Mineral Components in Foods

Chemical and Functional Properties of Food Components Series

SERIES EDITOR

Zdzisław E. Sikorski

Mineral Components in Foods
Edited by Piotr Szefer and Jerome O. Nriagu

Chemical and Functional Properties of Food Components, Third Edition
Edited by Zdzisław E. Sikorski

Carcinogenic and Anticarcinogenic Food Components
Edited by Wanda Baer-Dubowska, Agnieszka Bartoszek and Danuta Malejka-Giganti

Methods of Analysis of Food Components and Additives
Edited by Semih Ötleş

Toxins in Food
Edited by Waldemar M. Dąbrowski and Zdzisław E. Sikorski

Chemical and Functional Properties of Food Saccharides
Edited by Piotr Tomasik

Chemical and Functional Properties of Food Lipids
Edited by Zdzisław E. Sikorski and Anna Kolakowska

Chemical and Functional Properties of Food Proteins
Edited by Zdzisław E. Sikorski

Mineral Components in Foods

EDITED BY

Piotr Szefer
Medical University of Gdańsk
Gdańsk, Poland

Jerome O. Nriagu
University of Michigan
Ann Arbor, Michigan

CRC Press is an imprint of the
Taylor & Francis Group, an informa business

CRC Press
Taylor & Francis Group
6000 Broken Sound Parkway NW, Suite 300
Boca Raton, FL 33487-2742

First issued in paperback 2019

ISBN-13: 978-0-367-45333-6 (pbk)
ISBN-13: 978-0-8493-2234-1 (hbk)

Library of Congress Cataloging-in-Publication Data

Mineral components in foods / [edited by] Piotr Szefer and Jerome O. Nriagu.
p. ; cm. -- (Chemical and functional properties of food components series)
Includes bibliographical references and index.
ISBN-13: 978-0-8493-2234-1 (hardcover : alk. paper)
ISBN-10: 0-8493-2234-0 (hardcover : alk. paper)
1. Nutrition. 2. Food--Analysis. 3. Food--Consumption. 4. Food--Toxicology.
I. Szefer, Piotr. II. Nriagu, Jerome O. III. Series.
[DNLM: 1. Food Analysis--methods. 2. Minerals--adverse effects. 3. Food Contamination. 4. Metals--adverse effects. 5. Trace Elements--adverse effects. QU 130 M664 2006]

RA1258.M56 2006
363.19'2--dc22 2006015225

Visit the Taylor & Francis Web site at
http://www.taylorandfrancis.com

and the CRC Press Web site at
http://www.crcpress.com

Preface

Understanding of the mineral composition of foodstuff is important from nutritional and toxicological points of view. According to the Paracelsian maxim, dose makes poison and every element can become a human health hazard if its concentration in food is high enough. On the other scale, lack of inadequate quantities of essential trace elements can cause health problems for consumers. Therefore, the concern of food manufacturers and processors is to make sure that food products do not breach the essentiality/toxicity duality embodied in the various legal requirements or codes of practice for metals in food.

Nutritional or essential elements required for normal bodily function are classified according to their relative amounts and requirement. Magnesium (Mg), calcium (Ca), potassium (K), and sodium (Na) are classified as *macronutrients* while chromium (Cr), cobalt (Co), copper (Cu), molybdenum (Mo), nickel (Ni), and selenium (Se) are grouped as *micronutrients*. Chemical elements regarded as possibly essential micronutrients include arsenic (As), boron (B), and vanadium (V), whereas the *toxic metals* include beryllium (Be), cadmium (Cd), lead (Pb), and mercury (Hg). Among the *nontoxic, nonessential metals* are aluminium (Al) and tin (Sn); these two elements are sometimes called "packaging metals," in recognition of their unique application in industry and commerce. In the past, the term *trace element* was used for elements that were present in biological systems in minute quantities (from nanogram [10^{-9} g] to picogram [10^{-12} g]), near or below the detection limits of the analytical methods of the time. In recent years, this term is commonly applied to an element that typically occurs at a level of < 10 mg/100 g in biological systems. All the elements detectable at this concentration are sometimes referred to as *minor elements*. The less common term *ultratrace element* pertains to elements found at concentrations less than 0.001 mg/100 g.[1]

The essential micro- and macroelements in food are important for the development and maintenance of life functions. They affect all aspects of growth, health, and reproduction, from the formation of cells, tissues, and organs to the initiation and development of host defense by the immune system in response to foreign microbes and viruses. To deal with the essentiality/toxicity mirage, biological systems have developed sophisticated genomic recognition and transport mechanisms to ensure delivery of the right metal ion to its target. Quite often, a toxic metal is able to use an established metabolic pathway for an essential element to reach a critical target to induce undesirable effects. A number of factors may influence the absorption, recognition, and transport of essential elements including their bioavailability, interelement interactions, dose-response effects, oxidation state (of the element and cellular environment), and associations with biomolecules. These factors also affect the ability of toxic metals to mimic the essential elements in their cellular and genomic functions.

Recent studies on the biological mechanisms of life and the likely influence of trace elements in the increasing incidence of chronic diseases, cancers, and degenerative diseases have given rise to growing concerns about the health effects of dietary exposure to trace elements. It is estimated that poor diet is responsible for trace element deficiencies (mainly related to iron, iodine, zinc, selenium, and copper) in about 40% of the world's population, especially in the developing countries. In poor countries, food supply is usually limited and monotonous; for instance, *ugali*, a porridge from white maize (corn), or *gari* (made of cassava) is eaten in many African countries three times a day. Besides the geo-environmental factors,[2] the mineral composition of diet is dependent frequently on the sociocultural and religious habits which can reduce the bioavailabilty of several trace elements; for instance, if one were to rely exclusively on vegetables for iron and calcium. Blood losses, parasitic infections, sweating (especially under tropical condition), menstruation, frequent pregnancies, and prolonged lactation tend to increase the mineral requirements and losses, hence exacerbating the deficiency syndrome.

Not only deficiency but also excess in the intake of mineral components may result in endocrine, cardiovascular, skeletal, gastrointestinal, kidney, genetic, and opthalmologic disorders in various individuals. A comprehensive evaluation of the linkage between diet and cancer in 1997 by the World Cancer Research Fund and the American Institute for Cancer Regulation concluded that diets with very low or elevated levels of iodine may lead to increased risk for thyroid cancer whereas selenium may decrease the risk of lung cancer. The development of oxidative stress due to increased cellular levels of reactive oxygen species in human immunodeficiency virus (HIV) infection has been linked to depletion of essential mineral components, especially zinc and selenium, and to altered iron metabolism as well as to loss of magnesium and calcium. Such changes in body stores of minerals in the HIV host may affect the course of HIV disease and progression to acquired immune-deficiency syndrome (AIDS). The balance of essential elements is particularly important for immune defense, and the moderating influence of mineral components in pathogenesis of HIV would seem like a fertile field for further investigation.[2]

Any advancement in the science of metallomics begins with improvements in our ability to measure a given elements and its forms in the exposure matrices (mainly foods and drinks) with high sensitivity, selectivity, and precision. Analytical methods used in quantification of mineral components in foods are described in Chapter 1. Knowledge of the total concentration of an element in food is not necessarily sufficient for evaluating how well this element will be absorbed or how it will be metabolized when consumed.[1] Concerted effort has therefore been made to establish the chemical forms of trace elements in foodstuffs as helpful adjuncts in establishing dietary requirements and related legislation. Chapter 2 provides an overview of analytical methods for the speciation of mineral components in food products. In spite of important recent advancements in analytical techniques for metal speciation in foods, there is obligation to validate the quality of the data obtained using various quality assurance programs. Chapter 3 discusses the analytical quality controls in analyses of foods for trace elements.

The application of statistical multivariate approaches to data on mineral composition of foods to ascertain the geographic origin, authenticity, and level of contamination of trace metals is reviewed in Chapter 4. The functional role of some minerals present in foods, interactions with other mineral components in food, effects of their deficiency and excess, as well as influence of storage and processing on mineral components, are summarized in Chapter 5. Even very small amounts of chemical elements may influence food quality negatively, causing undesirable changes in the composition during production, technological processing, packing, storage, transport, and cooking. Chapter 6 and Chapter 7 are devoted to the occurrence and distribution of mineral components in animal and plant foods including, honeys, confectioneries, and alcoholic beverages. Trace elements in wines are described in detail in Chapter 8.

Food processing plants and containers may be serious sources of metal contamination in foods. Specific problems associated with food contamination from containers are discussed in Chapter 9. Contamination of foods with trace elements from various environmental pathways is reviewed in Chapter 10 whereas Chapter 11 deals with the bioaccumulation of various radionuclides (including ^{137}Cs, ^{239}Pu, ^{90}Sr, ^{131}I, and ^{35}S) in the human food chain. Chapter 12 provides an overview on the assessment of exposure, dietary intakes, standards, risk, and benefits of chemical contaminants in food and water. There has been a recent explosion in the availability and prophylactic consumption of metal-containing supplements. Chapter 13 provides an overview of metal contamination of food supplements and discusses the health risk from this exposure pathway.

This volume is mainly directed to food chemists, nutrition and public health educators, analytical chemists, quality control professionals, food manufacturers, environmental scientists and students, and practicing professions in food chemistry and food safety. The critical reviews and reference lists should guide the reading of anyone with interest in food sciences and dietary risk assessment. The book represents the dedicated effort of our distinguished group of authors and we thank the book department of the publisher, for invaluable editorial assistance.

[1] Reilly, C., *Metal Contamination of Food: Its Significance for Food Quality and Human Health*, Blackwell Science, Oxford, 2002.

[2] Bogden, J. D, and Klevay, L. M., Eds., *Clinical Nutrition of the Essential Trace Elements and Minerals*, Humana Press, Totowa, New Jersey, 2000.

Editors

Piotr Szefer received his M.S., Ph.D., and D.Sc. degrees from the Medical University of Gdansk (MUG), Poland. He was granted a full professorship at MUG in 2000. Between 1990 and 2002, he was vice dean and the dean of the Faculty of Pharmacy, MUG. Since 2000 he has been the head of the Department of Food Sciences, MUG. Prof. Szefer's research includes instrumental analysis of certain groups of foodstuffs for mineral components and metaloorganic compounds such as butyltins. He is involved in the development of novel multivariate statistical techniques that can be used to estimate the quality, geographical origin, or authenticity of food products. His research also covers the fate of environmental contaminants in the marine environments, speciation of chemical elements, and their bioavailability and biomagnification in the aquatic food chain. Prof. Szefer has published 160 scientific papers including 12 book issues. He has served on editorial boards for seven scientific journals, including *Science of the Total Environment* (Elsevier) and *Oceanologia*. Professor Szefer has been a visiting professor or visiting researcher at several universities and research institutions including Miyazaki University, Japan; University of Aden, Yemen; the Geological Survey of Finland; Karlsruhe University, Germany; Unidad Académica Mazatlán (UNAM), Mexico; Andhra University, Visakhapatnam, India; Instituto de Radioproteçao e Dosimetria (IRD), Rio de Janeiro, Brazil; French Research Institute for Exploitation of the Sea (IFREMER), Nantes, France; Environmental Science and Technology Department, Roskilde, Denmark; University of Plymouth, England; Fundación AZTI Technalia, San Sebastian, Spain; Seoul National University and School of Earth and Environmental Sciences, Seoul National University and Korea Research and Development Information (KORDI), South Korea. He has supervised seven Ph.D. students and more than 100 M.Sc. students.

Jerome O. Nriagu is professor in the Department of Environmental Health Sciences and School of Public Health and Research professor, Center for Human Growth and Development, University of Michigan. For many years, he was a research scientist with Environment Canada, National Water Research Institute, Burlington, Ontario. He is the editor-in-chief of the journal *Science of the Total Environment* and editor of 28 books on various environmental topics including *Thallium in the Environment*, *Arsenic in the Environment*, and *Nickel and Human Health*. Dr. Nriagu received his B.Sc. and D.Sc. degrees from the University of Ibadan, Nigeria, an M.S. from the University of Wisconsin, and a Ph.D. from the University of Toronto. He has published extensively and is listed among the most highly cited scientists in the fields of environmental studies and ecology. He has received a number of awards for his work and is a Fellow of the Royal Society of Canada.

List of Contributors

Marek Biziuk
Department of Analytical Chemistry
Gdansk University of Technology
Gdansk, Poland

Marcelo Enrique Conti
Dipartimento di Controllo e Gestione delle Merci e del loro Impatto sull'Ambiente
Università La Sapienza
Rome, Italy

Smaragdi M. Galani-Nikolakaki
Analytical and Environmental Chemistry Laboratory
Technical University of Crete
Crete
Chania, Greece

Małgorzata Grembecka
Department of Food Sciences
Medical University of Gdansk
Gdansk, Poland

Nikolaos G. Kallithrakas-Kontos
Technical University of Crete
Analytical and Environmental Chemistry Laboratory
Crete
Chania, Greece

Joanna Kuczyńska
Department of Analytical Chemistry
Gdansk University of Technology
Gdansk, Poland

Michał Nabrzyski
Department of Food Sciences
Medical University of Gdansk
Gdansk, Poland

Jerome O. Nriagu
Department of Environmental Health Sciences
School of Public Health
University of Michigan
Ann Arbor, Michigan

Zitouni Ould-Dada
Food Standards Agency
Radiological Safety Unit
Contaminants Division
London, England

Peter J. Peterson
Division of Life Sciences
King's College
University of London
London, England

Aleksandra Polatajko
Group of Bioinorganic Analytical Chemistry
Pau, France

Conor Reilly
Chipping Norton
Oxfordshire, United Kingdom

Melissa J. Slotnick
Department of Environmental Health Sciences
School of Public Health
University of Michigan
Ann Arbor, Michigan

Piotr Szefer
Department of Food Sciences
Medical University of Gdansk
Gdansk, Poland

Joanna Szpunar
Group of Bioinorganic Analytical Chemistry
Pau, France

Barbara Szteke
Department of Food Analysis
Institute of Biotechnology of Food and Agricultural Industry
Warszawa, Poland

Table of Contents

1 Mineral Components in Food — Analytical Implications

Marek Biziuk and Joanna Kuczyńska

CONTENTS

1.1 INTRODUCTION

The problem of the presence of mineral components in food is complicated. Some metals are necessary for human life and play very important roles in bodily functions. Macronutrients such as carbon, hydrogen, oxygen, nitrogen, potassium, sulfur, magnesium, calcium, sodium, phosphorus, and chlorine are the main components of organic matter but also have important metabolic functions. So-called macroelements are the main cellular and structural building materials but also take part in osmotic pressure and acid/base regulation. Micronutrients such as B, Co, Cr, Cu, F, Fe, I, Li, Mo, Mn, Ni, Rb, Se, Si, V, and Zn participate in the metabolic functions and are the constituents of enzymes, hormones, vitamins, etc.[1] The contents of the main components in an adult man are estimated to be 1200 g Ca, 800 g P, 200 g K, 160 g S, 95 g Na, 95 g Cl, 30 g Mg, 4 g Fe, 3.5 g F, 2 g Zn, 0.007 g Cu, and 0.003 g I.[2] A deficiency of these elements can cause adverse effects not only in humans but in other animal species and plants. The main sources of mineral components, recommended daily requirements, and diseases caused by the deficiency of selected elements are listed in

TABLE 1.1
Recommended Daily Intake, Main Exposure Routes, and Diseases Caused by the Deficiency of Selected Macro- and Micronutrients[2]

Element	Recommended Daily Intake (mg)	Main Exposure Routes for Humans	Effects of Deficiency
Ca	900–1200	Milk, cheese, fish products, vegetables, cereal	Osteoporosis, rickets, neurological disorders
Fe	14–18	Eggs, offal, meat, fish, potatoes, nuts, whole wheat, mushrooms	Anemia
P	700–900	Fish, cheese, eggs, leguminous plants	Muscular weakness, bone aches, lack of appetite
Mg	300–370	Cereal, leguminous plants, chocolate	Nausea, vomiting, muscular cramps
Na	About 1000	Salt, meat, vegetables, cheese	Headache, lack of appetite, weakness, diarrhea, nausea
K	2–3.5	Potatoes, green vegetables, nuts	Muscular weakness, neurological disorders
Zn	13–16	Meat, cereal, leguminous plants	Hair loss, skin lesions, reduced resistance to infection
Cu	2–2.5	Green vegetables, seafood, cereal	Anemia, heart arrhythmia, blood vessels ruptures
F	1.5–4	Tea, cereal, cheese, fish	Tooth decay
I	0.15–0.16	Fish, seafood, vegetables, milk	Underactive thyroid, goiter
Se	0.06–0.075	Brazil nuts, fish, cereals, eggs	Cancer, heart diseases, reduced resistance to microbiological and viral infection, reduced antibody production

Table 1.1. Some of these elements (Cu, Cr, F, Mo, Ni, Se, or Zn) are toxic at higher concentrations, so their levels in a human body and in food must be kept below the toxic threshold, which should be higher than the level necessary for organisms. Very often the difference between an indispensable and toxic dose is very small. Some of these elements are antagonistic to organisms, and their presence can cause very harmful effects.[3–7] Most notorious among these are the so-called Big Four elements, namely, Cd, Hg, As, and Pb.[8,9] Intensive industrialization of the world has resulted in an increased input of metals in the environment, drastically altering the quality of surface- and groundwater as well as agricultural land and food.

Metals can be transported by water and air over long distances. Anthropogenic pollution of drinking water supplies and resources indispensable for food production has become a fact of life. The main route of introduction of metals to the human organism is through ingestion of food and drinking water, but the inhalation route can sometimes be significant. Metals are persistent in the environment and tend to

TABLE 1.2
Selected Metals Input, Their Levels in Main Food Sources, Organs and Tissues in which Metals Are Accumulated, and Daily Dietary Intake (Adults in U.S.)[12–16]

Metal	Total Input to Freshwater, (10^3 metric tons/year)	Main Food Sources Concentration (mg/kg)	Estimated Daily Dietary Intake (μg)	Tissue and Organs Accumulating Metals
As	41	Seafood, 3–37; wine, 0.02– 0.11	10–20	Liver, kidney, skin, hair, nails
Cd	2.1–17	Meat, fish, fruits, 0.005–0.010; peanuts, spinach, 0.06; pork kidney, 0.18–1.0	30–40	Liver, kidney, bones, prostate[17]
Cr	45–239	Nuts, 0.14; egg yolk, 0.2; cheeses, 0.13	42–78	Bones
Hg	0.3–88	Fish, fish products, 0.09–1.2	0.6	Kidney, bones, parathyroid gland, brain
Ni	33–194	Cacao products, up to 9.8; nuts, up to 5.1; other products < 0.5	100–800	Kidney, bones
Pb	97–180	Dairy products, 0.003–0.083; vegetables, 0.005–0.65; meat, fish, poultry, 0.002–0.16; cereals, cheese, 0.03	80	Liver, kidney, bones, brain, aorta

bioaccumulate in plants and organisms, even becoming biomagnified in the food chain.[10] People are at the top of the food chain, increasing their exposure to already enriched toxicants. Selected pathways for metal flow into environmental media, levels of metal in the main food sources, daily dietary intakes, and organs and tissues in which metals are accumulated are listed in Table 1.2. Certain levels of metals are associated with serious human health effects.[11] Table 1.3 lists the main sources of metals in the environment and their adverse health effects in humans.

There is a need for continuous monitoring of the degree of pollution of food and potable and surface waters by inorganic (and organic) compounds from anthropogenic sources. Excessive concentrations of pollutants above permitted maximum contaminant levels often signal violations of environmental laws (such as uncontrolled discharge of wastes, improperly operating treatment plants, lack of enforcement of legislation dealing with food production and water management, etc.).

1.2 SAMPLING AND SAMPLE PROCESSING

1.2.1 General Requirements

The sampling and analysis of food should be carried out according to the standards published by the National Standard Committee, the International Standard

TABLE 1.3
Main Sources of Selected Metals in the Environment and Main Health Effects Associated with Each Metal[12,17,18]

Metal	Sources	Health Effects
As	Pesticides, feed additive for poultry and swine production, domestic wastewater, sewage sludge, phosphate fertilizers, fossil-fuel combustion, pulp, and paper industry, metallurgic (mining and smelting), and steam electrical production	Skin, liver, prostate and lung cancer,[14,19] central nervous system effects leading to coma and death, diseases of respiratory and gastrointestinal tract, neurological disorders, muscular weakness, nausea, peripheral vascular diseases, hyperpigmentation, keratosis, etc.
Cd	Domestic wastewater, sewage sludge, smelting and refining, manufacturing processes, steam electrical production, smoking cigarettes[20]	Acute: nausea, vomiting, salivation, muscular cramps, liver injury, renal failure, convulsion, shock; chronic: lung cancer, renal toxicity (proteinuria)
Cr	Domestic wastewater, sewage sludge, manufacturing processes (metals, chemicals, pulp and paper, petroleum products), smelting and refining, and base metal mining and dressing	Cr(VI) is extremely toxic and genotoxic; induces DNA crosslinks and breaks; causes nausea, diarrhea, liver and kidney damage, dermatitis, respiratory problems, tissue damage, and cancer risk increases in the presence of arsenic and nickel[21]
Pb	Production of cement, metals, chemicals, pulp and paper, petroleum, fertilizers, combustion of leaded petrol	Anemia, insomnia, hypertension, renal dysfunction, sperm count suppression and damage to the peripheral nervous system, cognitive and behavioral disorders in children, carcinogenity
Ni	Coal combustion, domestic wastewater, sewage sludge, petroleum products, smelting and refining, and base metal mining and dressing	Dermatoses, allergic sensitization; nickel is carcinogenic in humans; among nickel refinery workers statistically significant elevation in the incidence of respiratory cancers was found[22]
Hg	Coal burning power plants, atmospheric fallout, manufacturing processes, chlorine and caustic soda production	Acute: nausea, vomiting, pharyngitis, bloody diarrhea, nephritis, annuria, hepatitis, followed by death from gastrointestinal and kidney lesions; chronic: carcinogenic for animals and humans, attacks nervous system (Minamata disease) and brain
Se	Natural, phosphate fertilizers, sewage sludge, coal burning, smelting and refining	Hair and nails brittleness and loss, skin lesions, liver enlargement; Deficiencies can cause: cancer, heart diseases, reduced resistance to microbiological and viral infection, reduced antibody production, endemic cardiomyopathy, and osteoarthropathy

Organization or the European Committee of Standardization. The laboratory should pass necessary accreditation process and all procedures (sampling, sampling pretreatment, analysis, and data evaluation) should be validated. Good Laboratory Practice (GLP) should be introduced and followed. Only then can the results of analysis be considered seriously by consumers. These results, sometimes affecting decisions concerned with life and death issues, must be reasonably accurate and require detailed quality assurance programs. Quality assurance is the procedure that includes certifications of reagents, materials, laboratory facilities and even analysts, statistical procedures for data evaluation, standard operating procedures (SOPs; tested and approved for particular determinations), instrumentation validation, and specimen/sample tracking. The analyst is responsible officially, disciplinarily, and even financially for the reliability of results of analysis. It is especially important in the case of food analysis. These analyses are very difficult because of the complicated nature of the matrices. Mineral components occur in the analyzed matrices at very small levels which often include organic compounds. Many biological samples are nonhomogenous and need additional steps of preparation. The sample taken for analysis should be representative so that its composition should correspond to the average composition of the analyzed object and cannot change during transport and storage. Most important, the sample should not be contaminated during collection, handling, and analysis. Very important is the purity of reagents and water used for pretreatment and analysis, cleanliness of equipment, and even indoor air and furniture in the laboratory. In the case of trace analysis, a clean laboratory with filtered air is often necessary. In terms of validation of methods, the blank determinations should be carried out and repeated after each change of reagents, water, equipment, washing procedure of vessels, analyst, etc. In general, vessels made from polytetrafluoroethylene (PTFE) or perfluoroalkoxy (PFA) are recommended instead of conventional borosilicate glass.

1.2.2 Sampling

Sampling design depends strongly on the matrix of an analyzed object and its size. Taking primary samples from large-sized objects (daily production, such as a load of van, vehicle tracks, car, tank, etc.) and preparing representative samples are described in various standard methods. Heterogeneous samples have to be homogenized by cutting, mixing, chopping, milling, grinding, mincing, and blending, using hand or mechanical equipment.[23] This is necessary not only for taking a representative sample, but also for better dissolution and digestion. After mixing, dividing, and making the sample smaller, the aliquot for analysis is taken. It should be divided in three parts: one for analysis, one for the client, and the third for archiving. Any reagents, vessels, or equipment used during operations must not introduce or absorb/adsorb the elements or compounds to be analyzed. This is why agate grinders, porcelain or glass mortars, and equipment with titanium blades instead of stainless steel are recommended.[24] For some food samples, cryogenic grinding is necessary.[25] Any analysis of biological samples should be carried out as soon as possible, especially in the case of samples that can deteriorate, rot, or decompose (meat, vegetables, fruits, fish, etc.). When this is impossible to do immediately, samples should be preserved under the proper conditions of temperature, moisture, light, contact with oxygen, etc. The simplest way is to seal samples in a sterile and dry container and then store them in a refrigerator or deep-freeze. All steps of sample preparation should be described in the final report.

The most widely used and convenient form of the sample to be analyzed for determination of mineral components would be as a water solution, but in the case of food there are not too many samples in such a form. When nebulization is part of the analytical procedure, the solution has to be characterized by low surface tension and low viscosity. Such requirements are fulfilled only, in the case of food, by samples of mineral water. Such samples do not need, in most cases, any pretreatment steps or any homogenization, and can be directly analyzed using appropriate methods, e.g., spectrophotometric, electrochemical, or even gravimetric or titrimetric techniques. This applies also to samples that can be dissolved readily in water, such as sugar, salt, water extracts, etc. When suspension appears in the water solution, and we are interested only in the soluble form, a filtration step is necessary. This also permits speciation and determination of soluble forms of analytes in the sample. After filtration, mineralization and dissolution of suspended matter, the insoluble forms may be quantified. In the case of other fluids rich in organic matrices, such as juices, milk, wine, and beer or liquids with suspended matter, and with solid or biological samples, there is need for a mineralization step.

Extraction techniques such as liquid extraction (LE), accelerated solvent extraction (ASE), supercritical fluid extraction (SFE), and solid phase extraction (SPE) have not found significant applications in the isolation of inorganic analytes but are often used for speciation analysis and determination of metallo–organic compounds. ASE, used for the isolation of analytes from solid samples, is based on the same principles as traditional solvent extraction but uses elevated temperature and pressure during the extraction process with liquid solvents (water, aqueous, or organic solvent). This allows for a better rate of extraction. High temperature and pressure influence the surface tension of the solvent and its penetration into the sample. The extraction of analytes from the solid samples is often carried out in a cell heated in an oven. SFE is based on the same principles as liquid–solid extraction. Supercritical fluids (substances above their critical temperature and pressure) have the properties of gases and liquids, so they can diffuse through solids like gases and dissolve analytes like liquids enhancing the extraction rates. Carbon dioxide alone or with some percent of organic solvents is most often used as supercritical fluid. SPE (column filled with ion-exchange resins) can be used for separation of inorganic analytes (anions and cations) from liquid foods including water, mineral water, wine, and beer. Analytes are trapped (and preconcentrated) in the column according to their affinities for the column packing material. After passing the aqueous sample and cleaning the column, the ions of interest are eluted by suitable solution. This facilitates the collection of higher concentration of analytes, thereby decreasing the limit of detection.

1.2.3 Mineralization of Sample

Only a few procedures of mineral component determination, such as neutron activation analysis, do not require any pretreatment in the analysis of solid and biological samples. In most procedures, the step of sample mineralization is indispensable. It can be carried out by dry ashing and wet digestion.

The cheap, simple, and rapid dry ashing method is often used in an elemental analysis of organic and inorganic compounds. It can be performed by heating the sample in the atmosphere (static) or with a stream (dynamic) of air or oxygen in a muffle furnace or in a quartz tube with a programmable temperature that destroys all organic compounds. The process transforms a huge quantity of the biological sample (food) into a small quantity of ash made up of inorganic compounds (mainly oxides). Most samples require drying or lyophilization before ashing. The samples are weighed with ashing aid (for example, magnesium, calcium, or aluminum nitrates) in a crucible or boat made of porcelain, quartz, or platinum. Because large samples (up to 25 g) can be easily processed, dealing with problems connected with homogenization of the sample is not very critical. Ashing aids can catalyze or moderate the process, help retain volatile analytes, and transform all analytes into soluble nitrates. The mineralized samples are dissolved after ashing in diluted inorganic acids. The dry ashing method in an open system cannot be used for the determination of volatile elements — for example, mercury, selenium, arsenic, chlorine, or iodine. Semivolatile elements such as cadmium, lead, and zinc can form volatile compounds (for example, chlorides [$FeCl_3$]) by reacting with some anions. For such samples, mineralization in closed glass flasks (Schöniger flasks) is recommended. A weighed aliquot of the solid sample is placed in the flask filled with oxygen and burned. The combustion products are absorbed in a suitable solution in the flask and can be measured titrimetrically, as is often the case for I, Cl, and Br.

The wet digestion procedures can be subdivided into four basic types:

- Open vessel digestion using acid or a mixture of acids with conductive heating
- Closed vessel digestion using acid or a mixture of acids with conductive heating
- Closed vessel digestion using acid or a mixture of acids with microwave heating
- Alkali fusion

Alkali fusions (with lithium metaborate, lithium tetraborate, sodium carbonate, sodium hydroxide, sodium peroxide, etc.) are widely used for decomposition of geological and metallurgical samples. They are not recommended for the decomposition of biological samples because of the high level of total dissolved solids introduced during sample preparation.[26]

Wet digestion with acids is currently the most common technique used for decomposition of organic matter, especially food material. The strong mineral acids or their appropriate combinations are the most convenient to decompose most types of food samples. The weighed sample with the addition of acids and digestion aids can be heated in the appropriate vessel (test tube, beaker, digestion flask, etc.) using a burner, heating plate, or an aluminum block with a programmed temperature. Nitric acid alone or as a mixture with perchloric or sulfuric acids is the most popular reagent used for sample decomposition, as it is a strong oxidizing agent and forms soluble nitrates with metals. Environmental chemists strongly recommend that all acid digestion should be done in Teflon or similar plastic containers to minimize sample contamination. To completely destroy the organic matrices, the addition of stronger oxidizing agents, such as hydrogen peroxide is sometimes necessary. Perchloric acid combined with metals can form spontaneously explosive compounds and can react explosively with undigested fats and oils after evaporation of HNO_3. Extreme caution must be exercised when using this acid. Hydrochloric acid and hydrofluoric acid are rarely used alone for the digestion of organic matrices. Some volatile chlorides and fluorides (with As, Sb, Sn, Se, Ge, Hg, and B) formed during digestion can cause the loss of analytes. Sulfuric acid, traditionally used in combination with nitric acid, cannot be used during determination of elements (particularly Ba, Ca, Pb, and Sr) which form sulfates with low water solubility or when volatilization of trace elements (Ag, As, Ge, Hg, Re, and Se) may occur during digestion.[26] The use of phosphoric acid results in the formation of phosphate ions, and difficulties in mass spectrometric determination (such as the production of polyatomic species) should be avoided. To sum up, the choice of acid or mixture of acids for sample digestion strongly depends on the type of matrix and the elements to be determined.

Closed vessel digestion was introduced into elemental analysis by Carius in 1860; he described the digestion of samples with concentrated nitric acid in sealed, strong-walled glass vials (Carius tube), but this method of mineralization was time-consuming and dangerous. Modern devices for closed acid digestion with conductive heating, called *digestion bombs* or *Parr bombs*, consist of nickel or a PTFE baker (inside) and a stainless steel jacket (outside). After the introduction of weighed samples and acids (mainly nitric acid) to the PTFE baker and sealing of the bomb, it is placed in an oven and then heated to 110–250°C for periods of one to several hours. The analyst should take necessary precautions during this step because sample digestion under supercritical conditions can generate high pressure, leading to explosive rupturing of the vessel. Organic samples should not be mixed with strong oxidizing agents, and the PTFE baker should not be filled to more than a 10–20% of its volume. After digestion, it is recommended that the bombs be cooled to room temperature before they are opened — and then with great caution. The digestates are next transferred to a volumetric flask (filtered, if necessary) and diluted to volume with deionized water.

Closed vessel digestion has many advantages. The high temperature and pressure enable a better and quicker decomposition of the sample to be achieved. Volatile elements (As, B, Cr, Hg, Sn, Se, and Sb) do not escape from the vessel and are retained in solution. The release of toxic fumes into laboratory air is avoided. Smaller volumes of acids can be used, and the mixture is isolated from air, thereby reducing contamination of the sample.

The advantages of closed-vessel acid digestion are equally applicable to processes based on microwave heating. During irradiation, an aqueous phase is heated

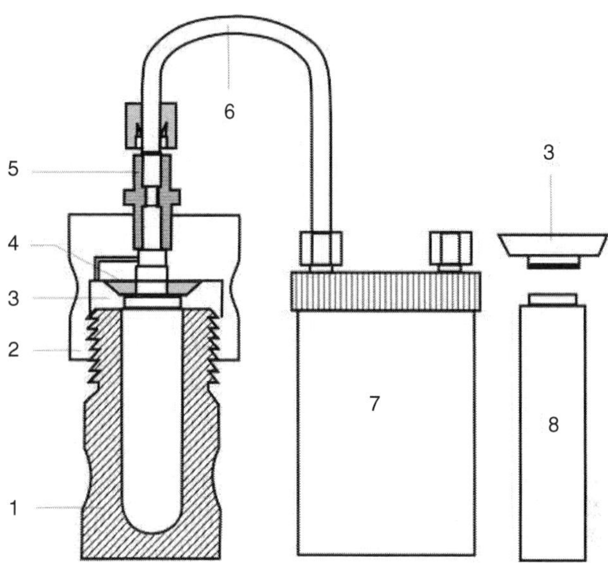

FIGURE 1.1 Basic set for microwave digestion: (1) vessel underpart, (2) screw cap, (3) closing cap, (4) rupture disc, (5) clumping cup, (6) connection tubing, (7) collecting vessel, and (8) forming tool.

rapidly because polar molecules and ions are energized through the mechanism of dipole rotation and ionic conductance.[27] The energy is also absorbed by sample molecules causing an increase in kinetic energy, internal heating, and differential polarization that expand, agitate, and rupture surface layers of the solid materials, exposing fresh surfaces to an acid attack.[26] The efficiency of sample digestion is much higher compared with conductive heating, and the time of digestion is much shorter. A schematic drawing of vessels used in this method is shown in Figure 1.1. The vessels are made from PTFE; fluorocarbon resins cannot be used because they cannot be heated convectively with their low heat-conducting capacity. PTFE is a totally inert material thermally stable up to 250°C and has high tensile strength. The vessels are generally thick-walled (Figure 1.1 [1]), and are suitable for use under higher pressure and temperature with aggressive reagents. An aluminum rupture disc (Figure 1.1 [4]) in the closing cap (Figure 1.1 [3]), form on a tranducer (Figure 1.1 [8]), which provides a safe relief from excessive pressure load within the digestion vessel (Figure 1.1 [1]) to the collecting vessel (Figure 1.1 [7]). Nitric acid alone or its combination with hydrogen peroxide, hydrofluoric acid, or hydrochloric acid are most often used, but sulfuric acid must be avoided because its high boiling point (338°C) is above PTFE's melting point. Due to high microwave transparency of the digestion vessel material, the vessel contents are heated directly and the vessel itself can be heated only by energy radiation from vessel contents. The thermal stress of the vessel within the first minutes is low but increases with time, so that a maximum digestion time should not exceed 5–10 min. Opening of the vessel after digestion is possible only after allowing enough time for cooling (5–15 min). Microwave acid digestion is widely used in food analysis and standard

TABLE 1.4
Advantages and Disadvantages of Wet Digestion Methods[30]

Digestion Technique	Possible Way of Losses	Source of Blank	Sample Size (g)	Digestion Time	Degree of Digestion	Economical Aspects
			Open Systems			
Conventional heating	Volatilization	Acids, vessels, air	< 5	Several hours	Incomplete	Inexpensive, needs supervision
Microwave heating	Volatilization	Acids, vessels, air	< 5	< 1 h	Incomplete	Inexpensive, needs supervision
UV digestion	None		Liquid	Several hours	High	Inexpensive, needs supervision
			Closed System			
Conventional heating	Retention	Acids (low)	< 0.5	Several hours	High	Needs no supervision
Microwave heating	Retention	Acids (low)	< 0.5	< 1 h	High	Expensive, needs no supervision

Source: Matusiewicz, H., Wet Digestion Methods, in *New Horizons and Challenges in Environmental Analysis and Monitoring*, Namieśnik, J., Chrzanowski, W., Żmijewska, P., Eds., Centre of Excellence in Environmental Analysis and Monitoring, Gdansk, 2003, p. 224, chap. 13. With permission.

methods for the determination of elements using microwave–assisted digestion are available.[24,28,29]

Table 1.4 shows the advantages and disadvantages of the wet digestion techniques with respect to losses of analytes, blank levels, contamination problems, sample size, digestion time, degree of digestion, and costs.[30]

1.3 METHODS OF FINAL DETERMINATION

1.3.1 Spectrophotometric Methods

The final determination of inorganic components in foods is done by spectrophotometric methods including atomic absorption spectrometry (AAS) and atomic emission spectrometry (AES) using a flame, inductively coupled plasma, or an electrothermal device for atomization.[24,31,32] These instrumental methods are based on the specific radiation emitted by excited atoms (the intensity of emitted characteristic light is measured proportionally to the concentration of the element in the sample), or absorbed by atoms in a ground state (the absorbance of characteristic light, passing the sample containing the atoms, which vary according to Beer–Lamberts law, is measured). Spectrophotometric methods are the most specific of all analytical methods (strictly connected with atomic structure) and can frequently be completed in a few minutes. More than 70 elements including alkali, alkaline earth, transition, and heavy metals can be determined in various samples with sensitivities from μg/g to ng/g range using spectrophotometric methods.

In flame atomic emission spectrophotometry (FAES) or flame photometry (FP) the sample is converted to atomic vapor by applying thermal energy (a flame). The samples have to be transformed into dissolved forms, nebulized using a pneumatic nebulizer, and in the form of aerosol, transported to the flame, where the solvent and the sample are vaporized and atomized. As energy continues to be applied, a number of atoms are changed to an excited state, in which they remain for a short time (10^{-7} to 10^{-4} sec). After the outer electron occupies more highly excited energy levels (orbitals), the excited atoms return to the ground state. In this process, the energy difference between the excited level and electron ground state is released. The atomic line radiation emitted is characteristic for each element. The characteristic wavelength is selected using a filter or monochromator (for example, grating or prism with entrance and exit slits) and detected using a suitable method. The detector converts electromagnetic radiation into an electric signal that can be measured after amplification if needed. There are single-element detectors, such as photovoltaic cells, solid-state photodiodes, photoemissive tubes, and photomultiplier tubes, or multiple-element detectors such as solid-state array detectors. Detectors can be characterized by precision, spectral sensitivity, limit of detection, and response time. If the flame burns evenly and the substance is fed into the flame at a constant rate over the period of the measurement, the intensity of the spectral lines observed provides a measure of the concentration of substances. This method is now mainly used for the determination of alkali metals and alkaline earth metals.

Atomic emission spectrometry with inductively coupled plasma excitation (ICP-AES) has much wider applications and can be used, because of high temperature of atomization, for the determination of most elements. At temperatures above 6000°C, a state that can be described as *plasma* occurs. Plasma is characterized by the fact that electrically charged particles are created by the breaking up of gas molecules — a process referred to as *ionization*. Plasma is thus a gas whose atoms or molecules have, to greater or lesser extent, broken up into positively or negatively charged carriers. For atoms to ionize, a specific energy, the so-called ionization potential, must first be applied. This energy, furthermore, must be fed to the atom itself. The ionization potential is expressed in electron volts (eV). The eV values for the majority of elements lie between 4 and 25 eV. In emission spectrometry, plasma is used to excite the atoms to be measured. In principle, plasma is an electrically conductive, fluid system in predominantly gaseous matter. The properties of plasma are determined by the charge carriers and by its quasi-neutrality, as the plasma has an overall electrically neutral effect. Plasma is produced by transferring electrical energy to a gas flow. In the case of inductively coupled plasma, a high frequency generator with an induction coil is used to supply energy for the ionization potential required. The energy is directly proportional to the density and temperature of the plasma. It is particularly important for practical analytical application of plasma that high temperature speeds up thermal ionization. This effect is used in ICP. A gas is fed through a system of quartz tubes whose shape facilitates flow, and to the end of these tubes a strong current of high frequency is applied. After ionization of the gas, homogenous plasma of high thermal intensity is produced. Minutely atomized particles of a solution are fed via an added carrier gas into the plasma core and ionized as a result of the residence time of these elemental particles in plasma. This causes the elements to emit spectra, which can be recorded qualitatively and quantitatively with the usual spectrometric system. ICP methods, due to the large dynamic range, enable simultaneous analysis of both major and trace elements in a single analytical run.

Mass spectrometry is a modern technique that can be used to measure ionic masses with high accuracy. Mass spectrometry with inductively coupled plasma excitation (ICP-MS) offers better precision than ICP-AES.

Since the beginning of the twentieth century, MS has been a powerful tool used to obtain some essential information about the analytes. The fundamental principle of a mass spectrometer is the ionization of particles of a given sample and the measurement of the ratio of mass to electric charge of the ion (m/z). In a typical spectrometer, there are several subassemblies that are represented in the block diagram (Figure 1.2).

As mentioned previously, plasma is used for ionization in the ICP technique. The heterogeneous ion (positive or negative) stream is transferred into a vacuum system containing a mass filter (most often a quadruple mass analyzer), and the ions are then separated according to their mass to charge (m/z) ratio. In the case of positively charged ions, the mass of the sample measured in the spectrometer is enhanced by the proton mass or protons that are attached to the particle of the sample. For the negatively charged ions, the mass of the sample is reduced by the proton mass or protons that are separated from the particle of the analyte. The ions with the specific ratios of mass to charge are transmitted to a detector. The mass-to-charge ratio is used to identify the isotopes of the elements, whereas the concentration is determined on the basis of the intensity of a specific peak in the mass spectrum, that is proportional to the amount of the given isotope.

Both ICP-AES and ICP-MS have found many applications in various fields, for instance, in the determination of nutritional and toxic inorganic elements in foods. Table 1.5 lists several surveys of elements in food based on this analytical method. Analytical problems associated with the interferences observed in ICP-MS have been overviewed by Capar and Szefer.[24]

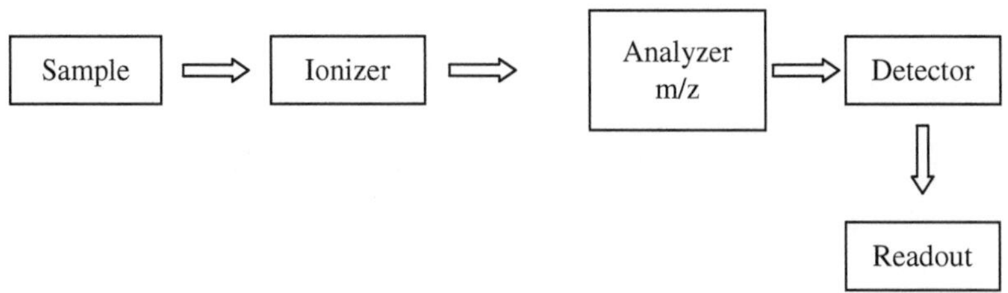

FIGURE 1.2 Schematic diagram of principle of operation of MS.

TABLE 1.5
Analyses of Food Samples Using ICP-AES and ICP-MS

Elements	Kind of Food	References/Methods
Al, Ba, B, Ca, Co, Cu, Fe, Li, Lu, Mg, Mn, Mo, Ni, P, K, Rb, Si, Na, Sr, Sn, Ti, V, Zn	Orange juices, peel extracts, deacidifed juices	33 ICP-AES and ICP-MS
As, Ca, Cd, Cr, Cu, Fe, K, Mg	Seaweeds	34 ICP-AES
Na, K, Ca, Mg, P, Fe, Cu, Mn, Zn, Cr, Al, B, Se, Mo	Meat, fish, chicken, cereal, sweet dishes	35 ICP-AES
I	Fish, food (from plants)	36 ICP-MS
Ca, Co, Cu, Cr, Fe, Mg, Mn, Mo, Na, Zn, Ag, Al, As, Ba, Be, Cd, Hg, Ni, Pb, Sb, Sn, Sr, Ti, Tl, U, V	Milk	37 ICP-OES
Al, Ba, Ca, Cu, Fe, K, Mg, Mn, Na, Sr, Zn	Tea beverages	38 ICP-AES
Pb	Wine	39 ICP-MS
As, Cd, Cr, Cu, Fe, Mn, Ni, Pb, Sn, V, Zn	Honey	40 ICP-MS
B	Hazelnut	41 ICP-OES

Atomic Absorption Spectroscopy (AAS), developed by Alan Walsh, has revolutionized quantitative analysis since the 1950s. An important milestone in the evolution of this technique was the development of the stable light sources that can emit sharp atomic spectral lines of the elements. Nowadays, it is one of the most frequently used methods for determining concentrations of a wide range of elements in materials such as food, water, metals, pottery, and glass.

Atomic absorption spectrometry is a highly sensitive technique, that is able to measure down to parts per billion (μg/l) of an element in a sample. This method has found many important applications in various areas such as food analysis, biochemistry, medicine, industry, mining, and environmental control.

Atomic absorption spectroscopy is based on the measurement of the absorption of the characteristic wavelengths of light by atoms in the gaseous state. Figure 1.3 shows the schematic diagram of an AAS.

As the diagram indicates, one of the main features of the instrument is the source of light, characteristically a hollow cathode tube, that is constructed out of a tungsten anode and a cylindrical metal cathode containing the element to be determined. The potential difference between the cathode and the anode as a consequent of the ionization of noble gas atoms (neon or argon) causes electron bombardment and

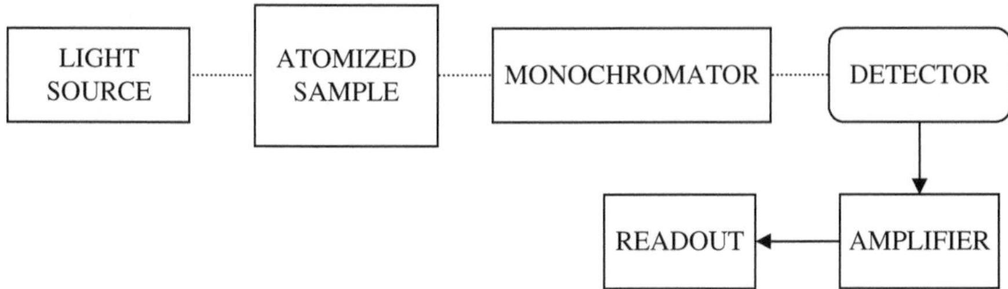

FIGURE 1.3 Schematic diagram of an atomic-absorption experiment.

ejection of metal atoms from the cathode. As a consequence, some atoms are excited and subsequently emit a radiation characteristic of the metal.

In order to perform a measurement the analyte must be vaporized and atomized. There are two types of methods, that can be used to atomize the sample: flame atomic absorption spectrometry (FAAS) and graphite furnace atomic absorption spectrometry (GF-AAS or ET-AAS). In the first case, the sample is brought into a flame using a nebulizer, whereas in the second method the samples are located in the graphite tube with electrical heating, which provides the thermal energy to dry the specimen, the ash organic component, and produces free ground–state atoms. This technique is preferred for measuring concentrations of most elements in different samples for the reason that the graphite tube is a much more efficient atomizer than the flame. The thermal energy, obtained by using one of the two methods above, causes the transition of the atom from the ground state to the first excited state; the absorption of some of the light (in the ultraviolet or visible wavelength) is generated by a lamp that is specific for a target metal. The selection of the specific light, which is absorbed by the sample, is done using a monochromator. This allows for the determination of one element in the presence of others. The light selected is directed to a detector, where an electrical signal proportional to light intensity is produced. The amount of absorbed radiation is proportional to the concentration of the element of the interest in solution. This dependence is described by the Beer–Lambert law, which is expressed as:

$$A = \varepsilon \cdot c \cdot l$$

where:

A, is the measured absorbance,
ε, is a wavelength-dependent molar absorptivity coefficient,
c, is the analyte concentration and
l, is the path length.

Concentrations are usually determined from a calibration curve, which is derived from known concentrations of the elements of interest. Comparing the amount of the standard absorption with the working curve enables the calculation of the concentration of an element in the unknown sample.

TABLE 1.6
Analyses of Food Samples Using AAS

Elements	Kind of Food	References
Ca	Cereal	43
Cr	Seafood, cereals, vegetables, olive oils	44
Se	Foodstuffs	45
Cd	Wine	46
Ca, Cl, K, Mg, N, P, Al, As, Au, Br, Cd, Co, Cr, Cs, Cu, Fe, Hg, I, Mn, Mo, Ni, Pb, Rb, Sb, Sc, Se, Sn, Sr, V, Zn, W	Total diet	47
Cd	Flour	48
As, Se	Foods	49
Zn, Mg	Maize	50
Ca, Cu, Fe, Mg, Mn, K, Na, Zn	Fruits	51
Hg, Se	Seafood	52
Pb, Cd, Fe, Cu, Mn, Zn	Fish	53
Fe, Cu	Peanuts	54
Pb, Cu, Zn, Mg, Ca, Fe	Eggs	55
Cd, Co, Cu, Fe, Hg, Mn, Mo, Ni, Pb, Zn	Fish	56
Ag, Ba, Cd, Cr, Cu, Fe, Mn, Pb, Zn	Fish, shrimp	57

As mentioned earlier, AAS is a versatile and powerful tool in nutritional studies.[42] An important aspect of AAS is that the sample preparation is usually simple. Moreover, the chemical form of the element is often unimportant. Depending on categories of food, the treatment consists mostly of cleaning, weighing, ashing, extraction, and acid digestion. Table 1.6 lists some analyses of food samples performed with AAS.

Unfortunately, AAS is a single-element analyzer and, thus, the technique is rather slow to use. Recent developments have led to configurations of AAS that can be used for multi-element analysis.

For elements that can be easily converted into volatile form (i.e., mercury, and iodine), a cold vapor atomic absorption spectrometry (CV-AAS) can be used in the analysis.

Hydride generation (HG) is a convenient technique for the determination of some elements (i.e., antimony, arsenic, bismuth, germanium, lead, selenium, tellurium, and tin) that are difficult to analyze by conventional methods. The analysis is based on the separation of these elements from the matrix by formation of their volatile hydrides after reaction with a reducing agent (sodium borohydride). The process, as shown in the following example, proceeds in acid media[58]:

$$BH_4^- + 3H_2O + H^+ \rightarrow H_3BO_3 + 8H$$

$$E^{+m} + 8H \rightarrow EH_n + H_2\left(excess\right)$$

TABLE 1.7
Analyses of Food Samples Using Hydride Generation Method

Elements	Kind of Food	References
As, Sb, Se	Green tea	59
As, Se	All foods	49
Zn	Food	60
As	Seafood	61
Se	Food (especially breads and flours)	62
Hg, Se	Seafood	52
Te	Milk	63
As, Sb	Milk	64
Se	Food	65

In principle, the HG procedure consists of several distinct processes such as the generation of the hydrides and their collection, the transport of the gaseous hydrides and by-products (i.e., H_2O, CO_2, and H_2) into the excitation source, and the atomization of the hydrides. The manner of atomization hinges on the detection system cojoined with HG. This technique can be combined with different methods such as AAS, AES, and ICP-MS; the combination with atomic spectrometer can be used to attain very low limits of detection. For this reason, HG-AAS is a technique that is applied to the determination of elements in a wide range of matrices, for instance, in food. Various examples of the application of a hydride generation technique in the analyses of food samples are shown in Table 1.7.

Determination of mercury in food products can be carried out using mercury specific atomic fluorescence spectrometry after mineralization of the samples and with or without preconcentration of mercury on a gold trap (by amalgamation). Samples for this process do not need any pretreatment step. After excitation of mercury vapor, atomic fluorescence is obtained by the emission at a wavelength of 254 nm, which can be measured by a detector. This technique is simple, usually 100 times more sensitive than atomic absorption, and is relatively free from interferences. Very small samples are sufficient for analysis (1–100 mg) and the detection limit is very low (0.1 ng/l, 0.05 mg/kg wet weight). Mercury can be analyzed in fruits, vegetables, meat, fish, seafood, and other solid or liquid food products using this method.[24,66,67]

Analysis of mercury in aqueous solutions entail a number of steps including acid digestion (sulfuric acid/nitric acid), oxidation of all mercury forms to Hg(II) with permanganate/persulfate, reduction of the excess oxidant with hydroxylamine, reduction of Hg(II) to Hg^0 using $SnCl_2$, purging of elemental Hg to gold amalgamation traps, and thermal desorption of the Hg followed by CV-AAS detection. Quantification is achieved using a calibration curve.

Organic mercury is much more toxic to plants, invertebrates, fish, and people than the inorganic forms of the element. Methylmercury can be measured by GC-ECD; however, this detector is not selective to Hg. The selectivity of atomic fluorescence eliminates this problem. The organomercury derivatives (methylmercury, dimethylmercury, and methylethylmercury) are separated on a chromatographic column, decomposed in a quartz tube at a high temperature, and then detected as elemental Hg.[24,68,69] The critical step in measuring MeHg is its extraction.

One of the most widely used and the cheapest (inexpensive instrumentation) methods for the determination of inorganic analytes (cations and anions) is the ultraviolet and visible absorption spectrometry method.[70] This method is simple and selective with moderate sensitivity and flexibility due to the many reagents available. The method is based on the formation of colored compounds with appropriate and usually specific reagents, and on the absorption of characteristic electromagnetic wavelength by this compound. Formations of metal–organic complex are well known. The characteristic light passing the cuvette containing aqueous or organic solvent solution with the analyte–reagent complex is absorbed by this complex, as per Beer–Lamberts law. The absorbance depends on the concentration of the analyte.[32] Solvents used in these methods must fulfill certain requirements. The solvents must be able to dissolve the sample, be compatible with cuvette material, nonvolatile, nontoxic, and be relatively transparent in the spectral region of interest. The selected examples of metals and anions determination using molecular absorption spectrophotometry in visible/UV light are listed in Table 1.8. The method is often used for the determination of major transition elements but is not often applied to the determination of alkali and alkaline earth metals because of lack of suitable reagents.

1.3.2 Radiochemical Methods

Radiochemical methods such as neutron activation analysis and X-ray fluorescence spectrometry are also strictly connected with atomic structure.

X-ray fluorescence analysis, one of the oldest nuclear techniques, is based on subjecting the sample to electromagnetic radiation of sufficient energy to remove electrons from the inner orbitals. The primary X-ray irradiation is produced by means of an X-ray tube or by a radioactive source (usually, e.g., ^{57}Co, ^{109}Cd, and ^{241}Am). After absorption of photons and removal of electron, the electron from an outer orbital is transferred to fill the "hole" formed, and the energy difference between two orbitals is emitted as secondary or fluorescent X-radiation. The radiation is diffracted and reflected by a movable analyzer crystal, thereby breaking it down into the various wavelengths. The intensity of the radiation is measured in the receiver at a wavelength specific to the element. The fluorescence radiation stops immediately when the primary radiation is halted. The fluorescence X-radiation is characteristic for each element, well-defined (based on the atomic structure), and thus enables qualitative and quantitative determination of elements with high selectivity. This radiation, however, has a low energy, that easily can be absorbed by the sample matrix; therefore, this technique is more suitable for very thin, very flat, and homogenous samples. The analysis of water can be realized using a dry residue of dissolved solids. The sensitivity of X-ray spectrometry is lower than that of the neutron activation method. Fluorescence radiation can also be obtained after bombardment of atoms with protons or charged particles produced by accelerator (PIXE). Commercially available equipment, even hand-held systems for field measurement, can be used.

TABLE 1.8
Selected Examples of Metals and Anions Determinations Using Molecular Absorption Spectrophotometry[70,71]

Analyte	Reagent	Wave-Length (nm)	Analyte	Reagent	Wave-Length (nm)
Aluminium	Chromazurol S	545	Iron	1,10–phenantroline	512
	8–hydroxychinoline	390		Thiocyanate	495
Ammonium	Hypochlorite/phenol	625	Iodide	Bromine/starch	590
Antimony	Iodide	425	Lead	Dithizone	520
Arsenic	Diethyldithiocarbamate	535	Mercury	Dithizone	485
Bismuth	Dithizone	490	Molybdenum	Thiocyanate	470
	Iodide	465	Nitrate	3,4-xylenol phenyldisulphonic acid	410
	Xylenol orange	450			
Bromide	Hypochloric/phenol red	580	Nickel	Dimethylglioxime	400
Cadmium	Cadion	480	Phosphate	Molybdate/vanadate	400
	Dithizone	520		Molybdenum blue	780
Chloride	Mercury Thiocyanate/Fe(III)	480	Selenium	Diaminobenzidine	420
	Diphenylcarbazone	560			
Chromium	Diphenylcarbazide	545	Sulphate	Barium chloranilate	530
	EDTA	540		Barium chromate	436
Cobalt	Nitrosonaphtol	415	Sulphide	Dimethyloaminoaniline/Fe(III)	662
	Thiocyanate	620			
Copper	Diethyldithiocarbamate	436	Tellurium	Bismuthiol II	330
	Ditizone	550	Vanadium	8–hydroxychinoline	550
Cyanide	Chloramine-T/pyridine/ barbituric acid	580	Zinc	Dithizone	538
				Pyridylazonaphtol	560

Neutron activation analysis (NAA) was introduced in 1936 by G. Hevesy with H. Levi but became more useful after the development of the nuclear reactor as a neutron source. Equally important was the introduction of the NaI scintillation γ-ray detector and of the high resolution Ge (Li) γ-ray detector. Neutron activation analysis (NAA) is a very powerful method applicable for both quantitative and qualitative measurements of many elements in different types of samples. This method has stimulated research in different fields such as food analysis,[72] chemistry, biology, geology and geochemistry, environmental science, and related fields. In Table 1.9 several surveys of elements in food using NAA are listed.

In general, NAA is the most suitable technique for materials that are difficult to convert into a solution for analysis and where only milligram quantities are available. The procedure for NAA can be divided into several steps, which are presented in Figure 1.4. The required amount of samples (maximally, 200 mg) is simply packaged in an irradiation container (quartz, polyethylene, or Al foil), sealed, and irradiated with neutrons for a time determined by the half-life of the radionuclide or the composition of the sample, in addition to radiological considerations.

TABLE 1.9
Analyses of Food Samples Using NAA

Elements	Kind of Food	References
Br, Ca, Cl, Co, Cu, I, K, Mg, Mn, Na, Rb, S, Ti, V	Cereals, oils, sweeteners, vegetables	72
Ag, Br, Ca, Cl, Co, Cr, Cs, Fe, Hg, K, Mg, Mn, Na, Rb, Sb, Se, Sr, Zn	Diets of children and adolescents	73
Cs	Daily diet	74
Co, Cr, Fe, Rb, Se, Zn	Cheese	75
Ca, Mg, Mn, Cu, Cr, Zn, Se, I	Daily diet	76
Se, Cr, Rb, Zn, Fe, Co, Sb, Ba, Ni, Ag, Hg, Sn	Flour and bran	77
Ca, Cl, K, Mg, Al, As, Au, Br, Cd, Co, Cr, Cs, Cu, Fe, Hg, I, Mn, Mo, Na, Rb, Sb, Sc, Se, Sn, Sr, Zn	Total diet	47
Al, As, Br, Cl, I, K, Mg, Mn, Na, Sb, Sm, Cr, Fe, Rb, Sc, Zn	Health food	78
Al, Ca, Cl, Cu, K, Mg, Mn, Na	Tea leaf	79

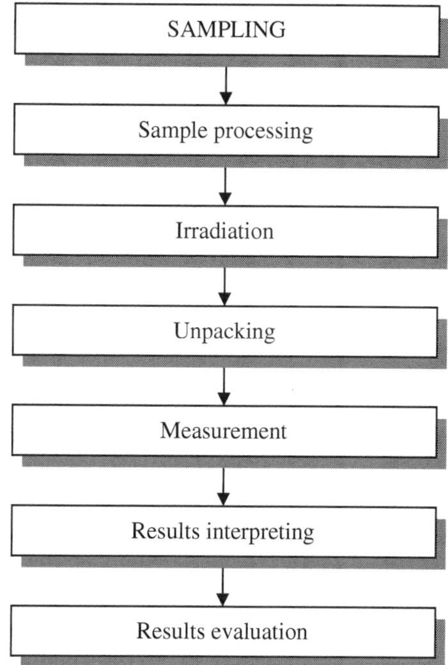

FIGURE 1.4 Schematic diagram of NAA analysis.

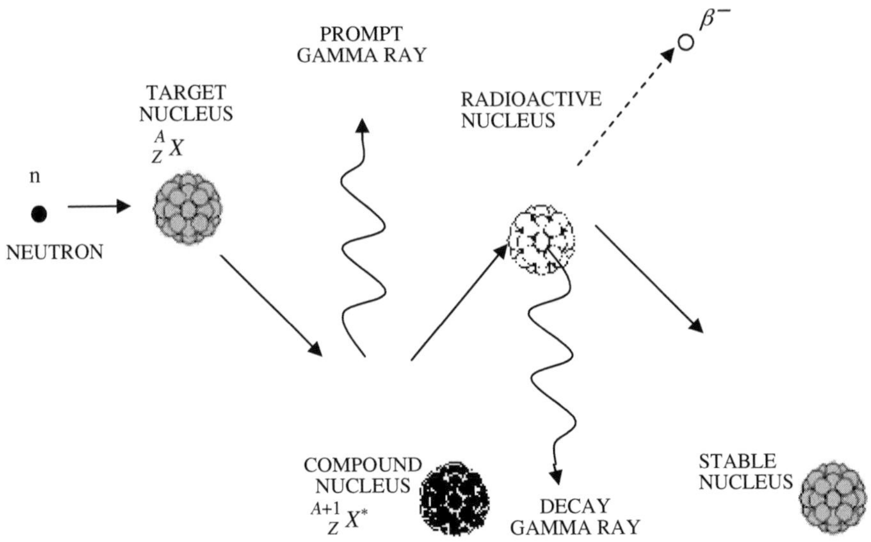

FIGURE 1.5 Schematic diagram illustrating the successions of events for a typical (n,y) reaction.

The physical effects on which NAA is based are the interaction of radiation with matter, the properties of the nucleus, and radioactivity. The sequences of events are illustrated in Figure 1.5.

When a particle (neutron, proton, and α-particle) or photon interacts with the atomic nucleus in collision, radioactive nuclides are produced in a highly excited state. The compound nucleus very quickly decays into a more stable configuration through emission of one or more characteristic prompt gamma rays. The distinct energy that is created during the process is unique to each element and provides positive identification of the targeted element in the sample, when the quantification of the element can be determined by relative standardization.[80] In this method, an unknown sample is irradiated with a calibration sample with known amounts of the elements. Both the real and the calibration samples are treated under the same conditions. The quantity of the element is determined by a comparison of the peak areas of the measured spectra for the sample with reference material. In order to evaluate the concentration of an individual element, the activities of one or more specific isotopes have to be determined. This may be done in two different ways. The first method depends on chemical separation and is generally referred to as radiochemical neutron activation analysis (RNAA), whereas the second technique is known as instrumental neutron activation analysis (INAA), refers to selective counting of radiation emitted by the appropriate radionuclide in the mixture of radionuclides.[81] In this case, some information about the activity of an individual radionuclide is obtained by using suitable measurement and calculation techniques.

The main advantage of NAA is that it is nondestructive and thus can be used in unique and very precious samples such as moon rocks and archaeological artifacts. The main component of the sample matrices, such as H, C, O, N, P, and Si, hardly

TABLE 1.10
Detection Limit of Activation Analysis Using a Thermal Neutron Flux of 10^{13} n/cm^{2}*s[82]

Sensitivity 10^{-12} g	Elements
1	Dy, Eu
1–10	Mn, In, Lu
$10–10^2$	Co, Rh, Ir, Br, Sm, Ho, Re, Au
$10^2–10^3$	Na, Ge, Sr, Nb, Cs, La, Yb, U, Ar, V, Cu, Ga, As, Pd, Ag, I, Pr, W
$10^3–10^4$	Al, Cl, K, Sc, Se, Kr, Y, Ru, Gd, Tm, Hg, Si, Ni, Rb, Cd, Te, Ba, Tb, Hf, Ta, Os
$10^4–10^5$	Pt
$10^5–10^6$	P, Ti, Zn, Mo, Sn, Xe, Ce, Nd, Mg, Ca, Tl, Bi
10^7	F, Cr, Zr, Ne, S, Pb, Fe

form any radioactive isotopes, so they seem to be invisible. This makes the method highly sensitive for measuring trace elements, because the main constituents do not cause any interference.

Activation analysis, under the appropriate experimental parameters, simultaneously can determine concentrations for nearly 70 elements with a high degree of sensitivity and accuracy (see Table 1.10).

The facility for sample treatment before analysis is another significant advantage of NAA compared with other techniques. Depending on the type of samples to be analyzed, sample preparation procedures may be very different. In most cases, neither chemical treatment nor addition of reagent is required to prepare the sample for analysis. Therefore, activation analysis is relatively free of sources of contamination. In fact, the only steps where contamination may occur and have an impact on the final result are sampling, sample handling and storage, and sample preparation. Most often, before the sample analysis, there is a need to wash, peel, separate seeds and stones, dry (oven drying or lyophilization),[83] and to homogenize (with or without cryogenic techniques). The aim of sample preparation is to reduce the sample size suitable for encapsulation prior to conducting the irradiation procedure. Generally, a suitable sample mass may vary between a few milligrams and several grams; its size depends on the sensitivities required and on the specific nature of the analyzed elements.

Despite all its advantages, NAA also has some limitations. The biggest disadvantage is the requirement for access to a nuclear research reactor. Also there is a huge problem with turnaround time, which may be somewhat longer than in other techniques, varying from a few days to one or more weeks. The cause of this drawback is the different half-lives for radioactive isotopes, which can be divided into three categories: short-lived, moderately-lived, and long-lived nuclides. Furthermore, some elements (Pb, Be, N, O, P, and Bi) cannot be determined by this method, which also is not suitable for analysis of liquid (water) samples without extensive pretreatment.

1.3.3 Electrochemical Methods

Two other important electrochemical methods for elemental analysis are voltamperometry and ion selective electrodes. Both methods can be used only for water solutions; therefore, food products must first be mineralized and dissolved before analysis.

Anodic stripping voltammetry (ASV) is characterized by compact equipment and reasonable cost, and does not require preliminary sample concentrating. Voltammetric measurements are performed with a voltamperometric analyzer equipped, for example, with three electrodes — a stationary mercury electrode (hanging mercury drop electrode or mercury film electrode) as a working electrode, an Ag/AgCl/KCl electrode as a reference electrode, and a platinum wire used as a reference and auxiliary electrode. In the first step, the metal ions are collected in the mercury by reduction to a metallic state and amalgamation with Hg. The deposition is followed by a voltammetric scan towards more positive potentials, collected metal is oxidized, and the current produced is determined.[84–87] The ASV method, due to preconcentrating metal in the mercury film, makes it possible to determine very low concentrations (down to about 0.01 μg/l). Quantitative analysis is performed by adding standards directly to the electrochemical cell containing the analyzed sample. The results are calculated by a comparison of signals for real samples and samples spiked with known standards. This method eliminates the influence of the matrix on the results, an important consideration in the case of food. Only dissolved metals (in the form of ions) are detected by ASV,[66] however.

Ion-selective electrode (ISE) methods are very useful for the determination of major cations and anions in aqueous solutions.[88] Ion-selective electrodes for different analytes are made and commercialized by many companies. The most popular applications are in the determination of fluorides. An example of a measuring device with an ion-selective electrode used for fluorides-determination is presented in Figure 1.6. The main part of the device is the membrane made of europium-doped (to lower its electrical resistance and to facilitate ionic charge transport) LaF_3 crystal incorporated in the measuring electrode body. The membrane is connected to an internal electrolyte solution (typically, NaCl and NaF each 0.1 mol/l) and the sample solution to be analyzed. The device is equipped also with internal (typically, Ag/AgCl wire) and external (constant-potential) reference electrodes. The electrochemical cell can be described as an Ag/AgCl, Cl^- (0.1 mol/l), F^- (0.1 mol/l)/LaF_3 crystal/analyte solution//external reference electrode.

The measurement is carried out potentiometrically at zero current according to the Nernst-type relation. Both ISE and external reference electrodes are dipped into the measuring cell containing the sample solution with the addition of an appropriate buffer, and the potential difference between these electrodes is measured as the analytical signal.

ISE methods are often used for the determination of chlorides with the active measuring membrane made from silver chloride or a mixture of silver chloride and silver sulfide, nitrates using porous or nonporous membrane electrodes, and cyanides with membrane made of AgI and Ag_2S. These methods also can determine different cations such as: Na^+, K^+, Ca^{2+}, Mg^{2+}, Cu^{2+}, Fe^{3+}, Cd^{2+}, and Zn^{2+}.[88] During analysis

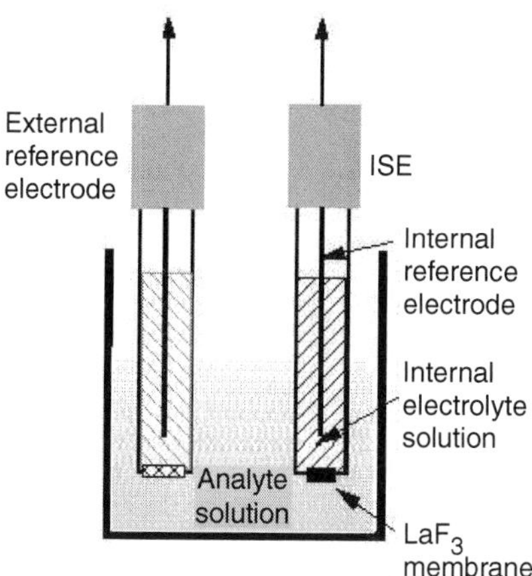

FIGURE 1.6 An example of a measuring device with an ion-selective electrode used for fluoride determination.

some interferences can occur. These interferences can result from the presence in the sample solution of some components, which can prevent the probe from sensing the ion of interest (for example, complexion) or other ions from being measured together with the analyte.

1.3.4 Ion Chromatography

Chromatographic techniques generally are used for the separation of organic components, gaseous, or liquid mixtures as a part of the procedure of their determination. However, some organometallic compounds can be determined — for example, methylmercury or organotin compounds. Only ion chromatography (IC) connected with conductivity detectors is suitable for the determination of inorganic analytes — both anions (such as fluoride, bromide, chloride, iodide, chlorate, iodate, bromate, nitrite, nitrate, and sulfate) and cations (such as lithium, sodium, ammonium, potassium, alkaline earth metals, and transition metals). Aqueous solutions, which may need filtration, dilution, and cleaning to remove interferences, are required for analysis. Ion chromatography is a special type of liquid chromatography in which an ion separator column or ion-exchange resins (as immobile phase) are used to separate atomic or molecular ions. The ions of interest are separated by means of differing affinities for the column packing material, whereas the ions are swept along by flowing eluent. The anions are most often analyzed in this manner and can be determined with high precision simultaneously. Also, many cations can be determined simultaneously in a single run using cation-exchange resins as a separator column packing material. Such a system consists of a pump, an injector, a guard column, a separator column, a suppressor device (only for suppressed IC, the majority of applications), and a flow-through detector. An eluent suppressor, consisting of an ion-exchange column or membrane, situated between an analytical column and detector, converts the mobile phase ions to neutral forms. In the case of anion analysis, when the mobile phase is often NaOH or $NaHCO_3$, the eluent suppressor supplies H^+ to neutralize anions and retain or remove the Na^+. During cation analysis (HCl or HNO_3 are often in the mobile phase), an eluent suppressor supplies OH^- anions for neutralizing an eluent. The suppressor improves the detection limits, increasing the analyte conductivity signal while decreasing eluent background noise and sample counter-ion interferences, but single column (only a separator column) IC is also in wide use. Sometimes the third column, between the suppressor and the detector (which replaces the analyte ions with other ions in order to improve the detection sensitivity) can be used. Electric conductivity detectors are most often used, but specific electrometric detectors such as solute-specific amperometric detectors or colorimetric, fluorescence, ICP-AES, and UV-VIS spectrometric detectors are also in use.[88,89] In Table 1.11 some selected examples of inorganic ions determination in food using ion chromatography are listed.[89]

TABLE 1.11
Selected Examples of Inorganic Ions Determination in Food Using Ion Chromatography

Analyte	Types of Food	Eluent	Detector	Reference
NO_3^-, IO_3^-	Milk, popcorn	Sodium octanesulphonate	Conductivity	90
NO_2^-	Spinach	Phtalic acid–10% acetone	Coulometry	91
NO_3^-, PO_4^{3-}	Fruit juice	NaOH, CH_3OH, C_2H_5OH	Conductivity	92
NO_3^-, NO_2^-	Cereal baby food	KH_2PO_4-Na_2HPO_4	UV/VIS	93
SO_4^{2-}	Tea, wine	Potassium–hydrogen-phtalate–phtalic acid	Conductivity	94, 95
SO_3^{2-}	Beer	Na_2CO_3-$NaHCO_3$, formaldehyde	Conductivity	96
SO_3^{2-}	Different food	Sulphuric acid	Amperometry	97, 98
PO_4^{3-}, Cl^-, F^-, I^-	Rice, tea, liver, bovine, fruits, vegetables	Na_2CO_3-$NaHCO_3$	Conductivity, ICP/MS, ICP/AES	99, 100, 101, 102
Br^-, BrO_3^-	Bakery products	NaOH	ICP-MS	103
CrO_4^{2-}, Ca^{2+} MnO_4^-, Mg^{2+}	Orange juice, potato chips	2,5-dihydroxy-1,4-benzenedisulphonic acid	UV/VIS	104
CN^-	Chopped fruits	H_3BO_3-NaOH-Na_2CO_3	Amperometry	105
Pb^{2+}, Cd^{2+}, Fe^{3+}, Cu^{2+}, Ni^{2+}, Co^{2+}, Zn^{2+}	Rice flour, tea leaves,	H_2SO_4-HCl-KCl pyridine-2-6-dicarboxylic acid CH_3COOH-CH_2COONa	UV/VIS	99
Pb^{2+}, Cd^{2+}, Fe^{3+}, Cu^{2+}, Ni^{2+}, Co^{2+}, Zn^{2+}, Fe^{2+}, Mn^{2+}	Porridge oats, wine	Tartaric acid	UV/VIS	106

TABLE 1.11 (CONTINUED)
Selected Examples of Inorganic Ions Determination in Food Using Ion Chromatography

Analyte	Types of Food	Eluent	Detector	Reference
II group, NH_4^+	Fruit, juice, puree	Methanesulphonic acid	Conductivity	107
Ca^{2+}, Sr^{2+}, Ba^{2+}	Milk, skimmed powder	KNO_3 – lactic acid	UV/VIS	108
I & II group, NH_4^+	Bread crumbs, cheese, pretzels	EDTA-HNO_3	Conductivity	109
NH_4^+, Ca^{2+}, Na^+, K^+, Rb^+, Ba^{2+}, Mg^{2+}	Food extracts	Lithium hydrogenphtalate–ethylendiamine, lithium 4–hydroxybenzoate	Conductivity	110

ABBREVIATIONS

AAS Atomic absorption spectrometry
AES Atomic emission spectrometry
ASE Accelerated solvent extraction
ASV Anodic stripping voltammetry
CV-AAS Cold vapor atomic absorption spectrometry
ET-AAS Electrothermal atomic absorption spectrometry
FAAS Flame atomic absorption spectrometry
FAES Flame atomic emission spectrometry
FP Flame photometry
GC-ECD Gas chromatography with electron capture detector
GF-AAS Graphite furnace atomic absorption spectrometry
GLP Good Laboratory Practice
HG Hydride generation
HG-AAS Atomic absorption spectrometry with hydride generation
DNA Deoxyribonucleic acid
IC Ion chromatography
ICP Inductively coupled plasma
ICP-AES Atomic emission spectrometry with inductively coupled plasma excitation
ICP-MS Mass spectrometry with inductively coupled plasma excitation
ICP-OES Optical emission spectrometry with inductively coupled plasma excitation
INAA Instrumental neutron activation analysis
ISE Ion-selective electrode
LE Liquid extraction
MeHg Methylmercury
MS Mass spectrometry

NAA	Neutron activation analysis
PFA	Perfluoroalkoxy
PTFE	Polytetrafluoroethylene
QA	Quality assurance
RNA	Ribonucleic acid
RNAA	Radiochemical neutron activation analysis
SFE	Supercritical fluid extraction
SPE	Solid-phase extraction
UV	Ultraviolet
VIS	Visible

REFERENCES

1. Synowiecki, J., Składniki mineralne, in *Chemia Żywności* (Food Chemistry), Sikorski, Z., Ed., WN-T, Warszawa, 2000.
2. Gawęcki, J. and Mossor-Pietraszewska, T., Red, *Kompendium Wiedzy o Żywności, Żywieniu i Zdrowiu*, PWN, Warszawa, 2004.
3. Eichler, W., *Gift in unserer Nahrung,* Kilda, Greven, 1982.
4. Bull, R.J., Carcinogenic and mutagenic properties of chemicals in drinking water, *Sci. Total Environ.,* 47, 385, 1985.
5. Hartwig, A., Carcinogenicity of metal compounds: possible role of DNA repair inhibition, *Toxicol. Lett.*, 102, 235, 1998.
6. Kasprzak, K.S., Oxidative DNA and protein damage in metal-induced toxicity and carcinogenesis, *Free Radic. Biol. Med.*, 32, 958, 2002.
7. Pourahmad, J. et al., Carcinogenic metal induced sites of reactive oxygen species formation in hepatocytes, *Toxicol. Vitro*, 17, 803, 2003.
8. Schwerdtle, T., Walter, I., and Hartwig, A., Arsenite and its biomethylated metabolites interfere with the formation and repair of stable BPDE-induced DNA adducts in human cells and impair XPAzf and Fpg, *DNA Repair*, 2, 1449, 2003.
9. Tully, D.B. et al., Effects of arsenic, cadmium, chromium, and lead on gene expression regulated by a battery of 13 different promoters in recombinant HepG2 cells, *Toxicol. Appl. Pharmacol.*, 168, 79, 2000.
10. Connell, D.W., *Bioaccumulation of Xenobiotic Compounds*, CRC Press, Boca Raton. FL, 1990.
11. Ferguson, L.R., Natural, man-made mutagens and carcinogens in the human diet, *Mutat. Res.*, 443, 1, 1999.
12. Moore, J.W., *Inorganic Contaminants of Surface Water: Research and Monitoring Priorities*, Springer-Verlag, New York, 1991.
13. Nriagu, J.O. and Pacyna, J., Quantitative assessment of worldwide contamination of air, water and soils by trace metals, *Nature*, 333, 134, 1988.
14. Rojas, E., Herrera, L.A., Poirier, L.A., and Ostrosky-Wegman, P., Are metals dietary carcinogens?, *Mutat. Res.*, 443, 157, 1999.
15. Kabata-Pendias, A., *Trace Elements in Soil and Plants*, CRC Press, Boca Raton, FL, 2001; Kabata-Pendias, A., *Biogeochemia Pierwiastków Śladowych*, PWN, Warszawa 1999.
16. Waalkes, M.P. and Misra, R.R., Cadmium carcinogenicity and genotoxicity, in *Toxicology of Metals*, Chang, L.W., Ed., CRC Press, Boca Raton, FL, 1996, p. 231.

17. Cohen, M.D. et al., Mechanisms of chromium carcinogenicity and toxicity, *Crit. Rev. Toxicol.*, 23, 255, 1993.
18. McLaughlin, M.J., Parker, D.R., and Clarke, J.M., Metals and micronutrients — food safety issues, *Field Crops Res.*, 60, 143, 1999.
19. Pershagen, G., The carcinogenity of arsenic, *Environ. Health Perspect.*, 40, 93, 1981.
20. Pocock, S.J. et al., Blood cadmium concentration in the general population of British middle-aged men, *Hum. Toxicol.*, 7, 95, 1988.
21. Verougstraete, V., Lison, D., and Hotz, P., Cadmium, lung a prostate cancer: a systematic review of recent epidemiological data, *J. Toxicol. Environ. Heallth*, Part B, 6, 227, 2003.
22. Coogan, T.P. et al., Toxicity and carcinogenicity of nickel compounds, *CRC Crit. Rev. Toxicol.*, 19, 341, 1989.
23. Rains, T.C., Application of atomic absorption spectrometry to the analysis of foods, in *Atomic Absorption Spectrometry; Theory, Design and Applications*, Haswell, S.J., Ed., Elsevier, Amsterdam, 1991.
24. Capar, S.G. and Szefer, P., Determination and speciation of trace elements in foods, in *Methods of Analysis of Food Components and Additives*, Otles, S., Ed., CRC Press, Boca Raton, FL, 2005, p. 111.
25. Santos, D. et al., Determination of Cd and Pb in food slurries by GFAAS using cryogenic grinding for sample preparation, *Anal. Bioanal. Chem.*, 373, 183, 2002.
26. Jarris, I., Sample preparation for ICP-MS, in *Handbook of Inductively Coupled Plasma Mass Spectrometry*, Jarris, K.E., Gray, A.L., and Houk, R.S., Eds., Glasgow, 1993, p. 172.
27. Gilman, L.B., and Engelhart, W.G., Recent advances in microwave sample preparation, *Spectroscopy*, 4, 14, 1989.
28. Environmental Protection Agency, SW-846 EPA Method 3052, Microwave assisted acid digestion of siliceous and organically based matrices, in *Test Methods for Evaluating Solid Waste*, U.S. E.P.A., Washington, D.C., 1996.
29. Official Methods of Analysis of AOAC International, Official Method 999.10, Lead, Cadmium, Zinc, Copper, and Iron in Foods — Atomic Absorption Spectrophotometry after Microwave Digestion, AOAC, Gaithersburg, 2002.
30. Matusiewicz, H., Wet Digestion Methods, in *New Horizons and Challenges in Environmental Analysis and Monitoring*, Namieśnik, J., Chrzanowski, W., Żmijewska, P., Eds., Centre of Excellence in Environmental Analysis and Monitoring, Gdańsk, 2003, p. 224, chap. 13.
31. Taylor, L.R., Pap, R.P., and Pollard, B.D., *Instrumental Methods for Determining Elements,* VCH, New York, 1994.
32. Willard, H.H. et al., *Instrumental Methods of Analysis*, Wadsworth, Belmont, CA, 1981.
33. Simpkins, W.A. et al., Trace elements in Australian orange juice and other products, *Food Chem.,* 71, 423, 2000.
34. Munilla, M.A. et al., Determination of metals in seaweeds used as a food by inductively coupled plasma atomic-emission spectrometry, *Analusis,* 23, 463, 1995.
35. Sawaya, W.N. et al., Nutritional profile of Kuwaiti composite dishes: minerals and vitamins, *J. Food Compos. Anal.,* 11, 70, 1998.
36. Eckhoff, K.M. and Maage, A., Iodine content in fish and other food products from East Africa analyzed by ICP-MS, *J. Food Compos. Anal.,* 10, 270, 1997.
37. Ikem, A. et al., Levels of 26 elements in infant formula from USA, UK, and Nigeria by microwave digestion and ICP–OES, *Food Chem.,* 77, 439, 2002.

38. Fernández, P.L. et al., Multi-element analysis of tea beverages by inductively coupled plasma atomic emission spectrometry, *Food Chem.,* 76, 483, 2002.
39. Marisa, C. et al., Determination of lead isotope ratios in port wine by inductively coupled plasma mass spectrometry after pre-treatment by UV-irradiation, *Anal. Chim. Acta*, 396, 45, 1999.
40. Caroli, S. et al., A pilot study for the production of a certified reference material for trace elements in honey, *Microchem. J.*, 67, 227, 2000.
41. Şimşek, A. et al., Determination of boron in hazelnut (*Corylus avellana* L.) varieties by inductively coupled plasma optical emission spectrometry and spectrophotometry, *Food Chem.,* 83, 293, 2003.
42. Bulska, E., Analytical advantages of using electrochemistry for atomic spectrometry, *Pure Appl. Chem.,* 73, 1, 2001.
43. Bazzi, A., Kreuz, B., and Fischer, J., Determination of calcium in cereal with Flame Atomic Absorption Spectroscopy: an experiment for a quantitative methods of analysis course, *J. Chem. Educ.*, 81, 1042, 2004.
44. Lendinez, E. et al., Chromium in basic foods of the Spanish diet: seafood, cereals, vegetables, olive oils and dairy products, *Sci. Total Environ.*, 278, 183, 2001.
45. Smrkolj, P. et al., Selenium content in selected Slovenian foodstuffs and estimated daily intakes of selenium, *Food Chem.*, 90, 691, 2005.
46. Jaganathan, J., Reisig, A.L., and Dugar, S.M., Determination of cadmium in wines using Graphite Furnace Atomic Absorption Spectrometry with Zeeman background correction, *Microchem. J.,* 56, 221, 1997.
47. Iyengar, G.V. et al., Content of minor and trace elements, and organic nutrients in representative mixed total diet composites from the U.S., *Sci. Total Environ.*, 256, 215, 2000.
48. De-qiang, Z. et al., Determination of cadmium in flour by atom trapping flame atomic absorption spectrometry using derivative signal processing, *Anal. Chim. Acta*, 405, 185, 2000.
49. Mindak, W.R. and Dolan, S.P., Determination of arsenic and selenium in food using a microwave digestion-dry ash preparation and flow injection hydride generation atomic absorption spectrometry, *J. Food Compos. Anal.*, 12, 111, 1999.
50. Anzano, J.M. et al., Zinc and Manganese Analysis in Maize by Microwave Oven Digestion and Flame Atomic Absorption Spectrometry, *J. Food Compos. Anal.*, 13, 837, 2000.
51. Miller-Ihli, N.J., Atomic Absorption and Atomic Emission Spectrometry for the determination of the trace element content of selected fruits consumed in the U.S., *J. Food Compos. Anal.,* 9, 301, 1996.
52. Plessi, M., Bertelli, D., and Monzani, A., Mercury and selenium content in selected seafood, *J. Food Compos. Anal.*, 14, 461, 2001.
53. Tüzen, M., Determination of heavy metals in fish samples of the middle Black Sea (Turkey) by graphite furnace atomic absorption spectrometry, *Food Chem.*, 80, 119, 2003.
54. Anzano, J.M. and Gónzalez, P., Determination of iron and copper in peanuts by flame atomic absorption spectrometry using acid digestion, *Microchem. J.*, 64, 141, 2000.
55. Kiliç, Z. et al., Determination of lead, copper, zinc, magnesium, calcium and iron in fresh eggs by atomic absorption spectrometry, *Food Chem.*, 76, 107, 2002.
56. Karadede, H. and Unlu, E., Concentration of some heavy metals in water, sediment and fish species from the Ataturk Dam Lake (Euphrates), Turkey, *Chemosphere,* 41, 1371, 2000.

57. Vazquez, F.G. et al., Metals in fish and shrimp of the Campeche Sound, Gulf of Mexico, *Bull. Environ. Contam. Toxicol.*, 67, 756, 2001.
58. Pohl, P., Hydride generation — recent advances in atomic emission spectrometry, *Trends Anal. Chem.*, 23, 87, 2004.
59. Yang, J. et al., Evaluation of continuous hydride generation combined with helium and argon microwave induced plasmas using a surfatron for atomic emission spectrometric determination of arsenic, antimony and selenium, *Spectrochim. Acta* Part B, 50, 1351, 1363, 1995.
60. Sun, H., Suo, R., and Lu, Y., Determination of zinc in food using atomic fluorescence spectrometry by hydride generation from organized media, *Anal. Chim. Acta*, 457, 305, 2002.
61. Boutakhrit, K. et al., Open digestion under reflux for the determination of total arsenic in seafood by inductively coupled plasma atomic emission spectrometry with hydride generation, *Talanta*, 66, 1042, 2005.
62. Murphy, J. and Cashman, K. D., Selenium content of a range of Irish foods, *Food Chem.,* 74, 493, 2001.
63. Ródenas-Torralba, E., Morales-Rubio, A., and de la Guardia, M., Multicommutation hydride generation atomic fluorescence determination of inorganic tellurium species in milk, *Food Chem.*, 91, 181, 2005.
64. Cava-Montesinos, P. et al., Determination of arsenic and antimony in milk by hydride generation atomic fluorescence spectrometry, *Food Chem.*, 60, 787, 2003.
65. Klapec, T. et al., Selenium in selected foods grown or purchased in eastern Croatia, *Food Chem.,* 85, 445, 2004.
66. Biziuk, M., Namieśnik, J., and Zaslawska, L., Heavy metals in food products and biological samples from the Gdansk district, in *Radionuclides and Heavy Metals in Environment,* Frontasyeva, M.V., Perelygin, V.P., Vater, P., Eds., NATO Science Series, IV. Earth and Environmental Sciences, Vol. 5, Kluwer Academic Publ., Dordrecht, Holland, 2001, p. 209.
67. Dabeka, R., Bradley, P., and McKenzie, A.D., Routine, high-sensitivity, cold vapor atomic absorption spectrometric determination of total mercury in foods after low-temperature digestion, *J. AOAC Int.*, 85, 1136, 2002.
68. Palmieri, H.E.L., and Leonel, L.V., Determination of methylmercury in fish tissue by gas chromatography with microwave-induced plasma atomic emission spectrometry after derivatization with sodium tetraphenylborate, *Fresenius J. Anal. Chem.*, 363, 466, 2000.
69. Grinberg, P. et al., Solid phase microextraction capillary gas chromatography combined with furnace plasma emission spectrometry for speciation of mercury in fish tissue, *Spectrochim. Acta* Part B, 58, 427, 2003.
70. Marczenko, Z., *Separation and Spectrophotometric Determination of Elements*, Ellis Harwood, London, 1986.
71. Cresser, M.S. and Marr, I.L., Optical spectrometry in the analysis of pollutants, in *Instrumental Analysis of Pollutants*, Hewitt, C.N., Ed., Elsevier, London 1991, p. 99, chap. 3.
72. Soliman, K. and Zikovsky, L., Determination of Br, Ca, Cl, Co, Cu, I, K, Mg, Mn, Na, Rb, S, Ti, and V in cereals, oils, sweeteners and vegetables sold in Canada by neutron activation analysis, *J. Food Compos. Anal.*, 12, 85–89, 1999.
73. Zaichick, V. et al., Instrumental neutron activation analysis of essential and toxic elements in child and adolescent diets in the Chernobyl disaster territories of the Kaluga Region, *Sci. Total Environ.*, 192, 269, 1996.

74. Akhter, P., Orfi, S.D., and Ahmad, N., Cesium concentration in the Pakistani diet, *J. Environ. Radioact.*, 67, 109, 2003.
75. Gambelli, L. et al., Minerals and trace elements in some Italian dairy products, *J. Food Compos. Anal.,* 12, 27, 1999.
76. Pokorn, D. et al., Elemental composition (Ca, Mg, Mn, Cu, Cr, Zn, Se, and I) of daily diet samples from some old people's homes in Slovenia, *J. Food Compos. Anal.,* 11, 47–53, 1998.
77. Bonafaccia, G. et al., Trace elements in flour and bran from common and tartary buckwheat, *Food Chem.*, 83, 1, 2003.
78. Chien-Yi, C., Trace elements in Taiwanese health food, Angelica keiskei, and other products, *Food Chem.*, 84, 545, 2004.
79. Jonah, S.A. and Williams, I.S., Nutrient elements of commercial tea from Nigeria by an instrumental neutron activation analysis technique, *Sci. Total Environ.*, 258, 205, 2000.
80. Bode, P. and De Goeij, J.J.M., Activation analysis, in *Encyclopedia of Environmental Analysis and Remediation*, Meyers, R.A., Ed., John Wiley & Sons, New York, 1998, p. 68.
81. Bode, P., Nuclear analytical techniques for environmental research, in *New Horizons and Challenges in Environmental Analysis and Monitoring,* Namienik, J., Chrzanowski, W., and Mijewska, P., Eds., Centre of Excellence in Environmental Analysis and Monitoring, Gdańsk, 2003, chap. 1, p. 1.
82. IAEA-TECDOC-564, Practical Aspects of Operating a Neutron Activation Analysis Laboratory, IAEA, Vienna 1990.
83. Howe, A. et al., Elemental composition of Jamaican foods 1: a survey of five food crop categories, *Environ. Geochem. Health*, 27, 19, 2005.
84. Achterberg, E.P. and Braungardt, Ch., Stripping voltammetry for the determination of trace metal speciation and in-situ measurement of trace metal distribution in marine waters, *Anal. Chim. Acta,* 400, 381, 1999.
85. Bersier, P.M., Howell, J., and Bruntlett, C., Advanced electroanalytical techniques versus Atomic Absorption Spectrometry, Inductively Coupled Plasma Mass Spectrometry in environmental analysis, *Analyst*, 119, 219, 1994.
86. Locatelli, C. and Torsi, G., Cathodic and anodic stripping voltammetry: simultaneous determination of As-Se and Cu-Pb-Cd-Zn in the case of very high concentration ratios, *Talanta*, 50, 1079, 1999.
87. Wang, J., *Stripping Analysis: Principles, Instrumentation and Applications*, VCH, Deerfield Beach, 1985.
88. Nollet, L.M.L., Ed., *Handbook of Water Analysis*, Marcel Dekker, Basel, New York, 2000.
89. Buldini, P.L., Cavalli, S., and Trifiro, A., State-of-art ion chromatographic determination of inorganic ions in food, *J. Chromatogr. A,* 789, 529, 1997.
90. Kaine, L.A., Crowe, J.B., and Wolnik, K.A., Forensic application of coupling non-suppressed ion-exchange chromatography with ion-exclusive chromatography, *J. Chromatogr. A*, 673, 141, 1992.
91. Bosch-Bosch, N. et al., Determination of nitrite levels in refrigerated and frozen spinach by ion chromatography, *J. Chromatogr. A*, 706, 221, 1995.
92. Saccani, G. et al., Use of ion chromatography for the measurement of organic acids in fruit juices, *J. Chromatogr. A*, 706, 395, 1995.
93. Nieto, M.T. et al., Nitrite and nitrate contents in cereal-based baby foods, *Alimentria*, 265, 71, 1995.

94. Ding, M.Y., Chen, P.R., and Luo, G.A., Simultaneous determination of organic acids and inorganic anions in tea by ion chromatography, *J. Chromatogr. A*, 764, 341, 1997.
95. Mongay, C., Pastor, A., and Olmmos, C., Determination of carboxylic acids and inorganic anions in wines by ion-exchange chromatography, *J. Chromatogr. A*, 736, 351, 1996.
96. Villasenor, S.K., Matrix elimination in liquid chromatography using heart-cut column switching techniques, *Anal. Chem.*, 63, 1362, 1991.
97. Kim, H.J., Determination of sulfite in food and beverages with ion-exclusive chromatography with electrochemical detection: collaborative study, *J. Assoc. Off. Anal. Chem.*, 73, 216, 1990.
98. Kim, H.J. and Conca, K.R., Determination of sulfur dioxide in grapes, comparison of Monier-Williams method and two ion exclusion chromatographic methods, *J. Assoc. Off. Anal. Chem.*, 73, 983, 1990.
99. Buldini, P.L., Cavalli, S., and Mevoli, A., Sample pretreatment by UV photolysis for the ion chromatographic analysis of plant material, *J. Chromatogr. A*, 739, 167, 1996.
100. Novic, M. et al., Determination of chlorine, sulphur and phosphorus in organic materials by ion chromatography using electrodialysis sample pretreatment, *J. Chromatogr. A*, 704, 530, 1995.
101. Smolders, E., van Dael, M., and Merckx, R.J., Simultaneous determination of extractable sulfate and malate in plant extracts using ion-chromatography, *J. Chromatogr. A*, 514, 371, 1990.
102. Schnetger, B. and Muramatsu, Y., Determination of halogens, with special reference to iodine, in geological and biological samples using pyrohydrolysis for preparation and inductively coupled plasma mass spectrometry and ion chromatography for measurement, *Analyst*, 121, 1627, 1996.
103. Heitkemper, D.T. et al., Practical application of element-specific detection by inductively coupled plasma atomic emission spectroscopy and inductively coupled plasma mass spectrometry to ion chromatography of food, *J. Chromatogr. A*, 671, 101, 1994.
104. Mehra, M.C. and Kandil, M., Ion chromatographic determinations of ionic analytes with benzendisulfonic acid eluent using indirect ultraviolet detection, *Analusis*, 24, 17, 1996.
105. Chadha, K. and Lawrence, J. F., Ion chromatographic determination of cyanide contamination in fruit and evaluation of a colorimetric test kid, *Int. J. Environ. Anal. Chem.*, 44, 197, 1991.
106. Yan, D. and Schwedt, G., Trace determination of aluminum and iron by ion chromatography and post chromatographic derivatization, *Fresenius Z. Anal. Chem.*, 320, 252, 1985.
107. Trifiro, E. et al., Determination of cations in fruit juices and purées by ion chromatography, *J. Chromatogr. A*, 739, 175, 1996.
108. Jones, R., Foulkes, M., and Paull, B., Determination of barium and strontium in calcium containing matrices using high-performance chelation ion chromatography, *J. Chromatogr. A*, 673, 173, 1994.
109. Morawski, J., Alden, P., and Sims, A., Analysis of cation nutrients from food by ion chromatography, *J. Chromatogr. A*, 640, 359, 1993.
110. Saari-Nordhaus, R. and Anderson, J.M., Jr., Simultaneous analysis of anions and cations by single-column ion chromatography, *J. Chromatogr. A*, 549, 1991.

2 Speciation of Mineral Components in Food – Analytical Implications

Aleksandra Polatajko and Joanna Szpunar

CONTENTS

2.1 INTRODUCTION

The safety and nutritional quality of food are determined by both the total level and the speciation (distribution among different chemical forms) of trace elements. Because bioavailability and metabolism are often highly variable between different chemical forms of the same element, information on trace element speciation is essential. It provides an understanding of how the absorption and bioavailability of elements can be reduced for toxic elements or improved for essential nutrients. As a result, accurate, reproducible, and precise methodologies are needed for food sampling, sample pretreatment, and analysis to maintain the initial species intact and to determine them at the levels required.

According to the International Union of Pure and Applied Chemistry (IUPAC) definition, metal speciation in foods comprises analytical activities of identifying and/or measuring the quantities of one or more individual chemical species in a sample.[1] The analytical instrumental approach is usually based on the combination of a chromatographic separation technique with an element-specific detection technique (see Chapter 1 this book). The former ensures that the compound of interest leaves the column unaccompanied by other species of the analyte element, whereas the latter enables a sensitive and specific detection of the target element. Coupled (hyphenated) techniques have become a fundamental tool for speciation analysis and have been discussed in many review publications[2–5] and in Chapter 1.

2.2 SPECIES OF INTEREST

The source of elemental species in foodstuffs can be either anthropogenic or natural. In the first case it is a result of external contamination (environmental; occurring during processing, or leaching from packaging materials). In the second case it results from an endogenous synthesis by a plant or an animal (methylmercury, organoarsenic, or organoselenium species).

From the chemical point of view, metal and metalloid species found in foodstuffs can be divided into one of those two categories:

1. Those which contain a covalent bond between a carbon atom and a metal or metalloid atoms such as
   - tetraalkylated lead compounds used an antiknock additives to gasoline, which are degraded to trialkyl or dialkyl species
   - ingredients of antifouling paints, such as butyl, octyl, and phenyl tin species, released into the aquatic environment and found in seafood products
   - methylmercury, a product of biomethylation of mercury by marine organisms
   - products of the metabolism of arsenic by marine biota leading to the formation of the carbon–arsenic bond, e.g., as in arsenobetaine or arsenosugars
   - selenoaminoacids, selenopeptides, and proteins, biosynthesized by plants and animals
2. Coordination complexes in which a metal is coordinated by a bio-ligand present in a biological matrix (e.g., metal-binding peptides, which are enzymatically synthesized in living organisms exposed to heavy-metal stress)

Arsenic and selenium species of interest for food analysis are listed in Table 2.1 and Table 2.2, respectively.

2.3 INSTRUMENTAL APPROACHES

Analytical technique for species-selective analysis should fulfill three major requirements: (1) selectivity of the separation technique, which should allow the target

TABLE 2.1

Arsenic Species of Interest for Food Analysis

Species	Formula
Arsenous acid (arsenite)	$OH\text{-}As(OH)_2$
Arsenic acid (arsenate)	$O{=}As(OH)_3$
Monomethylarsonic acid	$CH_3AsO(OH)_2$
Dimethylarsinic acid	$(CH_3)_2AsO(OH)$
Tetramethylarsonium	$(CH_3)_4As^+$
Arsenobetaine	$(CH_3)_3As\text{-}CH_2\text{-}COOH$
Arsenobetaine-CH_2	$(CH_3)_3\text{-}CH_2As\text{-}CH_2\text{-}COOH$
Arsenocholine	$(CH_3)_3As\text{-}CH_2\text{-}CH_2\text{-}OH$
DMAsEt	$(CH_3)_2AsO(C_2H_5)$
DMAsAc	$(CH_3)_2AsO(CH_3CO)$

Arsenosugar 392

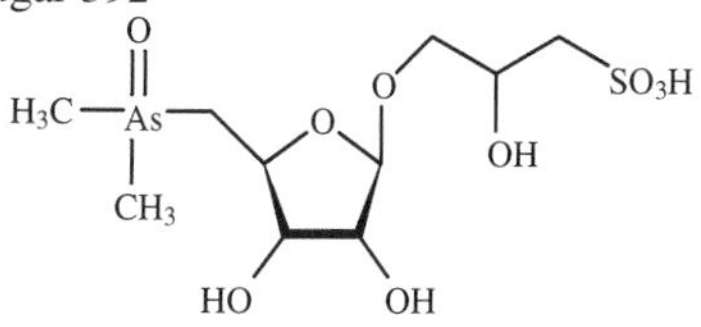

Arsenosugar 328

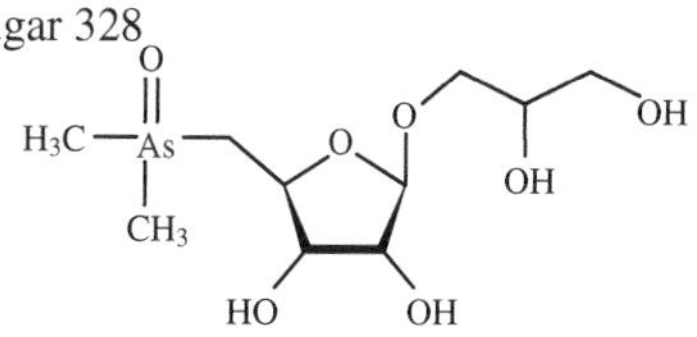

Arsenosugar 482

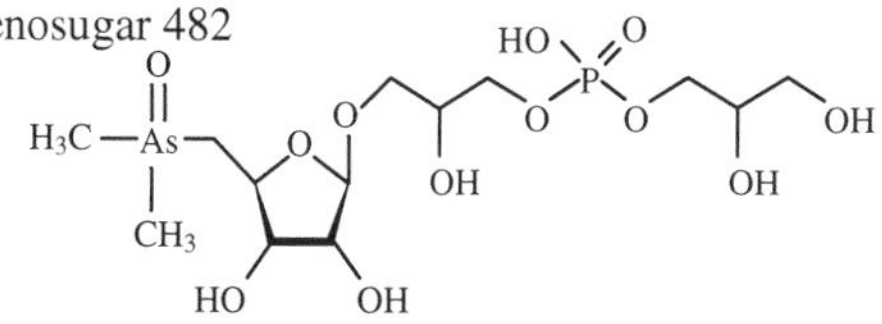

Arsenosugar 408

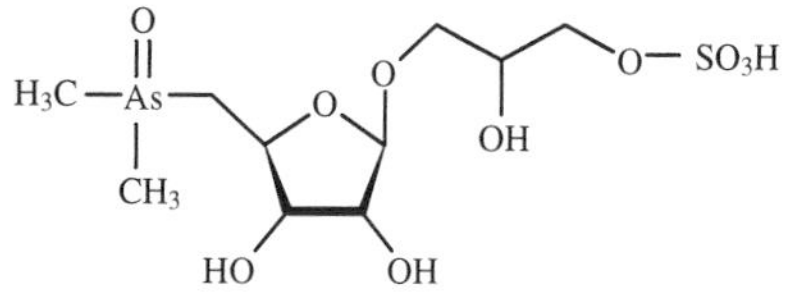

TABLE 2.2

Selenium Species of Interest for Food Analysis

Species	Formula
Selenite	SeO_3^{2-}
Selenate	SeO_4^{2-}
Trimethylselenonium	$(CH_3)_3Se^+$
Selenocysteine	H_3N^+-CH(COO^-)-CH_2-SeH
Selenocystine	H_3N^+-CH(COO^-)-CH_2-Se-Se-CH_2-CH(COO^-)-NH_3^+
Selenomethionine	H_3N^+-CH(COO^-)-CH_2-CH_2-SeH
Se-methylselenocysteine	H_3N^+-CH(COO^-)-CH_2-Se-CH_3
y-glutamylmethylselenocysteine	H_3N^+-CH_2-CH_2-CO-NH-CH(COO^-)-CH_2-Se-CH_3
Selenocystathionine	H_3N^+-CH(COO^-)-CH_2-CH_2-Se-CH_2-CH(COO^-)-NH_3^+
Selenohomocysteine	H_3N^+-CH(COO^-)-CH_2-CH_2-SeH
Selenocystamine	H_3N^+-CH_2-CH_2-Se-Se-CH_2-CH_2-NH_2
Adenosyl derivatives	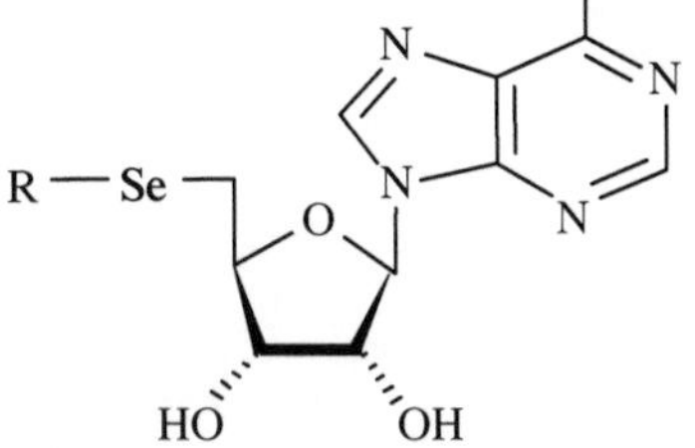

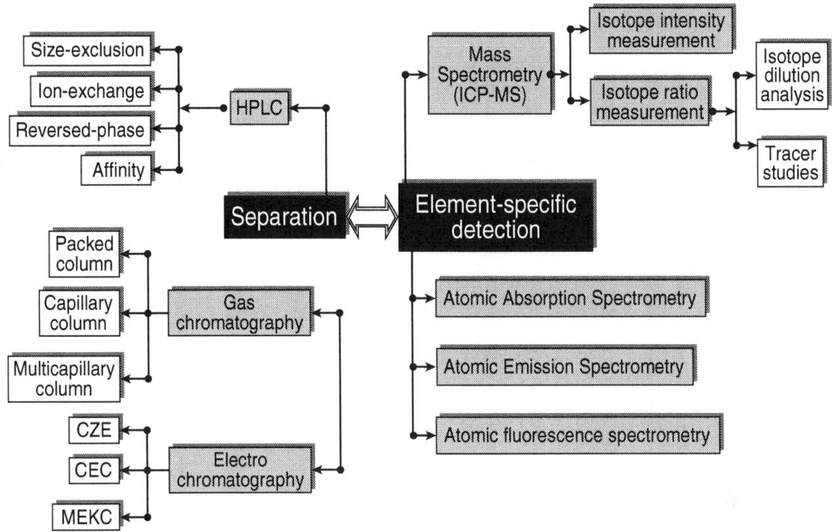

FIGURE 2.1 Instrumental approaches to speciation analysis.

analyte species to arrive at the detector well separated from potential matrix interferents and from other species of the same element, (2) sensitivity of the element selective detection technique as the already low concentrations of trace elements are usually distributed among several species, (3) capability of species identification; for this purpose retention-time matching with standards is usually employed. When standards are not available, the use of a molecule-specific detection technique (molecular mass spectrometry) is necessary.

The hyphenated techniques available for speciation analysis in foods are schematically shown in Figure 2.1. In the most frequent cases, a separation technique, notably chromatography (gas or liquid) is combined with inductively coupled plasma-mass spectrometric (ICP-MS) detection.

The separation technique used becomes of particular concern when the target species have similar physicochemical properties. Gas chromatography offers high separation efficiency, and very low detection limits can be achieved in the absence of the condensed mobile phase. For nonvolatile species, liquid phase column separation techniques, such as high performance liquid chromatography (HPLC) and cation exchange (CE), are the usual choice. They can be easily coupled online with ICP-MS. A variety of separation mechanisms and mobile phases can be used to allow the preservation of the species identity.

For element-specific detection in gas chromatography, a number of dedicated spectrometric detection techniques can be used, e.g., quartz furnace atomic absorption or atomic fluorescence for mercury, or microwave-induced plasma atomic emission for lead or tin. However, ICP-MS is virtually the only technique capable of coping, in the online mode, with the trace element concentrations (the femtogram level absolute detection limits) in liquid chromatography (LC) and capillary electrophoresis (CE) effluents. The isotope specificity of ICP-MS offers a still underexploited potential for tracer studies and for improved accuracy via isotope dilution analysis.

2.3.1 Gas Chromatography–Based Techniques

The beginnings of element-specific detection in chromatography are associated with the use of flame atomic absorption spectrometry (AAS) followed by flame photometric detection (FPD) with a bandpass filter selective to tin, electron capture detector (ECD), and especially electron impact MS. The wider use of ICP-MS and the availability of a commercial interface have contributed to the expansion of this technique for element-specific detection in gas chromatography (GC). Today, the practical applications for the determination of volatile organometallic species are dominated by the three hyphenated techniques: GC-MIP with atomic emission detection (AED), GC-ICP-MS, and gas chromatography-electron impact-mass spectrometry (GC-EI MS). The only exception is the determination of methylmercury in the environment where the position of GC-AAS and gas chromatography-atomic fluorescence spectrometry (GC-AFS) is still remarkably strong. A similar statement applies, however to a lesser degree, to organotin analysis that can still be successfully performed by GC-FPD, especially when the improved pulsed flame photometric detector is used.

An electrothermally heated silica tube is usually used as the atomization cell in GC-AAS systems. Despite some problems related to eliminating flame background and hydrocarbon interferences, FPD enjoyed a strong position as a tin-selective detector, and recent introduction of the pulsed flame photometric detector (PFPD) has offered a 10-fold gain in sensitivity for organotin detection.[2]

Plasma detectors are characterized by high sensitivity, and a large number of elements can be efficiently determined using them. The coupling of GC-MIP AED has been extremely popular in speciation analysis of anthropogenic environmental contaminants and products of their degradation. This is due to the versatility of this hyphenated system and the sub-picogram detection limits that could be matched only by ICP-MS.[6] GC-MIP AED offers sufficiently attractive figures of merit to be applied on a routine basis to speciation of organotin and organolead compounds in food matrices.[7] It is being gradually replaced by GC-ICP-MS, whose lower detection limits allow a simpler sample preparation procedure, work with more dilute extracts, and especially provides a sensitive speciation analysis of complex organic matrices. The position of ICP-MS has recently become stronger owing to the availability of appropriate commercial interfaces.

The basic requirement for an interface is that the analytes should be maintained in the gaseous form during transport from a GC column to the detector in a way that prevents any condensation.[6] New perspectives for GC-based hyphenated techniques in a variety of research areas including food sciences are expected both on the level of sample introduction and mass spectrometry. They include sample preparation methods using microwave-assisted, solid phase microextraction or purge; capillary trap automated sample introduction systems; miniaturization of GC hardware, allowing the time-resolved introduction of gaseous analytes into an element-sensitive detector (e.g., based on microcolumn multicapillary GC) or wider availability of collision cell technology; and time-of-flight detection in ICP-MS. The most popular application of GC-ICP-MS in food speciation analysis is the determination of methylmercury in seafood samples.[8–11]

An atomic fluorescence spectrometry (AFS) detector coupled with a gas chromatograph is a commercially available hyphenated system allowing speciation of mercury, but the risk of artefacts due to the presence of hydrocarbons prevents this technique from being successfully applied to more complex matrices.

Mass spectrometry of molecular ions, which is a common detection technique in GC of organic compounds, is relatively seldom used in speciation analysis. The widest popularity was enjoyed by electron impact mass spectrometers due to their availability as benchtop systems at relatively low cost. For quantitative analysis, the systems are operated in the single-ion monitoring mode; for most organometallic compounds, detection limits at the low picogram level can be achieved. The unquestionable advantage of molecular ion MS is the confirmation of the identity of the detected compound. This is achieved in the analysis by GC with element-selective detection by retention-time matching. The drawback is the high background due to the ionization of other (not necessarily metal-containing) species coeluted with the analyte compound.

2.3.2 Liquid Chromatography–Based Techniques

Most species of elements of interest in food are nonvolatile and cannot be converted into such by means of derivatization. They include virtually all the coordination complexes of trace metals but also many truly organometallic (containing a covalently bound metal or metalloid) compounds. For all these species, an HPLC is the principal separation technique prior to element-selective detection. Alternatively, capillary electrophoresis can be used, but its practical significance is limited. In spite of the elevated acquisition and running costs, ICP-MS remains the sole element-selective detector in HPLC of real-world food samples. The coupling of HPLC and ICP-MS offers an unmatched performance for the detection and determination of nonvolatile metallospecies. Technically it is very simple, consisting of connecting the exit of the chromatographic column to the nebulizer. The optimization of the interface is limited to the choice of the nebulizer matching the flow rate from the column and assuring the stability of the plasma in the presence of the mobile phase. HPLC-ICP-MS offers multielement capability, low detection limits, and the possibility of online isotope dilution. Indeed, with its sub-ng l^{-1} detection limits offered by the (most popular) quadrupole analyzers, ICP-MS allows the detection of HPLC signals from as little as 0.1 ng ml^{-1} of an element in the injected solution. HPLC-ICP-MS with enriched stable isotopes is a unique analytical method by which speciation of both endogenous elements and exogenous tracers can be achieved in a single experiment.

The principal HPLC separation mechanisms used in speciation analysis in food matrices include size-exclusion, ion-exchange, and reversed-phase chromatography. Size-exclusion chromatography (SEC) is based on the molecular sieve effect and enables species to be separated according to their size. This mechanism is usually valid for fairly large proteins and polysaccharides. For smaller species, especially ions with a high charge-to-mass ratio, secondary adsorption and ion-exchange effects can affect the separations. These phenomena, initially considered as a nuisance, are becoming more and more often employed for the separation of organometalloid species such as organoselenium,[12,13] organoarsenic compounds[14–17] and metal complexes with macromolecules.[18]

Both cation- and anion-exchange chromatography are widely used for the separation of organoarsenic[19] and organoselenium[20] compounds. Reversed-phase HPLC seems to be superior to ion-exchange for the separation of metal biomolecule complexes but is less popular due to problems in interfacing with ICP-MS when high concentrations of organic modifiers are necessary.

The potential of LC-ICP-MS for trace metal speciation has been widely discussed.[21–23]

2.4 SAMPLE PREPARATION APPROACHES

Foodstuffs represent a chemically complex matrix. Therefore, sample preparation for species-selective analysis is often troublesome, and developments in this area are attracting considerable attention.

2.4.1 Liquid Samples

In terms of preparation of liquid samples, the filtration of the sample using a 0.45-μm or, better, a 0.22-μm filter before injection onto the chromatographic column is mandatory. A guard column should be used to protect the analytical column particularly from effects of lipids that would otherwise degrade the separation. A number of techniques, including ultracentrifugation and microdialysis, have been proposed for sample pretreatment.

2.4.2 Solid Matrices

The major limitation of HPLC-ICP-MS is the need for analytes to be present in an aqueous solution, a requirement that makes impossible the straightforward analysis of biological samples. With some exceptions, the recovery of metal complexes by leaching is usually low, in the order of 10 to 50%. The majority of species is incorporated in or simply stick to the water-insoluble tissue residue and are therefore not accessible to direct analysis. Contrary to compounds with a metal (metalloid)-carbon bond, the pH required in alkaline or acid hydrolysis would affect the complexation equilibria, thereby changing speciation in an irreversible way. Selective degradation of the matrix at pH natural for a sample being studied appears therefore to be the only appropriate solution. The most popular methods used for the recovery of element species from solid food matrices are listed in Table 2.3.

2.4.2.1 Low Molecular Species

Many compounds containing a metal–carbon bond (organotin, methylmercury, alkyllead) are stable enough to survive the alkaline hydrolysis of biological matrices, e.g., with tetramethylammonium hydroxide.[24] It is the most popular protocol of sample preparation for organometallic contaminant species determination, usually followed by derivatization for the subsequent GC separation and element-selective detection. An example of the determination of organotin compounds in mussels is

TABLE 2.3
Most Popular Sample Preparation Procedures Used for Speciation Analysis in Food Samples

Sample Preparation Procedure	Species Recovered
Leaching with water, water–methanol mixtures, or neutral buffers	Low molecular species of arsenic and selenium and metal complexes from solid foodstuffs
Organic solvents (methanol)	Arsenolipids
Proteolysis	Alkylmetals, amino acids resulting from protein hydrolysis (including selenoamino acids)
Pectinolysis	Metals complexed with polysaccharides
Leaching with SDS, CHAPS	Denaturated proteins (including proteins containing selenium)
Digestion with artificial gastric and/or intestinal juice	Bioavailable species
Alkaline digestion (e.g., with TMAH)	Alkylmetals

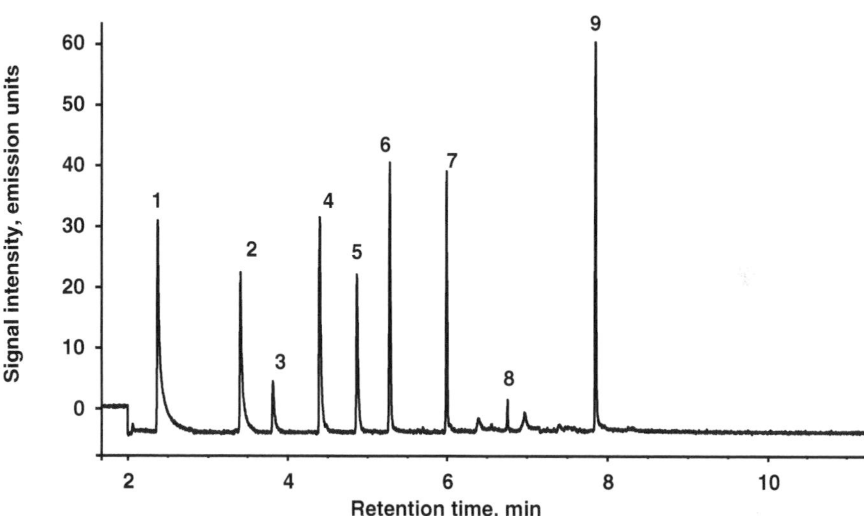

FIGURE 2.2 Speciation analysis of organotin compounds in mussels by gas chromatography-atomic emission spectrometry; 1 — inorganic tin, 2 — monobutyltin, 3 — tripropyltin (internal standard), 4 — dibutyltin, 5 — monophenyltin, 6 — tributyltin, 7 — tetrabutyltin (internal standard), 8 — diphenyltin, 9 — triphenyltin.

shown in Figure 2.2. Speciation analysis for mercury in seafood has recently been reviewed by Carro et al.[25]

Leaching with water, water–methanol mixtures, and neutral buffers is an established technique for the recovery of metal complexes and low molecular species of arsenic and selenium from solid foodstuffs. Sonication with a water–methanol

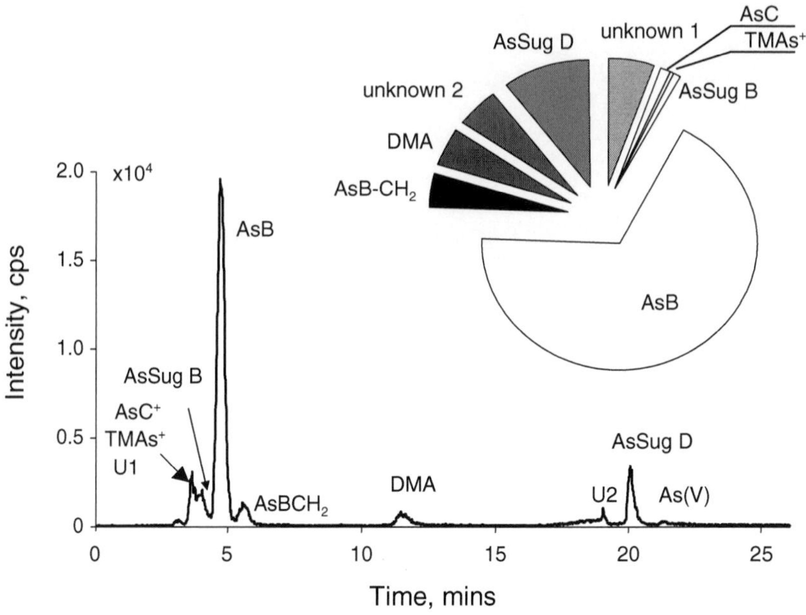

FIGURE 2.3 Determination of arsenic species in oyster tissue by anion-exchange HPLC-ICP-MS.

solution is the most popular method used for extraction of arsenic species from rice powder,[26,27] algae,[28,29] chicken meat,[30] oyster tissue,[31,32] and baby foods.[33] A typical HPLC-ICP-MS chromatogram of arsenic species in a seafood sample is presented in Figure 2.3.

Because selenoaminoacids are water-soluble, leaching with hot water has been judged sufficient to recover selenium species not incorporated into larger molecules. The typical recoveries of selenium extracted in this way from selenized yeast samples are usually ca. 10 to 20%.[34–37] A total of 27 different procedures employing buffering systems between pH 1 and 9 were compared for extraction of selenium compounds from high-selenium broccoli.[38] In extractions using nonbuffered solvents, more than 40% of the spiked Se-methylselenocysteine was not recovered. When buffer solutions were used for extraction, losses for Se-methylselenocysteine ranged from 10 to 20%. It was found that ca. 30% of the selenium naturally occurring in broccoli samples was volatilized and lost to the atmosphere in the course of extractions with buffer solutions.[38]

2.4.2.2 High Molecular Species

The most widely investigated example is the recovery of the selenium-containing protein fraction from selenized yeast.[37,39] The selenium-containing proteins were isolated from nut samples by dissolution in 0.1 *M* sodium hydroxide, precipitated with acetone, and then dissolved in a phosphate buffer of pH 7.5 prior to HPLC-ICP-MS analysis.[40]

The recovery of selenium-containing proteins from selenized yeast-based food supplements can be increased to above 95% by degrading the species originally present with a mixture of proteolytic enzymes.[37,39] The enzymes used in most digestion protocols were pure protease[37,39,41] or a mixture of proteolytic enzymes.[13] The use of protease XIV and proteinase K provided extraction yields of ca. 70% in plant tissues.[42] A dramatic activity enhancement of two proteolytic enzymes (protease XIV and subtilisin) was reported in a high-energy ultrasonic field.[43]

Different enzymatic protocols were compared to elucidate the possible role of the cell-wall digesting enzymes for the improvement of extraction efficiency of selenium from edible mushrooms.[44] A three-step enzymatic procedure gave the highest extraction efficiency (89%).[45] A three-step protocol using water extraction and two proteolytic enzymes (pepsin and trypsin) was proposed for Se-enriched edible mushroom after comparing five different sample extraction methods.[45] Hydrolysis carried out by boiling the sample with methanesulfonic acid was claimed to offer higher recovery of selenomethionine in comparison with the Proteinase K-protease XIV procedure.[46] During chromatography of enzymatic digests, the recoveries from the column were not complete (40–90%). This was attributed to the presence of strongly hydrophobic peptides resulting from the incomplete decomposition of proteins to amino acids.[41,47,48]

Four different extraction procedures were evaluated for a quantitative recovery of Se species during extraction from cod muscle. The highest Se recoveries were obtained in the presence of enzymes, whereas only 5% of total Se in cod was extracted when a "soft" extraction procedure (MeOH/HCl) was used.[49]

Enzymatic procedures were also reported for the recovery of organoarsenic species. A two-step procedure for extraction of arsenic from freeze-dried apples using a treatment with α-amylase enzyme followed by sonication with 40% acetonitrile was found to provide good extraction efficiency.[50]

Cellulose and complex pectic polysaccharides are the main matrix compounds of the water-insoluble residue after centrifugation of fruit and vegetable homogenates. The use of pectinolytic enzymes is therefore necessary to solubilize solid samples. Pectinolysis is known to degrade efficiently large pectic polysaccharides, but some of them, e.g., rhamnogalacturonan-II, are considered to be resistant to pectinolytic enzymes.[51] A mixture of commercial preparations — Rapidase LIQ™ and Pectinex Ultra-SPL™ — was reported for release of metal complexes from the solid parts of edible plant, fruits, and vegetables.[51]

Water-soluble polysaccharide species with higher molecular weights can be readily degraded by enzymic hydrolysis with a mixture of pectinase and hemicellulase to release the dRG-II complex.[51] The same mixture was found to be efficient to extract the dRG-II-metal complexes from the water-insoluble residue of vegetables owing to the destruction of the pectic structure.[51]

It is evident that even a delicate enzymatic attack changes the initial speciation, but this is often the only way to access, at least partially, information on metal species present in high-molecular-weight water-insoluble constituents of plant and animal tissues. The challenge is always to release into an aqueous phase the largest possible moiety containing the trace element of interest.

Figure 2.4 shows two examples dealing with the recovery of high molecular species from food samples. Speciation of zinc in infant formula (Figure 2.4a) after the centrifugation of the sample homogenized with a Tris buffer led to a chromatogram with one major signal corresponding to ca. 20 to 30% of the total zinc initially present. After the addition of sodium dodecylsulfate (SDS), a quantitative recovery of zinc into the 0.2 μm filtrable fraction was achieved. In addition to the increase in the concentration of the compounds excluded from the column, an additional compound could be released from the sample. The proteolytic hydrolysis of the proteins excluded from the column resulted in the formation of a mixture of oligopeptides that still contained firmly bound zinc. The recovery of zinc from the sample did not increase in comparison with the extraction with SDS. In Figure 2.4b, speciation of lead in an apple homogenate sample was investigated. The supernatant obtained after ultracentrifugation of the sample contained ca. 40% of the lead initially present, which was eluted as a 60-kDa polysaccharide compound (as confirmed by the refractometric method). The rest of the lead was retained in the solid residue that consisted of cellulose and complex water-insoluble pectic polysaccharides. An extraction using a mixture of pectinolytic enzymes that are known to degrade efficiently large pectic polysaccharides[51] allowed the partial destruction of the residue to release a dimer of Rhamnogalacturonan II. The lead fraction in the residue was insignificant (ca. 5%).

2.4.2.3 Evaluation of the Bioavailability of Metals in Foodstuffs by Sequential Enzymatic Extractions

Some attention has been paid to the analysis of enzymatic digests of foodstuffs in the quest for molecular information to contribute to the knowledge of the bioavailability of some elements. Sequential enzymolyses in simulated gastric and gastrointestinal juice were proposed for an estimation of bioavailability of heavy metal species from meat[52] and cocoa samples.[53,54] The soluble fractions of the stomach and upper intestinal contents of guinea pigs on different diets were investigated for estimating the bioavailable forms of Al, Cu, Zn, Mn, Sr, and Rb.[55]

An *in vitro* model simulating enzymatic activity in the gastrointestinal tract was developed for the assessment of the potential bioavailability of Cd and Pb in cocoa powder and cocoa liquor of different geographical origins. The model was based on the sequential extraction with simulated gastric and intestinal juices. The residue after the latter extraction was further investigated by using, in parallel, solutions of phytase and cellulose.[53]

The chemical stability of four arsenosugars isolated from seaweed extracts was investigated. Four arsenosugars were subjected to simulated gastric juice and acidic artificial stomach degradation, which resulted in a single common product tentatively identified by electrospray ionization tandem mass spectrometry (ESI-MS/MS), and an acid hydrolysis mechanism was proposed for its formation from each of the native arsenosugars.[56]

The *in vitro* gastric and intestinal digests of selenized yeast food supplements were investigated by successive HPLC–ICP-MS and HPLC–ESI-MS/MS. The main compound extracted by both gastric and intestinal juice was Se-methionine, which

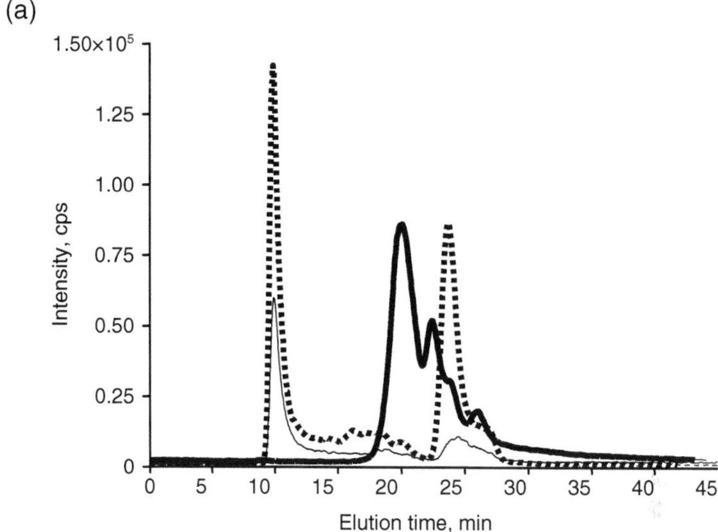

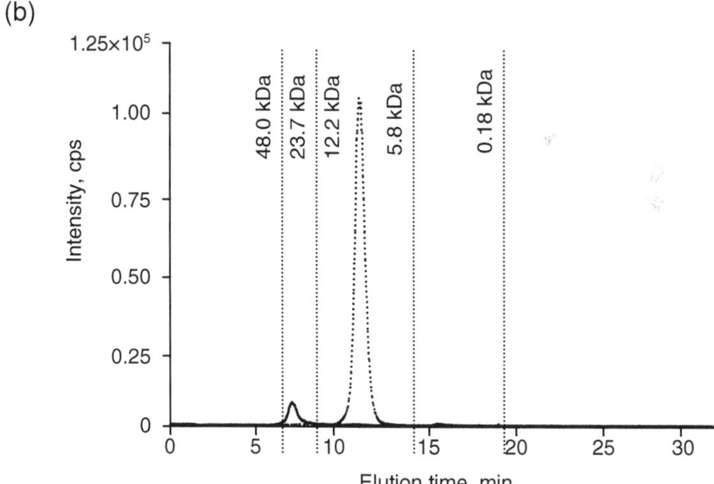

FIGURE 2.4 The effect of sample preparation procedure on size-exclusion HPLC-ICP-MS chromatographic profile in speciation analysis in food matrices. (a) Zinc species in infant formula (bold line — original sample, dashed line — sample incubated with SDS, thin line — sample digested with a mixture of protease and lipase enzymes). (b) Lead species in an apple homogenate sample (bold line — original sample juice, dashed line — supernatant after treatment with pectinolytic enzymes).

TABLE 2.4
Standard Reference Materials for Species-Selective Analysis in Food Matrices

Element	Standard Reference Material Name	Matrix	Species Certified	Concentration
Sn	BCR 477	Mussel tissue	Monobutyltin Dibutyltin Tributyltin	1.50 ± 0.28 mg/kg 1.54 ± 0.12 mg/kg 2.20 ± 0.19 mg/kg
Sn	NIES 11	Sea bass	Tributyltin	1.3 ± 0.1 mg/kg
Hg	BCR 463	Tuna fish	CH_3Hg^+	2.85 ± 0.16 mg/kg
Hg, As	NRCC DORM-2	Dogfish muscle	CH_3Hg^+ Arsenobetaine Tetramethylarsonium	4.47 ± 0.32 mg/kg as Hg 16.4 + 1.1 mg/kg as As 0.248 + 0.054 mg/kg as As
Hg	NIST 1566b	Oyster tissue	CH_3Hg^+	0.0132 ± 0.00007 mg/kg as Hg
Hg	NIST 2976	Mussel tissue	CH_3Hg^+	61.0 mg/kg
Hg	NRCC TORT-2	Lobster hepatopancreas	CH_3Hg^+	0.152 ± 0.013 mg/kg as Hg
As	BCR 627	Tuna fish	Arsenobetaine DMA	52 ± 3 μmol/kg 2 ± 0.3 μmol/kg
Hg	NRCC LUTS-1	Nondefatted lobster hepatopancreas	CH_3Hg^+	0.0094 ± 0.0006 mg/kg as Hg
Se	NRCC SELM-1	Yeast	Selenomethionine	3431 ± 157 mg/kg

was also the main Se-compound extracted by proteolytic digestion from the yeast supplements. Two other minor compounds were identified as Se-cystine and Se(*O*)-methionine, a degradation product of Se-methionine.[57]

2.5 QUALITY ASSURANCE

Validation of the analytical methods for speciation of organometallic anthropogenic contaminants was comprehensively discussed by Quevauviller.[58] A number of interlaboratory comparison round-robin exercises were carried out that contributed to the awareness of the problems encountered in quantitative speciation analysis and resulted in the availability of certified reference materials (CRMs) for organomercury, organotin, organoselenium, and organoarsenic species in food matrices (Table 2.4).

In contrast to pollutant metals, the problem of validation of speciation analyses concerning endogenous metal species in biological materials is still open. The chromatographic purity of peaks and the identity of the species that produce signals in chromatography often remain unknown. Also, it is usually unknown whether the species observed at the detector had existed in the original sample or was simply created during the analytical procedure, e.g., by oxidation and ligand exchange.

In order to cope with the lack of CRMs for endogenous species, the practice of laboratory internal quality control materials is developing. An example may be a laboratory reference material (LRM) prepared from Brazil nuts for quality control of the selenomethionine determinations. The homogeneity and stability of this candidate reference material passed the relevant tests recommended by the Measurement and Testing Programme.[59]

2.6 STABILITY STUDIES

A number of stability studies related to elemental species have been reported. They are concerned with selenium[60] and arsenic species including arsenobetaine in baby foods,[61] and monomethylarsonic acid and As(V) in rice.[26] Arsenobetaine was the only arsenic compound present in the analyzed baby foods, and its levels remained unchanged when the samples were stored for different times or when the samples were freeze dried. The study allowed the confirmation of the stability of arsenobetaine and the absence of the formation of other arsenic species by interconversion during storage.[61]

The stability of As species in water–methanol rice extract stored at + 4°C for at least 1 month was demonstrated. However, once the rice grains were ground, monomethylarsonic acid and As(V) were not stable under any storage conditions, probably due to microbiological activity unless gamma irradiation was applied.[26] Arsenic species in chicken meat candidate reference material were demonstrated to be stable for at least 12 months.[30]

The stability of SeMet and $TMSe^{+}$ in freeze-dried oyster and in its enzymatic extracts stored at different temperatures was studied. The results obtained for the freeze-dried sample showed that SeMet and $TMSe^{+}$ were stable for at least 12 months, under all the conditions tested. However, Se species in the enzymatic extracts were only stable for 10 d if stored at 4°C.[60]

REFERENCES

1. Templeton, D. et al., Guidelines to speciation analysis, *Pure Appl. Chem.*, 72, 1453, 2000.
2. Szpunar, J. and Lobinski, R., *Hyphenated Techniques for Speciation Analysis*, Royal Society of Chemistry, Cambridge, MA, 2003.
3. Szpunar, J., Bioinorganic speciation analysis by hyphenated techniques, *Analyst*, 125, 963, 2000.
4. Cornelis, R. et al., *Handbook of Speciation Analysis*, John Wiley & Sons, New York, 2002.
5. Lobinski, R., Elemental speciation and coupled techniques, *Appl. Spectrom.*, 51, 260A, 1997.

6. Bouyssiere, B., Szpunar, J., and Lobinski, R., Gas chromatography with inductively coupled plasma mass spectrometric detection in speciation analysis, *Spectrochim. Acta, B* 57, 805, 2002.
7. Lobinski, R. and Adams, F.C., Speciation analysis by gas chromatography with plasma source spectrometric detection, *Spectrochim. Acta*, 52B, 1865, 1997.
8. García Fernandez, R.G. et al., Comparison of different derivatization approaches for mercury speciation in biological tissues by gas chromatography/inductively coupled plasma mass spectrometry, *J. Mass Spectrom.*, 35, 639, 2000.
9. Mester, Z. et al., Determination of methylmercury by solid-phase microextraction inductively coupled plasma mass spectrometry: a new sample introduction method for volatile metal species, *J. Anal. At. Spectrom.*, 15, 837, 2000.
10. Slaets, S. et al., Optimization of the coupling of multicapillary GC with ICP-MS for mercury speciation analysis in biological materials, *J. Anal. At. Spectrom.*, 14, 851, 1999.
11. Tseng, C.M. et al., Field cryofocussing hydride generation applied to the simultaneous multi-elemental determination of alkyl-metal(loid) species in natural waters using ICP-MS detection, *J. Environ. Monit.*, 2, 603, 2000.
12. Casiot, C. et al., Sample preparation and HPLC separation approaches to speciation analysis of selenium in yeast by ICP MS, *J. Anal. At. Spectrom.*, 14, 645, 1999.
13. Casiot, C. et al., An approach to the identification of selenium species in yeast extracts using pneumatically-assisted electrospray tandem mass spectrometry, *Anal. Commun.*, 36, 77, 1999.
14. Morita, M. and Shibata, Y., Speciation of arsenic compounds in marine life by high-performance liquid chromatography combined with inductively coupled argon-plasma atomic-emission spectrometry, *Anal. Sci.*, 3, 575, 1987.
15. Shibata, Y. and Morita, M., Exchange of comments on identification and quantification of arsenic species in a dogfish muscle reference material for trace elements, *Anal. Chem.*, 61, 2116, 1989.
16. Shibata, Y. and Morita, M., Speciation of arsenic by reversed-phase high-performance liquid chromatography inductively coupled plasma mass spectrometry, *Anal. Sci.*, 5, 107, 1989.
17. Yoshinaga, J. et al., NIES certified reference materials for arsenic speciation, *Accred. Qual. Assur.*, 2, 154, 1997.
18. Makarov, A. and Szpunar, J., The coupling of size-exclusion HPLC with ICP MS in bio-inorganic analysis, *Analusis*, 26, M44, 1998.
19. Francesconi, K.A. and Kuehnelt, D., Determination of arsenic species: a critical review of methods and applications, 2000–2003, *Analyst*, 129, 373, 2004.
20. Uden, P.C. et al., Selective detection and identification of Se containing compounds-review and recent developments, *J. Chromatogr. A*, 1050, 85, 2004.
21. Caruso, J.A., Sutton, K.L., and Ackley, K.L., Elemental speciation: new approaches for trace element analysis, in *Comprehenisve Analytical Chemistry*, Barcelo, D., Ed., Elsevier, Amsterdam, 2000.
22. Michalke, B., The coupling of LC to ICP-MS in element speciation — Part II: recent trends in application, *Trends Anal. Chem.*, 21, 154, 2002.
23. Michalke, B., The coupling of LC to ICP-MS in element speciation: I. General aspects, *Trends Anal. Chem.*, 21, 142, 2002.
24. Szpunar, J. et al., Rapid speciation of butyltin compounds in sediments and biomaterials by capillary gas chromatography-microwave induced plasma atomic-emission spectrometry after microwave-assisted leaching-digestion, *J. Anal. At. Spectrom.*, 11, 193, 1996.

25. Carro, A.M. et al., Different extraction techniques in the preparation of methylmercury biological samples: classic extraction, supercritical-fluid and microwave extraction, *Int. Lab.* 28, 23, 1998.
26. Pizarro, I. et al., Evaluation of stability of arsenic species in rice, *Anal. Bioanal. Chem.*, 376, 102, 2003.
27. D'Amato, M., Forte, G., and Caroli, S., Identification and quantification of major species of arsenic in rice, *J. AOAC. Int.*, 87, 238, 2004.
28. McSheehy, S. et al., Identification of dimethylarsinyl-riboside derivatives in seaweeds by pneumatically assisted electrospray tandem mass spectrometry (ESI MS/MS), *Anal. Chim. Acta*, 410, 71, 2000.
29. McSheehy, S. and Szpunar, J., Speciation of arsenic in edible algae by bi-dimensional size-exclusion anion exchange HPLC with dual ICP MS and electrospray MS/MS detection, *J. Anal. At. Spectrom.*, 15, 79, 2000.
30. Polatajko, A. and Szpunar, J., Speciation of arsenic in chicken meat by anion-exchange liquid chromatography with inductively coupled plasma-mass spectrometry, *J. AOAC Int.*, 233, 2004.
31. McSheehy, S. et al., Investigation of arsenic speciation in oyster test reference material by multidimensional HPLC-ICP MS and electrospray tandem mass spectrometry (ES MS/MS), *Analyst*, 126, 1055, 2001.
32. Velez, D., Ybanez, N., and Montoro, R., Determination of arsenobetaine in manufactured seafood products by liquid chromatography, microwave-assisted oxidation and hydride-generation atomic-absorption spectrometry, *J. Anal. At. Spectrom.*, 12, 91, 1997.
33. Vela, N.P. and Heitkemper, D.T., Total arsenic determination and speciation in infant food products by ion chromatography-inductively coupled plasma-mass spectrometry, *J. AOAC Int.*, 244, 2004.
34. McSheehy, S. et al., Identification of selenocompounds in yeast by electrospray quadrupole-time of flight mass spectrometry, *J. Anal. At. Spectrom.*, 17, 507, 2002.
35. McSheehy, S. et al., Speciation of selenocompounds in yeast aqueous extracts by three dimensional liquid chromatography with ICP MS and electrospray MS detection, *Analyst*, 127, 223, 2002.
36. McSheehy, S. et al., Analysis for selenium speciation in selenized yeast extracts by two-dimensional liquid chromatography with ICP MS and electrospray MS/MS detection, *J. Anal. At. Spectrom.*, 16, 68, 2001.
37. Ruiz Encinar, J. et al., Methodological advances for selenium speciation analysis in yeast, *Anal. Chim. Acta*, 500, 171, 2003.
38. Roberge, M.T., Borgerding, A.J., and Finley, J.W., Speciation of selenium compounds from high selenium broccoli is affected by the extracting solution, *J. Agric. Food Chem.*, 51, 4191, 2003.
39. Polatajko, A. et al., A systematic approach to selenium speciation in selenized yeast, *J. Anal. At. Spectrom.*, 19, 114, 2004.
40. Kannamkumarath, S.S. et al., HPLC-ICP MS determination of selenium distribution and speciation in different types of nut, *Anal. Bioanal. Chem.*, 373, 454, 2002.
41. B'Hymer, C. and Caruso, J.A., Evaluation of yeast-based selenium food supplements using HPLC-ICP MS, *J. Anal. At. Spectrom.*, 15, 1531, 2000.
42. Montes-Bayon, M. et al., Selenium in plants by mass spectrometric techniques: developments in bio-analytical methods, *J. Anal. At. Spectrom.*, 17, 1015, 2002.
43. Capelo, J.L. et al., Enzymatic probe sonication: enhancement of protease-catalyzed hydrolysis of selenium bound to proteins in yeast, *Anal. Chem.*, 76, 233, 2004.

44. Dernovics, M., Stefanka, Z., and Fodor, P., Improving selenium extraction by sequential enzymatic processes for Se-speciation of selenium-enriched *Agaricus bisporus*, *Anal. Bioanal. Chem.*, 372, 473, 2002.
45. Stefánka, Z. et al., Comparison of sample preparation methods based on proteolytic enzymatic processes for Se-speciation of edible mushroom (*Agaricus bisporus*) samples, *Talanta*, 55, 437, 2002.
46. Wrobel, K. et al., Hydrolysis of proteins with methanesulfonic acid for improved HPLC-ICP MS determination of seleno-methionine in yeast and nuts, *Anal. Bioanal. Chem.*, 375, 133, 2003.
47. Bird, S.M. et al., Speciation of selenoamino-acids and organoselenium compounds in selenium-enriched yeast using high-performance liquid chromatography inductively coupled plasma mass spectrometry, *J. Anal. At. Spectrom.*, 12, 785, 1997.
48. Kotrebai, M. et al., Selenium speciation in enriched and natural samples by HPLC-ICP-MS and HPLC-ESI-MS with perfluorinated carboxylic acid ion-pairing agents, *Analyst*, 125, 71, 2000.
49. Diaz Huerta, V., Fernandez Sanchez, M.L., and Sanz-Medel, A., Quantitative selenium speciation in cod muscle by isotope dilution ICP MS with a reaction cell: comparison of different reported extraction procedures, *J. Anal. At. Spectrom.*, 19, 644, 2004.
50. Caruso, J.A., Heitkemper, D.T., and B'Hymer, C., An evaluation of extraction techniques for arsenic species from freeze-dried apple samples, *Analyst*, 126, 136, 2001.
51. Szpunar, J. et al., Speciation of metal-carbohydrate complexes in fruit and vegetable samples by size-exclusion HPLC-ICP MS, *J. Anal. At. Spectrom.*, 14, 639, 1999.
52. Crews, H.M. et al., Application of HPLC-ICP MS to the investigation of cadmium speciation in pig kidney following cooking and in vitro gastro-intestinal digestion, *Analyst*, 114, 895, 1989.
53. Mounicou, S. et al., Development of a sequential enzymolysis approach for the evaluation of the bioaccessibility of Cd and Pb from cocoa, *Analyst*, 127, 1638, 2002.
54. Mounicou, S. et al., Bioavailability of cadmium and lead in cocoa: comparison of extraction procedures prior to size-exclusion fast-flow liquid chromatography with inductively coupled plasma mass spectrometric detection (SEC-ICP-MS), *J. Anal. At. Spectrom.*, 17, 880, 2002.
55. Owen, L.M.W. et al., Determination of copper, zinc and aluminum from dietary sources in the femur, brain and kidney of guinea pigs and a study of some elements in vivo intestinal digestion by size-exclusion chromatography inductively coupled plasma mass spectrometry, *Analyst*, 120, 705, 1995.
56. Gamble, B.M. et al., An investigation of the chemical stability of arsenosugars in simulated gastric juice and acidic environments using IC-ICP-MS and IC-ESI-MS/MS, *Analyst*, 127, 781, 2002.
57. Dumont, E., Vanhaecke, F., and Cornelis, R., Hyphenated techniques for speciation of Se in *in vitro* gastrointestinal digests of *Saccharomyces cerevisiae*, *Anal. Bioanal. Chem.*, 379, 504, 2004.
58. Quevauviller, P., *Method Performance Studies for Speciation Analysis*, Royal Society of Chemistry, Cambridge, MA, 1998.
59. Bodo, E.T. et al., Preparation, homogeneity and stability studies of a candidate LRM for Se speciation, *Anal. Bioanal. Chem.*, 377, 32, 2003.
60. Moreno, P. et al., Stability of total selenium and selenium species in lyophilized oysters and in their enzymatic extracts, *Anal. Bioanal. Chem.*, 374, 466, 2002.
61. Vinas, P. et al., Stability of arsenobetaine levels in manufactured baby foods, *J. Food Prot.*, 66, 2321, 2003.

3 Criteria of Evaluation of Food Elements Analysis Data

Barbara Szteke

CONTENTS

3.1 INTRODUCTION

Food quality is one of the most important factors determining the consumer's perception and acceptance, attraction to, and purchase of the product. As food is the main source of major and trace elements for humans, their levels are among the factors that characterize the quality and safety of foodstuffs. Consumer pressure for safer, cleaner, and better quality foods cannot be overemphasized. For that reason, regulation in many countries define recommended levels of nutritive and essential elements, and permissible levels of toxic or potentially toxic elements in food products, raw materials, and food additives.

The determination of levels of both major and trace constituent elements in food and food products is becoming increasingly important in nutritional and food safety considerations. What might be a major (%) constituent in one food product can often be at a trace (milligram or microgram) concentration in another. For example, calcium is a major constituent in dairy product but found in relatively low concentrations in cereal grains.[1]

Food analysis, especially to satisfy the requirements of the labeling laws, is far more complicated than traditional chemical analysis. Determinations of elements present in food can be very complex as a task and complicated in execution. They can range from measuring simple elements in simple matrices (drinks) to complex compounds in complex materials (solid food). Such measurements may be complex and may require rather sophisticated methods and techniques, but there is no way they can escape the requirement for any measurement: that the result must be what it claims to be.

Because the potential toxicity of many elements and the effect of minerals and supplements for humans depend strongly on their concentrations in food, food safety is related to the accurate knowledge of the quantities of these elements. Product quality control depends on analytical data and must therefore ensure the accuracy and reliability of such data in order to represent its true condition. This condition is essential in sample determination to ensure control of product quality, prevention of product contamination, and protection of human health.

Analytical chemistry is at the center of programs to protect the public from toxic chemical contaminants that may enter the food chain. Regulatory limit enforcement, trend monitoring, surveys to determine the extent of contamination, and exploratory investigations all depend on laboratory analysis. Reliable data on the incidence and levels of contaminants are essential for recognition and control of potential food safety problems and for maintaining public confidence in food safety. Consumers, farmers, food manufacturers, and analytical chemists all have an interest in this issue but each from a different point of view. Yet, all must rely on data generated in laboratory analyses.

In contemporary human nutrition, many important technological, administrative, and legal decisions may depend on the results of analytical determinations. This is true not only for macro constituents but also for trace elements. In the case of some essential elements (e.g., Se, Cr, Zn), if there is only a minimal deviation in either direction from the optimal concentrations fixed by nature (e.g., during production, storage, or subsequent treatment), serious impact on health may result. The concentrations must, therefore, be rigidly controlled.[2]

Proper development of food production — pertaining especially to agriculture and the food industry — depends on the ability to meet the needs of consumers and the trade for products that are accepted and recognized as wholesome and of the expected quality. Food control and associated standards are needed to guide the agro-industry on sound scientific lines to make its products better accepted within the country as well as in export trade.

Considering both the health and economic aspects of unsafe or inadequate quality food, governments are paying increasing attention to the prevention and control of food contamination by strengthening and implementing national, regional, and local programs for food control. International terms of reference are much needed in order to improve the quality and safety of food supplies moving at global levels. For the food control system to contribute to the achievement of these goals, there must be a basic food law, supplemented by detailed regulations and administered by a capable food control organization. Almost all countries give high priority to the objectives mentioned above, and an increasing number are giving consideration to food control services as a tool for reaching them.[3]

Harmonization of analysis methods at the international level is a necessity if international trade in food products is to be facilitated. This harmonization can be reached only in an international forum, such as the FAO/WHO Codex Alimentarius

TABLE 3.1
Some Examples of Results of Interlaboratory Comparisons

Year	Material	n[a]	Element	Concentration Range (mg/kg)	"True" Content (mg/kg)	References
1985	Cabbage	27	Pb	0.10–3.9	0.23–0.41[b]	4
1992	Oriental tobacco	43	Cr	0.038–11.6	2.59 ± 0.32	5
	Leaves	17	Cs	0.117–11.6	0.177 ± 0.022	
	CTA-OTL-1	43	Na	48.2–8083	345	
		40	Pb	0.051–19.5	4.91 ± 0.80	
1993	Milk powder	92	Pb	0.198–7.692	1.483	6
		93	Cd	0.017–1.264	0.489	
		74	Hg	0.000–1.423	0.400	

[a] Number of participating laboratories.
[b] Acceptable range.

Commission, AOAC (originally the Association of Official Agricultural Chemists), the International Organization for Standardization (ISO), and International Laboratory Accreditation Cooperation (ILAC).

3.2 PROBLEM OF RELIABILITY OF ANALYTICAL RESULTS

Laboratory comparative studies, initiated by national and international institutions, have raised the alarm that analytical results, in spite of good reproducibility in the individual laboratories, may diverge widely and hence differ from the true value.[2] In spite of the impressive advancements in analytical chemistry, new developments in methodology, more stable equipment, computers, and access to more reference materials for method validation and calibration, the accuracy in element analysis of real food samples still leaves much to be desired. Such a conclusion can be drawn from the review of results of numerous interlaboratory comparisons where differences of orders of magnitude for certain elements are not a rarity. As an example, one may cite the dispersion of analytical results observed in interlaboratory comparisons for the determination a few elements in some foodstuffs and biological materials (Table 3.1).

According to Horwitz,[7]

> we can only speculate as to the cause of the poor performance of analytical chemists as exhibited by the large number of outliers apparent in collaborative studies. Within a single generation of scientists, analytical chemistry has changed precipitously from being a science of macroanalysis to one of trace analysis. In macrochemistry, gross errors are relatively easy to identify by the principle of consistency. A misplaced decimal point in concentration usually means that the physical, chemical, biological or sensory properties of substances were changed sufficiently to arouse suspicions as to accuracy of the analysis. For example, a moisture content of 8% is typical of rather

dry solid, but a moisture content of 80% is more typical of a fruit. A misplaced decimal point at the parts-per-million level can be discovered only by a completely independent analysis.

There are a number of analytical techniques from which the analyst of today can choose when conducting an elemental assay. The choice of analytical procedure depends on a number of factors, but it is necessary to consider the potential sources of error and all that may affect the final result, e.g., improper sampling and storage of the sample, sample preparation, calibration, instrumentation, analysis, and interpretation.

In sampling, a fraction or portion of a greater quantity of some bulk material is taken in order to facilitate its handling for the analysis. This procedure of reduction of bulk by selection must not change the ratio of matrix to element constituents. The relationship must be the same in the sample drawn and in the original bulk. All samples of fresh foods (vegetables, fish, fruits, meat, etc.) are liable to change in matrix and element concentrations. Almost always, the sampling will include some enrichment step. Obviously, the homogeneity of the bulk material is the most critical parameter of the entire sampling procedure. The food matrices met in elemental analysis are often inhomogeneous.

Sample preparation is still a difficult problem for the analyst as he or she must convert the substance into a suitable form for analysis without contaminating it or causing a loss of dry weight or a loss of any elements of interest. Drying, grinding, and solubilization techniques are frequently difficult and uncertain in their final result. Progress made analytically in recent years has far exceeded progress made towards improving sample preparation procedure. Usually, in the first stage, the solid sample has to be transferred to a homogenous liquid phase. With organic matrices of food, this always includes a sample destruction. Sample preparation often includes separation or preconcentration of the analytes, and at these stages one can make fatal errors. The analyte may be lost by adsorption (surfaces of tools and vessels) or volatilization (elements — Hg, As, Se; compounds — oxides, halogenides, hydrides), and it can also be added to the sample from the air, laboratory glassware, acids, reagents, etc.

In such multistage procedures, the actual step of determination for which many methods are available today can easily be calibrated and is, in general, least subject to systematic errors. Hence, the causes of systematic errors inherent to multistage procedures lie, in the first instance, in all the steps of sample preparation that have to precede the actual determination. The main concern is to avoid contamination from vessels and containers, reagents, and the laboratory air. It is also important to keep the blank as low and constant as possible. So, sample preparation is probably still the weakest link in the determination of the elemental content of biological substances.

Among others potential sources of error are chemical reactions: change of valency of ions, precipitation reaction, ion exchange, complex formation of volatile and nonvolatile compounds, and interference of the signal, such as matrix effects, target, overlapping of signals, signal background, and incorrect calibration and evaluation.

Other risks include incorrect standards, instable standard solutions, blanks, errors during measurements, false calibration functions, inadmissible extrapolation, etc.

Sherlock and others stated:[4]

> Users of analytical data should be made aware by those who provide it, that customer gets what he pays for. If a customer wants accurate and reliable data, then they will not come cheaply. Since inaccurate and unreliable data will rarely meet the customer's needs, their continued generation will be both a waste of consumers' money and of analysts' time.

3.3 REQUIREMENTS FOR METHODS OF ANALYSIS

As mentioned, many analytical methods can be applied to the determination of chemical elements in food. There are many requirements for analytical methods, and the main criteria determining their quality are based on adequate international standards and documents.

For a laboratory to produce consistently reliable data, it must implement a program of quality assurance procedures that gives objective evidence of its technical competence and the reliability of results and performance, and then seek formal accreditation. Quality assurance of elemental analysis is of crucial importance for the proper assessment of food quality. Reliable and accurate data are hence necessary to verify compliance with regulations or other specifications and to make sound decisions. Such data can be acquired by implementation of a suitable quality system through laboratories involved in the analysis of food.

According to ISO Standard,[8] quality measurement system, quality assurance and, as related with it, quality control are defined as follows:

- **Quality measurement system** — management system to direct and control an organization with regard to quality
- **Quality assurance** — part of quality management focused on providing confidence that quality requirements will be fulfilled

Quality assurance describes the overall measures that a laboratory uses to ensure the quality of its operations. Examples of typical items are: quality control, suitable and reference materials, traceability, proficiency testing, nonconformance management, and internal audits, statistical analysis. The quality assurance activities should be embedded in a managerial quality system.

- **Quality control** — part of quality management focused on fulfilling quality requirements

Quality control procedures relate to ensuring the quality of specific samples or batches of samples and include the following: analysis of reference materials, blind samples blanks, spiked samples, duplicates, and other control samples.

An essential part of good laboratory practice and a requirement of ISO 17025[9] for in-house methods also include the following:

- **Validations of methods** — is the confirmation, through the provision of objective evidence, that the requirements for a specific intended use or application have been fulfilled.[8]

Analytical methods must be thoroughly validated before use. Because in practice most analytical procedures used are not standard methods, the laboratory should demonstrate that the method used is suitable for the intent before routinely introducing the procedure. It means that the method meets the requirements of the client or generally accepted criteria at an international level (e.g., the levels of the minimum detectable amount). These methods must be well documented, staff adequately trained in their use, and control charts should be established to ensure proper statistical control. Where possible, all reported data should be traceable to reliable and well-documented standard materials, preferably certified reference materials. Accreditation of the laboratory requires applying sound quality assurance principles.

There is increasing hope that authorities (be it national or international, e.g., the FAO/WHO Codex Alimentarius Commission or the European Union) will not prescribe a single method of analysis but a method criteria. This is because stipulating a single method means that:

1. The analyst is denied freedom of choice, and thus may be required to use an inappropriate method in some situations.
2. The procedure inhibits the use of advanced methodology, especially if the "official" method is dated.
3. The procedure inhibits the use of automation.
4. It is administratively difficult to change a method found to be unsatisfactory or inferior to another currently available.

Laboratories may use any method of analysis provided it has been validated as performing to the adequate standards. Before the decision on use any analytical method, the following should also be taken into consideration:

- **The method's applicability** — the analytes, matrices, and concentrations for which a method of analysis may be used satisfactorily.

Development of quality system is initially a slow process; awareness building easily may require many months. The laboratory shall have quality control procedures for monitoring the validity of tests and calibration undertaken. Monitoring shall be planned and reviewed, and may include but not be limited to the following:

- Regular use of certified reference material and internal quality control using secondary reference materials

- Participation in interlaboratory comparison or proficiency-testing programs
- Replicate tests or calibrations using the same or different methods[9]

Unfortunately, the terms "quality control" and "quality assurance" are frequently not correctly interpreted, and sometimes even abused. As an example, it sometimes occurs that the analysis of a sample of a reference material is considered to be the laboratory's quality assurance. It is not.

Organizations that provide standard methods of analysis, such as AOAC, the American Society for Testing and Materials (ASTM), ISO, and the International Union of Pure and Applied Chemistry (IUPAC) have developed their own interlaboratory procedures to predict how a given method of analysis will perform in actual practice. Such a study requires analyses of identical test samples by a number of laboratories over the concentration and commodity ranges of interest.

Analytical research laboratories are often involved in other activities such as routine analyses and teaching. In this situation, making such duties compatible introduces an additional difficulty in implementing a quality system. There is a pressing need for analytical laboratories involved in research and development activities to implement gradually quality systems with a view to improving performance.[12]

For elemental analysis in food, appropriate methods should be assessed by the laboratory from the following main criteria defined by international standards:[10,11]

- **Accuracy** — closeness of the agreement between the result of a measurement and true value of the measurand*

The term accuracy, when applied to a set of test results, involves a combination of random components and a common systematic error or bias component.

- **Bias** — difference between the expectation of the test results and an accepted reference value

Bias is the total systematic error as contrasted to random error. There may be one or more systematic error components contributing to bias. A larger systematic difference from the accepted reference value is reflected by a larger bias value.

Estimation of bias (the difference between the measured value and the true value) is one of the most difficult elements of method validation, but appropriate reference materials (RMs) can provide valuable information within the limits of uncertainty of the RM's certified value(s) and the uncertainty of the method being validated.

- **Precision** — closeness of the agreement between the independent test results and true value of the measurand and obtained under prescribed conditions

* Particular quantity subjected to measurement.

Precision depends only on the distribution of random errors and does not relate to the true value or the specified value. The measure of precision is usually expressed in terms of imprecision and computed as a standard deviation of the test results. Less precision is reflected by a larger standard deviation. "Independent test results" means results obtained in a manner not influenced by any previous result on the same or similar test object. Quantitative measures of precision depend critically on the stipulated conditions. Repeatability and reproducibility conditions are particular sets of extreme conditions:

- **Repeatability** — precision under repeatability conditions
- **Repeatability conditions** — conditions where independent test results are obtained with the same method on identical test items in the same laboratory by the same operator using the same equipment within short intervals of time
- **Reproducibility** — precision under reproducibility conditions
- **Reproducibility conditions** — conditions where test results are obtained with the same method on identical test items in different laboratories with different operators using different equipment

When different methods give test results that do not differ significantly, or when different methods are permitted by the design of the experiment, as in a proficiency study or a material-certification study for the establishment of a consensus value of a reference material, the term "reproducibility" may be applied to the resulting parameters. The conditions must be explicitly stated.

- **True value** — the value that characterizes a quantity perfectly defined in the conditions existing when the quantity is considered; the true value of a quantity is a theoretical concept and, in general, cannot be known exactly
- **Trueness** — closeness of the agreement between the average value obtained from a large series of test results and an accepted reference value

The measure of trueness is usually expressed in terms of bias.

Moreover, the following definitions of other important criteria are currently in use:

- **Traceability** — property of the result of the value of a standard whereby it can be related to stated references, usually national or international standards, through an unbroken chain of comparison having stated uncertainties.[13]

To achieve comparability of results over space and time, it is essential to link all the individual measurement results to some common, stable reference or measurement standard. The results can be compared through their relationship to that reference. This strategy of linking results to a reference is termed *traceability*. The above definition implies a need for effort at national and international levels to provide widely accepted reference standards, and at the individual laboratory level to demonstrate the necessary links to the standards.[14]

- **Limit of detection (LOD)** — conventionally defined as field blank + 3σ, where σ is the standard deviation of the field blank value signal
- **Limit of quantification (LOQ)** — the same as for detection limit except that 6σ or 10σ is required rather than 3σ (the quantification limit is strictly the lowest concentration of analyte that can be determined with an acceptable level of repeatability, precision, and trueness)
- **Selectivity/Specificity** — the extent to which a method can determine particular analytes in mixtures or matrices without interferences from other components. Selectivity is the extent to which a method can determine particular analytes under given conditions in mixtures or matrices, simple or complex

Selectivity is the recommended term in analytical chemistry to express the extent to which a particular method can determine analytes in the presence other components. Selectivity can be graded. The use of the term specificity for the same concept is to be discouraged as this often leads to confusion.

- **Sensitivity** — change in the response divided by the corresponding change in the concentration of the standard (calibration) curve; that is, the slope s_i of the analytical calibration curve (it means that a method is said to be sensitive if a small change in concentration, **c**, or quantity, **q**, causes a large change in the measure, **x**; that is, when derivative **dx/dc** or **dx/dq** is large)
- **Uncertainty (of measurement)** — parameter associated with the result of a measurement that characterizes the dispersion of the values that could reasonably be attributed to the measurand

The parameter may be, for example, a standard deviation (or a given multiple of it), or the half-width of an interval having a stated level of confidence. Uncertainty of measurement comprises, in general, many components. Some of these components may be evaluated from the statistical distribution of results of a series of measurements and can be characterized by experimental standard deviations. The other components, which can also be characterized by standard deviations, are evaluated from assumed probability distributions based on experience or other information.

Important to and required by standard ISO 17025[9] is that analysts be aware of the uncertainty associated with each analytical result and that the uncertainty be estimated. The measurement uncertainty may be derived by a number of procedures. Food analysis laboratories are required to be in control, use collaboratively tested methods when available, and verify their application before taking them into routine use. Such laboratories, therefore, have available to them a range of analytical data that can be used to estimate their measurement uncertainty.

Most quantitative analytical results take the form of a ± 2u or a ± U, where "a" is the best estimate of the true value of the measurand's concentration (the analytical result) and "u" is the standard uncertainty, and "U" (equal to 2u) is the expanded uncertainty. The range a ± 2u represents a 95% level of confidence where the true value would be found. The value of "U" or 2u is the value that is normally used and reported by analysts and is hereafter referred to as *measurement uncertainty*, and may be estimated in a number of different ways.

The parameter may be, for example, a standard deviation (or a given multiple of it), or the half-width of an interval having a stated level of confidence. The measurement uncertainty of an analytical result may be estimated by a number of procedures, notably those described by ISO[15] and EURACHEM.[16]

The measurement uncertainty and its level of confidence must, on request, be made available to the user (customer) of the results.

Summarizing, according to Hu and Liu,[17] suitable analytical methods for trace elements in food should have the following features:

1. The uncertainty of the methods should be minimized (good precision). In general, the relative standard deviation of method should be lower than ± 5%.
2. The sensitivity and detection limits of the methods should meet the needs of the standard (high sensitivity and low detection limit). In general, the detection limit of the method should be lower than the permitted content in sample by at least one order of magnitude.
3. A fair agreement between the true content and the expected content is sufficient (good accuracy).

3.4 CERTIFIED REFERENCE MATERIALS (CRMs)

More and more decisions based on chemical measurements are having global effects. Reference materials are important tools to obtain global comparability of results of chemical measurements. All quantities and properties on the macroscopic scale of analytical chemistry can be represented by standards. In many cases the standards are divisible without changing the properties. They are then called reference materials. Reference materials can be:

- Simulates or artifacts
- Spiked and unspiked real life samples

The range of reference materials covers a three-dimensional space of coordinates:

- Analytes
- Matrices
- Applications[18]

Certified reference materials (CRMs) and reference materials (RMs) are widely used in food analysis. The basic purpose of all CRMs is to improve the comparability of measurements, but CRMs can be used at two different stages of the measurement process:

- For calibration, i.e., as a tool to transfer traceability
- For quality control, i.e., as a tool to verify the performance of a method as applied; and the required type of material depends on that intended use[19]

According to Vocabulary International in Metrology (VIM),[13] reference materials are defined as follows:

Reference material — a material or substance, one or more of whose property values are sufficiently homogeneous and well established to be used for the calibration of an apparatus, the assessment of a measured method, or assigning values to materials

Certified reference material — a reference material, accompanied by a certificate, one or more of whose property values are certified by a procedure that establishes its traceability to an accurate realization of the unit in which the property values are expressed, and for which each certified value is accompanied by an uncertainty at a stated level of confidence

CRMs are being used for the validation of new analytical methods and also for checking the performance of new laboratory, new analyst, etc. They are also employed as control samples for routine monitoring of laboratory performance with the use of control charts.

A targeted use of certified reference materials is one approach to the quality control of food analysis. CRMs are used to evaluate the accuracy of analytical procedures and compare different preparation procedures and techniques of determinations. They are also used for internal quality control of routine food analysis. In order to demonstrate the credibility of analytical results in all these areas, several CRMs are used, chosen with respect to the analyte, levels of certified values, and matrix composition.

CRMs serve in elemental analysis as a means of transferring measuring quality. When the determination of an element in CRM gives the correct result (i.e., the one that is in agreement with the certificate), it may be assumed that the results for the same element in an unknown sample is also correct, provided the overall composition of both materials and the concentrations levels of the analyte are not too different. So, the use of CRMs assures the transfer of accuracy and achievement of measurement compatibility on a global scale.[20]

The need that drives the production of certified reference materials is where regulation exists in terms of tolerance, and there is scope for disagreement between parties as to the true values. Reference materials provide an unequivocal benchmark against which the abilities of laboratories or contrasting methodologies can be compared. Moreover, where there are unusual analytical difficulties (and this is particularly the case with trace elements), reference materials offer the way to demonstrate internally that all is well. Required limits of determination are frequently very low, which may be a source of difficulty.

Reference materials provide a means of assessing whether this is a potential problem. Frequently the tried-and-tested methods, such as the official AOAC method, may not employ the latest technology; reference materials enable one to demonstrate that the new technologies can give the same results as the established procedure.

For elemental food analysis, many specific CRMs are available, among them are NIST SRMs: nonfat milk powder, oyster tissue, wheat flour, rice flour, spinach leaves, bovine liver, baking chocolate, slurried spinach or BCR CRMs: skim milk powder, bovine muscle, pig kidney, cod muscle, wholemeal flour, single cell protein, white cabbage, rye flour, wheat flour, haricots verts (French Beans), and pork muscle.

There is frequently the misconception that reference materials should be used for quality assurance purposes for routine use in laboratories. CRMs are in fact too expensive for this routine purposes and participation in proficiency testing (PT) or interlaboratory comparisons (ILC), and the purchase and use of surplus material from these exercises is a better and cheaper alternative.

The determination of the bulk content of elements in food and other materials is often not sufficient for the estimation of, for example, health hazards, and therefore the determination of individual species of various elements, for instance, methylmercury, organotins, As species, Cr(III), and Cr(VI), is requested. Very few CRMs certified for individual species of elements is available so far. The problems with their production and certification are more complex than the problems of production of CRMs for total element content analysis.[20]

3.5 PROFICIENCY TESTING

A regular independent assessment of the technical performance of a laboratory is recommended as an important means of assuring the validity of analytical measurement and as part of an overall quality strategy. A common approach to this assessment is the use of independent proficiency testing (PT) schemes (evaluating measurement performance by interlaboratory comparisons). A PT scheme is defined as a system for objectively evaluating laboratory results by an external agency. It includes regular comparison of a laboratory's results at intervals with those of other laboratories, the main object being the establishment of trueness. This is achieved by the scheme coordinator regularly distributing homogeneous test samples to participating laboratories for analysis and reporting of the data. Each distribution of test samples is referred to as a round. The main purpose of a PT scheme is to help the participating laboratory to assess the accuracy of its test results. It is now recommended by ISO that accreditation agencies require laboratories seeking accreditation to the ISO 17025[9] to participate in an appropriate proficiency testing scheme before accreditation is gained. Indeed, in some sectors such participation is mandatory.

Laboratories must demonstrate that they are in control and operate proficiently. This may be demonstrated by proficiency testing. Participation in proficiency testing schemes provides laboratories with a means of objectively assessing and demonstrating the reliability of data they produce. A proficiency (laboratory-performance) study determines the performance of laboratories or analysts in conducting an analysis. All laboratories, which in a proficiency study are self-chosen, must analyze the same material but by any method they wish to use.

The running of a PT scheme will be the responsibility of the coordinating laboratory. Sample preparation will either be contracted out or undertaken in-house and distributed to participants for analysis, who then are to return the results within a given time. Participants can use the analytical method of their choice, but it is recommended that all methods should be properly validated before use. Participants will be identified only by code to preserve confidentially. Reports issued to participants should include data on the results from all laboratories together with the participant's performance score.

Proficiency can be measured in a number of ways. Some schemes calculate Z scores for enumeration data in the international protocol.[21] Z scores are derived from the reported results, the assigned (true) result calculated using the robust or consensus mean of all data, and the target value for standard deviation derived from collaborative trial data, and the calculated from:

$$Z = (x\hat{X})/\sigma$$

where x is the reported value of analyte concentration in the test material; $\hat{X}$ is the assigned value, the best estimation of the true concentration of the analyte; and σ is the target value for standard deviation of value of x.

As the Z score is standardized, it is comparable for all analytes and methods. The Z scores for all participants can then be plotted on a graph or performance limit can be interpreted as follows:

$|\text{Z score}| \leq 2$ Satisfactory
$2 < |\text{Z score}| < 3$ Questionable
$|\text{Z score}| \geq 3$ Unsatisfactory

Proficiency testing schemes may be operated either by laboratory accreditation bodies or by other organizations. As the results of laboratories' performance in proficiency testing schemes are used in judging their technical competence, it is critical that proficiency testing schemes be operated competently, effectively, and fairly.[22]

Most PT schemes involve comparisons of participants' results with an assigned value that has been derived from test results by different method (e.g., IPE — International Plant Analytical Exchange, organized by the Wageningen Agricultural University, Netherlands, or FAPAS — Food Performance Assessment Scheme, organized by the Food Science Laboratory MAFF, U.K.

The EC–JRC IRMM International Evaluation Programme (IMEP) is one of the few interlaboratory comparison programs worldwide that are not based on consensus values as reference values (i.e., derived from the participants' results). IMEP participants can compare their results to a "reference range" that is traceable to the SI. This way, the demonstrated performance can be compared to the claimed performance. IMEP is owned and coordinated by the Institute for Reference Materials and Measurements (IRMM), Belgium.[23]

Proficiency testing is gaining increasing importance as a quality assurance tool for laboratories carrying out analytical measurements. The performance of laboratories in PT schemes is also being increasingly used, particularly by accreditation bodies, as a measure of the competence and quality of laboratories. It is important for laboratories to have comprehensive information on the scope and availability of PT schemes in the areas in which they work. This will enable them to make appropriate decisions about which schemes they should participate in.

3.6 ACCREDITATION OF LABORATORIES

The growing international trade is driving the increased need for conformity. As so many nationally accepted calibration and test methods differ, there must be acceptable means of determining their conformity. The credibility of test results depends on accuracy, precision, repeatability, and reproducibility. In turn, these depend on the competence of the tester and the validity of the methods used. Bodies that have to accept goods must have confidence that the laboratories conducting tests and calibrations are competent and their results are valid. So, the accreditation of a laboratory is a formal recognition of its technical ability to perform defined analyses and to ensure the quality of these analyses.

It follows that a laboratory seeking accreditation must:

- From the technical point of view, posses the necessary means, expertise and methods
- Be able to provide assurance of the quality of its services, i.e., has a well-developed quality assurance system

The accreditation concept is now internationalized. To meet these needs (concept), a few international organizations were created, and among them were ILAC (International Laboratory Accreditation Cooperation), IAF (International Accreditation Forum) EAC (European Accreditation of Certification), and IAS (International Accreditation Service). In order to conform to international regulations, the accreditation organizations must themselves possess a quality assurance system for their services in order to:

- Provide proof of the validity of accreditation granted
- Give international sanction (accreditation) to the analysis reports of the accredited laboratories

Accreditation should be a priority objective for laboratories. It fits into the scheme of international quality systems and constitutes an essential complement to certification of a company's products and quality systems.

The requirements to which the accrediting organizations must conform are contained in ISO Standard 17025.[9] This standard has been produced as the result of extensive experience in the implementation of ISO Guide 25 and EN 45001, both of which it now replaces. It contains all of the requirements that testing and calibration laboratories have to meet if they wish to demonstrate that they operate a quality system, are technically competent, and are able to generate technically valid results. The Standard[9] specifies the general requirements for the competence to carry out tests and calibrations, including sampling. It covers testing and calibration performed using standard method, nonstandard methods, and laboratory-developed methods. The standard is for use by laboratories in developing their quality, and administrative and technical systems that govern their operations. The acceptance of testing results between countries should be facilitated if laboratories comply with this standard and if they obtain accreditation from bodies that have entered into mutual recognition agreements with equivalent bodies in other countries using this international standard.

Many countries around the world have one or more organizations responsible for the accreditation of their laboratories. Most of these accreditation bodies have now adopted an international standard, ISO 17025,[9] as the basis for the accreditation of their country's testing and calibration laboratories. Adoption of this international standard has helped countries formulate a uniform approach to determining laboratory competence. Such a uniform approach allows countries with similar accreditation systems to establish agreements between themselves, based on mutual evaluation and acceptance of each other's accreditation systems.

Such international agreements, usually called *mutual recognition arrangements*, are crucial in enabling test data to be accepted between countries. In effect, each partner in such an arrangement recognizes the other partner's accredited laboratories as if they themselves had undertaken the accreditation.

The provision for accreditation is thus globally the same for all accrediting organizations respecting these requirements. They may, however, differ in certain practical details from one country to another.

However, even if the task of obtaining and maintaining accreditation remains difficult over the first years, it clearly constitutes a minimum level of organization and an intermediate step. The laboratory must continue to improve its quality system and not consider ISO 17025[9] as simply a series of requirements and constraints but consider it as a true management model.

Laboratory accreditation provides a means of determining the competence of laboratories to perform specific types of testing, measurement, and calibration. As laboratory accreditation involves an independent assessment by technical experts, it also allows a laboratory to determine whether it is performing its work correctly and to appropriate standards. Importantly, laboratory accreditation provides formal recognition to competent laboratories, thus providing a ready means for customers to access reliable testing and calibration services.

This developing system of mutual recognition between accreditation bodies has enabled accredited laboratories to achieve a form of international recognition and allowed test data accompanying exported goods to be more readily accepted in overseas markets. This effectively reduces costs for both the manufacturer and the importers, as it reduces or eliminates the need for products to be retested in another country.

Countries without viable accreditation systems can also seek to have their laboratories certified by those with established accreditation systems, so that their test data and associated goods can be accepted in foreign markets. These countries can also endeavor to develop their own accreditation system based on the structure and experience of those in other countries.

3.7 GEMS/FOOD PROGRAM (FOOD CONTAMINATION MONITORING)

The use of validated and harmonized analytical methods is increasingly required in order to compare monitoring and exposure data in various countries. There is a strong need for the development and harmonization of reliable, validated and, if possible, simple methods. Furthermore, the establishment and enlargement of data banks and material banks and the use of reference materials are necessary for improvement of the efficiency of food control.

The greatest existing program in the food area is the Joint UNEP/FAO/WHO Food Contamination Monitoring Program, commonly known as GEMS/Food, being a component of the Global Environment Monitoring System (GEMS) established in 1976 by the United Nations Environment Program (UNEP). Reliable and comparable analytical results of measurements of contaminants in foods and diets are the prerequisite for a realistic assessment of the contamination situation of food in the world.

Although the project came to an end in 1994, WHO has continued to implement GEMS/Food for the benefit of its contributing institutions located in over 70 countries around the world.[24] The purpose of the program is to inform governments, intergovernmental institutions such as the Codex Alimentarius Commission, other relevant institutions, and the public on the levels and trends of contaminants in food (e.g., cadmium, lead, and mercury), their contribution to the total human exposure, and significance with regard to public health and trade. The program is conducted globally, regionally, and nationally, as well as in local areas of special concern that serve as models — for remedial actions, food control, and resource management. Supporting components of the program involve technical cooperation, training, analytical quality assurance studies, and information exchange.

The main objectives of the GEMS/Food program are:

- To collect data on levels of certain chemicals in individual foods and in total diet samples and to evaluate these data, review trends, and produce and disseminate summaries, thus encouraging appropriate food control and resource management measures
- To obtain estimates of the intake via food of specific chemicals, with a view to combining these data with those from other sources and thus enabling the total intake of the contaminant to be estimated
- To provide technical cooperation with the governments of countries wishing to initiate and strengthen food contaminant monitoring programs
- To provide the joint FAO/WHO Codex Alimentarius Commission with information on the level of contaminants in food to support and accelerate its work on international standards for contaminants in foods

GEMS/Food maintains links with the FAO, UNEP, and other international organizations, such as the International Atomic Energy Agency, as well as with relevant international nongovernmental organizations such as the Association of Official Analytical Chemists International on specific food contamination monitoring and surveillance matters. In particular, GEMS/Food remains most active in the work of the Codex Alimentarius Commission and its scientific advisory bodies such as the Joint FAO/WHO Meeting on Pesticide Residues (JMPR) and the Joint FAO/WHO Expert Committee on Food Additives (JECFA).

In the past, contaminant data have been submitted to WHO by the participating institution or collaborating center via written forms transcribed from laboratory reports. In 1996, WHO started the development of a new data structure for food contaminant data and protocols for the electronic submission of such data. Electronic submission of data on dietary intakes is the subject of a separate manual. In addition to protocols for electronic data submission, WHO has also developed a computer system to allow the direct entry of data into — as well as the retrieval of data and creation of reports from — the GEMS/Food program database.

The laboratory (or the majority of the contributing laboratories) which successfully participates in relevant proficiency tests during the sampling and analysis period should be officially accredited.

For the analytical method used in the program, it is necessary to document the following: identification of method, validation, limit of detection/ limit of quantification, recoveries, measurement uncertainty, calibrants, laboratory accreditation, and interlaboratory comparison studies. GEMS/Food-EURO has sponsored two workshops on this topic. Details of recommendations of the second GEMS/Food-EURO workshop may be found in the report of the workshop titled "Reliable Evaluation of Low-level Contamination of Food."[25]

REFERENCES

1. Benton Jones, J. Jr., Developments in the measurement of trace metal constituents in foods, in *Analysis of Food Contaminants*, J. Gilbert, Ed., Elsevier, London and New York, 1984, p. 157.
2. Tölg, G., Assets and deficiencies in elemental analysis of food-stuffs, in *Recent Development in Food Analysis,* I Euro Food Chem., Verlag Chemie, Weinheim, Deerfield Batch, Florida-Basel 1982, p. 335.
3. Canet, C., Importance of international cooperation in food safety, *Food Addit. Contam.*, 10, 97, 1993.
4. Sherlock, J.C. et al., Analysis — accuracy and precision?, *Chemistry in Britain*, 1019, November 1985.
5. Dybczyński, R., Reference materials and their role in quality assurance in inorganic trace analysis, in *Problemy Jakości Analizy Śladowej w Badaniach Środowiska Przyrodniczego,* Kabata-Pendias A. and Szteke, B., Eds., Żak, Warsaw, 41, 1998 (in Polish).
6. GEMS/Food-Euro, Report to participants in Proficiency Testing Exercise 93/01, Ministry of Agriculture, Fisheries and Food, Food Science Laboratory, U.K., 1994.

7. Horwitz, W., International coordination and validation of analytical methods, *Food Addit. Contam.,* 10, 61, 1993.
8. ISO/IEC 9000:2000, Quality management systems — Fundamentals and vocabulary.
9. ISO/IEC Standard 17025:1999, General requirements for the competence of calibrating and testing laboratories.
10. ISO/IEC 3534-1:1993, Statistics – Vocabulary and symbols — Part 1: Probability and general statistical terms.
11. ISO/IEC 5725-1:1994, Accuracy (trueness and precision) of measurement methods and results. Part 1. General principles and definitions.
12. Valcárcel, M. and Rios, A., Quality assurance in analytical laboratories engaged in research and development activities, *Accred. Qual. Assur.*, 8, 78, 2003.
13. ISO, International vocabulary of basic and general terms in metrology, Geneva, 1993.
14. EURACHEM/CITAC Guide, Traceability in Chemical Measurement: A Guide to Achieving Comparable Results in Chemical Measurement, 2003.
15. ISO, Guide to the Expression of Uncertainty in Measurement, Geneva, 1993.
16. EURACHEM/CITAC Guide Quantifying Uncertainty in Analytical Measurement, 2nd ed., EURACHEM Secretariat, BAM, Berlin, 2000.
17. Hu, Z. and Liu, L., Quality assurance for analytical data of microelements in food, *Accred. Qual. Assur.*, 7, 106, 2002.
18. Zsunke, A., The role of reference materials, *Accred. Qual. Assur.*, 5, 441, 2000.
19. Guide for the Production and Certification of BCR Reference Materials, EC, DG XII, Doc. BCR/48/93, December 15, 1994.
20. Dybczyński, R., Preparation and use of reference materials for quality assurance in inorganic trace analysis, *Food Additiv. Contam.*, 19, 928, 2002.
21. Thompson, M. and Wood, R., The international harmonised protocol for the proficiency testing of (chemical) analytical laboratories, *Pure Appl. Chem.*, 65, 2123, 1993; *J. AOAC Int.*, 76, 926, 1993.
22. ISO/IEC — Guide 43: 1997, Proficiency testing by interlaboratory comparisons: Part 1: Development and operation of proficiency testing scheme; Part 2: Selection and use of proficiency testing schemes by laboratory accreditation bodies.
23. Quétel C.R. et al., Certification of the lead content in wine to provide reference values for international laboratory comparisons (IMEP-16 and CCQM-P12), 2nd International IUPAC Symposium Trace Elements in Food, Brussels, Belgium, 114, 2004.
24. WHO 2003, Global Environment Monitoring System. Food Contamination Monitoring and Assessment Programme (GEMS Food). Instructions for electronic submission of data on chemical contaminants in food and diets: www.who.int/foodsafety/publications/chem/gems_instructions/en/.
25. http://www.who.int/fsf/Chemicalcontaminants/index2.htm.

4 Chemometric Techniques in Analytical Evaluation of Food Quality

Piotr Szefer

CONTENTS

4.1 INTRODUCTION

The aim of multivariate data analysis is to break down mixed data structure into its components. In order to reduce relatively large number of variables to a smaller number of orthogonal factors, the data are treated by multivariate statistical methods, i.e., principal components analysis (PCA), factor analysis (FA), linear discrimination analysis (LDA), canonical discriminant analysis (DA), end-member analysis, cluster analysis (CA), neural network (NN), etc.[1] Defernez and Kemsley[2] have reported the use and misuse of chemometrics for treating classification problems. Multivariate data analysis has been presented extensively by several authors.[3–8]

Because of PCA, it is possible to create "new" dimensions of the data[9] and evaluate a reduced number of independent factors or principal components describing the information included in a system of characteristic but partly dependent variables. PCA and FA are techniques for finding a few components or factors that explain the major variations within the data matrix. Each component or factor in PCA or FA is a weighted linear combination of the original variables. Components or factors only with eigenvalues higher than unity should be preferably considered.[10,11] The factor loading characterizes quantitatively the contribution of individual variables to the corresponding factors. The ranking of the factors is characterized by the amount of variances they explain.[12] However, the difficulties in interpreting the components may sometimes be observed because of the lack of information about their meaning in either a physical or chemical sense. Moreover, in the process of reducing numerous original variables to a few orthogonal factors or components (mostly 3), some information is omitted.[11] However, this unexplained variance can be taken into account, resulting in an improvement in the reliability of this approach.[13]

Cluster analysis consists of various techniques.[14] The objects are grouped so that "similar" objects fall into the same class. Objects in one cluster should be homogenously distributed relative to some characteristics explaining within cluster properties, and they should also be clearly separated from other object score groupings.[11] Cluster analysis assigns particular variables with similar courses to clusters of variables.[12] Clustering techniques are divided into two basic groups, namely hierarchic and nonhierarchic methods. It is important to decide which clustering procedure is the most suitable. According to Sharma,[14] Wards's minimum variance technique was superior because it gives a larger amount of correctly classified observations as compared to most other methods, although it is not always better than average linkage clustering. This finding was supported by Massart and Kaufman.[15] One of the major difficulties and criticisms of the technique is defining objectivity.[11,16] Clustering always produces some clusters even if the results are completely random. Most methods are biased towards finding spherically and elliptically shaped clusters. When clusters with another shape are obtained, these are not always found to result in a loss of information and sometimes even misleading data.[5,16,17]

Discriminant analysis makes possible the determination of variations between groups of nominal elements characterized by numerical variables. Discriminant functions (DFs) that depend linearly on the element concentration studied are formed. The numerical values are the coordinates of the locations in a plane described by the two discriminant functions.[12] The particular endmember analysis (EA) is described by Renner et al.[18] An objective definition of external endmembers in the analysis of mixtures was presented by Full et al.[19] In general, there are indefinitely many sets of extreme points for a particular set of exact mixtures. However, since associations between elements of a geochemical dataset are not arbitrary, a conservative strategy is to seek extreme compositions (datapoints) that are geometrically close to the data and therefore close to observed reality.[19–21] A detailed examination of the multivariate analysis was performed by Renner[21–24] and Renner et al.[25]

Multivariate analysis has been applied to estimate food quality in view of its quantitative characteristic of mineral composition. They have been useful in processing elemental data of abiotic components, e.g., atmospheric deposits and marine aerosols,[26–28] marine suspended matter,[29] bottom sediments,[18,30–36] and iron–manganese concretions.[37] To biotic compartments, categorized chemometrically in relation to their elemental composition, belong seaweeds,[12,38] plankton,[26] mussels,[12,39–41] fish,[41–43] and sea mammals.[44–47]

4.2 CHEMOMETRIC EVALUATION OF FOOD QUALITY

4.2.1 Meat

Sporadic information is available on the application of chemometric techniques in evaluating meat product quality. Pork liver pastes canned in glass, plastic, and ceramic containers were analyzed for Cd, Co, Cr, Cu, Fe, Mn, Ni, Pb, and Zn by atomic absorption spectrometry (AAS).[48] The data obtained were processed statistically using a combination of multivariate and univariate techniques. Among the former, LDA, PCA, and FA were carried out. A total of 80 observations and 12 variables (metal contents and weight of sample, country of origin, factory and type of material of the container) were considered. LDA and PCA appeared to be useful in the differentiation of samples according to the material of the container (Figure 4.1). For instance, a cluster of object samples belonging to type II represent a high load from the country of origin and brand purchased. However, it should be emphasized that 6 scores of type I (marked with an asterisk) are separated from the rest of those attributable to that group. It means that these samples correspond to the pork composition. The other object samples were clustered relative to the container material.[48] Statistical techniques such as a single-factor Multivariate Analysis of Variance (MANOVA), univariate tests of normality (Kolmogorov–Smirnov), and equality of variances (Levene) were used based on the data of natural abundance stable-isotope ratios of C ($^{13}C/^{12}C$), N ($^{15}N/^{14}N$) and S ($^{34}S/^{32}S$) measured in samples of Belgian, Dutch, French, German, Indian, and Brazilian beef by continuous-flow isotope ratio mass spectrometry (CFIRMS).[49] As can be seen in Figure 4.2, object samples of beef from the U.S. and Brazil are distinguished from those attributable to northern European mainly due to the different proportion of plants with C_3 and C_4 photosynthetic pathways in the diet of the cattle. The chemometric data obtained appeared to be useful in assessing meat authentication.[49]

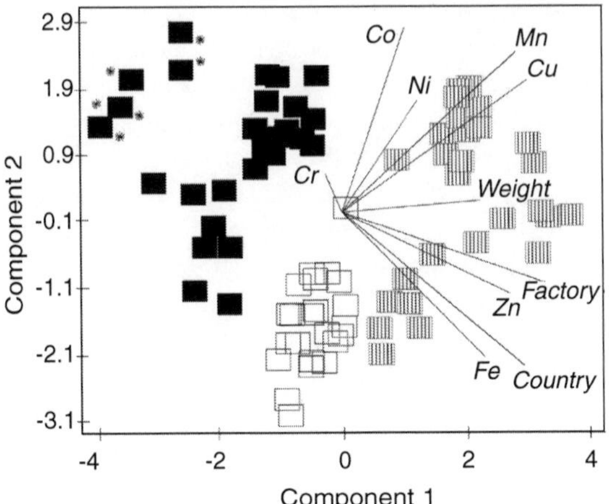

FIGURE 4.1 Biplot defined by PC1 and PC2 (black squares Type I, open squares Type II, grey squares Type III). (Adapted from Brito, G. et al., Differentiation of heat-treated pork liver pastes according to their metal content using multivariate data analysis, *Eur. Food Res. Technol.*, 218, 584, 2004. With permission.)

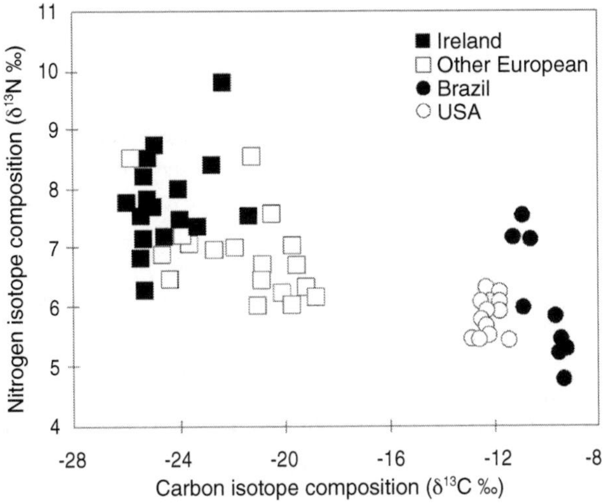

FIGURE 4.2 C and N stable isotope ratios in defatted bovine muscle tissue (International study). (Adapted from Schmidt, O. et al., Inferring the origin and dietary history of beef from C, N, and S stable isotope ratio analysis, Elsevier, *Food Chem.*, 91, 545, 2005. With permission.)

4.2.2 Fish

The application of PCA and FA to the data matrix (Ca, Cu, Fe, K, Mg, Na, and Zn determined in the muscle samples of the fish *Chromis limbatus* by Flame (F)-AAS and Graphite Furnace (GF)-AAS was able to differentiate between two groups of elements and between object samples belonging to different seasons. This distribution pattern was confirmed by CA (*k*- nearest neighbors method, KNN).[50] Szefer et al.[42] have reported multivariate distribution of loadings such as Hg, Cd, Pb, Cu, and Zn in muscle and hepatic samples of perch from the Pomeranian Bay, southern Baltic. Concentrations of the metals were analyzed by F-AAS and the data obtained processed by means of FA. A biplot of the object scores shows a grouping of the muscle samples (Figure 4.3A). Seasonal differences, similarly to hepatic objects (Figure 4.4A), are also clearly visualized. Muscle samples corresponding to the summer season are separated from those attributable to winter; however age-related differences (Figure 4.3B) show no such regular distribution pattern as in the case of hepatic samples (Figure 4.4B). As can be seen in the distribution pattern of loadings (Figure 4.3C), the winter muscle samples are generally loaded with Cd and Pb, whereas both muscle Zn and Cu are mainly determinants of summer objects.

PCA has been a useful method for processing data concerning concentration of chemical elements (Al, As, B, Ca, Cd, Cu, Fe, Li, Mg, Mn, Pb, Rb, Se, Si, V, Zn) determined by ICP-MS in the yellowtail flounder.[43] As can be seen in Figure 4.5A, muscle scores (Numbers 1–8) described by the lower values of PC1 are distinctly isolated from hepatic scores (Numbers 15–20) identified by higher PC1 values. To elements responsible for such clustering belong Ca and Fe, Mn, Cd, Cu, and Se, respectively (Figure 4.5B). PC2 (Figure 4.5A) differentiates muscle scores of off-shore fish (Numbers 1–6 with higher PC2 values) from those corresponding to inshore flounder (Numbers. 7 and 8 with lower PC2 values). On the other hand, hepatic scores (with higher PC2 values) attributable to males are distinguished from the remaining, e.g., females (with lower PC2 values). The distribution of the loadings (Figure 4.5B) indicates that muscle Ca is the main identifier of inshore flounder (low Ev2), whereas hepatic Cu is attributable to males. According to Hellou et al.,[43] an increase in the concentration of Al, Ca, Li, and Pb is registered between the muscle of inshore and offshore fish (two times greater than mean concentration and above the offshore concentration range).

4.2.3 Seafood

PCA has been successfully applied in the analytical evaluation of the degree of seafood pollution by toxic trace elements. Popham and D'Auria[39] have used PCA for deciding if blue mussels have been collected from the coastal waters of British Columbia, polluted with trace metals such as Pb, Zn, Cu, Mn, and Fe. These metals were determined by X-ray energy spectroscopy (XES). The authors identified areas in which mussels contained very high levels of tissue heavy metals reflecting industrial pollution of the ambient water. An interesting aspect of their study is that it was possible to identify regions characterized by less Zn and Pb pollution or a much higher extent in relation to the reference area. These areas have been identified in spite of the fact that specimens of the mussels analyzed were different in size (age) and were collected in different times of the year.

In order to characterize and differentiate polluted and unpolluted edible mussels taken from the waters of the Adriatic Sea, two multivariate techniques, i.e., LDA and LPCA (linear principal component analysis), have been successfully applied by Favretto et al.[51] after the determination of Mn, Fe, Co, Ni, Cu, Zn, Cd, Hg, and Pb

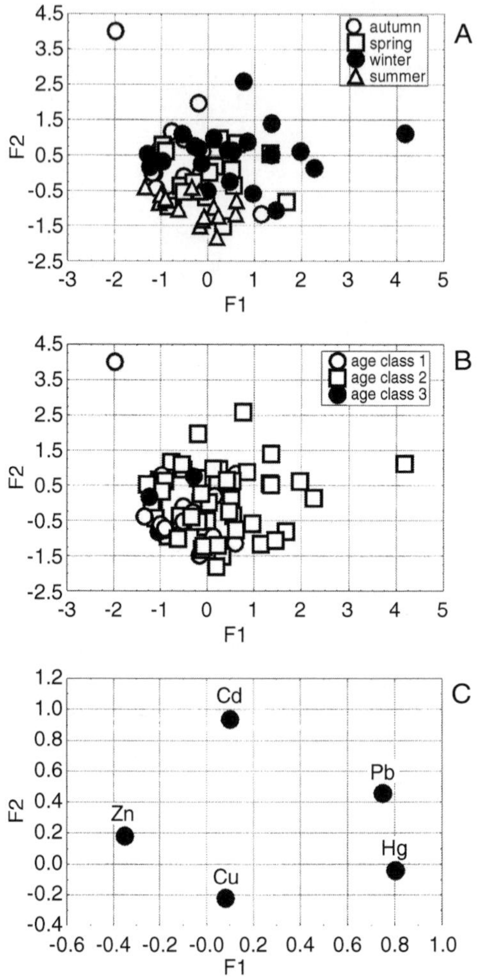

FIGURE 4.3 Biplot of scores showing seasonal (A) and age (B) variations of concentrations of trace metals (C) in the muscle of *Perca fluviatilis* from the Pomeranian Bay and Szczecin Lagoon, Baltic Sea. (Adapted from Szefer, P. et al., Distribution and relationships of mercury, lead, cadmium, copper and zinc in perch (*Perca fluviatilis*) from the Pomeranian Bay and Szczecin Lagoon, southern Baltic, Elsevier, *Food Chem.*, 81, 73, 2003. With permission.)

in mussel samples by atomic absorption spectrometry (AAS). Using LPCA it was also possible to distinguish some elements of terrigenous origin (Pb, Cd). PCA allowed a clear separation of polluted mussels from unpolluted ones from Muggia Bay, identified by Zn, Cd, Pb, and Hg, and Mn, Fe, Cu, Co, and Ni, respectively (Figure 4.6).

Edible mussels from different Spanish markets have been used to interpret a matrix data consisted of metal concentrations such as Cu, Fe, K, Mn, Na, Ni, V (AAS), Cd, Co, Pb (ETA-AAS) and Hg (CV- and HG-AAS).[52] The use of Potential Curves and PCA appeared to be very helpful in the identification of several

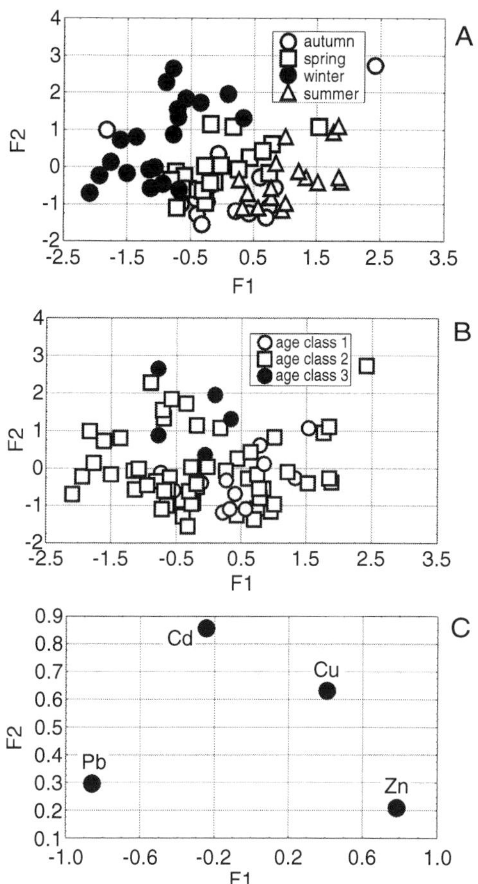

FIGURE 4.4 Biplot of scores showing seasonal (A) and age (B) variations of concentrations of trace metals (C) in the liver of *Perca fluviatilis* from the Pomeranian Bay and Szczecin Lagoon, Baltic Sea. (Adapted from Szefer, P. et al., Distribution and relationships of mercury, lead, cadmium, copper and zinc in perch (*Perca fluviatilis*) from the Pomeranian Bay and Szczecin Lagoon, southern Baltic, Elsevier, *Food Chem.*, 81, 73, 2003. With permission.)

noncontrolled factors during the period of harvest, packing, transport, etc., affecting the quality of the seafood studied. The simplified algorithm for potential curves permitted classifying of mussel groups and a simple understanding of quality control in commercial fresh and frozen products.[52]

Bechmann et al.[53] have reported spatial-related trends of metal concentrations in edible mussels from Limfjord, Denmark. High-resolution inductively coupled plasma–mass spectrometry (HR-ICP-MS) was used for the determination of metals such as Ba, Cd, Co, Cr, Cu, Ga, Ni, Pb, Rb, and V in the soft tissue of blue mussel (*Mytilus edulis*). PCA appeared to be a useful technique in cases when a decision had to be made regarding which regions were appropriate for mussel fishery. Figure 4.7A differentiates mussel scores with respect to their size and location of sampling

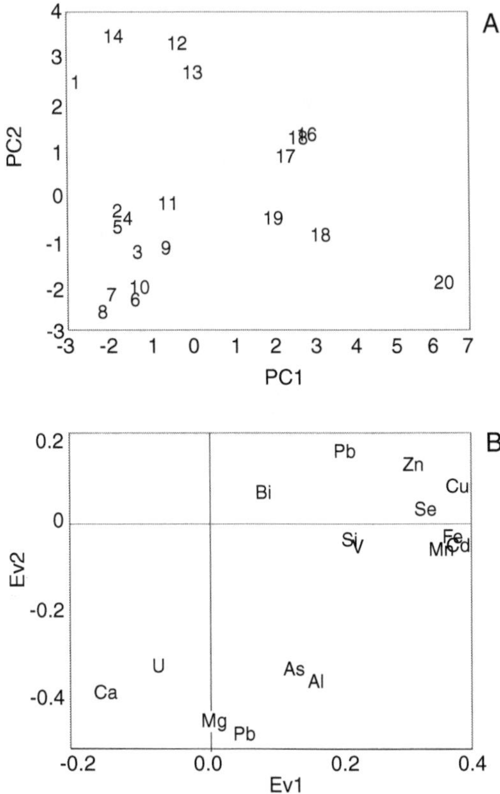

FIGURE 4.5 Biplot of distribution of scores (A) and the concentration of 16 elements (B) in the tissues of yellowtail flounder (*Pleuronectes ferruginea*). Numbers 1–6 represent the muscle of offshore fish, 7 and 8 represent the muscle of inshore flounder, 9–14 represent the gonads, and 15–20 the livers. The first three numbers of each subset are due to males (pc, principal component). (Adapted from Hellou, J. et al., Distribution of elements in tissues of yellowtail flounder *Pleuronectes ferruginea*, Elsevier, *Sci. Total Environ.*, 181, 137, 1996. With permission.)

sites. The distribution pattern of corresponding loadings is shown in Figure 4.7B. Object samples C (higher PC2 values) from the contaminated area differ significantly from others. One can infer from the above-mentioned studies that PCA is a powerful tool to identify polluted areas inhabited by edible mussels that can be simultaneously used as useful biomonitors.[54–56] Mussels transplanted from an aquaculture farm in a clean open bay to the Bays of northwest Mediterranean were analyzed for several biomarkers as well as Cd, Cu, and Zn concentrations. PCA was applied to discriminate the different transplantation sites and appeared to be a useful tool for assessing water quality in this area.[57] PCA has been used for the evaluation of effectiveness of seafood sample pretreatment methods for trace elements (As, Cd, Cr, Cu, Fe, Mg, Mn, Ni, Pb, Se, and Zn) determined by F-AAS and ET-AAS.[58]

The distribution of trace elements in the soft tissue of mollusks from different geographical locations has been discussed by Struck et al.,[12] Szefer and Wołowicz,[59]

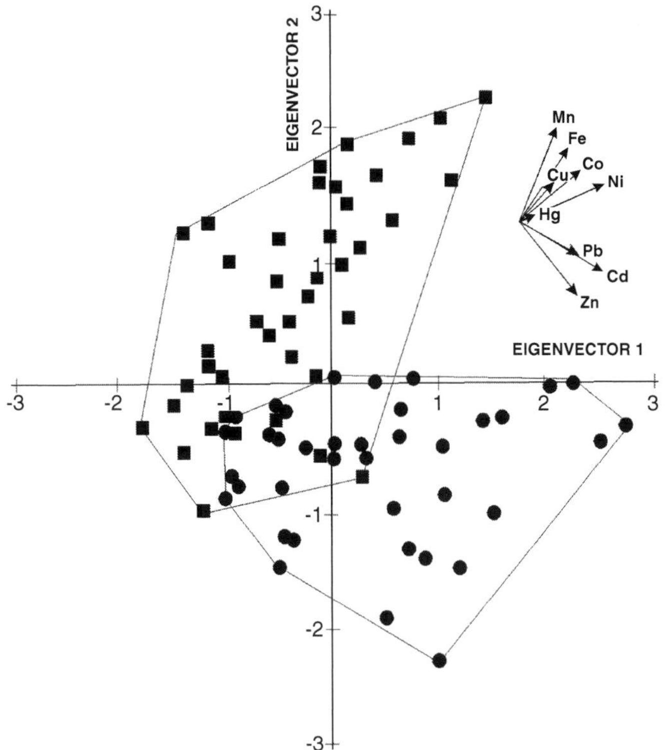

FIGURE 4.6 Eigenvector projection of component scores of mussels: circles, polluted mussels; triangles, unpolluted mussels. Scores were calculated from the unrotated matrix of correlations between variable and principal components. (Adapted from Favretto, L. and Favretto, L.G., Principal component analysis and pollution by trace elements in a mussel survey, John Wiley & Sons, *J. Chemometrics*, 3, 301, 1988. With permission.)

and Szefer et al.[40,60–63] Based on chemometrically processed data, it was possible to characterize and classify all the samples analyzed with respect to their species features and provenience. For example, the observed differentiation between two groups of object samples attributable to two species (*Acanthopleura haddoni* and *Ostrea cucullata*) of mollusks from the Gulf of Aden could be explained by the differences in their feeding habits.[60]

An example of spatial differentiation of mineral composition of mollusk tissue obtained by FA is an influence of industrial activity on heavy metal distribution in *Mytilus galloprovincials* from the coastal waters of the Korean Peninsula.[62] Interesting results have also been obtained for mollusks from different aquatic regions all over the world in terms of chemometric data. Szefer and Wołowicz[59] processed statistically the metal data (Cd, Cu, Fe, Mn, Ni, Zn) for the soft tissue of *Cerastoderma glaucum* from four geographical regions, i.e., the Gulf of Gdańsk (Baltic Sea), Marennes–Oleron Bay, Arcachon Bay (French Atlantic coast), and Embiez Islands (Mediterranean Sea). It was shown from comparison between the distribution of the object scores and the loading vector direction (Figure 4.8) that mainly Mn

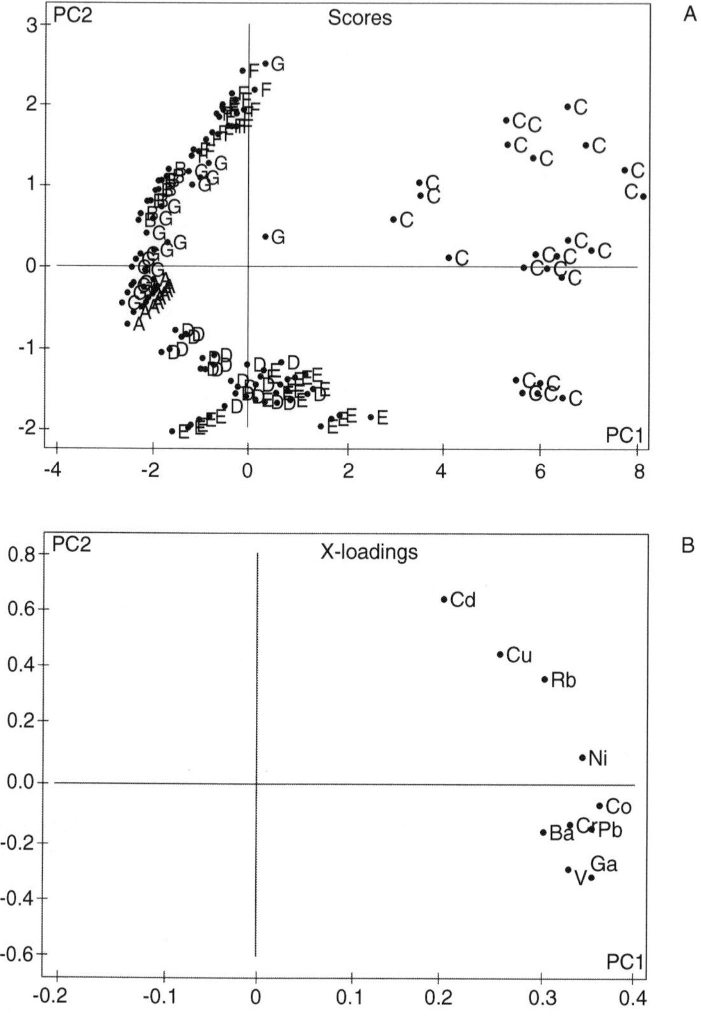

FIGURE 4.7 Scatterplot of scores (A) and loadings (B), i.e., 10 elements (V, Cr, Co, Ni, Cu, Ga, Rb, Cd, Ba, Pb) for all mussel samples analyzed. The letters A–G correspond to the seven sampling sites. (Adapted from Bechmann, I.E., Stürup, S., and Kristensen, L.V., High resolution inductively coupled plasma mass spectrometry (HR-ICPMS) determination and multivariate evaluation of 10 trace elements in mussels from 7 sites in Limfjorden, Denmark, Springer, *Fresenius J. Anal. Chem.*, 368, 708, 2000. With permission.)

and Fe concentrations in the mollusk analyzed were responsible for differentiation between populations from Marennes–Oleron Bay and Arcachon Bay. Zinc, Cd, and partly Ni, had a significant role in distinguishing the Gulf of Gdańsk cluster from the others. Such distribution pattern suggests that elevated levels of Cd and Zn may be attributable to anthropogenic sources densely distributed in this region. The PCA data show that both spatial and seasonal factors are of great importance in the distribution of the metals studied in mollusk tissues.[59]

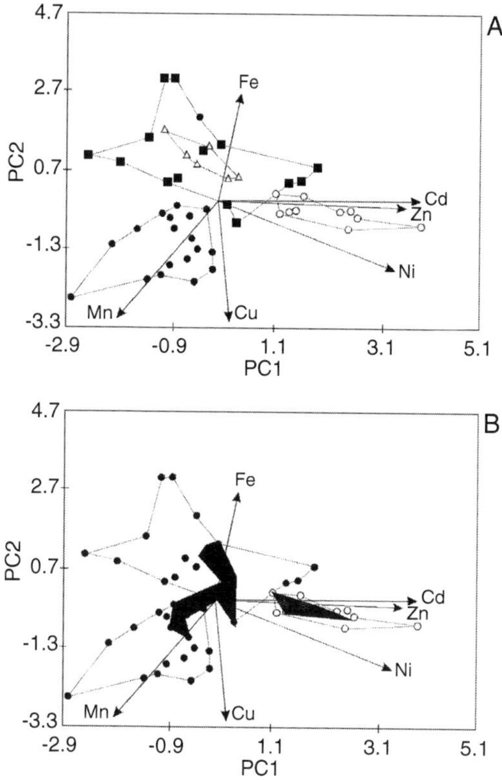

FIGURE 4.8 Biplot for object scores of the first two principal vectors of 50 mollusc samples: A — regional differences are illustrated by clusters of points corresponding to samples from the Gulf of Gdańsk (○), Marennes–Oleron Bay (●), Arcachon Bay (■), and Embiez Islands (△). Association between principal components (PCI × PC2) and variable (metal) vectors are also indicated; B — season-dependent variations are illustrated by clusters of points corresponding to samples collected during January — May. These groupings are indicated by shaded areas. (Adapted from Szefer, P. and Wołowicz, M., Occurrence of metals in the cockle *Cerastoderma glaucum* from different geographical regions in view of principal component analysis, *SIMO-Mar. Pollut.*, 64, 253, 1993. With permission.)

A special attention has been paid to applying multivariate techniques in the classification of Mytilids from different geographical zones.[40,60–65] The elemental data for the soft tissue of mussels collected in temperate, subtropical, and tropical zones were processed by FA. As seen in Figure 4.9, A and B, after removing extreme values (attributable to extremely contaminated samples in highly industrialized areas of Saganoseki, Japan, and Öxelosund, Sweden), it is possible to distinguish selected mussel populations with respect to their geographical provenance (Figure 4.9, C and D).

Interesting factorial data, with respect to mineral composition, have been obtained for the soft tissue of *Mytilus trossulus* taken from three areas of the southern Baltic.[40] In order to verify the regional influences of seawater on the biochemical

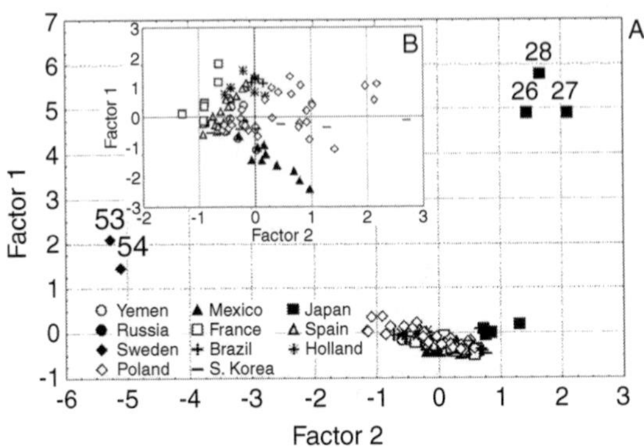

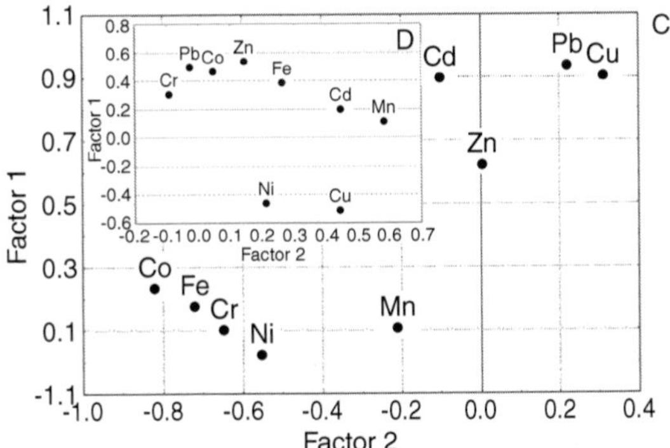

FIGURE 4.9 Two-dimensional scatterplot in the space determined by F1 and F2 corresponding to the metals in the soft tissue of Mytilidae. (A) Factorial distribution of all the scores (samples); (B) after excluding sample numbers 22–28, 53, and 54, which deviate significantly from the other scores; (C) factorial distribution of all the loadings (trace elements); (D) factorial distribution of loadings after excluding samples numbers 22–28, 53, and 54, which deviate significantly from the other scores. (Adapted from Szefer, P. et al., A comparative assessment of heavy metal accumulation in soft parts and byssus of mussels from subarctic, temperate, subtropical and tropical marine environments, Elsevier, *Environ. Pollut.*, 139, 70, 2006. With permission.)

composition of *Mytilus edulis* from the Baltic Sea and North Sea, which are independent of the presence of trace elements, DA was performed for macroelement concentrations in the mussel as variables.[12] This distribution pattern made it possible to distinguish Baltic and North Sea locations such as in the case of *F. vesiculosus* in spite of different food habits between these two zoobental organisms. Location groups based on the trace element concentration patterns showed a less distinctive geographical arrangement in comparison to the location clusters based on macroelement concentration patterns. This picture suggests modified conditions for the accumulation of trace elements in *M. edulis* as in *F. vesiculosus* compared to the uptake of macroelements.[12] The concentration of 15 elements in the seafood was determined by F-AAS (Ca, Co, Cr, Cu, Fe, K, Mg, Mn, Na, Ni, Zn), GF-AAS (Cd, Pb), CV-AAS (Hg) and HG-AAS (Se).[66] FA made it possible to distinguish different

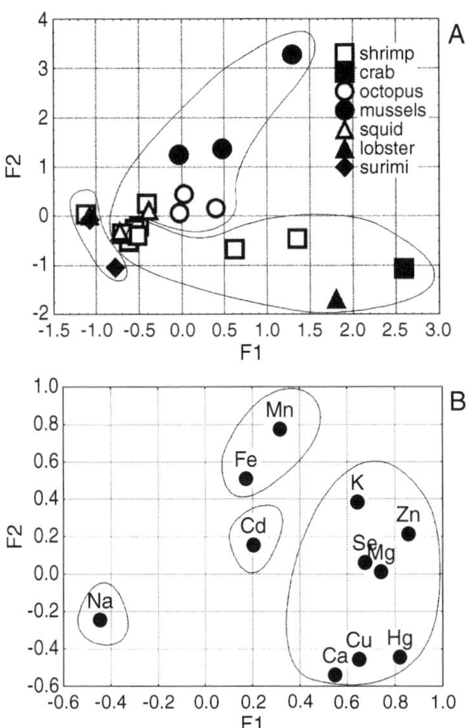

FIGURE 4.10 Biplot of the distribution of object scores (A) and loadings (B) in commercially distributed seafood from different countries such as India, Thailand, Canada, Philippines, New Zealand, Spain, Norway, and Great Britain. (Adapted from Kwoczek, M. et al., Essential and toxic elements in seafood available in Poland from different geographical regions, ACS, *J. Agric. Food Chem.*, 54, 3015, 2006. With permission.)

species of seafood as well as being useful for the identification of elements responsible for the separation of raw seafoods from technologically processed sea products. For instance, there is discrimination between mussels and octopus with higher values of F2 and the squids described by intermediate values of F2 (Figure 4.10A). It appeared that factor F2 is associated with taxonomic groups of the raw seafood distinguishing mussels, octopuses, squids, and crustaceans (shrimp, lobster, and crab). In the case of the loadings distribution, F1 achieves the lowest values for Na corresponding to seafood processed technologically, i.e., Kamaboko crab, Torpedo shrimp, surimi crab, and coated squid rings, which are clearly distinct from the group connected with Cd, Fe, Mn, Ca, K, Se, Cu, Mg, Hg, and Zn (Figure 4.10B). The latter group of elements is ascribed to the scores mostly associated with unprocessed, raw seafoods (Figure 4.10A), i.e., with higher values of F1. Presumably, a considerable pool of these essential elements reflects a natural elemental composition of the invertebrates studied. Hence, F1 is connected with the diverse ways in which some of the studied samples were processed, and some unprocessed. Factor F2 allows us to distinguish metals responsible for grouping object samples (Figure

4.10A) attributable to different features of particular taxonomic groups of the raw seafoods analyzed. The specific elemental composition for each of such group is the cause for the formation of two subclusters associated with Fe, Mn (mussels), and Cd (squids, octopus) described by higher values of F2 and with Hg, Cu, and Ca (crustaceans: shrimps, lobsters, crab) characterized by lower values of F2.

Multivariate techniques have been applied in the quantitative estimate of seafood quality.[53,67,68]

4.2.4 Milk and Dairy Products

4.2.4.1 Milk

In order to assess the authenticity of dairy products, it is recommended that the geographical origin of the milk and its final product be identified. Isotopic composition ($^{13}C/^{12}C$ and $^{15}N/^{14}N$) and concentrations of Ca, Li, Na, K, and Mg were determined by an isotope ratio mass spectrometry (IRMS) and high-performance ion chromatography (HPIC), respectively, whereas Al, Cr, Cu, Fe, Mn, Ni, Zn, Pb, Se, and Ba were quantified by ICP –OES.[69] Moreover, other parameters such as the contents of fat, lactose, protein, etc., were measured in milk samples. As can be seen in Figure 4.11, scores of milk samples from Caserta are characterized by negative values of PC1, whereas those from Foggia are described by positive F1 values. Such geographical distinguishing is connected with loadings of Na, Li, K, and protein content.

4.2.4.2 Cheeses

The results obtained by Brescia et al.[69] were evaluated using several chemometric techniques (PCA, HCA [hierarchical cluster analysis], DA) for the geographical

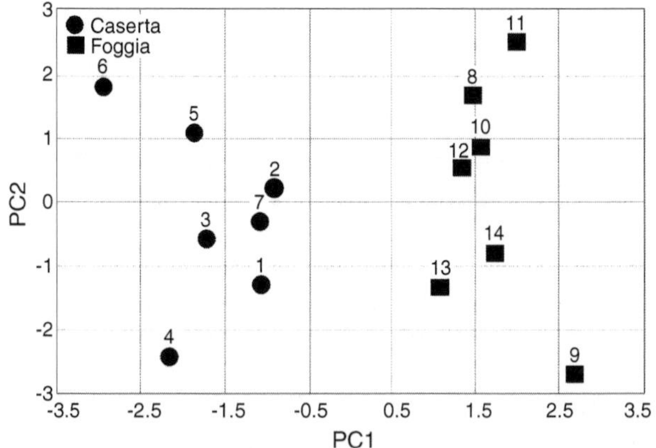

FIGURE 4.11 Scatterplot of the scores of milk samples from the first two principal components PC1 and PC2 obtained using analytical data. (Adapted from Brescia, M.A. et al., Characterisation of the geographical origin of Buffalo milk and mozzarella cheese by means of analytical and spectroscopic determinations, Elsevier, *Food Chem.*, 89, 139, 2005. With permission.)

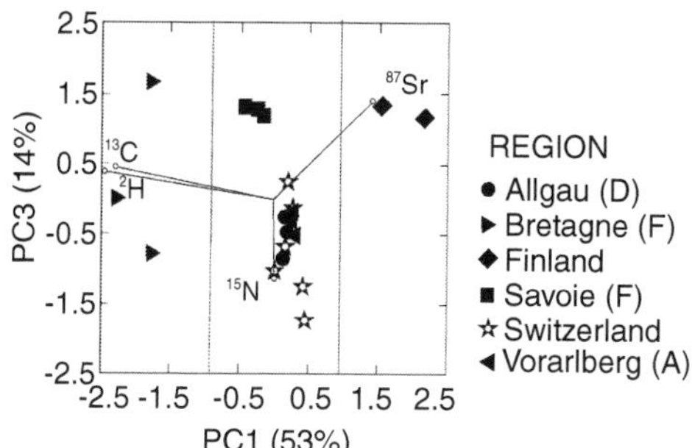

FIGURE 4.12 PCA of parameters $\delta^{13}C$, $\delta^{2}H$, $\delta^{15}N$, and $\delta^{87}Sr$. Separation of the groups "Finland," "Savoie," and "Bretagne." (Adapted from Pillonel, L. et al., Stable isotope ratios, major, trace and radioactive elements in emmental cheeses of different origins, Elsevier, *Lebensm.-Wiss. u.-Technol.*, 36, 615, 2003. With permission.)

classification of buffalo milk mozzarella cheeses produced in the Foggia and Caserta provinces, southern Italy. It is pointed out that the coupling of the isotopic ratios data with NMR data gave satisfactory results concerning geographical origin distinguishing. A variety of 20 emmenthal cheeses from six regions of four European countries (Germany, France, Austria, Switzerland) have been analyzed for concentrations of radioactive elements (^{90}Sr, ^{234}U, ^{238}U), stable isotope ratios ($^{13}C/^{12}C$, $^{15}N/^{14}N$, $^{18}O/^{16}O$, $^{2}H/^{1}H$, $^{87}Sr/^{86}Sr$), major elements (Ca, K, Mg, Na) and trace elements (Cu, I, Mn, Mo). Based on the stable isotope ratios and major or trace elements, it was possible to distinguish geographically distinct areas.[70] PCA of the data such as $\delta^{13}C$, $\delta^{15}N$, $\delta^{2}H$ and $\delta^{87}Sr$, as well as Ca, Cu, Mo, I, Mn, Na and Zn, played the most important role in the separation of clusters attributable to the geographically different regions (Figure 4.12). However, most of these parameters are affected by seasonal conditions because of variations in forage composition.[70]

Chemometric studies of goat's cheeses based on their mineral composition and concentration of fat, protein, pH, dry matter, and percentage of fat have been reported.[71] Ca, Cu, Fe, K, Mg, Na, Se, and Zn were determined by AAS. FA and DA appeared to be useful in the classification of samples with respect to their type (Figure 4.13A), season of production (Figure 4.13, B and C), and the type of goat's diet, although unsatisfactory separation was obtained in the case of criterion such as the region of production.

4.2.5 Rice

Isotopic ($\delta^{18}O$, $\delta^{13}C$) and elemental (B, Se, Rb, W, Gd, Ho, Mg) analyses using Isotope Ratio Mass Spectrometry (IRMS) and inductively coupled plasma mass spectrometry (ICP-MS), respectively, have been applied to determine the geographical origin of premium long-grain rice.[72] The maximum discrimination between rice samples from regions such as America, Europe, and India/Pakistan was clearly identified by canonical discriminant analysis (CDA) (Figure 4.14).

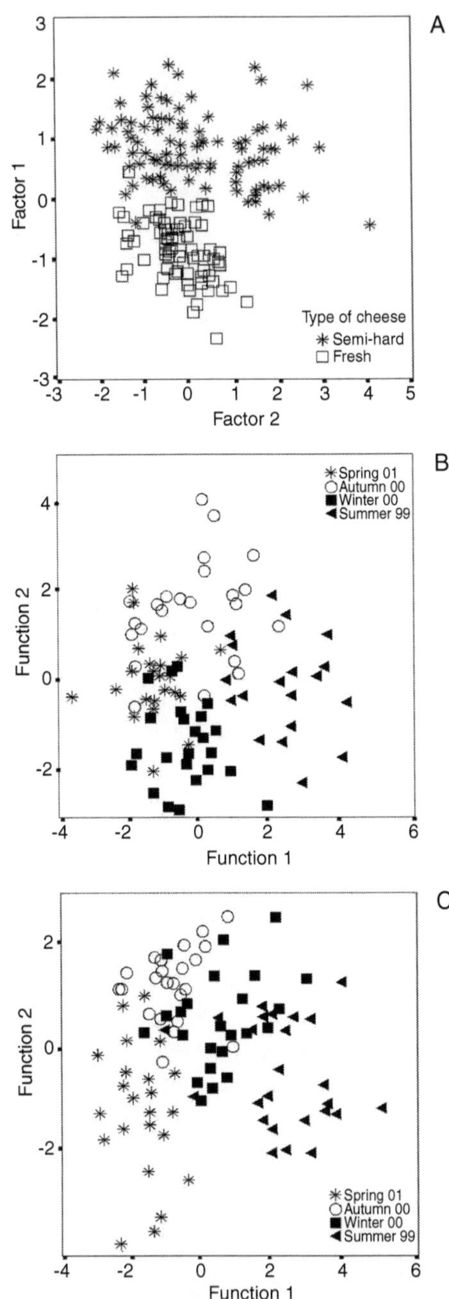

FIGURE 4.13 Scatter diagram of sample scores representing the first two factors differentiating the type of cheese (A); the first two-function discriminants are differentiated by the season of production of the fresh cheese samples (B) and the semi-hard cheese samples (C). (Adapted from Puerto, P. et al., Chemometric studies of fresh and semi-hard goats' cheeses produced in Tenerife (Canary Islands), Elsevier, *Food Chem.*, 88, 361, 2004. With permission.)

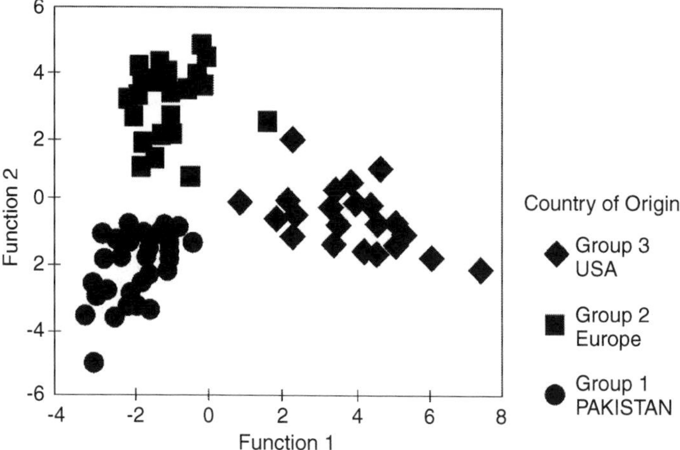

FIGURE 4.14 Graphical representation of the stepwise canonical discriminant analysis of isotopic and multielement data obtained from rice samples. (Adapted from Kelly, S. et al., The application of isotopic and elemental analysis to determine the geographical origin of premium long rice, Springer-Verlag, *Eur. Food Res. Technol.*, 214, 72, 2002. With permission.)

4.2.6 Vegetables

Chemomertic techniques appeared to be very useful in assessing the quality of different vegetables. For instance, 210 samples of onion and 190 samples of pea have been analyzed by HR–ICP-MS for 63 chemical elements (Ag, Al, Au, B, Ba, Be, Bi, Ca, Cd, Ce, Co, Cr, Cs, Cu, Dy, Er, Eu, Fe, Ga, Gd, Ge, Hf, Hg, Ho, In, Ir, K, La, Li, Lu, Mg, Mn, Mo, Na, Nb, Nd, P, Pb, Pr, Pt, Rb, Re, Ru, S, Sb, Sc, Si, Sm, Sn, Sr, Tb, Te, Th, Ti, Tl, Tm, U, V, W, Y, Yb, Zn, Zr) and 55 chemical elements (Ag, Al, Au, B, Ba, Be, Bi, Ca, Cd, Ce, Co, Cr, Cu, Dy, Er, Eu, Fe, Ga, Gd, Ge, Hf, Ho, Ir, La, Lu, Mn, Mo, Nb, Nd, P, Pb, Pd, Pr, Pt, Re, Rh, Sb, Sc, Se, Si, Sm, Sr, Ta, Tb, Te, Th, Ti, Tl, Tm, U, V, Y, Yb, Zn, Zr), respectively.[73] PCA split up the sites into two groups with respect to the cultivation methods for onions and peas, which affected the concentrations of some elements in the crops. Comparative statistical tests of element concentrations identified crop areas with organically and conventionally cultivated vegetables based on statistically different ($p < 0.05$) distribution pattern between Ca, B, Ba, Be, Bi, Co, Cu, Dy, Fe, Gd, Ge, La, Lu, Mo, Nd, Rb, Sc, Sr, Ti, U, and Y in onions and P, Gd, and Ti in peas (see, for instance, Figure 4.15). PCA has been also successfully applied for identification of tomato fruits cultivated in three different substrate systems.[74] Based on the different elemental composition of fruits (namely Ca, Cd, Fe, Mn, Mo, Na, Ni, Sr, Zn; and Ca, Cd, Fe, Mn, Na, Ni, Sr, Zn, Cu, K, Mg, P, Sn, and V), it was possible to identify crops from the different substrates and collected at three harvest times, respectively (Figure 4.16). According to Gundersen et al.,[74] the concentrations of Ca, Cd, Fe, Mn, Mo, Na, Ni, Sr, and Zn were significantly different ($p < 0.05$) for tomato fruits cultivated on different substrates. The use of PCA made it possible to distinguish the two groups attributable to soil-grown fruits and rockwool-grown fruits with the

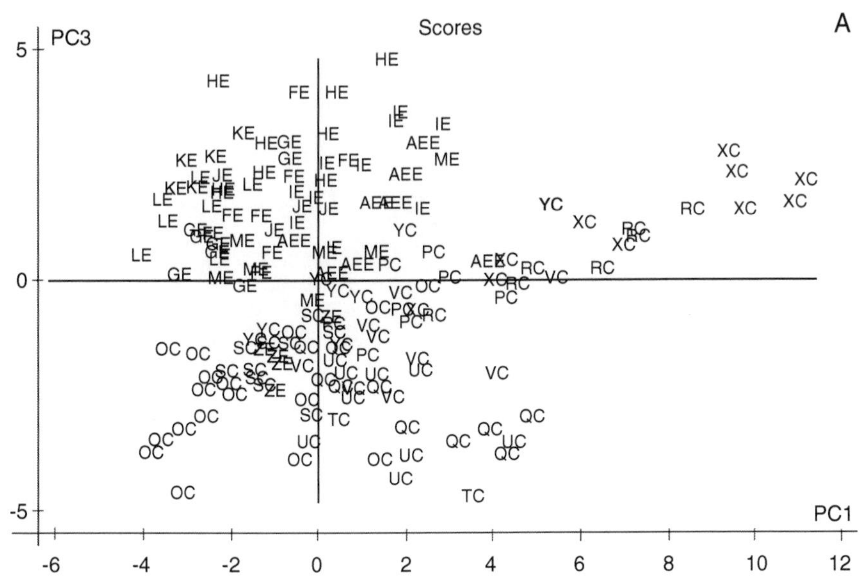

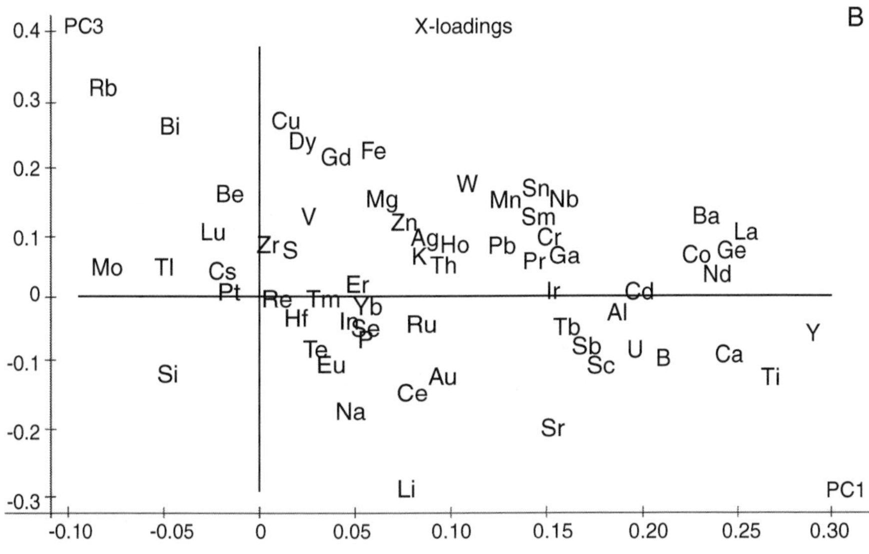

FIGURE 4.15 Biplot of scores (A) and loadings (B) for the first and third principal component of the PCA model for individual onion samples. Sites with second letter E are organic and sites with second letter C are conventional. (Adapted from Gundersen, V. et al., Comparative investigation of concentrations of major and trace elements in organic and conventional Danish agricultural crops. 1. Onions (*Allium cepa* Hysam) and Peas (*Pisum sativum* Ping Pong), ACS, *J. Agric. Food Chem.*, 48, 6094, 2000. With permission.)

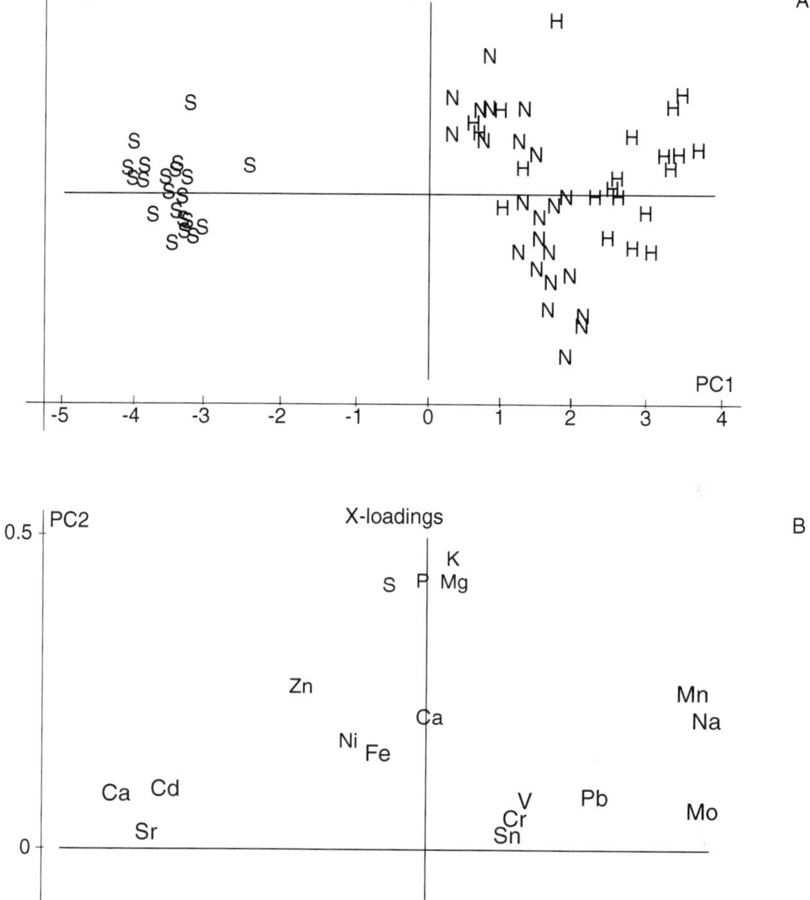

FIGURE 4.16 Biplot of scores (A) and loadings (B) for the first and second principal component of the PCA model for individual tomatoes samples. The letters in the plot refer to the treatment, (i.e., S, soil; N, rockwool with normal electrical conductivity (EC); H, rockwool with high EC treatment). (Adapted from Gundersen, V., McCall, D., and Bechmann, I.E., Comparison of major and trace element concentrations in Danish greenhouse tomatoes (*Lycopersicon esculentum* Cv. Aromata F1) cultivated in different substrates, *J. Agric. Food Chem.*, 49, 3808, 2001. With permission.)

two different nutrient solutions. Data for Ca, Cu, Fe, K, Mg, Mn, Na, Rb, and Zn (determined by AAS) in eight potato cultivars harvested in Tenerife (Spain) have been processed with FA and CA.[75] Traditionally and recently imported potatoes (*Solanum tuberosum*) were clearly separated by the application of CA. FA is the appropriate technique to identify potato samples according to the location of the farm (Figure 4.17A) and irrigation procedure (Figure 4.17B). It was also possible

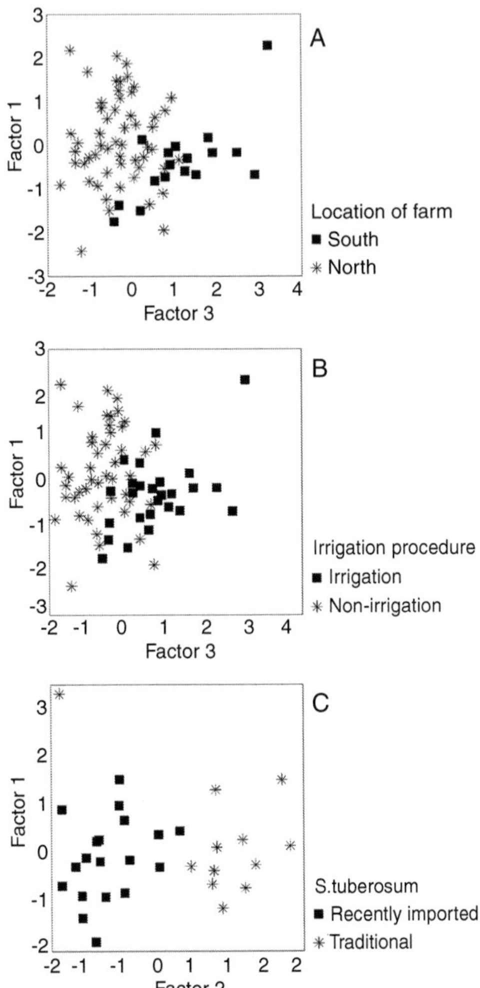

FIGURE 4.17 Scores of the potato samples on axes representing the two factors, differentiated by location of farms (A), irrigation procedure used (B), and recently imported and traditional vegetables (C). (Adapted from Rivero, R.C. et al., Mineral concentrations in cultivars of potatoes, Elsevier, *Food Chem.*, 83, 247, 2003. With permission.)

to separate traditional from recently imported potatoes (Figure 4.17C). A good agreement between FA data and CA data was obtained when concentrations of Zn, Mn, and Na (showing stronger correlations with the first three factors) were used in construction of the CA dendrogram (Figure 4.18).

Elemental composition data of 608 potato samples (Al, Ca, Cd, Co, Cu, Cr, Fe, K, Mg, Mn, Mo, Na, Ni, P, Pb, S, V, Zn determined by ICP-AES [atomic emission spectrometry]) have also been successfully processed by several statistical techniques, i.e., PCA, CDA, DFA (discriminant factor analysis), *k*-nearest neighbors, and artificial neural network (ANN).[76] Statistical classification may be widely

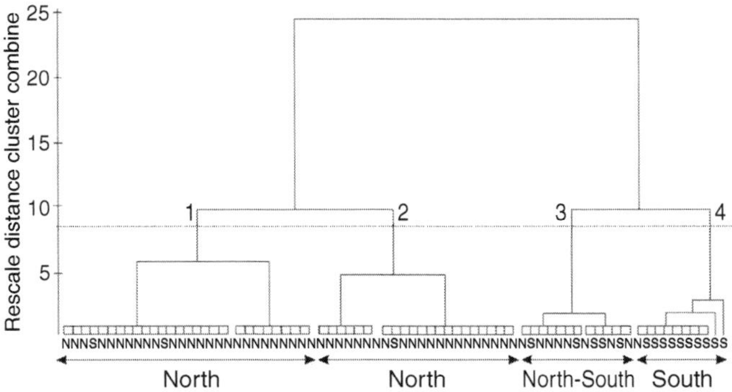

FIGURE 4.18 Dendrogram of the cluster analysis. (Adapted from Rivero, R.C. et al., Mineral concentrations in cultivars of potatoes, Elsevier, *Food Chem.*, 83, 247, 2003. With permission.)

applied to confirm the authenticity of Idaho-labeled potatoes as Idaho-grown potatoes. Assessment of Ca, Cd, Co, Cu, Fe, K, Mg, Mn, Na, Ni, Pb, and Zn determined by AAS in 12 different species of vegetables of Saudi Arabia (cucumber, vegetable marrow, tomato, potato, greenpepper, eggplant, carrots, parsley, lettuce, spinach, salq, onion, leek, watercress, and cabbage) has been made using CA, FA, and transfer factor analysis.[77] Concentrations of Cd, Cu, and Pb seemed to be appropriate descriptors in PCA, DA, and SIMCA (soft independent modeling-class analogy). A simplified mode of potential curves (SMPC) for grouping of edible vegetables to three predefined classes attributable to both agricultural conditions (private vs. commercial) and road traffic intensity has been recommended.[78] Padín et al.[79] have demonstrated that several pattern recognition techniques such as PCA, CA, LDA, KNN, SIMCA, and MLF-ANN (multilayer feed forward) are adequate to develop classification rules for authentication of 102 samples of Galician potatoes with a Certified Brand of Origin and Quality (CBOQ) based on their elemental profile (Ca, Cu, Fe, K, Li, Mg, Mn, Na, Rb, Zn) determined by F-AAS and F-AES (atomic emission spectrometry). The data obtained for LDA, KNN, and MLF-ANN are acceptable for the non-CBOQ class, whereas SIMCA appeared to have better recognition and prediction abilities for the CBOQ class. Self-organizing with adaptive neighbourhood network (SOAN) and MLF–ANN were applied to optimize the classification. The mineral compositional profiles provided suitable information allowing identification potatoes relative to their origin brand based on of SOAN-MLF-ANN. Owing to such a combination, it is possible to detect fraud, to preserve the quality name of the CBOQ product, and protect the consumer from overpayment and deception.[79] The dendrogram of CA (Figure 4.19) shows clear distinguishing Galician potatoes with a CBOQ from other potatoes that do not have this quality brand, i.e., Galician non-CBOQ and non-Galician non-CBOQ.

4.2.7 Market Basket Food Items

Torelm et al. and Torlem and Danielsson.[80,81] have reported variations in major nutrients and mineral components data in food items collected from three main

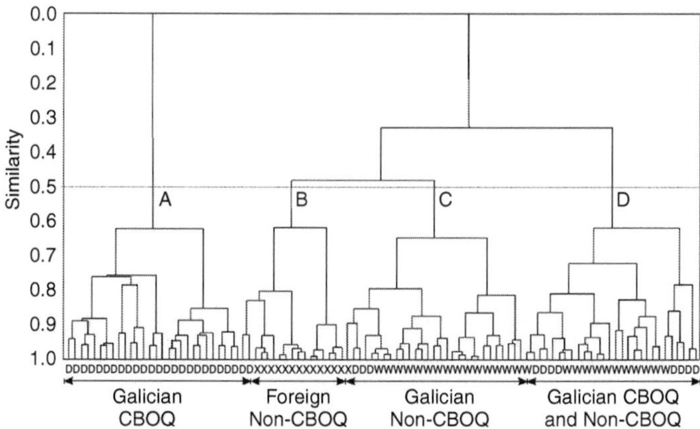

FIGURE 4.19 Dendrogram of cluster analysis. Sample codes: D, Galician CBOQ; W, Galician non-CBOQ; X, non-Galician non-CBOQ. (Adapted from Padín, P.M. et al., Characterization of Galician (N.W. Spain) quality brand potatoes: a comparison study of several pattern recognition techniques, *Analyst*, 126, 97, 2001. With permission.)

grocery chains in four geographical regions/cities in Sweden during four seasons of one year. Data on the following descriptors were obtained from the samples: Ca, Fe, K, N, Na, and P, as well as moisture, ash, and fat. A combination of PCA and analysis of variance (ANOVA) were used to explain the influence of some factors, i.e., geographic region, season, and store chain. For most vegetables, the seasonal variations were dominant, especially for tomato. In the case of animal products, the spatial differences appeared to have a greater contribution, and for hot dog it was the dominant factor. Pork, egg, and chicken exhibited somewhat pronounced influence of the different store chains. Torelm et al.[80] have also interpreted variations in the chemical elements in dishes prepared in ordinary households. PCA appeared to be useful for finding patterns in the manner of preparing dishes; the dominant variation pattern was associated with fat content, although enhanced by the closure effect (Figure 4.20).

4.2.8 Honey

Correspondence factor analysis (CFA), CA, and hierarchical cluster analysis (HCA) have been applied to the interpretation of data for trace elements in 86 honey samples sold in France. Concentrations of Ag, Al, Ca, Cd, Co, Cr, Cu, Fe, Hg, Li, Mg, Mn, Mo, Ni, P, Pb, S, and Zn were determined by ICP-AES.[82] Crude relationships were obtained between the distribution pattern of chemical elements in the honey samples and their botanical provenance. Based on PC data, it was possible to record the influence of metallic and nonmetallic elements on the factorial distribution of particular object samples in biplot PC1–PC2 (Figure 4.21). For instance, scores numbered 63 and 73, and 19 and 56, reflected the highest levels of Al and Fe respectively, implying that these elements are responsible for contamination of the honey samples located in the bottom and upper right part of the two-dimensional scatterplot. According to Devillers et al.,[82] honeys originating from caducous trees appear to be less

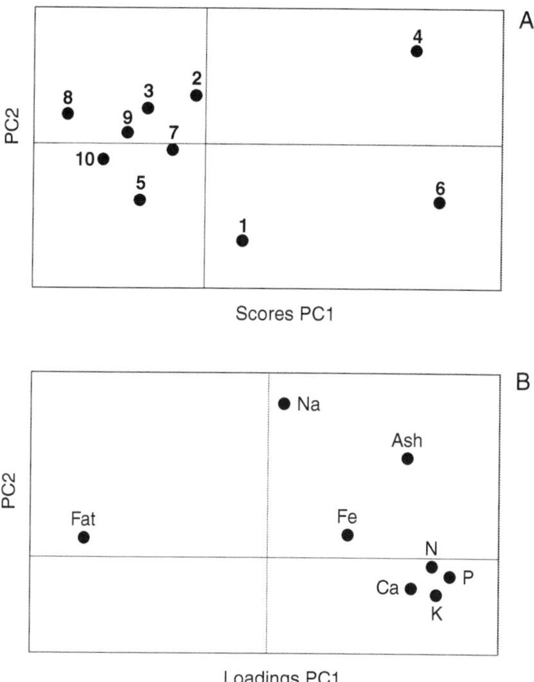

FIGURE 4.20 Principal component analysis of data on analytes in dishes prepared by 10 families. PC1 and PC2 analyzed, for each family, on analyte values when three dishes are combined. Score plot for families (A) and loading plot for analytes (B). (Adapted from Torelm, I. et al., Variations of major nutrients and minerals due to interindividual preparation of dishes from recipes, RSC, *J. Food Compos. Anal.*, 10, 14, 1997. With permission.)

polluted as compared to others. Metal contaminants are concentrated more easily in aromatic plants than in the herbaceous ones.[82]

Techniques such as PCA, LDA, KNN, and SIMCA have been used by Latorre et al.[83,84] to classify honeys according to their type and origin based on the concentrations of chemical elements determined using AES (K, Li, Na, Rb) and AAS (Co, Cu, Fe, Mg, Mn, Ni, Zn). Employing only three descriptors, i.e., Cu, Li, and Mn, a clear classification between two honey groups such as natural Galician honey and processed non-Galician honey was achieved (Figure 4.22). It demonstrates that the metal composition of honey combined with chemometric techniques can be used as an effective way for the detection of frauds — high-quality and expensive Galician honeys can be replaced with lower-priced non-Galician honeys as substrates for adulteration and falsification due to their comparable organoleptic properties. Terrab et al.[85] have determined the mineral content (Cu, Fe, K, Mg, Mn, Zn), ash content, and electrical conductivity of 98 honey samples from Northwest Morocco. The data have been processed by PCA, LDA, and multilayer perceptrons trained by back-propagation of error (MLP-BP). The latter is a very useful statistical tool based on ANN for classification purposes in chemometrics. LDA appeared to be an effective tool to discriminate two groups of samples, e.g., *Eucalyptus* honeys and honeydew

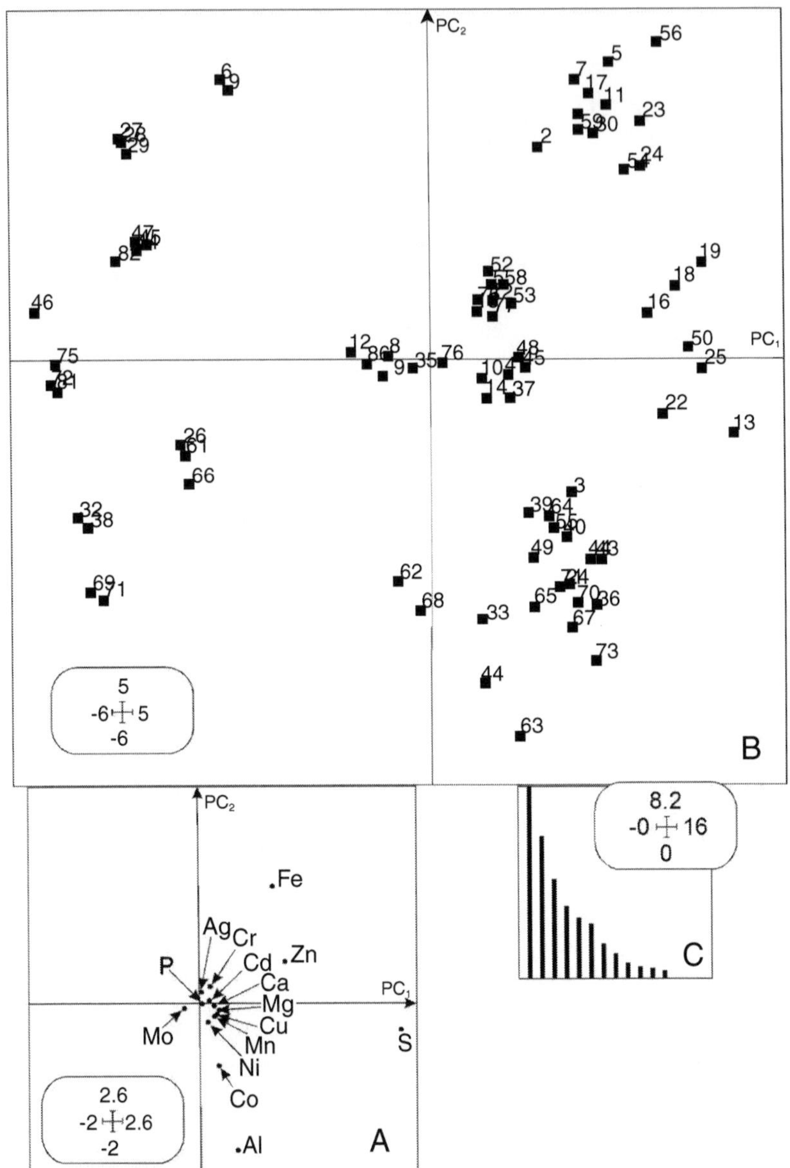

FIGURE 4.21 PC1-PC2 factorial maps for the 15 elements (A) and 86 honey samples (B). Graph of the eigenvalues (C). (Adapted from Devillers J. et al., Chemometrical analysis of 18 metallic and nonmetallic elements found in honeys sold in France, ACS, *J. Agric. Food Chem.*, 50, 5998, 2002. With permission.)

(Figure 4.23). According to the authors, the mineral content was too low to achieve much differentiation of the five unifloral honey classes considered, except the above mentioned *Eucalyptus* and honeydew. In order to improve the ability for distinguishing these object categories, it is recommended that other parameters be used,

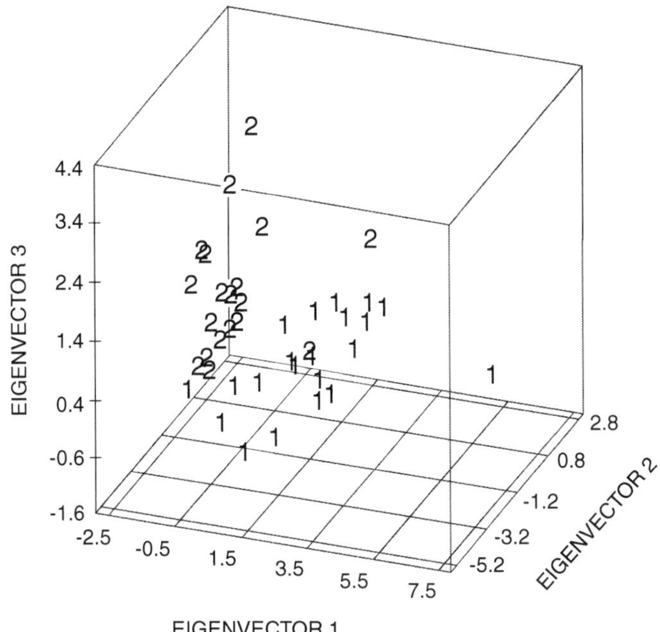

FIGURE 4.22 Eigenvector projection of honey samples. 1: natural Galician honey, 2: processed non-Galician honey. (Adapted from Latorre, M.J. et al., Chemometric classification of honeys according to their type: II. Metal content data, Elsevier, *Food Chem.*, 66, 263, 1999. With permission.)

especially organic compounds such as methyl anthranilate or hesperitin, sugars, as well as some physicochemical parameters. Hernandez et al.[86] classified honey from the Canary Islands based on its mineral composition (Fe, Cu, Zn, Mg, Ca, Sr, K, Na, Li, Rb). The chemometric processing of the spectroscopic results (AAS, AES) by various techniques (PCA, CA, DA, logistic regression) enabled an accurate categorization of the honey samples according to origin.

4.2.9 Tea

Based on measurements of different physical and chemical parameters, and after applying PCA and especially LDA, Fernández et al.[87] have concluded that the profiles of metals studied (Al, Ba, Ca, Cu, Fe, K, Mg, Mn, Na, Sr, Zn determined by ICP-AES) are a good descriptor for distinguishing the common types of tea beverages such as infusions, instant teas, and soft drinks (Figure 4.24). Statistical comparison shows that infusions prepared with black teas are enriched in Mn, Mg, Al, Ca, and K, and that there are significant differences between the mineral composition of infusions prepared with different extraction times. Moreda-Piñeiro et al.[88] have applied PCA, CA, LDA, and SIMCA for the classification of teas according to their region of origin based on data for 17 chemical elements determined by ICP-AES and ICP-MS. Descriptors such as Al, Ba, Ca, Cd, Co, Cr, Cu, Cs, Mg, Mn, Ni, Pb, Rb, Sr, Ti, V, and Zn have been determined in 85 tea samples from various Asian

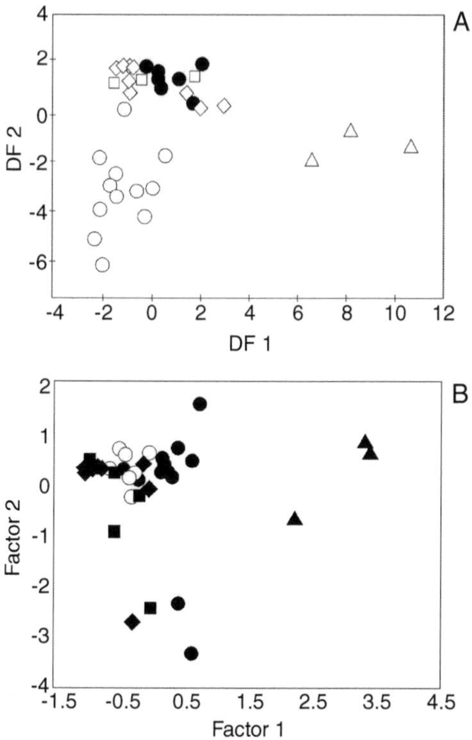

FIGURE 4.23 Linear discriminant analysis (A) and principal component analysis (B) of some Moroccan unifloral honeys, as shown by a scatter diagram representing the projection of the point of each sample on the plane formed by the first two discriminant functions (A) and the first two factors (B): open circles, *Eucalyptus*; diamonds, *Citrus*; squares, *Lythrum*; full circles, Apiacae; triangles, honeydew. (Adapted from Terrab, A. et al., Mineral content and electrical conductivity of the honeys produced in Northwest Marocco and their contribution to the characterization of unifloral honeys, *J. Sci. Food Agric.*, 83, 637, 2003. With permission.)

and African countries, as well as commercial blends and samples of unknown origin. The results obtained indicate that the differentiation and classification of tea samples from Africa and Asia (Figure 4.25) as well as identification of teas from China and India and Sri Lanka, in contrast to samples from Malaysia, Bangladesh, Japan, and Papua New Guinea, is possible (Figure 4.26). The authors concluded that it is also possible to assign unknown samples and tea blends with respect to their geographic origin. Marcos et al.[89] used the concentrations of Al, Ba, Ca, Cu, Fe, La, Mg, Mn, Sr, Ti, Zn (ICP-AES), and Cd, Co, Cr, Cs, Hg, La, Li, Nd, Ni, Pb, Pr, Rb, Se, Sn, Ti, V and Zr (ICP-MS) to achieve a distinction between African and Asian as well as between Chinese and other Asian teas using PCA (Figure 4.27A) and LDA (Figure 4.27B) methods. According to Herrador and González,[90] eight metals such as Al, Ca, Cu, Ba, Mg, Mn, K, and Zn determined by ICP-AES in 48 samples of commercial tea are good descriptors for the differentiation of green, black, and Oolong teas (Figure 4.28). Four statistical techniques, namely, PCA, CA, LDA, and ANN, have

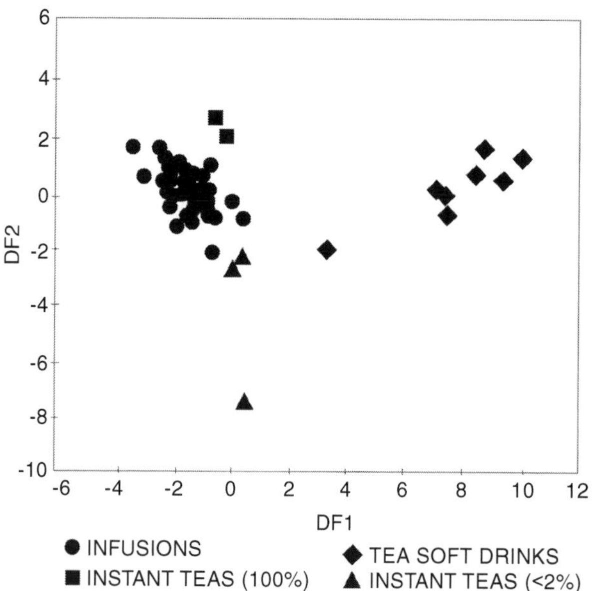

FIGURE 4.24 Scatterplot of tea beverages in the space of the two first discriminant functions. (Adapted from Fernández, P.L. et al., Multi-element analysis of tea beverages by inductively coupled plasma atomic emission spectrometry, Elsevier, *Food Chem.*, 76, 483, 2002. With permission.)

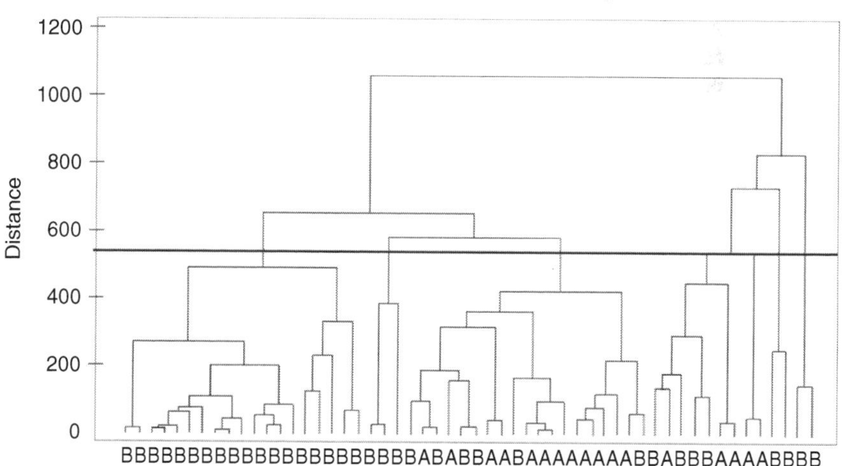

FIGURE 4.25 Dendrogram of cluster analysis. (A) African tea samples; (B) Asian tea samples. (Adapted from Moreda-Piñeiro, A., Fisher, A., and Hill, S.J., The classification of tea according to region of origin using pattern recognition techniques and trace metal data, Elsevier, *J. Food Compos. Anal.*, 16, 195, 2003. With permission.)

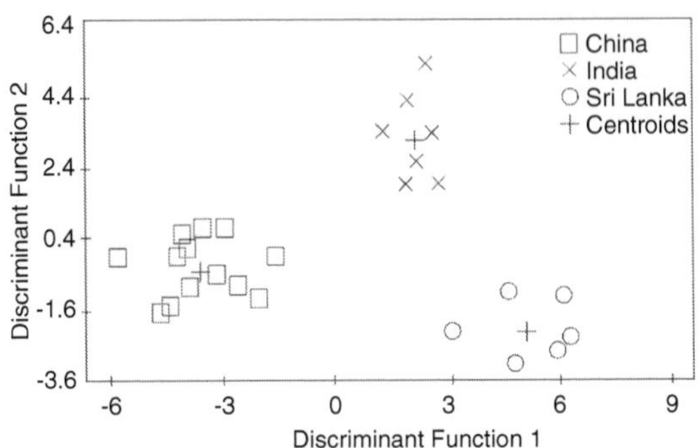

FIGURE 4.26 Projections of tea samples from China, India, and Sri Lanka in the space formed by the two discriminant functions after LDA. (Adapted from Moreda-Piñeiro, A., Fisher, A., and Hill, S.J., The classification of tea according to region of origin using pattern recognition techniques and trace metal data, Elsevier, *J. Food Compos. Anal.*, 16, 195, 2003. With permission.)

been applied successfully, with ANN being able to achieve a prediction ability of ~95% (Figure 4.29).

4.2.10 Coffee

Recently, there has been an increasing practice of selling coffees based on their varietal and geographic origin.[91–94] In order to characterize coffee varieties (Arabica and Robusta, roasted and green coffees from Brazil, Salvador, Costa Rica, Colombia, Honduras, Nicaragua, Guatemala, Cameroon, Uganda, Thailand, Indonesia, Vietnam, Ivory Coast) and resolve coffee mixtures, several authors have applied PCA, CA, and LDA, using as chemical descriptors concentrations of Ba, Ca, Cu, Fe, K, Mg, Mn, Na, P, Sr, and Zn (determined by ICP-AES) (Figure 4.30). The concentrations of P, Mn, and Cu appeared to be the most discriminating variables. These statistical techniques are useful in assessing the quality of the coffee product in case of fraudulent or accidental mislabeling.[94] The determination of mineral nutrients and nonessential metals including Al, Ca, Cd, Cr, Cu, Fe, K, Mg, Mn, Na, Ni, P, Pb, S, Sb, Sn, and Zn was performed in 21 samples of Brazilian soluble coffee by ICP-AES.[95] The PCA and AHC (agglomerative hierarchical clustering) data indicated differences in the process of industrial production and factors influencing the cultivation of the coffee plant, i.e., the geochemical type of soil, the use of different fertilizers, and the surrounding conditions. The ICP-AES method has also been used to determine 18 elements (Al, Ca, Cd, Co, Cu, Cr, Fe, K, Mg, Mn, Mo, Na, Ni, P, Pb, S, V, Zn) in 160 samples of coffee beans from Indonesia, East Africa, and Central/South America.[96] Applying the PCA, DFA, and NN models revealed significantly different groupings into the three major geographical regions that the coffee was grown in (Figure 4.31). The use of NN models and DFA successfully differentiated the coffees with respect to their subregional growing areas (70–86% successful classification).

Hasswell and Walmsley[97] have applied the PCA and CA models to data from multielemental (Br, Ca, Cr, Cu, Fe, K, Mn, Ni, Rb, Sr, Zn) analysis of coffees using total reflection X-ray fluorescence analysis (TXRF). Although the coffee samples

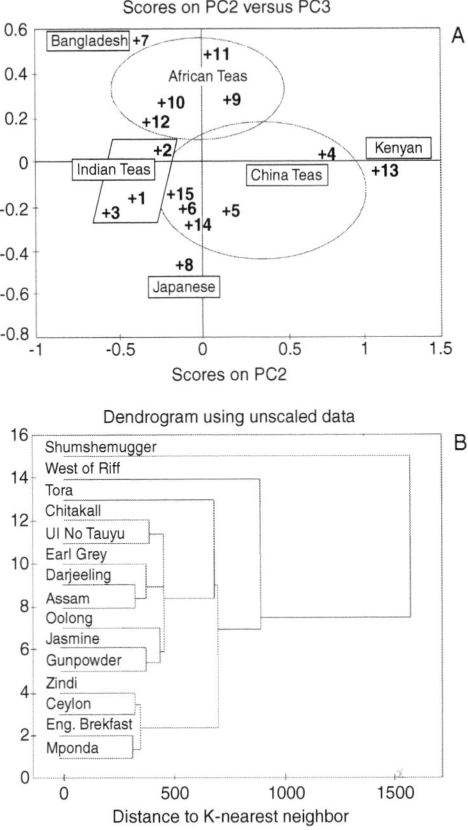

FIGURE 4.27 PCA plot of the ICP-AES data minus La results (A) and dendrogram using the untreated data (B). (Adapted from Marcos, A. et al., Preliminary study using trace element concentrations and a chemometrics approach to determine the geographical origin of tea, *J. Anal. At. Spectrom.*, 13, 521, 1998. With permission.)

were not adequately separated by the above techniques, most of the samples were grouped with respect to manufacturer rather than country of origin. The use of star plots has been a valuable means for identifying different coffee samples, and these plots, showing the importance of each variable in the sample, offer considerable potential in quality control applications.

According to Grembecka et al.,[98] loadings such as Co, Mn, Fe, Cr, Ni, Zn, Cu Ca, Mg, K, Na, and P (determined by F-AAS) are good descriptors for different kinds of coffees. Based on the mineral composition of coffee, it was possible to differentiate chemometrically particular types of coffee by distinguishing Arabica from Robusta, ground from soluble coffees, and their infusions.

As can be seen in Figure 4.32, F1 distinguishes Arabica group (lower values) from Arabica group (higher values). The elements responsible for identification of the latter include Mn and Mg, whereas the former cluster is attributable to Zn, Cu, Ca, Na, K, P, Fe, and Ni. Figure 4.32 shows that the ground coffee group, described

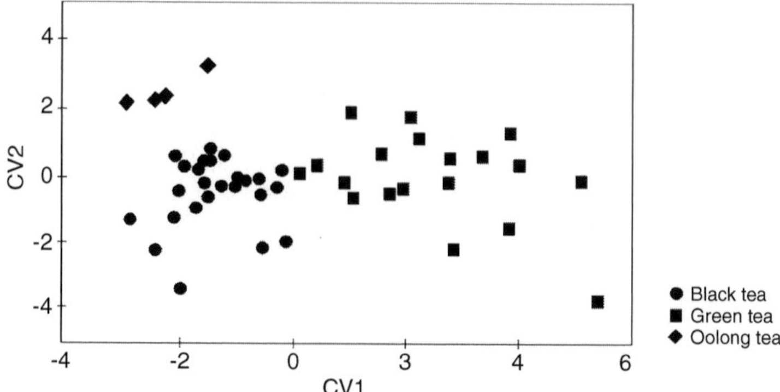

FIGURE 4.28 Scatterplot of tea samples using CV1 and CV2 as axes. (Adapted from Herrador, A.M. and González, A.G., Pattern recognition procedures for differentiation of Green, Black and Oolong teas according to their metal content from inductively coupled plasma atomic emission spectrometry, Elsevier, *Talanta*, 53, 1249, 2001. With permission.)

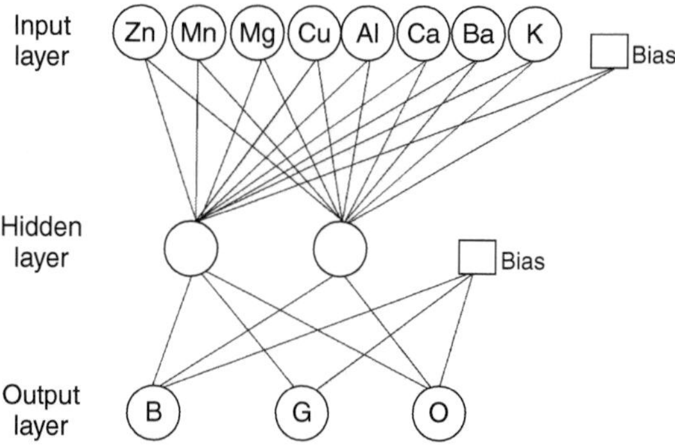

FIGURE 4.29 Architecture scheme of the multivariate neural network used. (Adapted from Herrador, A.M. and González, A.G., Pattern recognition procedures for differentiation of Green, Black and Oolong teas according to their metal content from inductively coupled plasma atomic emission spectrometry, Elsevier, *Talanta*, 53, 1249, 2001. With permission.)

by higher values of F2, is well distinguished from the soluble coffee group, described by lower values of F2, whereas the ground coffee infusion cluster is located in the middle part of the F2/F1 biplot. Ground coffee and soluble coffee scores are identified by Fe, Mn, and Zn, and Cd, Na, and Ni, respectively, whereas the other samples are identified by P, Mg, and K.

Multivariate estimation of the geographical origin of coffees has been performed successfully by Costa Freitas and Mosca.[99]

4.2.11 Mushrooms

Chemometrical techniques such as CA and DA have been used to analyze the data for fruiting bodies of mushroom *Xerocomus badius* (caps and stalks) and the underlying

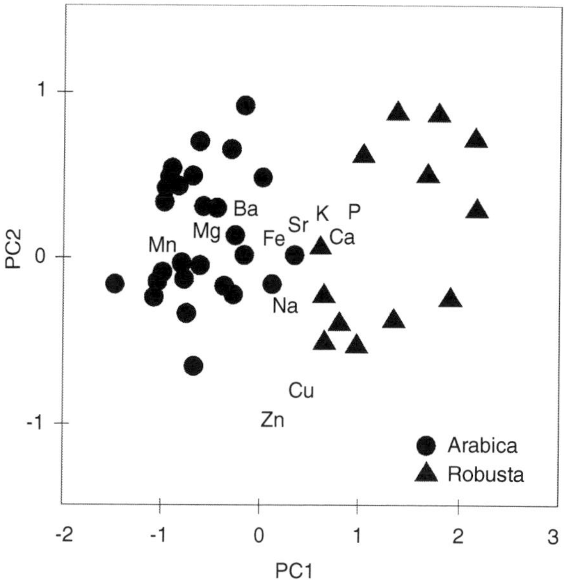

FIGURE 4.30 PCA biplot of the first two PCs illustrating distribution of scores and loadings for coffee samples Arabica and Robusta. (Adapted from Martin, M.J., Pablos, F., Gonzales, A.G., Characterization of green coffee varieties according to their metal content, Elsevier, *Anal. Chim. Acta*, Vol. 358, 177, 1998. With permission.)

soil substratum collected from northeastern Poland.[100] In the caps, three first functions explained 75.8% of the total variance. The lowest values of F1 identified caps samples coming from Trójmiejski Landscape Park, adjacent to the Tricity agglomeration (Figure 4.33). The CA data (hierarchical clustering, Ward's method) for the sampling sites as objects are shown in Figure 4.34. The dendrogram is built up of two main clusters. The first one contains two subclusters with the objects from Iławskie Lake district (C59–C68) and the adjacent area of Morag (C43–C58), the most similar regions by environmental conditions. A second cluster contains five subclusters with the objects from Augustowska Forest (C1–C16), Białowieska Forest (C17–C27), Borecka Forest (C28–C42), Wdzydzki Landscape Park (C69–C83), and the adjacent area of Kołobrzeg (C139–C145). The most similar areas were distinguished, e.g., Augustowska and Białowieska Forests are protected areas, deprived of industrial or urban influences (observed also for Trójmiejski Landscape Park). The principal factor governing the accumulation of trace elements in mushrooms is pollution via atmospheric deposition. Apart from the anthropogenic factor, natural effects such as geochemical composition of bed rock, pH, and granulometric structure of soil, genetic makeup of fungi, ectomycorrhizal occurrence, and the kind of undergrowth (mosses, lichens, ferns) have a real influence on some metal concentrations in mushrooms.[100] The dendrogram from the CA model made it possible to separate the most similar forest areas with respect to the chemical composition of both the underlying soil substratum and biomass overgrown with *X. badius* in the studied areas.[100]

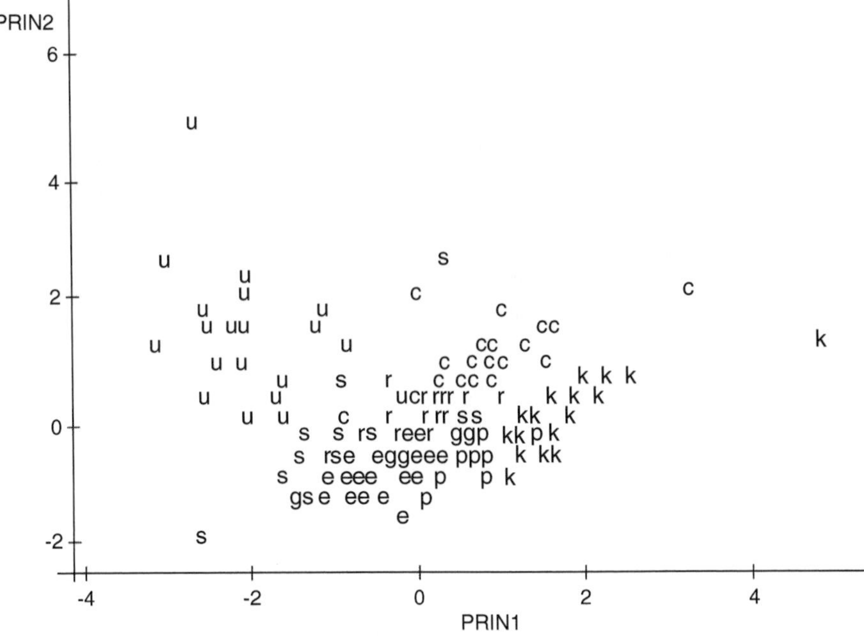

FIGURE 4.31 PC1 vs. PC2 for chemical profile of elements in roasted coffee beans from eight different growing regions: r, Costa Rica; c, Colombia; p, Panama; s, Sulawesi; u, Sumatra; k, Kenya; e, Ethiopia. There are 41 observations that are hidden. (Adapted from Anderson, K.A. and Smith, B.W., Chemical profiling to differentiate geographic growing origins of coffee, ACS, *J. Agric. Food Chem.*, 50, 2068, 2002.)

4.2.12 Fruit Juices

The PCA of trace elements determined by ICP-AES and ICP-MS (Al, Ba, B, Ca, Co, Cu, Fe, Li, Mg, Mn, Mo, Ni, P, K, Rb, Si, Na, Sr, Sn, Ti, V, Zn) in 482 samples of Australian and Brazilian orange juices, Australian peel extracts, and deacidified juices showed a distinct differentiation of the sample origin.[101] This distribution pattern could be attributable to elemental differences in the soil where the fruit was grown as well as to different types of rootstock (Figure 4.35). PCA showed that some 11 types of physicochemical measurements including mineral components are the most useful parameters for quality control analysis of pineapple juices and nectars.[102]

4.2.13 Sweets

Different cane sugar products (cane sugar plants, crude and syrup juices, molasses, the end products of consumer sugar) have been analyzed by instrumental neutron activation analysis (INAA), AAS, and ICP-AES for Al, Ca, Cl, Co, Cr, Fe, Mg, Mn, Na, Sc (INAA), Cu, Li, P, Sn, V and Zn (ICP-AES), and Pb and As (AAS).[103] The data obtained have been processed by CA applying Euclidean distance as a similarity coefficient and the group average. It is concluded that particular groups of multielement

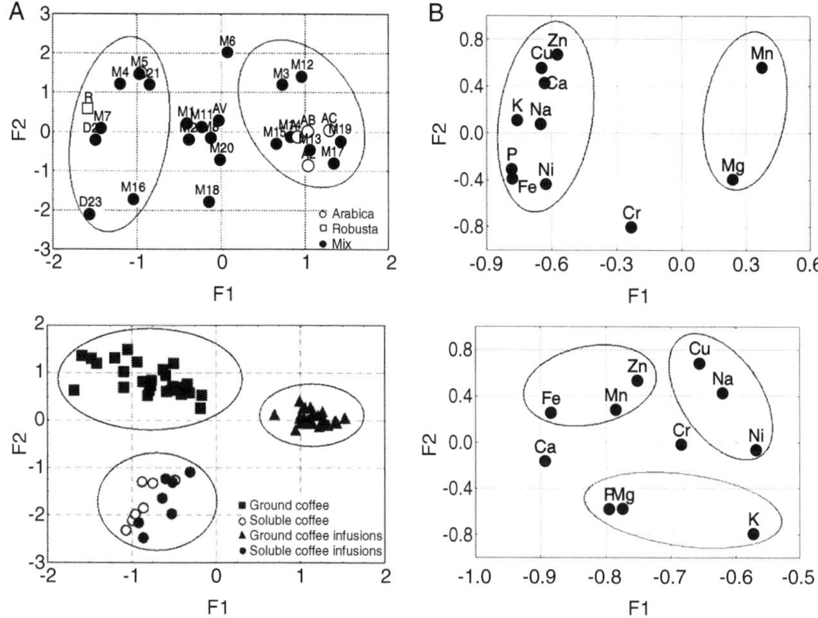

FIGURE 4.32 A biplot of the first two factors visualizing distribution of scores (A) and loadings (B) for coffee samples available in the Polish market. (Adapted from Grembecka, M. et al., Assessment of mineral composition of market coffee and its infusions, 2nd International IUPAC Symposium on Trace Elements in Food, Brussels, Belgium, 2004, p. 94. With permission.)

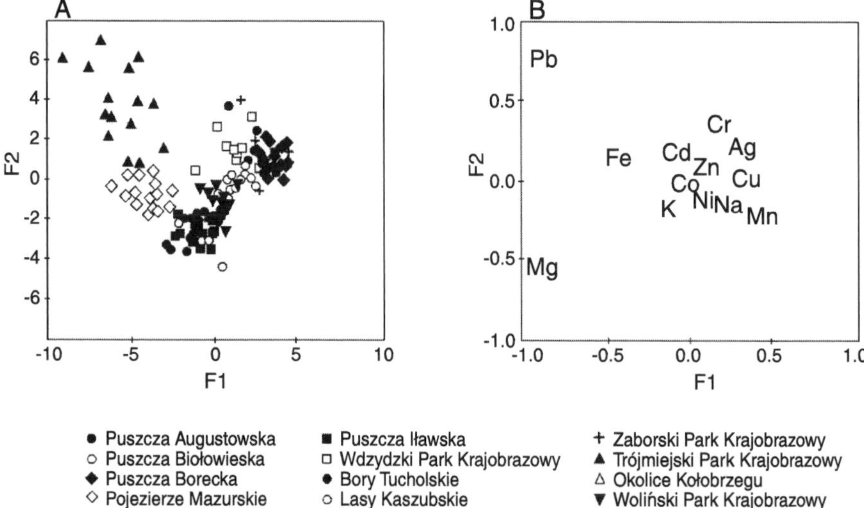

FIGURE 4.33 A) Scatterplot of object scores of the two first discriminant functions of 166 caps samples; (B) location of loadings for 14 metals in the cap. (Adapted from Malinowska, E., Szefer, P., and Falandysz, J., Metals bioaccumulation by bay bolete, *Xerocomus badius*, from selected sites in Poland, Elsevier, *Food Chem.*, 84, 405, 2004. With permission.)

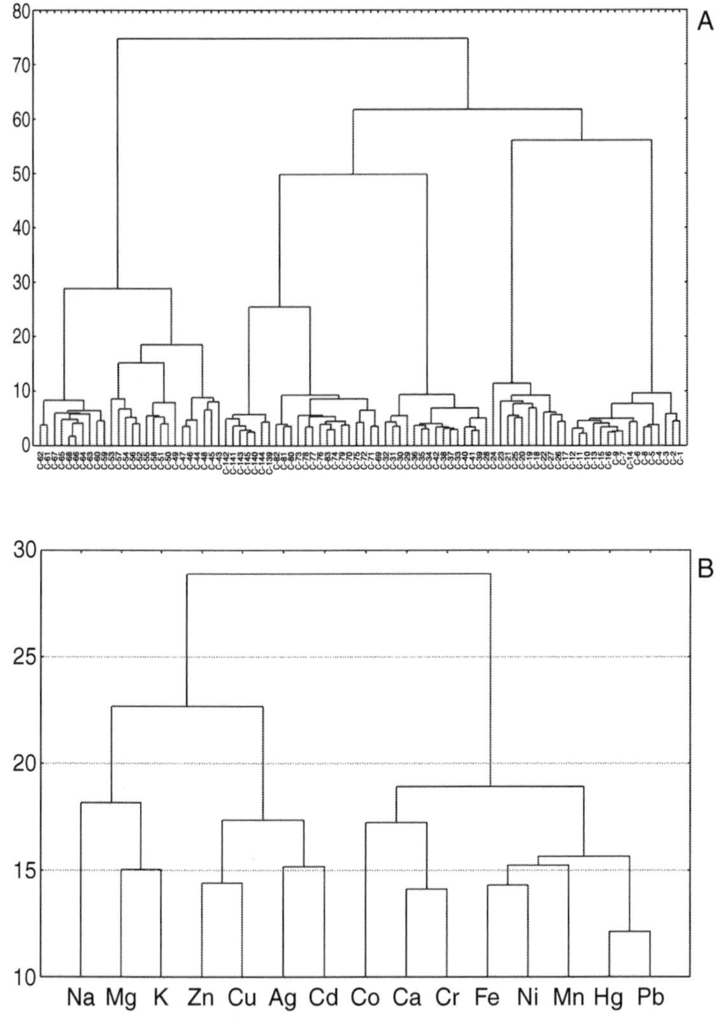

FIGURE 4.34 Hierarchical dendrogram for 90 objects samples (A) and chemical elements (B); sampling sites as (1–16) Augustowska Forest, 17–27 Białowieska Forest, 28–42 Borecka Forest, 43–58 Adjacent area of Morag, 59–68 Ilawskie Lake district, 69–83 Wdzydzki Landscape Park, 139–145 Adjacent area of Kołobrzeg. (Adapted from Malinowska, E., Szefer, P., and Falandysz, J., Metals bioaccumulation by bay bolete, *Xerocomus badius*, from selected sites in Poland, Elsevier, *Food Chem.*, 84, 405, 2004. With permission.)

clusters are attributable to specific sample composition, i.e., to their different texture, structure, and different soil types, reflecting differences or similarities in the processing steps in five Egyptian sugar-industry factories.[103] The dendrogram illustrates such clear differentiation between the above-mentioned groups (Figure 4.36).

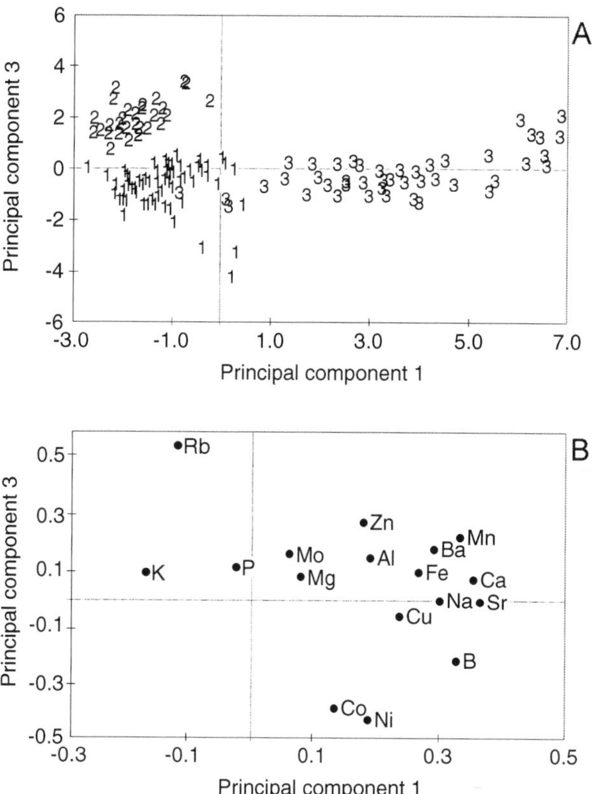

FIGURE 4.35 Scores (A) and loadings (B) of principal components analysis of trace elements in (1) Australian and (2) Brazilian reconstituted juices, and (3) Australian peel extract. (Adapted from Simpkins, W.A. et al., Trace elements in Australian orange juice and other products, Elsevier, *Food Chem.*, 71, 423, 2000. With permission.)

4.2.14 Alcoholic Beverages

4.2.14.1 Wines

Several authors[104–112] have processed the concentrations of chemical elements in different kinds of wines, including must, chemometrically.[105] Concentrations of Cu, Zn, Fe, Mn, Li, Rb, Sr, Ca, Mg, Na, and K were determined by AAS or AES,[106,107] and Al, B, Ca, Cr, Cu, Fe, I, K, Li, Mg, Mn, Na, P, Pb, Rb, Si, Sn, Sr, Zn, As, Ba, Be, Cd, Ce, Co, Cs, Ga, and Ge by ICP-MS.[108] Scatter diagram with the three most discriminating variables enables differentiation between five classes of wines. It is also possible to distinguish between two high quality wines from three less-fine wines (Figure 4.37). It has been reported[107] that PC1 is related to the place of wine origin, whereas PC2 is attributable to grape maturity (Figure 4.38). Better geographical differentiation with recognition ability amounting to 100% for each category has been attained using LDA (Figure 4.39). LCA and PCA appeared to be useful

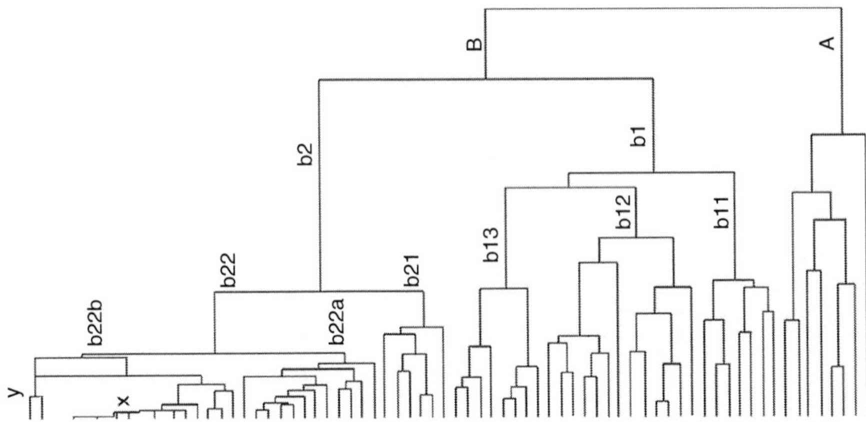

FIGURE 4.36 Cluster analysis of all sugar samples. (Adapted from Awadallah, R.M., Ismail, S.S., and Mohamed, A.E., Application of multi-element clustering techniques of five Egyptian industrial sugar products, *J. Radioanal. Nucl. Chem.*, 196, 377, 1995. With permission.)

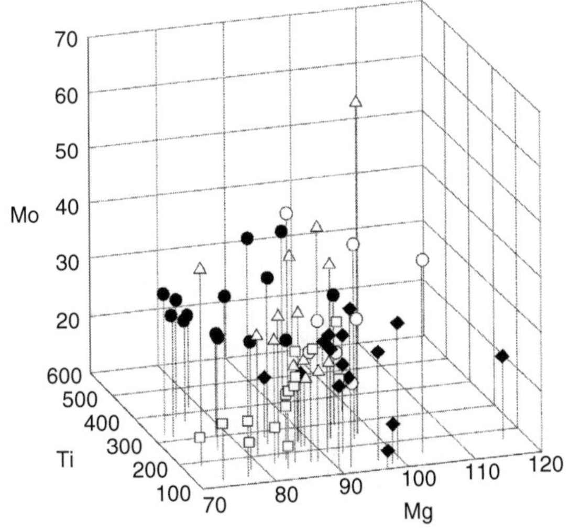

FIGURE 4.37 Scatter diagram with the three most discriminant variables. (○) Nebbiolo d'Alba, (❑) Langhe Nebbiolo, (◆) Barolo, (△) Barbaresco, (●) Roero. (Adapted from Marengo, E. and Aceto, M., Statistical investigation of the differences in the distribution of metals in Nebbiolo-based wines, *Food Chem.*, 81, 621, 2003. With permission.)

in the identification of Spanish wines relative to their geographical origin and the ripening state of the grapes.

A total of 34 chemical components including metals (Co, Fe, Li, Mn, Na, Ni, K, Rb, Zn), and volatile and phenolic compounds determined in 39 red wines from Galicia (northwest Spain) have been categorized into two different geographic groups based on CA, PCA, LDA, KNN, and SIMCA.[113] A satisfactory discriminant

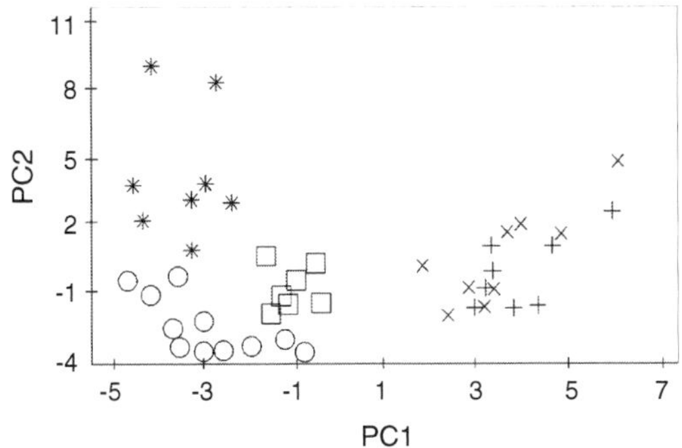

FIGURE 4.38 Scores of the wine samples on the first two principal components (57.5% of total variance). Codification: (❑) dry wines from El Hierro; (x) dry wines from Lanzarote; (○) dry wines from La Palma; (+) sweet wines from Lanzarote; (*) sweet wines from La Palma. (Adapted from Frías, S. et al., Classification of commercial wines from the Canary Islands (Spain) by chemometric techniques using metallic contents, Elsevier, *Talanta*, 59, 335, 2003. With permission.)

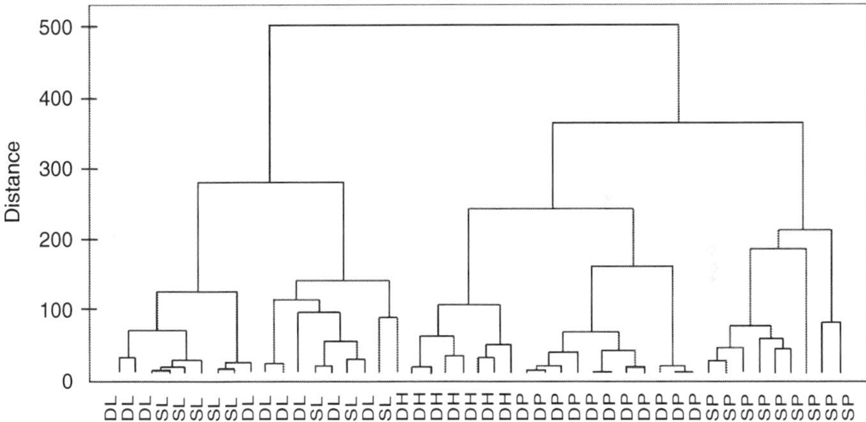

FIGURE 4.39 Dendrogram built with all the variables using Ward's method and Euclidean squared distance. Codification: SL = sweet wines from Lanzarote; DL = dry wines from Lanzarote; DH = dry wines from El Hierro; DP = dry wines from La Palma; SP = sweet wines from La Palma. (Adapted from Frías, S. et al., Classification of commercial wines from the Canary Islands (Spain) by chemometric techniques using metallic contents, Elsevier, *Talanta*, 59, 335, 2003. With permission.)

model has been presented to differentiate Spanish rosé PDO wines.[114] Ethanol and Ca content appeared to be the most important descriptors. The results reported by Marengo and Aceto[108] have indicated that the elemental composition of Italian wines could be a fast way to characterize and classify their samples and discriminate between two high-quality wines and three poor-quality wines.[108] PCA applied to the analytical data (Al, B, Ba, Ca, Cu, Fe, K, Mg, Mn, Na, Sr, Zn, Br, Cl, NH_4, NO_3, PO_4, SO_4, organic acids, alcohol content, acidity, etc.) showed that Italian wines could be divided into three groups according to their geographical origin.[115] The analysis of 48 elements, including the rare earths (REE), in 112 samples of Spanish and English wines by ICP-MS has been performed to identify the region of origin of wines using CA.[116] It was possible to identify the Spanish wines originated from three regions as well as differentiate between English and Spanish white wines. Application of PCA to elemental data (Ca, Cu, Fe, K, Mg, Mn, Na, P, Zn) and other descriptors (organic and sensor) resulted in satisfactory classification of Greek red wines in terms of their geographical provenance.[117] According to Frías et al.,[107] a high degree of sensitivity and specificity has been obtained for classification of commercial Spanish wines using SIMCA as a modeling multivariate technique. The metal content (Ca, Cu, Fe, K Mg, Na, Zn, determined by AAS) of Spanish red wines from the certified denomination of origin (DOC) has been processed with CA and LDA.[118] Sodium was identified as descriptor that best may differentiate wines according to the DOCs. González and Peña-Méndez[105] have used PCA, CA, BA (biplot analysis), and FA in the classification and differentiation of must and Canarian wine samples. It has been pointed out[106] that application of DA and especially ANNs are very useful in differentiating between dry and sweet wines from the Canary Islands according to their location on the island and the ripening state of grapes. ANNs appeared to be more useful techniques than LDA and showed 100% recognition and prediction abilities using only Rb, Sr, and Mn.[106] Back propagation artificial neural network (BP-ANN) has been successfully applied to classify wine samples in six different regions of Germany based on the concentrations of Al, B, Ba, Ca, Cu, Fe, K, Mg, Mn, Na, P, Rb, Sr, V, and Zn measured with ICP-OES.[119] The ANN data were compared to those obtained by CA, PCA, the Bayes discrimination method, and the Fisher discrimination method. Concentrations of different trace and ultratrace elements (As, Au, Ba, Be, Cd, Co, Cs, Cu, Ni, Re, Pb, Pt, Rb, Sb, Sn, Sr, Te, Ti, Tl, V, W, Zn, Zr, and 16 REEs) have been determined by ICP-MS in 153 samples of white, rosé, and red wines from the Canary Islands.[120,121] From the FA data, the 23 chemical elements, in contrast to the 16 REEs, presented significant discrimination according to type of wine and area of origin.[121] LDA and BP-ANN are recommended techniques to classify wines with respect to island of origin using the eight descriptors, i.e., Sr, Rb, Pb, Be, Ba, Tl, Ti, and Au. Similarly promising data have been obtained for 125 samples of bottled wines from the Tenerife Island (Spain) based on data of 11 chemical elements (Ca, Cu, Fe, K, Li, Mg, Mn, Na, Rb, Sr, Zn) using PCA and LDA.[122] Benítez et al.[123] have conducted a study on the influence of the Cu, Fe, and Mn content of fine sherry wine on its susceptibility to browning. Based on CA data, it was concluded that Fe and Mn appeared to influence the tendency of wine to undergo browning.

The SIMCA technique was used to assess eight denominations of origin (DOs) of wines bottled in the Canary Islands, based on their contents of trace elements grouped in three blocks (rare earths, lead isotopes ratios, and other metals such as As, Ba, Be, Au, Cd, Co, Cs, Cu, Ni, Pb, Pt, Rb, Re, Sb, Sn, Sr, Te, Ti, Tl, V, W, Zn, Zr determined by ICP-MS).[124] The model constructed with all the above element blocks was taken as the reference model and found to be adequately sensitive and specific. CA, using the Ward hierarchical clustering method, showed similar results with respect to the different DOs, and the models were similar with respect to their sensibility and specifity.[124] According to Murányi and Kovács,[125] knowledge of elemental composition of inorganic (Al, Ca, Cd, Cu, Fe, Mn, Pb) and organic components in Tokay wines from Hungary is very useful in establishing the area where the grapes were grown, and the type and the character of wines in view of FA.

According to Kment et al.,[126] the application of PCA and FA allowed a differentiation of the sources of chemical elements in wines with respect to the Czech wine/soil system. The authors used ICP-MS and AAS for measuring the 27 elements (Ag, Al, As, Ba, Be, Ca, Cd, Co, Cr, Cs, Cu, Fe, K, Li, Mg, Mn, Na, Ni, P, Pb, Rb, Sb, Sr, Tl, U, V, Zn) in samples of wine and Mehlich's soil extract. The group of lithophile elements (K, Mn, Cs, Cr, Ba, Sr) is associated with a single or predominant source in the vineyard soil, whereas Cu and As are attributable to the application of pesticides or similar operations. Variables such as Cd, Pb, and V can be anthropogenic in origin.

A combination of chemical analyses and chemometric approaches (PCA, LDA) appear to be a relevant tool for assessing of the influence of aging period, color, and sugar content of wines.[127] Concentrations of inorganic anions (Cl^-, NO_3^-, SO_4^{2-}) and heavy metals (Cd, Pb, Zn) were determined in DOC. Golden and Amber Marsala wines using ion exchange chromatography and derivative anodic chronopotentiometric stripping techniques (dASCP). PCA and, particularly, LDA data (Figure 4.40) clearly show a clear distinction between the four groups according to various ages of the wine.

Arvanitoyannis et al.[128] has reviewed the application of quality control methods for assessing wine authenticity by multivariate analysis. Diaz et al.[129] have reported chemometric data concerning differentiation of mineral composition of bottled wines with regard to origin in the Canary Island. Košir et al.[130] have used site-specific natural isotopic fractionation by ^2H nuclear magnetic resonance (SNIF-NMR) and isotope ratio mass spectrometry (IRMS) in combination with chemometric techniques for the determination of chaptalization and geographic origin of Slovenian wines.

4.2.14.2 Brandies

The 11 descriptors determined in 20 samples of Sherry brandies and 12 samples of Penedés by F-AAS (Ca, Cu, Fe, Mg, Mn, Zn), GF-AAS (Al, Cd, Pb) and AES (K, Na) have been analyzed chemometrically using PCA, agglomerative hierarchical CA, and MPN-BP.[131] The latter was found to be useful in the discrimination of Sherry and Penédes brandies based on the profile of Ca, Cu, Fe, and Mg descriptors (Figure 4.41). The use of classification procedures based on neural networks leads to ca. 90% hits in prediction ability.

4.2.14.3 Scotch Whisky

Concentrations of eight elements, i.e., Ca, Cu, Fe, Mg, Na, Ni, Pb, and Zn, determined in 35 Scotch Whiskies by GF-AAS, have been processed with CA.[132] It was shown that the fingerprinting of the data cannot be used as criteria to identify regions of whisky production. However, a simple analysis for Cu concentration could be used for distinguishing a malt whisky from a blended or grain whisky.

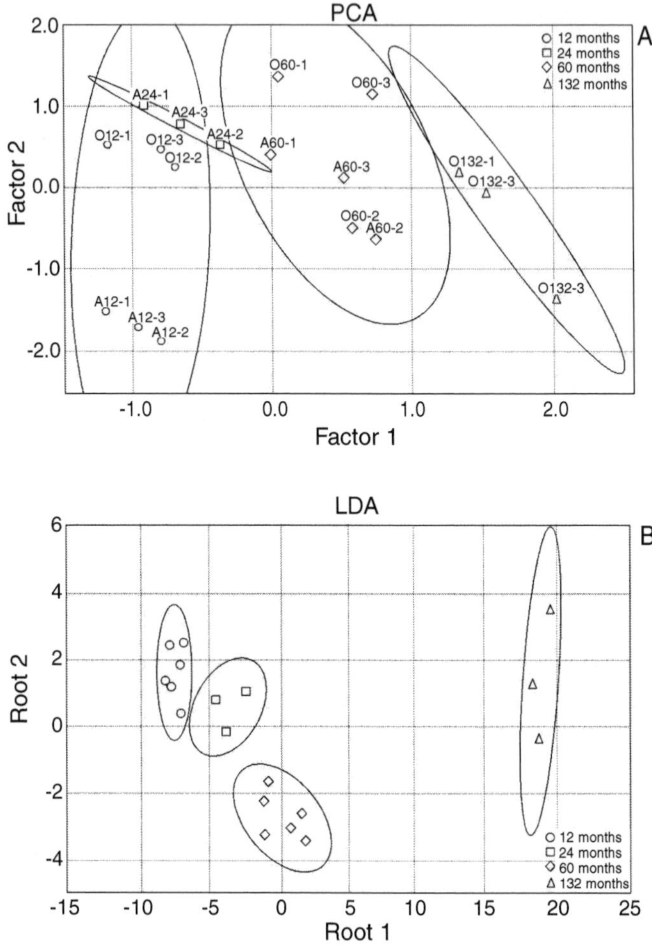

FIGURE 4.40 Principal component analysis (A) and linear discriminant analysis (B) biplots from the inorganic composition of 18 Marsala samples grouped according to their age (B) and their age and color (A). The used variables were Cd^{2+}, Cl^-, Cu^{2+}, NO_3^-, Pb^{2+}, SO_4^{2-}, and Zn^{2+} levels. (Adapted from Dugo, G. et al., Determination of some inorganic anions and heavy metals in D.O.C. Golden and Amber Marsala wines: statistical study of the influence of ageing period, colour and sugar content, Elsevier, *Food Chem.*, 91, 355, 2005. With permission.)

4.2.14.4 Spirituous Beverages

Elements such as Cu, Fe, and Zn have been determined by F-AAS in Cocuy de Penca firewater, a spirituous beverage very popular in Northwestern Venezuela, and the chemometric approach was followed to assess the geographic location of the manufacturers and the presence or absence of sugar in the end product.[133] Three chemometric approaches were applied, i.e., LDA, quadratic discriminant analysis (QDA), and various ANNs. The use of the classification procedure based on ANNs led to overall 97% hits of prediction ability.

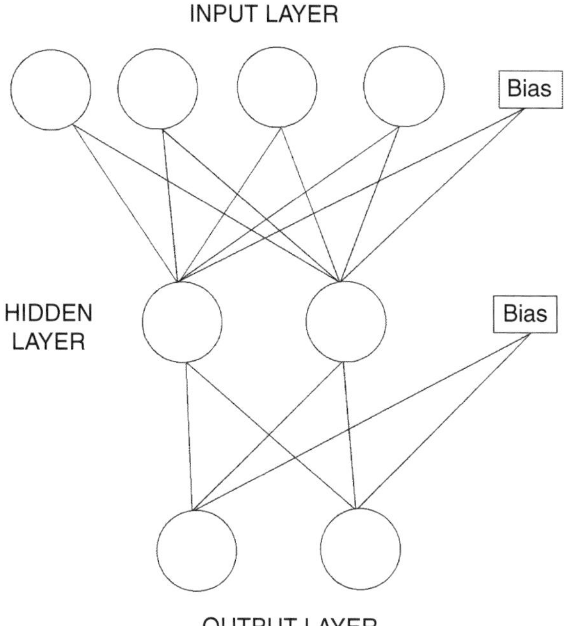

FIGURE 4.41 Architecture of the BPNN: 4 × 2 × 2 plus bias. Four metal descriptors for the input layer (Ca, Fe, Cu, and Mg), a hidden layer of two nodes (empirically established), and an output layer with two nodes (the two categories J and C). (Adapted from Cameán, A.M. et al., Differentiation of Spanish brandies according to their metal content, Elsevier, *Talanta*, 54, 53, 2001. With permission.)

4.2.14.5 Beers

Using a chemometric approach, the resulting mineral composition of beer has provided valuable information on how to distinguish the samples characterized as low alcoholic content beers, lagers, and dark beers (Figure 4.42).[134] Zinc, Mn, B, Fe, Al, Ba, P, Sr, Ca, Mg, Na, and K were considered as chemical descriptors and determined by ICP-AES in the 32 beer samples. The use of supervised learning PR methods, such as LDA, based on artificial neural networks-multilayer perceptions trained by back-propagation [BP-MLP]; Figure 4.43) was also found to be useful.

Differentiation and classification of beer samples have been achieved by Bellido-Milla et al.[135] with data obtained by flame AAS (Fe, Mn, Zn, Cu, Mg, Ca, Al) and flame AES (Na, K). The combination of the data obtained by flame AAS and the spectrum obtained by UV-VIS molecular absorption spectrometry provided information on the inorganic and organic components of the samples analyzed. The application of LDA appeared to be extremely useful for beer quality control and grouping the beers analyzed in spite of the new beer styles that had been introduced lately by the brewing industry. Based on the distribution of the grouping variables on the plot of the first discriminant functions (Df2/Df1), four types of beer were identified, namely, stout, ale, lager, and wheat beer (Figure 4.44). Three raw materials were also distinguished, i.e., barley, and other cereals as adjuncts of malt, as well

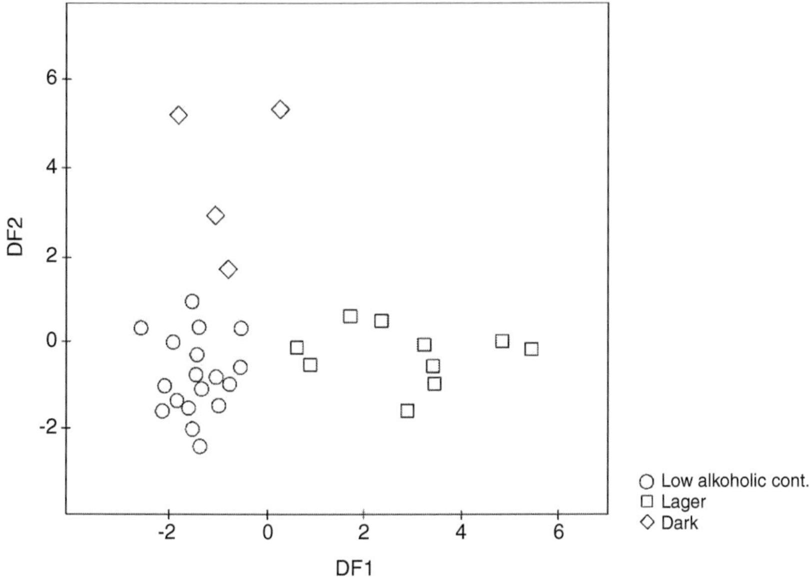

FIGURE 4.42 Plot of the two discriminant functions. (Adapted from Alcázar, A. et al., Multivariate characterisation of beers according to their mineral content, Elsevier, *Talanta*, 57, 45, 2002. With permission.)

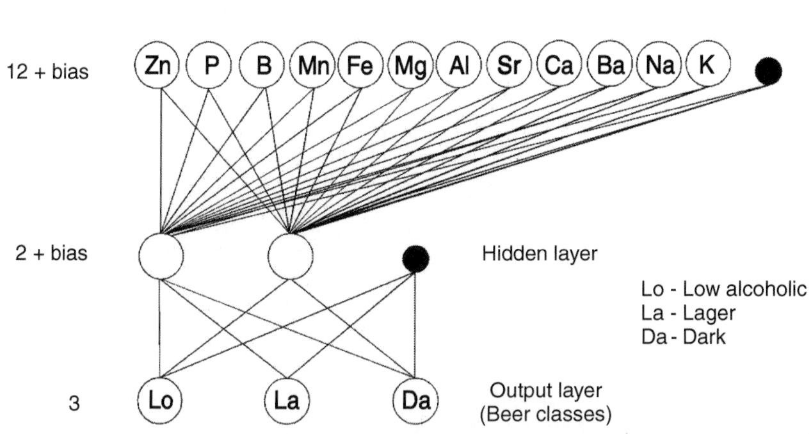

FIGURE 4.43 Architecture of the network. (Adapted from Alcázar, A. et al., Multivariate characterisation of beers according to their mineral content, Elsevier, *Talanta*, 57, 45, 2002. With permission.)

as wheat (Figure 4.45). Without the molecular absorption spectrum variable, the discrimination was poorer because of misclassification of stout with the lager.[135] The ANOVA model made possible the classification of beers based on the two types of

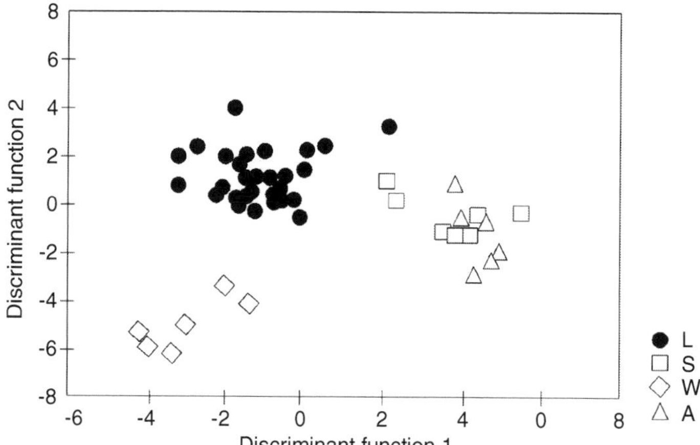

FIGURE 4.44 Biplot of the first DF vs. the second DF for types of beer: L, larger; S, stout; W, wheat beer; A, ale. (Adapted from Bellido-Milla, D., Moreno-Perez, J.M., and Hernandez-Artiga, M.P., Differentiation and classification of beers with flame atomic spectrometry and molecular absorption spectrometry and sample preparation assisted by microwaves, Elsevier, *Spectrochim. Acta B.* 55, 855, 2000. With permission.)

containers, namely bottle and can. A total of 23 elements (Ag, As, Bi, Cd, Co, Cr, Cs, Cu, Fe, Ga, Hg, In, Mn, Ni, Pb, Rb, Sb, Sn, Th, Tl, U, V, Zn) were determined in 35 bottled and canned Polish beers by ICP-MS.[136] Assessment of the data with PCA showed that the trace elements may originate from natural sources (soil, water, cereal, hops, yeast) and environmental pollution (fertilizers, pesticides, industrial processing, containers).

4.2.15 Wine Products

4.2.15.1 Wine Vinegars

According to Guerrero et al.,[137] the mineral composition (As, Ca, Cu, Fe, K, Mg, Mn, Na, and Zn determined by F-AAS, FES, HG-AAS) of 40 wine vinegars from Spain may be used for distinguishing quick and slow processed vinegars by applying the PCA (Figure 4.46), CA, SIMCA, and especially, the nonparametric KNN, LDA, and artificial neural networks trained by back propagation (BPANN). According to Benito et al.,[138] it is possible to characterize the vinegars obtained from wines with DOC Rioja (66 vinegars) and Jerez (18 vinegars) according to their chemical composition (e.g., Ca, Cl, Cu, Fe, K, Na).

4.2.15.2 Must

Multivariate data analyses (PCA, FA) have been successfully applied for differentiating between must and wines based on descriptors such as B, Ca, Cu, Fe, K, Mg, K, Na, Pb, S, and Zn, and other basic parameters such as ashes, density, reducing

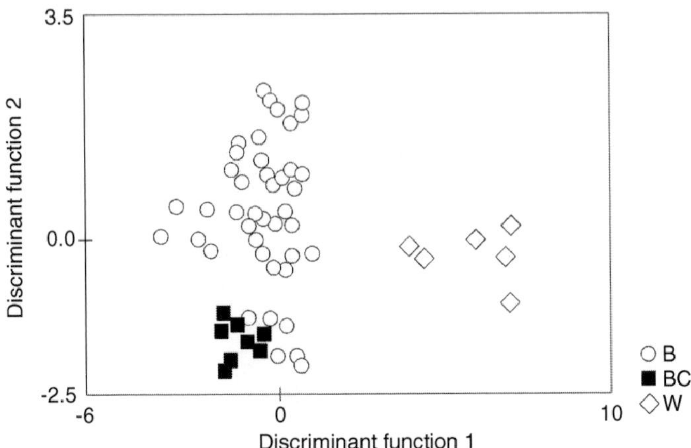

FIGURE 4.45 Biplot of the first DF vs. the second DF for raw materials: B, barley; BC, barley plus cereals; W, wheat. (Adapted from Bellido-Milla, D., Moreno-Perez, J.M., and Hernandez-Artiga, M.P., Differentiation and classification of beers with flame atomic spectrometry and molecular absorption spectrometry and sample preparation assisted by microwaves, Elsevier, *Spectrochim. Acta B*. 55, 855, 2000. With permission.)

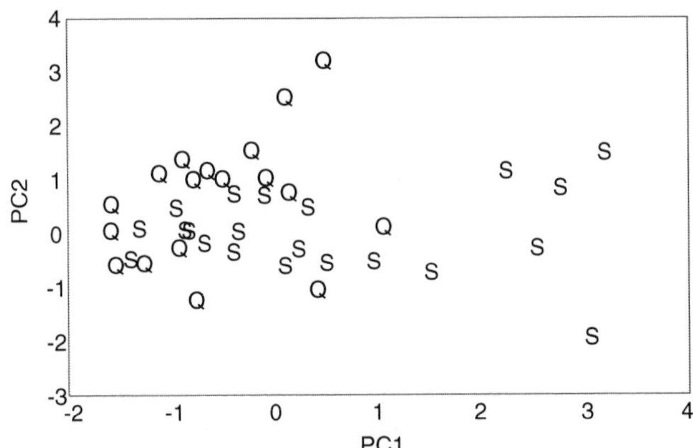

FIGURE 4.46 Scores plot of the vinegar samples using the two first PCs (Q and S stand for samples belonging to the "quick" and "slow" classes). (Adapted from Guerrero, M.I. et al., Multivariate characterization of vine vinegars from the south of Spain according to their metallic content, Elsevier, *Talanta*, 45, 379, 1997. With permission.)

sugars, alcohol strength, tartaric acid, and total acidity.[105] For instance, the successful use of PCA in differentiating between musts and wines is illustrated in Figure 4.47.

SUMMARY

The applicability of advanced statistical chemometric techniques such as principal component analysis (PCA), factor analysis (FA), linear discrimination analysis

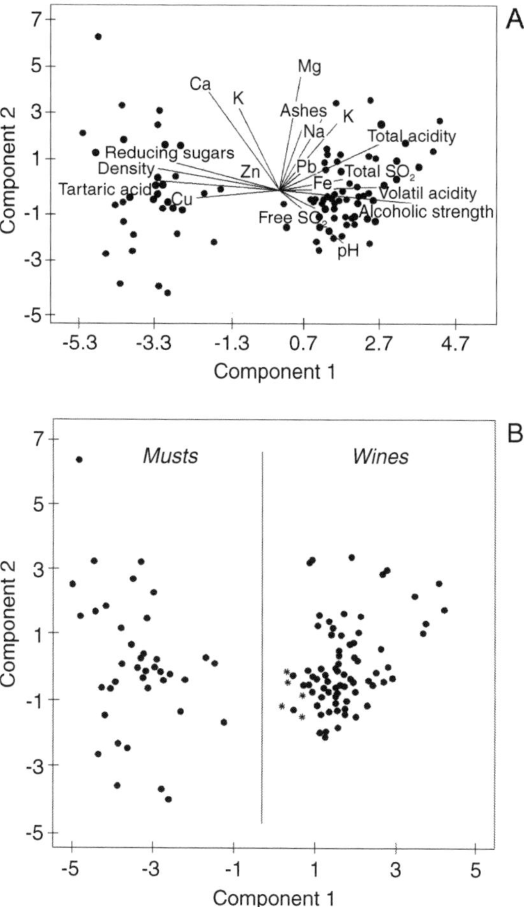

FIGURE 4.47 Biplot of loadings (A) and scores (B) for PCA. The first two scores, from the must and wine samples, on PCA. Outlier must samples. (Adapted from González, G. and Peña-Méndez, E.M., Multivariate data analysis in classification of must and wine from chemical measurements, *Eur. Food Res. Technol.*, 212, 100, 2000. With permission.)

(LDA), canonical discriminant analysis (DA), end-member analysis, cluster analysis (CA), and neural network (NN) and their capability in interpreting the complex data have been reviewed in this chapter. The analysis of a large volume of published data leads to the conclusion that chemometric methods are an effective statistical tool in analytical evaluation of food quality, as illustrated for meat, fish, seafood products, cheeses, honeys, vegetables, fruit juices and nectars, tea and coffee beverages, alcoholic beverages (i.e., commercial wines and must), Sherry wine vinegars, brandies, beer, and rice. The multivariate statistical techniques appear to be very useful in solving many problems and are helpful in both the classification of different features of the food products and the identification of existing pollution patterns. Their application is helpful for a deeper understanding of the distribution of selected metals and metalloids, including toxic trace elements, in foods. Multivariate methods can be recommended in monitoring the quality of foods and can help reduce the cost of such surveys. The chemometric approach is also a good tool in the analytical evaluation of foodstuff quality. The techniques may be useful, especially as combined techniques (e.g., SOAN and MLF), in detecting fraud to preserve the brand name of the original product and for protecting the consumer from being deceived and cheated financially.

REFERENCES

1. Szefer, P., Application of chemometric techniques in analytical evaluation of biological and environmental samples, in *New Horizons and Challenges in Environmental Analysis and Monitoring*, Namieśnik, J., Chrzanowski, W., and Żmijewska P., Eds., CEEAM, Gdańsk, chap. 18, p. 355.
2. Defernez, M. and Kemsley, E.K., The use and misuse of chemometrics for treating classification problems, *Trends Anal. Chem.*, 16, 216, 1997.
3. Cooley, W.W. and Lohnes, P.R., *Multivariate Data Analysis*, John Wiley & Sons, New York, 1971.
4. Morrison, D.F., *Multivariate Statistical Methods*, McGraw-Hill International, Auckland, 1978.
5. Mardia, K.V., Kent, J.T., and Bibby, J.M., *Multivariate Analysis*, Academic Press, London, 1989.
6. Deming, S.N. and Morgan, S.L., *Experimental Design: A Chemometric Approach*, Elsevier, Amsterdam, 1993.
7. Malinowski, E.R., *Factor Analysis in Chemistry,* A Wiley-Interscience Publication, John Wiley & Sons, New York, 1991.
8. Wackernagel, H., *Multivariate Geostatistic: An Introduction with Application*, 3rd ed., Springer-Verlag, Berlin, 2003.
9. Flury, B. and Riedwyl, H. *Multivariate Statistics — A Practical Approach*, Chapman & Hall, London, 1988.
10. Beebe, K.R., Pell, R.J., and Seasholtz, M.B., *Chemometrics: A Practical Guide*, John Wiley & Sons, New York 1998.
11. Danielsson, Å. et al., Spatial clustering of metals in the sediments of the Skagerrak/Kattegat, *Appl. Geochem.*, 14, 689, 1999.
12. Struck, B.D. et al., Statistical evaluation of ecosystem properties influencing the uptake of As, Cd, Co, Cu, Hg, Mn, Ni, Pb, Zn in seaweed (*Fucus vesiculosus*) and common mussel (*Mytilus edulis*), *Sci. Total Environ.*, 207, 29, 1997.
13. Kuik, P. et al., The use of Monte Carlo methods in factor analysis, *Atmos. Environ.*, 27A, 1967, 1993.
14. Sharma, S., *Applied Multivariate Techniques*, John Wiley & Sons, New York, 1996.
15. Massart, D.L. and Kaufman, L., *The Interpretation of Chemical Data by the Use of Cluster Analysis*, John Wiley & Sons, New York, 1983.
16. Danielsson, Å., *Spatial Modelling in Sediments*, Linköping Studies in Arts and Science, Sweden, 1998.
17. Everitt, B.S. and Dunn, G., *Applied Multivariate Data Analysis*, Edward Arnold, London, 1991.
18. Renner, R.M., Glasby, G.P., and Szefer, P., Endmember analysis of heavy-metal pollution in surficial marine sediments from the Gulf of Gdańsk and Southern Baltic Sea, *Appl. Geochem.*, 13, 293, 1998.

19. Full, W.E., Ehrlich, R., and Klovan, J.E., EXTENDED QMODEL — Objective definition of external endmembers in the analysis of mixtures, *Math. Geol.*, 13, 331, 1981.
20. Ehrlich, R. and Full, W.E., Sorting out geology — unmixing mixtures, in *Use and Abuse of Statistical Methods in the Earth Sciences*, Size, W.B., Ed., International Association for Mathematical Geology Studies in Mathematical Geology, Oxford University Press, Oxford, 1, 1987, p. 33.
21. Renner, R.M., The construction of extreme compositions, *Math. Geol.*, 27, 485, 1995.
22. Renner, R.M., An examination of the use of the log ratio transformation for the testing of endmember hypotheses, *Math. Geol.*, 23, 549, 1991.
23. Renner, R.M., The resolution of a compositional data set into mixtures of fixed source compositions, *Appl. Stat.*, *J. R. Stat. Soc.*, Ser. C 42, 615, 1993a.
24. Renner, R.M., A constrained least-squares subroutine for adjusting negative estimated element concentrations to zero, *Comp. Geosci.*, 19, 1351, 1993b.
25. Renner, R.M. et al., A partitioning process for geochemical datsets, in *Statistical Applications in the Earth Sciences*, Agterberg, F.P. and Bonham-Carter, G.F., Eds., Geological Survey of Canada Paper, 89-9, 1989, p. 319.
26. Li, Y.H., Geochemical cycles of elements and human perturbation, *Geochim. Cosmochim. Acta*, 45, 2073, 1981.
27. Hopke, P.K. et al., The use of multivariate analysis to identify sources of selected elements in the Boston urban aerosol, *Atmos. Environ.*, 10, 1015, 1976.
28. Heidam, N.Z., On the origin of the Arctic aerosol: a statistical approach, *Atmos. Environ.*, 15, 1421, 1981.
29. Bernard P.C., van Grieken, R.E., and Brügmann, L., Geochemistry of suspended matter from the Baltic Sea. 1. Results of individual particle characterization by automated electron microprobe, *Mar. Chem.*, 26, 155, 1989.
30. Szefer, P. et al., Distribution of selected heavy metals and rare earth elements in surficial sediments from the Polish sector of the Vistula Lagoon, *Chemosphere*, 39, 2785, 1999.
31. Selvaraj, K., Ram Mohan, V., and Szefer, P., Evaluation of metal contamination in coastal sediments of the Bay of Bengal, India: geochemical and statistical approaches, *Mar. Pollut. Bull.*, 49, 174, 2004.
32. Szefer, P., Distribution and behaviour of selected heavy metals and other elements in various components of the southern Baltic ecosystem, *Appl. Geochem.*, 13, 287, 1998.
33. Szefer, P. et al., Extraction studies of heavy-metal pollutants in surficial sediments from the southern Baltic Sea off Poland, *Chem. Geol.*, 120, 111, 1995.
34. Glasby G.P. et al., Heavy-metal pollution of sediments from Szczecin Lagoon and the Gdansk Basin, Poland, *Sci. Total Environ.*, 330, 249, 2004.
35. Hallberg, R.O., Environmental implications of metal distribution in Baltic Sea sediments, *Ambio*, 20, 309, 1991.
36. Virkanen, J., Effect of urbanization on metal deposition in the Bay of Töölönlahti, southern Finland, *Mar. Pollut. Bull.*, 36, 729, 1998.
37. Li, Y.H., Interelement relationship in abyssal Pacific ferromanganese nodules and associated pelagic sediments, *Geochim. Cosmochim. Acta*, 46, 1053, 1982.
38. Żbikowski, R., Szefer, P., and Latala, A., Distribution and relationships between selected chemical elements in green alga *Enteromorpha* sp. from the southern Baltic, *Environ. Pollut.*, 143, 435, 2006.

39. Popham, J.D. and D'Auria, J.M., Statistical approach for deciding if mussels *(Mytilus edulis)* have been collected from a water body polluted with trace metals, *Environ. Sci. Technol.*, 17, 576, 1983.
40. Szefer, P. et al., Distribution and relationships of trace metals in soft tissue, byssus and shells of *Mytilus edulis trossulus* from the southern Baltic, *Environ. Pollut.*, 120, 423, 2002.
41. Julshamn, K. and Grahl-Nielsen, O., Distribution of trace elements from industrial discharges in the Hardangerfjord, Norway: a multivariate data analysis of saithe, flounder and blue mussel as sentinel organisms, *Mar. Pollut. Bull.*, 32, 564, 1996.
42. Szefer, P. et al., Distribution and relationships of mercury, lead, cadmium, copper and zinc in perch (*Perca fluviatilis*) from the Pomeranian Bay and Szczecin Lagoon, southern Baltic, *Food Chem.*, 81, 73, 2003.
43. Hellou, J. et al., Distribution of elements in tissues of yellowtail flounder *Pleuronectes ferruginea*, *Sci. Total Environ.*, 181, 137, 1996.
44. Szefer, P. et al., Intercomparison studies on distribution and coassociations of heavy metals in liver, kidney and muscle of harbour porpoise, *Phocoena phocoena*, from southern Baltic Sea and coastal waters of Denmark and Greenland, *Arch. Environ. Contam. Toxicol.*, 42, 508, 2002.
45. Ciesielski, T. et al., Relationships and bioaccumulation of chemical elements in the Baikal seal (*Phoca sibirica*), *Environ. Pollut.*, 139, 372, 2006.
46. Julshamn, K. and Grahl-Nielsen, O., Trace element levels in harp seal (*Pagophilus groenlandicus*) and hooded seal (*Cystophora cristata*) from the Greenland Sea: a multivariate approach, *Sci. Total Environ.*, 250, 123, 2000.
47. Nyman, M. et al., Current levels of DDT, PCB and trace elements in the Baltic ringed seals (*Phoca hispida baltica*) and grey seals (*Halichoerus grypus*), *Environ. Pollut.*, 119, 399, 2002.
48. Brito, G. et al., Differentiation of heat-treated pork liver pastes according to their metal content using multivariate data analysis, *Eur. Food Res. Technol.*, 218, 584, 2004.
49. Schmidt, O. et al., Inferring the origin and dietary history of beef from C, N, and S stable isotope ratio analysis, *Food Chem.*, 91, 545, 2005.
50. Astorga-Espana, M.S., Pena-Mendez, E.M., and Garcia-Montelongo, F.J., Application of principal component analysis to the study of major cations and trace metals in fish from Tenerife (Canary Islands), *Chemometrics Intell. Lab. Syst.*, 49, 173, 1999.
51. Favretto, L. and Favretto, L.G., Principal component analysis and pollution by trace elements in a mussel survey, *J. Chemometrics*, 3, 301, 1988.
52. Machado Mesa, L. et al., Interpretation of heavy metal data from mussel by use of multivariate classification techniques, *Chemosphere*, 38, 1103, 1999.
53. Bechmann, I.E., Stürup, S., and Kristensen, L.V., High resolution inductively coupled plasma mass spectrometry (HR-ICPMS) determination and multivariate evaluation of 10 trace elements in mussels from 7 sites in Limfjorden, Denmark, *Fresenius J. Anal. Chem.*, 368, 708, 2000.
54. Favretto, L. and Favretto, L.G., Multivariate data analysis of some xenobiotic trace metals in mussels from the Gulf of Trieste, *Z. Lebensm. Unters. Forsch.*, 179, 201, 1984a.
55. Favretto, L. and Favretto, L.G., Principal component analysis as a tool for studying interedependences among trace metals in edible mussels from the Gulf of Trieste, *Z. Lebensm. Unters. Forsch.*, 179, 377, 1984b.

56. Favretto, L. et al., Terrigenous debris and mussel pollution — a differentiation based on trace element concentration by means of multivariate analysis, *Anal. Chim. Acta*, 344, 251, 1997.
57. Roméo, M. et al., Mussel transplantation and biomarkers as useful tools for assessing water quality in the NW Mediterranean, *Environ. Pollut.*, 122, 369, 2003.
58. Bermejo-Barrera, P., Moreda-Piñeiro, A., and Bermejo-Barrera, A., Sample pretreatment for trace elements determination in seafood products by atomic absorption spectrometry, *Talanta*, 57, 969, 2001.
59. Szefer, P. and Wołowicz, M., Occurrence of metals in the cockle *Cerastoderma glaucum* from different geographical regions in view of principal component analysis, *SIMO-Mar. Pollut.*, 64, 253, 1993.
60. Szefer, P. et al., Distribution and relationships of selected trace metals in molluscs and associated sediments from the Gulf of Aden, Yemen, *Environ. Pollut.*, 106, 299, 1999.
61. Szefer, P. et al., *Toxic metals in soft tissue and byssus of mollusc Mytilidae from different marine ecosystems*, III Conference on Trace Metals "Effects on Organisms and Environment," Polish Academy of Sciences, Sopot, Poland, 2000, 186/P2-14.
62. Szefer, P. et al., Distribution and coassociations of trace elements in soft tissue and byssus of *Mytillus edulis galloprovincialis* relative to the surrounding seawater and suspended matter of the southern part of Korean Peninsula, *Environ. Pollut.*, 129, 209, 2004.
63. Szefer, P., *Monitors of pollution in the Baltic Sea*, International Symposium Analytical Forum 2004, Book of Abstracts, Warsaw, Poland, 2004, p. 41.
64. Szefer, P. et al., Distribution of metallic pollutants in molluscs Mytilidae from the temperate, tropical and subtropical marine environments, First International Symposium, IEP '98 Issues in Environmental Pollution, The State and Use of Science and Predictive Models, Elsevier Science Ltd., Denver, CO, 1998, Section of Abstract Book 4.04.
65. Szefer, P. et al., A comparative assessment of heavy metal accumulation in soft parts and byssus of mussels from subarctic, temperate, subtropical and tropical marine environments, *Environ. Pollut.*, 139, 70, 2006.
66. Kwoczek, M. et al., Essential and toxic elements in seafood available in Poland from different geographical regions, *J. Agric. Food Chem.*, 54, 3015, 2006.
67. Szefer, P. and Grembecka, M., Applications of chemometry in analytical evaluation of foods quality based on their mineral composition, 2nd International IUPAC Symposium on Trace Elements in Food, Brussels, Belgium, 2004, p. 48.
68. Bermejo-Barrera, P. et al., The multivariate optimisation of ultrasonic bath-induced acid leaching for the determination of trace elements in seafood products by atomic absorption spectrometry, *Anal. Chim. Acta*, 439, 211, 2001.
69. Brescia, M.A. et al., Characterisation of the geographical origin of Buffalo milk and mozzarella cheese by means of analytical and spectroscopic determinations, *Food Chem.*, 89, 139, 2005.
70. Pillonel, L. et al., Stable isotope ratios, major, trace and radioactive elements in emmental cheeses of different origins, *Lebensm.-Wiss. u.-Technol.*, 36, 615, 2003.
71. Puerto, P. et al., Chemometric studies of fresh and semi-hard goat's cheeses produced in Tenerife (Canary Islands), *Food Chem.*, 88, 361, 2004.
72. Kelly, S. et al., The application of isotopic and elemental analysis to determine the geographical origin of premium long rice, *Eur. Food Res. Technol.*, 214, 72, 2002.

73. Gundersen, V. et al., Comparative investigation of concentrations of major and trace elements in organic and conventional Danish agricultural crops. 1. Onions (*Allium cepa* Hysam) and Peas (*Pisum sativum* Ping Pong), *J. Agric. Food Chem.*, 48, 6094, 2000.
74. Gundersen, V., McCall, D., and Bechmann, I.E., Comparison of major and trace element concentrations in Danish greenhouse tomatoes (*Lycopersicon esculentum* Cv. Aromata F1) cultivated in different substrates, *J. Agric. Food Chem.*, 49, 3808, 2001.
75. Rivero, R.C. et al., Mineral concentrations in cultivars of potatoes, *Food Chem.*, 83, 247, 2003.
76. Anderson, K.A. et al., Determining the geographic origin of potatoes with trace metal analysis using statistical and neural network classifiers, *J. Agric. Food Chem.*, 47, 1568, 1999.
77. Mohamed, A.E., Rashed, M.N., and Mofty, A., Assessment of essential and toxic elements in some kinds of vegetables, *Ecotoxicol. Environ. Saf.*, 55, 251, 2003.
78. Carlosena, A. et al., Classification of edible vegetables affected by different traffic intensities using potential curves, *Talanta*, 48, 795, 1999.
79. Padín, P.M. et al., Characterization of Galician (N.W. Spain) quality brand potatoes: a comparison study of several pattern recognition techniques, *Analyst*, 126, 97, 2001.
80. Torelm, I. et al., Variations of major nutrients and minerals due to interindividual preparation of dishes from recipes, *J. Food Compos. Anal.*, 10, 14, 1997.
81. Torelm, I. and Danielsson, R., Variations in major nutrients and minerals in Swedish foods: a multivariate, multifactorial approach to the effects of season, region, and chain, *J. Food Compos. Anal.*, 11, 11, 1998.
82. Devillers J. et al., Chemometrical analysis of 18 metallic and nonmetallic elements found in honeys sold in France, *J. Agric. Food Chem.*, 50, 5998, 2002.
83. Latorre, M.J. et al., Chemometric classification of honeys according to their type: II. Metal content data, *Food Chem.*, 66, 263, 1999.
84. Latorre, M.J. et al., Authentication of Galician (N.W. Spain) honeys by multivariate techniques based on metal content data, *Analyst*, 125, 307, 2000.
85. Terrab, A. et al., Mineral content and electrical conductivity of the honeys produced in Northwest Marocco and their contribution to the characterization of unifloral honeys, *J. Sci. Food Agric.*, 83, 637, 2003.
86. Hernandez, O.M. et al., Characterization of honey from the Canary Islands: determination of the mineral content by atomic absorption spectrophotometry, *Food Chem.*, 93, 449, 2005.
87. Fernández, P.L. et al., Multi-element analysis of tea beverages by inductively coupled plasma atomic emission spectrometry, *Food Chem.*, 76, 483, 2002.
88. Moreda-Piñeiro, A., Fisher, A., and Hill, S.J., The classification of tea according to region of origin using pattern recognition techniques and trace metal data, *J. Food Compos. Anal.*, 16, 195, 2003.
89. Marcos, A. et al., Preliminary study using trace element concentrations and a chemometrics approach to determine the geographical origin of tea, *J. Anal. At. Spectrom.*, 13, 521, 1998.
90. Herrador, A.M. and González, A.G., Pattern recognition procedures for differentiation of Green, Black and Oolong teas according to their metal content from inductively coupled plasma atomic emission spectrometry, *Talanta*, 53, 1249, 2001.
91. Martin, M.J., Pablos, F., and González, A.G., Application of pattern recognition to the discrimination of roasted coffees, *Anal. Chim. Acta*, 320, 191, 1996.
92. Martin, M.J., Pablos, F., and González, A.G., Characterization of green coffee varieties according to their metal content, *Anal. Chim. Acta*, 358, 177, 1998a.

93. Martin, M.J., Pablos, F., and González, A.G., Discrimination between arabica and robusta green coffee varieties according to their chemical composition, *Talanta*, 46, 1259, 1998b.
94. Martin, M.J., Pablos, F., and González, A.G. Characterization of arabica and robusta roasted coffee varieties and mixture resolution according to their metal content, *Food Chem.*, 66, 365, 1999.
95. dos Santos, E.J. and de Oliveira, E., Determination of mineral nutrients and toxic elements in Brazilian soluble coffee by ICP-AES, *J. Food Compos. Anal.*, 14, 523, 2001.
96. Anderson, K.A. and Smith, B.W., Chemical profiling to differentiate geographic growing origins of coffee, *J. Agric. Food Chem.*, 50, 2068, 2002.
97. Haswell, S.J. and Walmsley, A.D., Multivariate data visualisation methods based on multi-elemental analysis of wines and coffees using total reflection X-ray fluorescence analysis, *J. Anal. At. Spectrom.*, 13, 131, 1998.
98. Grembecka, M. et al., Assessment of mineral composition of market coffee and its infusions, 2nd International IUPAC Symposium on Trace Elements in Food, Brussels, Belgium, 2004, p. 94.
99. Costa Freitas, A.M. and Mosca, A.I., Coffee geographic origin — an aid to coffee differentiation, *Food Res. Int.*, 32, 565, 1999.
100. Malinowska, E., Szefer, P., and Falandysz, J., Metals bioaccumulation by bay bolete, *Xerocomus badius*, from selected sites in Poland, *Food Chem.*, 84, 405, 2004.
101. Simpkins, W.A. et al., Trace elements in Australian orange juice and other products, *Food Chem.*, 71, 423, 2000.
102. Camara, M., Diez, C., and Torija, E., Chemical characterization of pineapple juices and nectars: principal component analysis, *Food Chem.*, 54, 93, 1995.
103. Awadallah, R.M., Ismail, S.S., and Mohamed, A.E., Application of multi-element clustering techniques of five Egyptian industrial sugar products, *J. Radioanal. Nucl. Chem.*, 196, 377, 1995.
104. Moret I. et al., Multiple discriminant analysis in the analytical differentiation of Venetian white wines: 4. Application to several vintage years and comparison with the *k* nearest-neighbor classification, *J. Agric. Food Chem.*, 32, 329, 1984.
105. González, G. and Peña-Méndez, E.M., Multivariate data analysis in classification of must and wine from chemical measurements, *Eur. Food Res. Technol.*, 212, 100, 2000.
106. Frías, S. et al., Classification and differentiation of bottled sweet wines of Canary Islands (Spain) by their metallic content, *Eur. Food Res. Technol.*, 213, 145, 2001.
107. Frías, S. et al., Classification of commercial wines from the Canary Islands (Spain) by chemometric techniques using metallic contents, *Talanta*, 59, 335, 2003.
108. Marengo, E. and Aceto, M., Statistical investigation of the differences in the distribution of metals in Nebbiolo-based wines, *Food Chem.*, 81, 621, 2003.
109. Li, X.S., Danzer, K., and Thiel, G., Classification of wine samples by means of artificial neural networks and discrimination analytical methods, *Fresenius J. Anal. Chem.*, 359, 143, 1997.
110. Frías, S. et al., Metallic content of wines from the Canary Islands (Spain). Application of artificial neural networks to the data analysis, *Nahrung/Food*, 46, 370, 2002.
111. Šperkova, J. and Suchánek, M., Multivariate classification of wines from different Bohemian regions (Czech Republic), *Food Chem.*, 93, 659, 2005.
112. Nogueira, J.M.F. and Nascimento, A.M.D., Analytical characterization of Madeira wine, *J. Agric. Food Chem.*, 47, 566, 1999.
113. Rebolo, S. et al., Characterisation of Galician (NW Spain) Ribeira Sacra wines using pattern recognition analysis, *Anal. Chim. Acta*, 417, 211, 2000.

114. Pérez-Magariño, S. et al., Multivariate classification of rosé wines from different Spanish protected designations of origin, *Anal. Chim. Acta*, 458, 187, 2002.
115. Brescia, M.A. et al., Characterization of the geographical origin of Italian red wines based on traditional and nuclear magnetic resonance spectrometric determinations, *Anal. Chim. Acta*, 458, 177, 2002.
116. Baxter, M.J. et al., The determination of the authenticity of wine from its trace element composition, *Food Chem.*, 60, 443, 1997.
117. Kallithraka, S. et al., Instrumental and sensory analysis of Greek wines: implementation of principal component analysis (PCA) for classification according to geographical origin, *Food Chem.*, 71, 501, 2001.
118. Ortega, M. et al., Metal content of Spanish red wines from certified denomination of origin, *Quim. Anal.,* 18, 127, 1999.
119. Sun, L.-X., Danzer, K., and Thiel, G., Classification of wine samples by means of artificial neural networks and discrimination analytical methods, *Fresenius J. Anal. Chem.*, 359, 143, 1997.
120. Perez-Trujillo, J.-P., Barbaste, M., and Medina, B., Contents of trace and ultratrace elements in wines from the Canary Islands (Spain) as determined by ICP-MS, *J. Wine Res.*, 13, 243, 2002.
121. Perez-Trujillo, J.-P., Barbaste, M., and Medina, B., Chemometric study of bottled wines with denomination of origin from the Canary Islands (Spain) based on ultra-trace elemental content determined by ICP-MS, *Anal. Lett.*, 36, 679, 2003.
122. Conde, J.E. et al., Characterization of bottled wines from the Tenerife Island (Spain) by their metal ion concentration, *Ital. J. Food Sci.*, 14, 375, 2002.
123. Benítez, P. et al., Influence of metallic content of fino sherry wine on its susceptibility to browning, *Food Res. Int.*, 35, 785, 2003.
124. Barbaste, M. et al., Analysis and comparison of SIMCA models for denominations of origin of wines from the Canary Islands (Spain) builds by means of their trace and ultratrace metals content, *Anal. Chim. Acta*, 472, 161, 2002.
125. Murányi, Z. and Kovács, Z. Statistical evaluation of aroma and metal content in Tokay wines, *Microchem. J.*, 67, 91, 2000.
126. Kment, P. et al., Differentiation of Czech wines using multielement composition — a comparison with vineyard soil, *Food Chem.*, 91, 157, 2005.
127. Dugo, G. et al., Determination of some inorganic anions and heavy metals in D.O.C. Golden and Amber Marsala wines: statistical study of the influence of ageing period, colour and sugar content, *Food Chem.*, 91, 355, 2005.
128. Arvanitoyannis, I.S. et al., Application of quality control methods for assessing wine authenticity: use of multivariate analysis (chemometrics), *Trends Food Sci. Technol.*, 10, 321, 1999.
129. Diaz, C. et al., Chemometric studies of bottled wines with denomination of origin from the Canary Islands (Spain), *Eur. Food Res. Technol.*, 215, 83, 2002.
130. Košir, I.J. et al., Use of SNIF-NMR and IRMS in combination with chemometric methods for the determination of chaptalisation and geographical origin of wines (the example of Slovenian wines), *Anal. Chim. Acta*, 429, 195, 2001.
131. Cameán, A.M. et al., Differentiation of Spanish brandies according to their metal content, *Talanta*, 54, 53, 2001.
132. Adam, T., Duthie, E., and Feldmann, J., Investigations into the use of copper and other metals as indicators for the authenticity of Scotch whiskies, *J. Inst. Brewing*, 108, 459, 2002.

133. Hernandez-Caraballo, E.A. et al., Classification of Venezuelan spirituous beverages by means of discriminant analysis and artificial neural networks based on their Zn, Cu, and Fe concentrations, *Talanta*, 60, 1259, 2003.
134. Alcázar, A. et al., Multivariate characterisation of beers according to their mineral content, *Talanta*, 57, 45, 2002.
135. Bellido-Milla, D., Moreno-Perez, J.M., and Hernandez-Artiga, M.P., Differentiation and classification of beers with flame atomic spectrometry and molecular absorption spectrometry and sample preparation assisted by microwaves, *Spectrochim. Acta B*. 55, 855, 2000.
136. Wyrzykowska, B. et al., Application of ICP sector field MS and principal component analysis for studying interdependences among 23 trace elements in Polish beers, *J. Agric. Food Chem*., 49, 3425, 2001.
137. Guerrero, M.I. et al., Multivariate characterization of vine vinegars from the south of Spain according to their metallic content, *Talanta*, 45, 379, 1997.
138. Benito, M.J. et al., Typification of vinegars from Jerez and Rioja using classical chemometric techniques and neural network methods, *Analyst*, 124, 547, 1999.

5 Functional Role of Some Minerals in Foods

Michał Nabrzyski

CONTENTS

5.1 INTRODUCTION

Minerals are inorganic elements that occur in nature. They originate from the earth's crust, and in animal or plant cellules constitute only a small proportion of the body tissue. During dry or wet ash-destruction process, the organic matrix of the biological sample is completely decomposed and removed in the form of volatile carboneous material, and the remaining material, which is the nondestructive residue, contains only mineral compounds.

Many mineral elements are essential constituents of enzymes, regulate a variety of physiologic processes (e.g., maintenance of osmotic pressure, oxygen transport, muscle contraction, and central nervous system integrity), and are required for the growth and maintenance of tissues and bones. They are so potent and so important that without them the organism would not be able to use the other remaining constituents of foods.

The total daily diet contains minerals within the range of 0.2–0.3% of all nutrients. The main mass of these minerals is made up of six minerals (mentioned in the following paragraph), whereas dozens of others, present in trace quantities, constitute only a hundredth of a percent of the total mass of daily eaten nutrients. Because of their broad biochemical activity, many inorganic compounds are also intentionally used as functional agents in processing a variety of foods. On the other hand, cations in foods, such as Fe^{2+} and Cu^{2+}, may induce a diversity of undesirable effects that influence the nutritional quality of foods. Iron, for example, can actively catalyze lipid oxidation, and its presence even in trace amounts has long been recognized as potentially detrimental to the shelf-life of fats, oils, and fatty acids. Iron compounds can activate molecular oxygen by producing superoxide, which through dismutation and other steps of biochemical changes results in the formation of hydroxyl free radicals.

Minerals play an important role in plant life. They function as catalysts for many biochemical reactions, are responsible for changes in the state of cellular colloids, directly affect the cell metabolism, and are involved in changes in protoplasm turgor and permeability. They often become centers of electrical and radioactive processes in living organisms. Minerals are usually classified into two main groups on the basis of their relative amounts in the body. Those occurring in relatively large amounts and needed in quantities of 100 mg or more per day are called macroelements or macrominerals. This group includes calcium, magnesium, sodium, potassium, chloride, and phosphorus. Minerals occurring in small amounts and needed in quantities of a few milligrams or less per day are called microelements or microminerals or trace elements. The group known as essential microelements includes iron, iodide, zinc, copper, manganese, fluorine, selenium, cobalt, nickel, chromium, boron, and silicon. Recently, a few other elements have sometimes been considered as potentially essential, such as lithium and strontium.[1–5]

In body tissues, other trace minerals also occur, such as arsenic, aluminum, cadmium, lead, and mercury, which until now have been recognized as potentially harmful, although recently, Al, As, Cd, and Pb in very small quantities have been suggested to have some characteristics of essentiality but lacking a defined biochemical function.[4,6] Elevated levels of this group of elements are present in the organism mainly as a result of environmental contamination, with very small quantities often derived from natural sources. According to the WHO Experts[7] the trace elements are divided into three groups from the point of view of their nutritional significance in humans: (1) essential elements, (2) elements that are probably essential, and (3) potentially toxic elements, some of which as mentioned above, may nevertheless have some essential function.

The heavy metals have been studied intensively, and recent data suggest that insufficient intake of aluminum, lead, and cadmium in experimental animal diets may exhibit apparent deficiency signs, such as depressed growth, anemia, disturbed iron metabolism, and some other effects very specific for each animal and each trace element.[4,6,7] More experimental data is necessary to prove that trace amounts of lead, aluminum, and cadmium may have a defined positive biological function in the organism and thus be recognized as nutritionally necessary.

Toxicological studies of toxic metals such as aluminum, cadmium, lead, and mercury have been yearly summarized by the Joint FAO/WHO Expert Committee on Food Additives (JECFA) and published in the form of special reports within the International Programme on Chemical Safety.[8–11] Recommendation of their safe intake in diets is reported as provisional tolerably weekly intake (PTWI). The term is used to describe the endpoint of food contaminants such as heavy metals with cumulative properties. Its value, expressed in mg/kg of body weight, represents permissible human weekly exposure to those contaminants unavoidably associated with the consumption of otherwise wholesome and nutritious foods.

Generally, an element is considered essential if it is necessary to support adequate growth, reproduction, and health through the life cycle when all other nutrients are eaten daily at optimal levels. The proof of essentiality of an element in one animal species does not prove essentiality in another, but the probability of an essential function in any species, including humans, increases with the number of experiments with other species in which essentiality has been proved. In the case of some trace elements that are both essential nutrients and unavoidable constituents of food, two threshold doses of tolerance are expressed. The lower represents the level of essentiality and the upper represents the provisional maximum tolerably daily intake (PMTDI), expressed in mg/kg body weight. The threshold values for zinc and copper are 0.3 to 1.0 and 0.05 to 0.5 mg/kg bw, respectively. For iodine the joint FAO/WHO Expert Committee set, for the purpose of safety, a PMTDI of 1.0 mg/day or 0.017 mg/kg bw from all sources, and the level of dietary allowance for an adult person was set as 0.1–0.14 mg/day.[9,12,13]

The recommended daily intake (RDA) of essential minerals represents standards of nutrition set by the Food and Nutrition Board of the U.S. National Academy of Sciences in milligram per person.[14–16] It contains the levels of essential nutrients that are necessary to meet the nutritional needs of the normal, healthy person. Individuals may differ in their precise requirements, and to take these differences into account, the RDA provides a margin of safety, i.e., sets the allowances high enough to cover the needs of most healthy people. For additional nutrients that are necessary to keep the body healthy, for which the RDA has not yet been established, a safe and adequate daily intake (SAI) is approved.

The daily intake depends on many important external and internal factors such as chemical form of the minerals, their presence and levels in eaten foods, percentage of absorption from the gastrointestinal tract, as well as nutritional habits, weight, age, sex, and economic conditions of consumers. Table 5.1 contains data related to the mean daily intake, RDA or SAI, and the absorption percentage of minerals.

As can bee seen from Table 5.1, the absorption of most minerals from the gastrointestinal tract is variable and depends on many factors, some of which are mentioned in the following section entitled "Interaction between Minerals and Dietary Components," whereas others are given in the following text and include intrinsic and extrinsic factors. The intrinsic (physiological) factors include the developmental changes in infancy and senility. Infancy, in the immediate postnatal period, has poorly regulated absorption until homeostatic regulatory mechanism become established with increasing gut maturity. Senility appears to feature decline in the efficiency of absorption of some minerals. Homeostatic regulations play an important

TABLE 5.1
Data of Mean Daily Intake, Percentage of Absorption from the Gastrointestinal Tract, and RDA[b] or SAI[c][14–16,20] in Milligrams per Adult Person

Element	Macroelements					
	Calcium	Phosphorus	Chloride	Sodium	Potassium	Magnesium
Daily intake (DI)	960–1220	1760–2130	1700-5100	3000-7000	3300	145–358
Percent of absorption (PA)	10–50	High[a]	High[a]	High[a]	High[a]	20–60
Recommended daily intake (RDA[b]) or safe and adequate intake (SAI[c])	800–1200[b]	800–1200[b]	750[c]	500[c]	2000[c]	280–350[c]

Element	Microelements								
	Iron	Zinc	Manganese	Copper	Fluorine	Iodine	Molybdenum	Chromium	Nickel
DI	15	12, 18	5.6, 8	2.4	< 1.4	< 1.0	> 0.15	< 0.15	0.16–0.20
PA	10–40	30–70	40	25–60	High[a]	100	70–90	< 1.0 or 10–25 in form of GTF[f]	30–50
RDA[b] or SAI[c]	10–15[b]	12–15[b]	2–3[c]	1.5–3.0[c]	1.5–4.0[c]	0.15[b]	0.075–0.250[c]	0.05–020[c]	0.05, 0.3[c]

Element	Cobalt	Vanadium	Selenium	Silicon[d]	Boron[d]	Lithium[e]
DI	0.003–0.012	0.012–0.030	0.06–0.22	21–46, 200	1–3	< 0.001–0.99
PA	30–50	< 1.0	~ 70	3, 40	High[a]	60–100
RDA[b] or SAI[c]	0.002[g]	0.01–0.025[c]	0.055–0.07[b]	21–46	1–2	

[a] More than 40%; [b] recommended daily inake (RDA); [c] safe and adequate daily intake (SAI); [d] essential element; [e] potentially essential element; [f] GTF (Glucose Tolerance Factor); [g] 0.002 mg of cobalt in vitamin B_{12}.

role in adaptation to low trace element status or high demands by modifying the activity as well as the concentration of receptors and the level of their saturation, which can influence the uptake of minerals from the gastrointestinal tract during normal life, pregnancy, or disease states.

The extrinsic (dietary) factors include the solubility and molecular dimensions of trace element species within food, the mixture of digestive secretion, and the gut lumen conditions that can influence mucosal uptake (e.g., the nonavailable iron oxalates, copper sulfides, trace-element silicates, zinc, iron, and lead, and phytates associated with calcium). Synergists promoting, among other things, zinc absorption include citrate, histidine, ascorbate-modifying iron/copper antagonism, and those metabolic processes that maintain the transport and mobility of some elements (transferrins, albumins, and other plasma ligands). On the other hand, there are also antagonists that restrict gastrointestinal lumen solubility of components such as phytates of calcium and zinc, or minerals such as cadmium, zinc, and copper, which compete for receptor sites involved in absorption, transport, storage, and other less known mechanisms.

5.2 INTERACTIONS BETWEEN MINERALS AND DIETARY COMPONENTS

The interactions among minerals as well as between minerals and other substances in the diet are complex and numerous. The term *interaction* is used to describe interrelationships among minerals and other nutrients present in the diet and may be defined as the effect of one element (nutrient) on one or more others, and thus may cause some positive or negative biochemical or physiological consequences in the organism. At the molecular level, such interactions occur at specific sites on proteins such as enzymes, receptors, or ion channels. Different nutritional and nonnutritional components of the diet, other nutrients in vitamin–mineral supplements, contaminants, and also some medications can interact with minerals in the gastrointestinal tract and influence their absorption. For example, diets low in calcium promote significant increase in the absorption and retention of lead and cadmium. High levels of calcium or phytate in diets appear to restrict lead uptake.

There are clear indications that iron deficiency promotes cadmium retention and may thus decrease the tolerance for high environmental or dietary cadmium levels. Evidence from a variety of experiments on animals suggests that iron deficiency also promotes lead uptake and retention; the evidence for humans is controversial, but in some studies substantial increase in lead uptake are said to have occurred when the status of dietary iron and iron levels were low.

Some of the microelements, such as Fe, Mn, Co, Cu, Cr, Ni and Zn, generally function in an organism as cations complexed with organic ligands or chelators, i.e., proteins, porphyrins, flavines, pterins, etc. These elements belong to the first series of the Periodic Table of the Elements and have incompletely filled 3-d orbitals except for zinc. They are called the *transition metals*. One of the characteristic features of these metals is their ability to form complexes ions. Their electron configurations at the valence shell, in the elemental as well as in the cationic state, are shown below:

Element	Cr	Mn	Fe	Co	Ni	Cu	Zn
Elemental state	$3d^5\ 4s^1$	$3d^5\ 4s^2$	$3d^6\ 4s^2$	$3d^7\ 4s^2$	$3d^8\ 4s^2$	$3d^{10}\ 4s^1$	$3d^{10}\ 4s^2$
Divalent cationic state	$3d^4$	$3d^5$	$3d^6$	$3d^7$	$3d^8$	$3d^9$	$3d^{10}$

As can be seen from the above arrangement, copper may exist in either the monovalent or divalent ion state, and the sum of its biological importance is related to its ability to oscillate between the cuprous (I) or cupric (II) states, i.e., the ability to accept and donate electrons. The divalent zinc ($3d^{10}$) is not strictly a transition element because its orbital d is filled but its atomic structure is similar and thus ready to form complexes analogous to those formed by the transition metal ions. Some of the other divalent cations such as cadmium Cd^{2+} ($4d^{10}$) and lead Pb^{2+} ($6d^{10}$ $6s^2$), which are not transition metals, also have a pronounced tendency to form coordinate covalent bounds with ligands that contain electron donor atoms such as N, O, and S, found frequently in proteins. Some examples of similar tendencies to form coordination complexes by cations such as Zn^{2+}, Cd^{2+}, and Cu^+ are discussed in the following text.

There is a degree of selectivity of metal ions for electron donor atoms; copper tends to prefer nitrogen, whereas zinc prefers sulfur, but several different donor atoms will complex with each metal ion. The amino-acid residues in proteins serve as rich sources of electron donor atoms. For example, the imidazole group of histidine supplies nitrogen, the carboxyl groups of aspartic and glutamic acids supply oxygen, and the sulfhydryl group of cysteine supplies sulfur for complexation.

The interactions between a metal atom and ligands can be thought of as a Lewis acid-base reaction. Ligands as Lewis base are capable of donating one or more electron pairs, and a transition metal atom acts as Lewis acids accepting or sharing a pair of electrons from the Lewis bases. Thus, metal–ligand interactions that share pairs of electrons are coordinate covalent bound. Interesting data on competitive interaction in biological systems exist between the divalent cations such as zinc and cadmium, or zinc, copper, and iron. Zinc and cadmium, together with mercury, belong to the group 12 of the recent form of the Periodic Table and have very similar tendencies to form complexes with a coordination number of 4 and a tetrahedral disposition of ligands around the metal. They have, as divalent cations, very similar electronic structure to the monovalent cation of copper ($3d^{10}$), and a consequence of this is the possibility of isomorphous replacement between these elements in biological systems.[17]

The essential microelements prefer the coordination number of 4 or 6. Copper and zinc in the cationic form have filled their orbitals (they are d^{10} ions), and both Cu^+ and Zn^{2+} form tetrahedral complexes. Cu^+ ion may lose one of its d orbital electrons and become a d^9 ion (Cu^{2+}), and thus form square planar complexes. On the other hand, the ferrous d^6 ion (Fe^{2+}) forms octahedral configuration. According to the brief review above, it is possible to predict that Cu^+, having the same d orbital configuration as that of Zn^{2+} and Cd^{2+}, may interact with both of cations, and that Zn^{2+} may be antagonistic to Cu^+ rather than Cu^{2+}.

In systematic studies of the complex interactions between Cd, Zn, Cu, and Fe in chicks, rats, and mice, it has been demonstrated that increased dietary intake of cadmium can cause increased mortality, poor growth, and hypochromic microcytic anemia.[17] The growth rate could be restored by zinc supplementation, and the mortality and severity of anemia signs may be reduced by copper supplementation of the diet. Zinc also restored the activity of heart cytochrome oxidase and additional Cu prevented the degeneration of aortic elastin. According to Bremner,[17] it is evident that Cd can have antagonistic effects on zinc and copper metabolism.

Diets low in calcium promote significant increases in the absorption and retention of lead and cadmium in rats and promote the manifestation of pathologic changes associated with Cd toxicity. It is important to note that low intake of phosphorus had effect on Pb retention similar to Ca, and that the effects of Ca and P deficiencies are additive.[7,17] Many nutritional and nonnutritional components of dietary supplements and some medications can interact with minerals in the gastrointestinal tract and affect their absorption. Amino acids act as intraluminar binders for some trace minerals. Large complexes, and weakly digestible proteins, on the other hand, may bind minerals strongly and reduce their absorption.

Triacyloglicerols and long-chain fatty acid derived from triacyloglicerols may form soaps with calcium and magnesium and decrease the bioavailability of these two nutrients. Also pectins, cellulose, hemicellulose, and polymers produced by Maillard reaction during cooking, processing, and storage bind minerals in the lumen and thus reduce their biological availability. Lactose, on the other hand, increases calcium absorption from milk.

Interactions between and among minerals, or minerals with some anionic species such as ascorbic acid, have important effect on mineral absorption. Absorption of iron is hindered by fiber and phosphates and promoted by ascorbic acid, copper, and meat protein. Ascorbic acid also enhances absorption of selenium but reduces the absorption of copper. A high protein intake appears to increase the excretion of calcium, whereas vitamin D increases ingestion and retention of calcium.

Intestinal parasites, dietary fiber phylates, and excessive sweating interfere with zinc absorption. Phytates, oxalates, and tannates can interfere with the absorption of a number of minerals. Certain medications such as tetracycline can also inhibit absorption of minerals, whereas others such as jodoquinol function as antibacterial agent by chelating trace metals at bacterial surface, thereby inhibiting their growth. It may be concluded that chemically similar minerals share certain "channels" for absorption, and the simultaneous ingestion of two or more such minerals may result in competition for absorption. The possible competition mechanisms of some minerals have already been discussed. When nonphysiological imbalances among competitive nutrients exist as the result of excessive ingestion (occasioned by leaching from water pipes, storage in unlacquered tin cans, or improper formulation of vitamin–mineral supplements), nutritionally important consequences of the mineral–mineral interaction can be observed. Finally, to participate in a nutritionally relevant process for the organism as a whole, a mineral must be transported away from the intestine. The concentration of circulating binding proteins and the degree of saturation of their metallic binding sites may influence the rate and magnitude of transport of the absorbed minerals.

Minerals require a suitable mucosal surface across which they enter the body. Reaction in or diversion of a large portion of the small bowel obviously affects mineral absorption. Excessive mucosal damage due to mesenteric infarction or inflammatory bowel disease, or major diversion by jejunoileal bypass procedures, reduces the available surface area. Minerals whose absorption occurs primarily in the proximal intestine, e.g., copper or iron, are affected differently from those absorbed more distally, e.g., zinc. In addition, the integrity of the epithelium, uptake of fluids and electrolytes, intracellular protein synthesis, energy-dependent pumps, and hormone receptors must be intact.

Intrinsic diseases of the small intestinal mucosa may impair mineral absorption. Conditions such as celiac sprue, dermatitis herpetiformis, infiltrative lymphomas and, occasionally, inflammatory bowel disease produce diffuse mucosal damage. Protein energy malnutrition causes similar damage, and tropical enteropathy affects part of the population of developing countries living under adverse nutritional and hygienic conditions.[18]

In general, the function of minerals in the body may be divided into two categories, namely, building body tissue and regulating numerous processes. The roles of each group in the human body are summarized in Table 5.2.

Potassium, sulfur, phosphorus, iron, and other minerals are structural components of soft tissues. Sodium is principally found in extracellular fluid (bone is an exception), where it is the chief cation, and thus it is considered mainly as a primary determinant of body fluid osmolarity as well as the maintainer of body fluid pH. Intracellular fluid contains much smaller amounts of sodium though these stores, and perhaps the sodium of the bone also serve in this capacity. It is also important to mention that the energy for impulse transmission in the nerve and its action potential derive from the potential energy represented by the separation of sodium and potassium across the cell wall. On the other hand, calcium, together with phosphorus, magnesium, and fluorine, are components of bone and teeth. Deficiencies during the growing years cause growth to be stunted and bone tissue to be of poor quality.

A continual intake of minerals is essential for the maintenance of skeletal tissue in good condition. Minerals are an integral part of many hormones, enzymes, and other compounds that regulate physiological functions in the organism. For example, iodine is required to produce the hormone thyroxine, chromium is involved in the production of insulin, and hemoglobin is an iron-containing compound. Thus, the production of these substances in the organism depends on adequate intake of the involved minerals. Selected foods as source of some minerals are listed in Table 5.3.

Minerals can also act as catalysts. Calcium is a catalyst in blood clotting. Some minerals are catalysts in the absorption of nutrients from the gastrointestinal tract, the metabolism of proteins, fat, and carbohydrates, and the utilization of nutrients by the cell. Minerals dissolved in the body fluids are responsible for nerve impulses and the contraction of muscles, as well as for water and acid-base balance. Minerals play an important role in maintaining respiration, heart rate, and blood pressure within normal limits. The deficiency of minerals in the diet may lead to severe, chronic clinical signs of diseases, frequently reversible after their supplementation in the diet or following total parenteral nutrition. The influence of minerals on biochemical reactions in living systems also make it possible to use them intentionally in many food processes.

TABLE 5.2
Biological Role of Some Minerals in the Human Body[16,19–21]

Mineral	Function	Deficiency	Sources
		Macroelements	
Calcium	Bone and tooth formation, blood clotting, cell permeability, nerve stimulation, muscle contraction, enzyme activation	Stunted growth, rickets, osteomalacia, osteoporosis, tetany	Milk, hard cheese, salmon, and small fish eaten with bones, some dark green vegetables, legumes
Magnesium	Component of bones and teeth, activation of many enzymes, nerve stimulation, muscle contraction	Seen in alcoholism or renal disease, tremors leading to conclusive seizures	Green leafy vegetables, nuts, whole grains, meat, milk, seafood
Phosphorus	Bone and tooth formation, energy metabolism component of ATP and ADP, protein synthesis, component of DNA and RNA, fat transport, acid-based balance, enzyme formation	Stunted growth, rickets	Milk, meats, poultry, fish, eggs, cheese, nuts, legumes, whole grains
Potassium	Osmotic pressure, water balance, acid-based balance, nerve stimulation, muscle contraction, synthesis of protein, glycogen formation	Nausea, vomiting, muscular weakness, rapid heartbeat, heart failure	Meats, fish, poultry, whole grains, fruits, vegetables, legumes
Sodium	Osmotic pressure, water balance, acid-based balance, nerve stimulation, muscle contraction, cell permeability	Rare: nausea, vomiting, giddiness, exhaustion, cramps	Table salt, salted foods, MSG and other sodium additives, milk, meat, fish, poultry, eggs, bread
		Microelements	
Chromium	Trivalent chromium increases glucose tolerance and plays role in lipid metabolism, useful in prevention and treatment of diabetes; hexavalent chromium is toxic	Impaired growth, glucose intolerance, elevated blood cholesterol	Whole-grain cereals, condiments, meat products, cheeses, and brewer's yeast

TABLE 5.2 (CONTINUED)
Biological Role of Some Minerals in the Human Body[16,19–21]

Mineral	Function	Deficiency	Sources
	Microelements		
Cobalt	Cofactor of vitamin B_{12}, plays role in immunity	Rarely observed, but if exists: pernicious anemia with hematological and neurologic manifestations may be observed due to vitamin B_{12} deficiency	Organ meats (liver, kidney), fish, dairy products, eggs
Copper	Necessary for iron utilization and hemoglobin formation; constituent of cytochrome oxidase; involved in bone and elastic tissue development	Anemia, neutropenia, leucopenia, skeletal demineralization	Liver, kidney, oysters, nuts, fruits, dried legumes
Iron	Hemoglobin and myoglobin formation, essential component of many enzymes	Anemia, decrease in oxygen transport and cellular immunity, muscle weakness, etc.	Liver, lean meats, legumes, dried fruits, green leafy vegetables, whole grain, and fortified cereals
Manganese	Cofactor of large number of enzymes; in aging process has a role as an antioxidant (Mn-superoxide dismutase); important for normal brain function, reproduction, and bone structure	In animals: chondrodystrophy, abnormal bone development, reproductive difficulties; in humans: shortage of evidence	Tea, whole grain, nuts; moderate levels: fruits, green vegetables, organ meat, and shellfish contain well absorbable manganese
Molybdenum	Component of three enzymes: sulfite oxidase, xanthine dehydrogenase, aldehyde oxidase. These enzymes catalyse oxidative hydroxylations of substrate molecules at the molybdenum center and subsequently transfer electrons to other redox-active cofactors and ultimately to the electron acceptors cytochrome c, molecular oxygen, or NAD^+. Sulfite oxidase detoxifies in the body sulfiting agents used for food preservation, activates nitrate reductase (found in nitrogen-fixing bacteria), and promotes formation of assimilatory nitrite crucial to the global nitrogen cycle in biological systems, copper antagonist	Intolerance of sulfur-containing amino acids in patients fed intravenously reduces conversion of hypoxanthine and xanthine to uric acid, resulting in development of xanthine renal calculi; deficiency state may be potentiated by high copper intake	Organ meats, grain, legumes, leafy vegetables, milk, beans

Zinc	Constituent of many enzyme systems, carbon dioxide transport, vitamin A utilization	Delayed wound healing, impaired taste sensitivity; severe deficiency: retarded growth and sexual development, dwarfism	Oysters, fish, meat, liver, milk, whole grains, nuts, legumes
Fluoride	Resistance to dental decay	Tooth decay in young children	Drinking water rich in fluoride, seafood, teas
Iodine	Synthesis of thyroid hormones that regulate basal metabolic rate	Goiter, cretinism, if deficiency is severe	Iodized salt, seafood, food grown near the sea
Selenium	Protects against number of cancers	Cataract, muscular dystrophy, growth depression, liver cirrhosis, infertility, cancer, aging due to deficiency of selenoglutathione peroxidase, insufficiency of cellular immunity	Broccoli, mushrooms, radishes, cabbage, celery, onions, fish, organ meats
Boron (recently considered as essential)	Prevents osteoporosis in postmenopausal women, beneficial in treatment of arthritis, builds muscle; its essentiality still has to be proven	Probably impairs growth and development	Foods of plant origin

TABLE 5.3
Contents of Selected Minerals in Some Foods

Mineral	Food	Amount (mg/100g)	References
Calcium	Swiss cheese — low sodium	960	15
	Sardines in tomato sauce	437 [a]	
	Cod in tomato sauce	335 [a]	
	Yogurt natural	189 [a]	
	Milk	120 [a]	
	Orange	42	34
	Carrot	41	34
	Tuna in own sauce	25 [a]	
	Potato	10	34
Potassium	Wheat seeds	502	34
	Porcine liver	350	34
	Beef	342	34
	Oat, flaked	335	34
	Carrot	290	34
	Pork	260	34
	Orange	177	34
	Milk	157	34
	Wheat meal (550)	126	34
	Cheese (45% fat)	107	34
Magnesium	Sardine in tomato sauce	27 [a]	
	Tuna in own sauce	24 [a]	
	Yogurt natural	12 [a]	
	Milk	9 [a]	
Zinc	Oyster	up to 100	26
	Peas, yellow dried	4.2	27
	Oats flaked	3.1	27
	Liver chicken a. pig	3.6, 4.5	27
	Egg (yolk)	3.6	27
	Beef	3.8	27
	Whole milk powder	3.1	27
	Hard cheese	2.4	27
	Wheat meal	0.6	27
	Fish	0.4–1.2	23,27
	Bee honey	0.8–5.0	24,25
Iron	Porcine liver	22	34
	Egg (yolk)	7.2	34
	Oat, flaked	4.6	34
	Wheat seeds	3.3	34
	Pork	2.3	34
	Beef	2.6	34
	Wheat meal (550)	1.1	34
	Egg (white)	0.2	34

TABLE 5.3 (CONTINUED)
Contents of Selected Minerals in Some Foods

Mineral	Food	Amount (mg/100g)	References
Copper	Oysters	6, 17	26,32
	Liver, calf	7	32
	Liver, beef	3	32
	Wheat germ	0.9, 2	27,32
	Sunflower seeds	2	32
	Fish	0.11–0.5	23,32
	Ham	0.03, 0.08	27,32
	Bee honey	0.25–1.08	24,25
Chromium	Spices	> 0.1–0. 5	33
	Breakfast cereals	0.01–0.04	28
	Cacao	0.2	33
	Paprika, pepper, curry	~ 0.05	33
	Hawthorn	0.025	33
	Cheese, drowned	> 0.01	33
	Whole-meal bread	~ 0.02	33
	Beef	< 0.004	33
	Kidney, liver	< 0.0015	33
	Bee honey	0.0–0.1	24,25
Molybdenum	Beans white	> 0.1–0. 5	20
	Peas, whole grain	0.01–0.04	20
	Peas, green drained	0.2	20
	French bean, canned	~ 0.05	20
	Carrot, canned drained	0.025	20
	Strawberry	> 0.01	20
	Black currants	~ 0.02	20
	Red currants	0.004	20
	Plums	< 0.0015	20
	Fish	0.0–0.1	20
Fluoride	Black teas	3–34	29
	Fish, canned	0.09–0.8	21,31
	Shellfish	0.03–0.15	21
Iodide	Marine fish, oyster, shrimp, lobster	0.02–0.1	22
	Milk powder	0.06	30
Selenium	Tortilla chips	1.0	15
	Potato chips	0.97	15
	Pork kidney, braised	0.21	15
	Tuna, canned	0.12	15
	Salmon, canned	0.08	15
	Milk	0.002	15

[a] Author's unpublished data.

5.3 MINERALS IN FOOD PROCESSING

As discussed above, minerals are inherent in natural foodstuffs. They are also an integral part of many enzymes, and play important roles in food processing, e.g., in alcoholic and lactic fermentation, meat aging, and dairy production. Some processing of foods results in decreasing mineral content, and thus, addition of minerals to these products to the levels at which they occur naturally, or fortifying them even above the levels expected naturally, is a frequent technological practice. Many compounds used as food additives contain metallic cations in their structure. A number of these compounds function as antimicrobials, sequestrants, antioxidants, flavor enhancers, and leavening, texturizing, and buffering agents. Some minerals are also used as dietary supplements. Selected examples are gathered in Table 5.4.

Some heavy metal ions actively catalyze lipid oxidation. Their presence even in trace amounts has long been recognized as potentially detrimental to the shelf life of fats, oils, and fatty foods. They can activate molecular oxygen by producing superoxides and lead to the formation of hydroxyl free radicals. Three cations are involved in the activity of superoxide dismutase (SOD), an enzyme that has been patented as an antioxidant agent for foods. Three types of metalloenzymes of SOD exist in living organisms, such as Cu- and Zn-SOD, Fe-SOD, and Mn-SOD. All three types of SOD catalyze dismutation of superoxide anions to produce hydrogen peroxide *in vivo*. There is evidence of increased lipid oxidation in apple fruit during senescence. SOD activity may also been involved in reactions induced by oxygen, free radicals, and ionizing radiation and could help protect cells from damage by peroxidation products.[36]

Besides SOD, catalase, ceruloplasmin, albumin, appotransferrin, and chelating agents (e.g., ethylene-diaminetetraacetic acid (EDTA), bathocupreine, cysteine, and purine) are capable of inhibiting the oxidation of ascorbic acid by trace metals. Copper-induced lipid oxidation in ascorbic acid–pretreated cooked ground fish may be inhibited in the presence of natural polyphenolic compounds, the flavonoids, which are effective antioxidants able to prevent the production of free radicals.[37] In the presence of ADP-chelated iron and traces of copper, oxygen radicals are generated in the sarcoplasmic reticulum of muscles. Muscle contains notable amounts of iron, a known prooxidant, and trace amounts of copper, also able to catalyze peroxidative reaction.[38,39]

Iron occurs with heme compounds and as non-heme iron complexed to proteins of low molecular weight. Reactive non-heme iron can be obtained by the release of iron from heme pigments or from the iron storage protein, ferritin. Iron is part of the active site of lipoxygenase, which may participate in lipid oxidation. Reducing components of the tissue-like superoxide anion, ascorbate, and thiols can convert the inactive ferric ion to active ferrous ion. There are also enzymatic systems that use reducing equivalents from NADPH to reduce ferric iron. A number of cellular components are capable of reducing ferric to ferrous iron, but under most conditions the two major reductants are superoxide and ascorbate.[38]

In some cases reduction of ferric iron can be accomplished enzymatically using electrons from NADH and, to a lesser extent, NADPH through an enzymatic system associated with both the sarcoplasmic reticulum and mitochondria.

Ferrous iron can activate molecular oxygen by producing superoxide. Superoxide may then undergo dismutation spontaneously or by the action of SOD and produce the hydrogen peroxide that can interact with another atom of ferrous iron to produce the hydroxyl radical. The hydroxyl radical can initiate lipid oxidation. It is generally accepted that ferrous iron is the critical electron acceptor in oxidation reactions. Because it is likely that most iron ordinarily exists in the cell as ferric iron, the ability to reduce ferric to ferrous iron is critical. The development of rancidity and warmed-over flavor, a specific defect that occurs in cooked and reheated meat products following short-term refrigerated storage, has been directly linked to autooxidation of highly unsaturated, membrane-bound phospholipids and to the catalytic properties of non-heme iron.[38,40,41]

Dietary iron may influence muscle iron stores and thus theoretically also affect the lipid oxidation in muscle food, e.g. pork. There appears to be a threshold for dietary iron level (between 130 and 210 μg/g) above which muscle and liver non-heme iron, total iron, and muscle thiobarbituric acid–reactive substances begin to increase because of porcine muscle lipid oxidation.[42,43] The secondary oxidation products, mainly aldehydes, are the major contributors to warmed-over flavor and meat flavor deterioration because of their high reactivity and low flavor thresholds.

Ketones and alcohols have a high flavor threshold, thus causing off-flavors less often. Exogenous antioxidants can preserve the quality of meat products. Radical scavengers appear to be the most effective inhibitors of meat flavor deterioration. However, different substrates and systems respond in different ways. Active ferrous iron may be eliminated by chelation with EDTA or phosphates, or chemically by oxidation to its inactive ferric form. Ferroxidases are enzymes that oxidize ferrous to ferric iron in the presence of oxygen according to the formula:

$$4\ Fe^{2+} + O_2 + 4\ H^+ \rightarrow 4\ Fe^{3+} + 2H_2O$$

Ceruloplasmin, a copper protein of blood serum, is a ferroxidase. Oxidation of ferrous to ferric iron tends to be favored in extracellular fluids, whereas chelation is more likely in the intracellular environment.[38]

Sodium chloride has long been used for food preservation. Salt alters both the aroma and the taste of food. The addition of sodium chloride to blended cod muscle accelerates the development of rancidity.[44,45] This salt-induced rancidity is inhibited by chelating agents such as EDTA, sodium oxalate, and sodium citrate, and by nordihydroguaiaretic acid and propylgallate. Although sodium chloride and other metal salts act as prooxidants, they have a strong inhibiting effect on Cu^{2+}-induced rancidity in the fish muscle. The most effective concentration of NaCl for this antioxidant effect is between 1 and 8%. Castell and Spears[45] also showed that the other heavy metal ions were effective in producing rancidity when added to various fish muscles. The relative effectiveness was of the following decreasing order: $Fe^{2+} > V^{2+} > Cu^{2+} > Fe^{3+} > Cd^{2+} > Co^{2+} > Zn^{2+}$, whereas Ni^{2+}, Ce^{2+}, Cr^{3+} and Mn^{2+} had no effect at the concentrations used. Of the elements tested, Fe^{2+}, V^{2+}, and Cu^{2+} were by far the most active catalysts. There were, however, important exceptions. The comparative effectiveness of the metal ions was not the same for muscles taken from all the species tested.

TABLE 5.4
List of Selected Mineral Compounds Used as Food Additives According to FAO/WHO Expert Committee on Food Additives

Chemical Name of Compound and (INS)*	Synonyms or Other Chemical Name	Functional Class and Comments	ADI, TADI, PMTDI (mg/kg body weight)
Aluminum sodium sulphate (521)	Soda alum, Sodium alum	Buffering, neutralizing, and firming agent	No ADI allocated
Aluminum potassium sulphate (522)	Aluminum potassium sulphate dodecahydrate	Acidity regulator, stabilizer	No ADI allocated
Aluminum ammonium sulphate (523)	Aluminum ammonium sulphate	Stabilizer, firming agent, temporary acceptance, group ADI for aluminum	0–0.6
Calcium alginate (404)	Calcium alginate	Thickening agent, stabilizer	ADI "not specified"
Calcium ascorbate (302)	Calcium ascorbate dihydrate	Antioxidant	ADI "not specified"
Calcium benzoate (213)	Monocalcium benzoate	Antimicrobial agent, preservative	ADI 0–5.0
Calcium chloride (509)	Calcium chloride	Firming agent	ADI "not specified"
Calcium citrate (333)	Tricalcium citrate, tricalcium salt of beta hydroxytricarballylic acid	Acidity regulator, firming agent, sequestrant	ADI "not specified"
Calcium dihydrogen phosphate (341i)	Calcium dihydrogen tetraoxophosphate, monobasic calcium phosphate, monocalcium phosphate	Buffer, firming, raising, leavening, and texturing agents, used in fermentation processes	MTDI 70.0
Calcium disodium ethylene diaminetetraacetace (385)	Calcium disodium EDTA	Antioxidant, preservative sequestrant (no excess of disodium EDTA should remain in food)	ADI 0–2.5
Calcium glutamate (623)	Monocalcium DI-L-glutamate	Flavor enhancer, salt substitute	ADI "not specified"
Calcium hydroxide (526)	Slaked lime	Neutralizing agent, buffer, firming agent	ADI "not limited"
Calcium hydrogen carbonate (170ii)	Calcium hydrogen carbonate	Surface colorant, anticaking agent, stabilizer	ADI "not specified"
Calcium lactate (327)	Calcium dilactate hydrate	Buffer, dough conditioner	
Calcium sorbate (203)	Calcium sorbate	Antimicrobial, fungistatic, preservative agent	ADI 0–25.0 (as sum of calcium, potassium, and sodium salt)
Magnesium chloride (511)	Magnesium chloride hexahydrate	Firming, color retention agent	ADI "not specified"
Magnesium carbonate (504i)	Magnesium carbonate	Anticaking and antibleaching agent	ADI 0–50.0

Magnesium gluconate (580)	Magnesium gluconate dihydrate	Buffering, firming agents in yeast food.	ADI "not specified"
Magnesium glutamate DI-L (625)	Magnesium glutamate	Flavor enhancer, salt substitute	ADI "not specified" (group ADI for α glutamic acid and its monosodium, potassium, calcium, magnesium, and ammonium salts)
Magnesium hydrogen Phosphate (343ii)	Magnesium hydrogen ortophosphate trihydrate, dimagnesium phosphate	Dietary supplement	MTDI 70 (expressed as phosphorus from all sources)
Magnesium hydroxide (528)	Magnesium hydroxide	Alkali, color adjunct	ADI "not limited"
Magnesium hydroxide carbonate (504ii)	Magnesium carbonate hydroxide hydrated	Alkali, anticaking, color retention, carrier, drying agent	ADI "not specified"
Magnesium lactate D,L- (also magnesium lactate L) (329)	Magnesium DI-D,L-lactate	Buffering agent, dough conditioner, dietary supplement	ADI "not limited"
Magnesium oxide (530)	Magnesium oxide	Anticaking, neutralizing agent	ADI "not limited"
Magnesium sulfate (518)	Magnesium sulfate	Firming agent	ADI "not specified"
Potassium acetate (261)	Potassium acetate	Antimicrobial agent, preservative, buffer	ADI "not specified" (also includes the free acid)
Potassium alginate (402)	Potassium alginate	Thickening agent, stabilizer	ADI "not specified" (group ADI for alginic acid and its ammonium, calcium, and sodium salts)
Potassium aluminosilicate (555)	Potassium aluminosilicate	Anticaking agent	No ADI allocated
Potassium hydrogen sulfite (228)	Potassium hydrogen sulfite	Preservative, antioxidant	ADI 0–0.7 (group ADI for sulfur dioxide and sulfites, expressed as sulfur dioxide, covering sodium and potassium metabisulfite, potassium and sodium hydrogen sulfite, and sodium thiosulfate)
Sodium alginate (401)	Sodium alginate	Thickening agent, stabilizer	ADI "not specified"

TABLE 5.4 (CONTINUED)
List of Selected Mineral Compounds Used as Food Additives According to FAO/WHO Expert Committee on Food Additives

Chemical Name of Compound and (INS)*	Synonyms or Other Chemical Name	Functional Class and Comments	ADI, TADI, PMTDI (mg/kg body weight)
Sodium aluminum phosphate acidic (541i)	Salp. Sodium trialuminium tetradecahydrogen, octaphosphate tetrahydrate (A), trisodium dialuminum pentadecahydrogen octaphosphate (B)	Raising agent	ADI 0–0.6
Sodium aluminum phosphate basic (541ii)	Kasal. Autogenous mixture of an alkaline sodium aluminium phosphate	Emulsifier	ADI 0–0.6
Sodium ascorbate (301)	Sodium L-ascorbate	Antioxidant	ADI "not specified"
Sodium benzoate (211)	Sodium salt of benzenecarboxylic acid	Antimicrobial, preservative	ADI 0–5.0
Sodium dihydrogen phosphate (339i)	Monosodium dihydrogen monophosphate (ortophosphate)	Buffer, neutralizing agent, sequestrant in cheese, milk, fish, and meat products	MTDI 70.0
Disodium ethylenediaminetetraacetate (386)	Disodium EDTA Disodium edeteate	Antioxidant, sequestrant, preservative, synergist	ADI 0–2.5 (as calcium disodium EDTA)
Potassium ascorbate (303)	Potassium ascorbate	Antioxidant	ADI "not specified" (group ADI for ascorbic acid and its sodium, potassium, and calcium salts)
Potassium benzoate (212)	Potassium benzoate	Antimicrobial agent, preservative	ADI 0–5.0 (expressed as benzoic acid)
Potassium bromate (924a)	Potassium bromate	Oxidizing agent (previous acceptable level of treatment of flours for breadmaking withdrawn)	ADI withdrawn
Potassium carbonate (501i)	Potassium carbonate	Alkali, flavor	ADI "not specified"
Potassium chloride (508)	Potassium chloride, sylvine, sylvite	Seasoning and gelling agent, salt substitute	ADI "not specified" (group ADI for hydrochloric acid and its magnesium, potassium, and ammonium salts)

Potassium or sodium copper chlorophyllin (141ii)	Potassium or sodium chlorophyllin	Color of porphyrin	ADI 0–15
Potassium hydrogen carbonate (501ii)	Potassium bicarbonate	Alkali, leavening agent, buffer	ADI "not specified"
Sodium glutamate (621)	Monosodium L-glutamate, (MSG), glutamic acid monosodium salt monohydrate	Flavour enhancer	ADI 0–0.7 (group ADI for sulfur dioxide and sulfites expressed as SO_2, covering sodium and potassium salt)
Sodium phosphate (339iii)	Trisodium phosphate, trisodium mono-phosphate, ortophosphate, sodium phosphate	Sequestrant, emulsion stabilizer, buffer agent	MTDI 70.0
Sodium or potassium sorbate (201, 202)	Sodium or potassium sorbate	Antimicrobial, fungistatic agent	ADI 0–25.0

Notes:
TADI — Temporary ADI, a term established by the JECFA for a substance for which toxicological data are sufficient to conclude that use of the substance is safe over the relative short period of time required to evaluate further safety data but are insufficient to conclude that use of the substance is safe over a lifetime. A higher-than-normal safety factor is used when establishing a TADI, and an expiration date is established by which time appropriate data to resolve the safety issue should be submitted to JECFA.
MTDI — Maximum Tolerable Daily Intake or Provisional Maximum Tolerable Daily Intake (PMTDI), terms used for description of the end point of contaminants with no cumulative properties. Its value represents permissible human exposure as a result of the natural occurrence of the substance in food or drinking water. In the case of trace elements that are both essential nutrients and unavoidable constituents of food, a range is expressed, the lower value representing the level of essentiality and the upper value the PMTDI.
[*] INS — International numbering system, prepared by the Codex Committee for Food Additives for the purpose of providing an agreed international numerical system for identifying food additives in ingredient list as an alternative to the declaration of the specific name (*Codex Alimentarius*, Vol. 1, Second Edition, Section 5.1, 1992).
ADI — Acceptable Daily Intake is an estimate of the amount of a substance in food or drinking water, expressed on a body weight basis, for a standard human of 60 kg weight, that can be ingested daily over a lifetime without appreciable risk for health.
[a] -**ADI "not specified" or ADI " not limited"** — Terms applicable to a food substance of very low toxicity that, on the basis of the available data — chemical, biochemical, toxicological, and others, as well as of the total dietary intake of the substance arising from its use at the levels to achieve the desired effect and from its acceptable background in food — does not, in the opinion of the Joint FAO/WHO Expert Committee on Food Additives (JECFA), represent a hazard to health. For that reason, and for reasons stated in individual evaluations, the establishment of ADI in numerical form is not deemed necessary. An additive meeting this criterion must be used within the bound of good manufacturing practice, i.e., it should be technologically efficacious and should be used at the lowest level.
Source: From FAO/WHO Ed., *Summary of Evaluations Performed by the Joint FAO/WHO Expert Committee on Food Additives 1956–1993*, International Life Science Inst. Press, Geneva, 1994.

EDTA is reported to be an effective chelating agent for metal ions and is approved for use in the food industry as a preservative and antioxidant. It acts as an inhibitor of *Staphylococcus aureus* by forming stable chelates in the media with multivalent cations that are essential for cell growth. The effect is largely bacteriostatic and easily reversed by releasing the complexed cations with other cations for which EDTA has higher affinity.[46]

The addition of phosphates (pyro-, tripoly-, and hexametaphosphate) also protects cooked meat from autooxidation. Orthophosphate gives no protection. The mechanism by which phosphates prevent autooxidation appears to be related to their ability to sequester metal ions, particularly ferrous iron, which is the major pro-oxidant.[40] The addition of NaCl increases retention of moisture in meat and meat products.

5.4 EFFECT ON RHEOLOGICAL PROPERTIES

Rheology is the study of the deformation and flow of matter. Rheological properties should be considered a component of the textural properties of foods. Texture depends on the various constituents and structural elements of foods in which microstructure components are formed and then clearly recognized in terms of flow and deformation during different processing treatments. The interaction between metal ions and polysaccharides often affects the rheological and functional properties in food systems. In aqueous media, neutral polysaccharides have a little affinity for alkali metal and alkali earth metal ions. On the other hand, anionic polysaccharides have a strong affinity for metal counterions. This association is related to the linear charge density of the polyanions. The linear charge density is expressed as the distance between the perpendicular projection of an adjacent charged group on the main axis of the molecule. The higher the linear charge density, the stronger the interaction of counterions with the anionic group of the molecule. Such anionic hydrocolloids (0.1% solutions) as alginate, karaya, arabic, and ghati have higher calcium-binding affinity.[47] An important functional property of alginates is their capacity to form gels with calcium ions. This makes alginates well suited for preparing products such as fruit and meat analogs. They are also widely used in biotechnology as immobilization agents of cells and enzymes. The method involves diffusion of calcium ions through alginate and cross-linking reaction with the alginate carboxylic group to form the gel.[47–49] Carrageenans are reported to stabilize casein and several plant proteins against precipitation with calcium and are used to prepare texturized milk products.[50]

5.5 EFFECTS OF STORAGE AND PROCESSING ON THE MINERAL COMPONENTS IN FOODS

The effect of normal storage on mineral components is rather low and may be connected mainly with changes of humidity and biological or chemical contamination; however, high changes of mineral components may occur during canning, cooking, drying, freezing, peeling, and all the other steps involved in preserving, as well as in food processing. The highest losses of minerals are encountered in the milling and polishing process of cereals and groats. All milled cereals undergo a significant reduction of nutrients. The extent of the loss is governed by the efficiency with which the endosperm of the seed is separated from the outer seed coat (bran) and the germ. The loss of certain minerals and vitamins is deemed so relevant to health that in many countries a supplementation procedure has been introduced for food products so derived to enrich them with the lost nutrients, e.g., iron and calcium. In some countries, regulations have been issued for standards of identity for enriched bread. If bread is labeled "enriched," it must meet these standards. In white flours, the losses of magnesium and manganese may reach 90%. These minerals remain mainly in the bran — the outer part of cereals. For this reason, it seems reasonable to recommend consumption of bread baked from meals of whole grains instead of white breads. Sometimes, recommendation of steady consumption of bran alone, for dietary purpose, needs great care because it may contain many contaminants such as toxic metals and residue of organic pesticides.

During preparation for cooking or canning, vegetables should be thoroughly washed before cutting to remove dirt and traces of insecticide spray. Root vegetables should be scrubbed. The dark outer leaves of greens are rich in iron, calcium, and vitamins, so they should be trimmed sparingly. Peeling of vegetables and fruit should be avoided, whenever possible, because minerals and vitamins are frequently concentrated just beneath the skin. Potatoes should be baked or cooked in their skins even for hashed browns or potato salad. However, True et al.[53] showed that cooking potatoes by boiling whole or peeled tubers, as well as microwave cooking and oven baking, may have a negligible effect on the losses of Al, B, Ca, Na, K, Mg, P, Fe, Zn, Cu, Mn, Mo, I, and Se. Microwaved potatoes retain nutrients well, and contrary to popular belief, peeling potatoes does not strip away their vitamin C and minerals. Whenever practical, any remaining cooking liquid should be served with the vegetable or used in a sauce or gravy soup. To retain minerals in canned vegetables, one should pour the liquid from the can into a saucepan and heat at low temperature to reduce the liquid; then add vegetables to the remaining liquid and heat before serving. Low temperatures reduce shrinkage and loss of many other nutrients. Cooking and blanching lead to the most important nutrient losses. The liquid of cooked vegetables contain about 30–65% of potassium, 15–70% of magnesium and copper, and 20 to more than 40% of zinc in the fresh samples. Thus, it is reasonable to use these liquids for soup preparation and diminish the losses this way.[54,55] Mineral losses depend on both the kind of vegetables cooked and the course of the applied process. Steam blanching generally results in smaller losses of nutrients because leaching is minimized in this process. Frozen meat and vegetables thawed at ambient temperatures lose many nutrients, including minerals in the thaw drip. To avoid these losses, the drip should be added to the pot where the meal is prepared for consumption. Frozen fruits should be eaten without delay, fresh just after thawing, together with the secreted juice. Foods blanched, cooked, or reheated in a microwave oven generally retain about the same or even higher amounts of nutrients as those cooked by conventional methods.

5.6 OTHER EFFECTS

Sodium reduction in the diet is recommended as a means of preventing hypertension and subsequent cardiovascular disease, stroke, and renal failure. Reducing or substituting NaCl requires an understanding of the effects caused by the new factors introduced. Several methods have been proposed for reducing the sodium content in processed meat without an adverse effect on the quality (flavor, gelation) or shelf life of the products. This includes a slight sodium chloride reduction, replacing some of the NaCl with other chloride salt (KCl or $MgCl_2$) or non-chloride salt, or altering the processing methods.[51]

Calcium ion is a known activator of many biochemical processes. The calcium-activated neutral protease (CANP) plays an important role in postmortem tenderizing of meat. The function of the metal ion in such enzymes is believed to be either the neutralization of the charges on the surface by preventing electrostatic repulsion of subunits or effecting a conformational change required for association of the subunits. Thus, metal ions must be present in a specific state to perform this function.

The, cation in solution exists as aqua complex ions in equilibrium with hydroxyl ions:

$$M(H_2O)^{m+} <=> MOH^{(m-1)+} + H^+$$

aqua complex ion — hydroxy complex (weak base)

The acid ionization constant (pKa) of the aqua complex ion determines whether or not the ion would form complexes with a protein. This depends greatly on the pH of the medium. Because the ionization constant of low charge is 12.6, a metal ion would form a stable complex only with negatively charged protein in alkaline media. It cannot bind to cationic proteins because it does not share electrons to form a covalent bond. This consideration explains why the activity of Ca^{2+}-activated protease is optimum in the alkaline pH range. Thus a decrease in its activity at acidic pH values may partly be due to a change in the electronic state of Ca^{2+}.[51,52] Generally, sodium and potassium react only to a limited extent with proteins, whereas calcium and magnesium are somewhat more reactive. Transition metals, e.g., ions of Cu, Fe, Hg, and Ag, react readily with proteins, many forming stable complexes with thiol groups. Calcium ions and ferrous, cupric, and magnesium cations may be integral parts of certain protein molecules or molecular associations. Their removal by dialysis or sequestration appreciably lowers the stability of the protein structure toward heat and proteases.

5.7 BIOCHEMICAL AND TOXIC PROPERTIES OF ALUMINUM, ARSENIC, CADMIUM, MERCURY, AND LEAD

5.7.1 Introduction

A diet consisting of a variety of foods provides the best protection against potentially harmful chemicals in food. This is because the body tolerates very small quantities of many toxic substances but has only limited ability to cope with large quantities of any single one. Almost any chemical can have a harmful effect if taken in a large quantity. This is especially true for trace minerals and, to some degree, also for macroelements, as well as vitamins. For this reason, it is important to understand the difference between toxicity and hazard. Varieties of food contain toxic chemicals, but these chemicals do not present a hazard if consumed in allowable amounts.

A number of minerals can produce chronic toxicity when absorbed and retained in excess of the body's demands. The quantity of elements accumulated by the organism is different from the amount in the environment, and this results in bioconcentration of metals within the organism. Some of the elements are necessary to the organism for metabolic processes; others, however, that are accumulated in high proportions — sometimes specifically in some organs — do not have any metabolic significance for the organism (e.g., arsenic, cadmium, mercury, lead) and are recognized as toxic. Compounds of metals such as aluminum, arsenic, cadmium, lead, and mercury contaminate the environment and enter the food supply. However, among these elements, aluminum compounds are less toxic, and some of the aluminum compounds are even used as food additives, e.g., bread leavening, firming, and emulsifying agents. They may be also applied as drugs, e.g., antiacids, and as such they enter the gastrointestinal tract, along with aluminum compounds naturally present in foods.

The toxicity of these metals is a function of the physical and chemical properties as well as the dose that enters the body. Toxicity is also a function of the ability of metals to bioaccumulate in body tissues. For this reason, it is very important to have adequate information about the chemical forms of metals. Currently, this may be done by applying speciation analysis, which makes it possible to differentiate the chemical form of the element and assess the safety level of its residue in foods or drinking water.

5.7.2 Aluminum

Aluminum was previously considered to be a completely indifferent element in terms of metabolic function. Recent studies, however, have shown that the health risk of aluminum exposure is greatly increased in persons with impaired kidney function. Alfrey et al.[56,57] have found that dialysis encephalopathy in a large number of patients with renal failure, undergoing chronic dialysis, is attributable to the high aluminum content of water used for the preparation of dialysates. Patients on dialysis who died of a neurologic syndrome of unknown cause (dialysis encephalopathy syndrome) had brain matter concentration of 25 mg aluminum per kg dry weight, while in controls 2.2 mg/kg was measured. But whether the metal has a causative role in the pathogenesis of these diseases still remains unconfirmed.

Aluminum is a ubiquitous metal that constitutes about 8% of the Earth's crust — the third most abundant crustal element. No wonder that this element continuously enters the animal organism, both with the feed and drinking water, and also via the lung and skin. Aluminum is abundantly present in planets of the solar system and is generally found in biological organisms in trace levels. The concentrations of aluminum in human tissues from different geographic regions were found to vary widely, probably reflecting the geochemical environment of individuals and locally grown food products. In healthy human tissues, aluminum concentration is usually below 0.5 μg/g wet weight, but higher levels have been observed in liver (2.6 μg/g of ash), lung (18.2 μg/g of ash), lymph nodes (32.5 μg/g of ash), and bone (73.4 μg/g of ash).[9]

Studies on the essentiality of aluminum performed in the second half of the twentieth century have demonstrated the many important functions of this metal in the vital processes of both plants and animals. In plants, its negative effects have mainly been demonstrated especially on cell division and the metabolism of other required elements such as Mn, Mg, and P. However, aluminum at natural levels in animals has been reported to have several positive physiological effects. In some organisms, it seems to be involved in the reactions between cytochrome c and succinyl dehydrogenase, as well as being a necessary cofactor for the activation of guanine–nucleotide binding, which is important in protein metabolism. It has also been shown to play an important role in the development of potent immune response and in endogenous triacylgliceride metabolism. On the other hand, excess of aluminum in patients with renal failure and the syndrome of dialysis encephalopathy may also affect the skeleton by markedly reducing bone formation, resulting in osteomalacia. A further pathological manifestation of aluminum toxicity is a microcytic hypochromic anemia not associated with iron deficiency. Such problems have practically disappeared since the use of aluminum-free deionized water for dialysis became routine.

Aluminum is an extremely versatile metal with a wide variety of uses, e.g., in packaging and building materials, paint pigments, insulating materials, abrasives, cosmetics, food additives, and drugs (antacids). This results in a wide range of human contacts with the metal and a consequent potential impact on human populations. There is no substantial evidence that aluminum has any essential function in animals or humans. However, goats given long-term (4 years) low-aluminum semisynthetic ratios (162 μg/kg) showed significantly reduced life expectancy as compared with that of control goats receiving 25 mg/kg.[7]

In special studies on teratogenicity and reproduction, groups of mice were equally divided by sex and fed diets containing bread leavened with either yeast or aluminum phosphate or alum. The presence of aluminum-leavened bread in the diet resulted in decreased number of offspring as well as development of ovarian lesions. In another study, groups of mice were fed bread with yeast plus 4% physiological saline mixture or 13% saline mixture, or bread with alum phosphate baking powder (4% Al plus 4% saline mixture), or bread with alum phosphate powder (1.3% Al) for a period of 4 months. The presence of aluminum-treated bread resulted in a decreased number of offspring during the first week of life. The ovaries of these animals contained a large number of atritic follicles and were greatly reduced in size.[12]

In another study, groups of rats were maintained on diets containing a mixture of sodium aluminum sulphate (SAS) and calcium acid phosphate at dietary levels equivalent to approximately 0, 0.15, 1.8, or 0.44% Al. Some of the rats were bred for seven successive generations. This study found that the presence of either compound in the diet had no effect on birth weight, average weaning weight, and the number of weaned animals. Histopathological examinations of kidneys of rats that survived 21 months on the diet did not reveal any significant changes.[12]

Long-term administration to humans of aluminum containing antacids such as aluminum hydroxide results in decreased plasma concentration of phosphorous because of decreased absorption or increased deposition of phosphorus in bone as aluminum phosphate.[9] Aluminum antacids may cause an inhibition of intestinal absorption of phosphorus, and this may be followed by an increase in calcium loss. Large amounts of antacids have been reported to induce phosphorus depletion. This effect is probably due to the binding of dietary phosphorus in the intestine by the aluminum. This effect was not observed when phosphorus-containing aluminum salts were used or when the interfering anion and aluminum were taken separately. The same was observed with fluoride. Aluminum cations, such as calcium and magnesium cations, are able to form poorly soluble compounds with fluoride. Thus, the higher intake of aluminum by patients using aluminum antacids may reduce fluoride absorption to some extent. There is some data to shown that aluminum and fluoride in animal studies (with ruminants and poultry) are mutually antagonistic in competing for absorption in the gut. Thus, a higher intake of aluminum compounds in some circumstances may prevent fluorosis, and conversely, feeding animals with higher level of fluoride, present in drinking water, may prevent deposition of aluminum in the brain. Available evidence does not lead to any conclusion that consumption of drinking water containing high, but nontoxic, levels of fluorides will have a preventive effect on Alzheimer's disease.[58,59]

Because aluminum is ubiquitous in the natural environment, its levels in foods (especially in food crops) reflect the aluminum content of the soil and water as well as the ability of plants to absorb and retain its species from the adjacent environment (soil and water) during plant growth. Depending on a variety of environment and geographical factors and anthropogenic inputs, there can be considerable variation in levels of aluminum in cultivated and naturally growing plants and consequently in animal fodder and human foods. Besides the natural sources, some quantities (sometimes even high) of aluminum in diets may originate from food containing aluminum additives or may have leached out of aluminum containers, cookware, utensils, and food wrappings. As noted previously, an important contribution to aluminum intake comes from antacid medications that can provide up to several grams of the metal per day.[7]

Daily dietary aluminum intakes vary in different countries, but according to more recent data published in the WHO report,[9] they range from about 2 to 6 mg/day for children and to ca. 14 mg/day for teenagers and adults. The same levels of intake are given by Pennington in a paper published in 1987.[60] The author compiled the data on the basis of the Total Diet Study for teenagers and adults in 1984. The same paper also summarizes the data published for many countries. The data compiled by Pennington varied from very low to more than 30 mg of aluminum per day/person. It seems that intakes approximating 2–7 mg/day aluminum may be typical of people consuming diets low in herbs, spices, and tea, and without food with aluminum additives, and prepared with little contact with aluminum containers, cookware, utensils, or wrappings.[60]

Nabrzyski and Gajewska[29] examined the normal hospital diets and discovered that the mean daily intake of aluminum was about 21.3±12.4 mg/person. According to Pennington's Report,[60] the aluminum levels in some special hospital diets were markedly lower and ranged from 1.8 to 7.33 mg/day. The major food sources of aluminum in daily diets are grain products, which contain 36.5–69.4%, the next are milk, yoghurt, and cheese contributing 11.0–36.5%, and the contribution of all other products was small, in most cases less than 1%.[60]

Certain plants are able to absorb high levels of aluminum; an example are black tea leaves, which can contain 445–1552 mg Al/kg, whereas herbal or fruit teas contain about 45 mg/kg, and herbal teas that were partially supplemented with black teas can have up to 538 mg Al/kg.[29,61] Aluminum in wild mushrooms and cultivated *Agaricus bisporus* was determined by Müller et al.[62] They found that cultivated *Agaricus bisporus* had the lowest level of 14 ± 7 mg/kg of dry matter. Also, the most popular species such as *Boletus* and *Xerocomus* were low in aluminum (30–50 mg/kg dry matter). However, several other species of the genus *Suillus*, *Macrolepiota rhacodes*, *Hyoholoma capnoides,* as well as individual samples of *Russula ochroleuca* and *Amanita rubescens* contained high aluminum levels of up to 100 mg/kg dry matter. The conclusion is that none of the investigated species of mushrooms contributes significantly to the daily intake of aluminum by humans.

Edible parts of different seafood species such as fish, crustacean, and molluscan shellfish caught in the North Sea, the Barents Sea, the Baltic Sea, the Northeast Atlantic, the Greenland waters, as well as in the coastal waters of Norway near an aluminum smelting plant were analyzed for aluminum. Fillets of lean and fatty fish contained very low amounts of Al, below 0.2 mg/kg wet weight. Fillets of fish caught near the smelting plant were characterized by significantly higher levels of aluminum, especially in the case of haddock of up to 1 mg/kg wet weight. Presumably the contamination of the coastal waters by aluminum originating from smelting plants was responsible for the elevated tissue concentrations. Edible part of crustacean and molluscan shellfish were found to contain up to 5 mg Al/kg wet weight. A comparison between fillets and different organs of cod showed higher aluminum concentration in organs, especially in gills (above 0.6 mg/kg wet weight), which are in continuous contact with the ambient water, and in the brain and heart (above 0.4 mg/kg wet weight). Generally, comparison of the provisional daily intake of 1 mg aluminum/kg body weight/day, established by the Experts Committee of WHO, and its content in the edible parts of seafood indicated that the contribution of aluminum from the edible part of aquatic food does not pose a significant risk in daily diets.[9,63] It is worth mentioning that some quantities of the metal can migrate from aluminum packaging materials, storage containers, and cooking utensils into foods and beverages, especially when the food was in direct and prolonged contact during storage or boiling process.

It is suggested that the additional daily intake of aluminum resulting from preparing all foods in uncoated aluminum pans is approximately 3.5 mg. In recent times, however, most pans are made of stainless steel or Teflon-coated aluminum, which diminish the migration of aluminum into foods. In Switzerland, for example, where most pans are of stainless steel or Teflon-treated aluminum, the average contribution from the use of aluminum utensils to the daily intake of 2–5 mg in the diet is estimated to be less than 0.1 mg.[64] However, it is worth mentioning that the migrated quantities of the metal into foods depends to a certain extent on the acidity of the food and can markedly increase, especially where acidic beverages are stored or heated in aluminum cans. Generally, it may be said that the quantities of aluminum contributed to the daily diet from such adventitious sources are rather inconsistent and insignificant.

Some of the intentional aluminum-containing additives, such as aluminum ammonium sulphate (officially numbered E 523), aluminum potassium sulphate (E 522), aluminum sodium sulphate (E 521) and sodium aluminum phosphate, acidic and basic (E 541i and 541ii), are used as stabilizers, buffers, and neutralizing agents. Other compounds are used as anticaking agents, dough strengtheners, leavening agents, acid-reacting ingredients in self-rising flour or cornmeal, emulsifying agents for processed cheese, firming agents, processing aids, stabilizers, thickeners, curing agents, components in bleaching agents, and texturizers' agents. The acidic form of sodium aluminum phosphate reacts with sodium bicarbonate to cause a leavening action. It is used in biscuit, pancake, waffle, cake, doughnut, and muffin mixes, frozen rolls and yeast doughs, and canned biscuits and self-rising flours. The basic form of sodium aluminum phosphate is used in processed cheese and cheese foods as an emulsifying agent to give cheese products a soft texture and to allow easy melting. It is commonly used in individually wrapped, sliced, processed cheeses, and may result in about 50 mg of aluminum per slice. Sodium aluminum phosphate is also used as a meat binding agent. Sodium aluminum sulphate is an acidifying agent found in many household baking powders. It contains about 70 mg of aluminum per teaspoon. A slice of cake or bread made with an aluminum-containing baking powder may contain 5–15 mg of aluminum. Most commercial baking powders and some that are sold for household use contain monocalcium phosphate rather than aluminum salts. Recently, the use of aluminum salts as firming agents in pickled vegetables and some pickled fruits has largely been replaced by calcium oxide in both industrial and home pickling, although several brands of commercially packed pickles still may contain ammonium or potassium sulphates of aluminum. Aluminum silicates (sodium aluminum silicate, aluminum calcium silicate, and hydrated sodium calcium aluminosilicate) are found in anticaking agents in salt, nondairy creamers, and other dry, powdered products.[60] The most commonly used foods that may contain substantial amounts of aluminum from food additives are processed cheeses, baking powders, cake mixes, frozen dough, pancake mixes, self-rising flours, and pickled vegetables.

A Committee of the Life Sciences Research Office of the Federation of American Societies for Experimental Biology reviewed the safety of GRAS substances including aluminum compounds under contract with the FDA and estimated that the average daily intake of aluminum from aluminum compounds added to foods was about 20 mg, and that about 75% of this was in the form of sodium aluminum phosphate.[60]

Current estimates of aluminum intake range from about 2–6 mg/day for children, and 6–14 mg/day for teenagers and adults. Low total body burden of aluminum coupled with urinary excretion suggests that even at high levels of consumption, only small amounts, ca. 1% of the normally consumed doses, are absorbed by a healthy person from the gastrointestinal tract. These small amounts absorbed from diets are then excreted by healthy kidneys, so no bodily accumulation occurs. On the other hand, patients with renal failure are at greater risk of high doses of aluminum because administered aluminum may be retained consequent on the functional impairment of the kidneys. Aluminum is primarily localized in bone, liver, brain, heart, and spleen.[9]

In conclusion, there is unknown risk to healthy people from typical dietary intakes of aluminum. Risk may arise only from habitual consumption of gram-quantities of aluminum antacids over long period of time, and may increase substantially in persons with impaired kidney function. Long-term intravenous application also results in serious toxicity.

Factors influencing exposure to aluminum and its tolerance by human subjects have been extensively reviewed by the Joint FAO/WHO Expert Committee on Food Additives, and the provisional tolerable weekly intake of 7 mg/kg body weight has been established.[9]

5.7.3 Arsenic

Pentavalent and trivalent arsenicals react with biological ligands in different ways. The trivalent form reacts with the thiol protein groups, resulting in enzyme inactivation, structural damage, and a number of functional alterations. The pentavalent arsenicals, however, do not react with –SH groups. Arsenates can competitively inhibit phosphate insertion into the nucleotide chains of the DNA of cultured human lymphocytes, causing false formation of DNA because of instability of the arsenate esters. Dark repair mechanisms are also inhibited, leading to persistence of these errors in the DNA molecules. Binding differences of the trivalent and pentavalent forms arsenic to differences in the accumulation of this element. Trivalent inorganic arsenic is accumulated at a higher level than its pentavalent form. Organic arsenic compounds are considered less toxic or nontoxic in comparison to inorganic arsenic, of which trivalent arsenicals are the most toxic forms.

Dietary arsenic represents the major source of arsenic exposure for most of the general population. Consumers eating large quantities of fish usually ingest significant amounts of arsenic, primarily as organic compounds, especially those with structures similar to arsenobetaine and arsenocholine as well as various other arsenic derivatives. Fish of many species contain arsenic between 1 and 10 mg/kg. Arsenic levels at or above 100 mg/kg have been found in bottom feeders and shellfish. Both lipid and water-soluble organoarsenic compounds have been found, but the water-soluble forms, mainly the quaternary arsonium derivates, constitute the larger portion of the total arsenic in marine animals.[9,65] Studies in mice have demonstrated that after the administration of organoarsenic compounds, over 90% of the arsenobetaine and arsenocholine doses were absorbed, and about 98% of the administered dose of arsenobetaine was excreted unchanged in the urine, whereas 66 and 9% of the single oral dose of arsenocholine was excreted in the urine and feces, respectively, within 3 days. Most of the arsenocholine was oxidized in animal organism to arsenobetaine, and in this form was excreted in the urine. The retention of arsenocholine in the animal body, following administration, was greater than the retention of arsenobetaine. The fate of organic arsenicals in humans still has not been fully clarified.

The little information available on organoarsenicals present in fish and other seafood may indicate that these compounds appear to be readily excreted in the urine in an unchanged chemical form, with most of the excretion occurring within 2 d of ingestion. Volunteers who consumed flounder excreted 75% of the ingested arsenic in urine within 8 d after eating the fish. The excreted arsenic was in the same chemical form as it was in the fish. Less than 0.35% was excreted in the feces. There are no data on tissue distribution of arsenic in humans after ingestion of arsenic present in fish and seafood. Also, there have been no reports of ill effects among ethnic populations consuming large quantities of fish that result in organoarsenic intakes of about 0.05 mg/kg of body weight per day.[9] Inorganic tri- and pentavalent arsenicals are metabolized in humans, dogs, and cows to less toxic methylated forms such as monomethylarsonic and dimethylarsinic acids.[66]

5.7.4 Mercury

Contrary to organic arsenicals, organic mercury compounds, especially methylmercury, are recognized as the more dangerous for humans than the inorganic compounds. Most foods with the exception of fish contain very low amounts of total mercury (< 0.01 mg/kg), almost entirely in the form of inorganic compounds. Over 90% of mercury in fish and shellfish is in the form of methylmercury. This is so because the fish feed on aquatic organisms that contain this compound derived from the biomethylation of inorganic mercury by microorganisms. Marlin is the only pelagic fish known to have more than 80% of the total muscle mercury present as inorganic mercury.[67] The amount of methylmercury is especially high in large, old fish of predatory species such as the shark and swordfish. In freshwater fish, the mercury content depends on its concentration in water and sediment and on the pH of water. The concentration of methylmercury in most fish is generally less than 0.4 mg/kg, although predators such as swordfish, shark, and pike may contain up to several milligrams of methylmercury per kilogram in the muscles. The intake of methylmercury depends on fish consumption and the concentration of methylmercury in the fish consumed. Many people eat about 20–30 g of fish per day, but certain groups eat 400–500 g per day. Apparently, the daily dietary intake of methylmercury can range from about 0.2 to 3–4 μg/kg body weight.

Studies of the biokinetics of methylmercury after ingestion show that its distribution in the tissues is more homogenous than that of other mercury compounds, with the exception of elemental mercury. The most important features of the distribution pattern of methylmercury are a high blood concentration, a high ratio of erythrocyte/plasma concentrations (about 20), and a high deposition in the brain. Another important characteristic is slow demethylation, which is a critical detoxification step. Methylmercury and other mercury compounds have a strong affinity for sulphur and selenium. Although selenium has been suggested to provide protection against the toxic effect of methylmercury, no such antagonistic effect has been demonstrated in humans.

A variety of effects have been observed in animal experiments with toxic doses, but some of the adverse health effects, such as renal damage and anorexia, have not been observed in humans exposed to high doses. The primary tissues of concern in humans are the nervous system and particularly the developing brain, and these have stimulated a wide range of epidemiologial studies. Methylmercury passes about 10 times more readily through the placenta than other mercury compounds. The dermal absorption of methylmercury is similar to that of inorganic mercury salts.

The LD_{50} values after oral administration are 25 mg/kg of body weight in old rats (450 g of body weight) and 40 mg/kg in young rats (200 g). The clearance half-time of methylmercury for the human whole body is about 74 d, and for blood compartment it is about 52 d.[10]

5.7.5 Cadmium

Cadmium shares a number of chemical properties with zinc and mercury, but in contrast to mercury, it is incapable of environmental methylation due to the instability of the monoalkyl derivative. Similarities and differences also exist in the metabolism of zinc, cadmium, and mercury. Metallothioneins (MTs) and other Cd-binding proteins hold and transport Cd, Zn, and Hg within the body. MTs are metal-binding proteins of relatively low molecular mass with a high content of cysteine residues that have a particular affinity for cadmium, as well as for zinc and copper, and can affect their toxicity. The MT synthesis in organisms is induced by these metals, and MT is involved in their storage in the organs. Zinc metallothionein can detoxify free radicals. Cadmium-induced metallothionein is able to bind cadmium intracellularly and, in this way, protects the organism against the toxicity of this metal. Cadmium is transported in the plasma as a complex with metallothionein and may be toxic to the kidney when excreted in the glomerular filtrate.

Most cadmium in urine is bound to metallothionein. This protein occurs in the organism as at least four genetic variants. The two major forms, I and II, are ubiquitous in most organs, particularly in the liver and kidney, and also in brain. Metallothionein isolated from adult or fetal human livers contained mainly zinc and copper, whereas that from human kidney contained zinc, copper, and cadmium.[9,11] Cadmium, along with the other two metals, is bound to the peptide by mercaptide bonds and are arranged in two distinct clusters, namely, a four-metal cluster called the α *domain* and a three metal-cluster called *cluster* β, at the C terminal of the protein. The α cluster is an obligate zinc cluster, whereas the zinc in cluster β may be replaced by copper or by cadmium.

Reaction with metallothionein is the basis for metabolic interactions between these metals. Metallothionein III is found in the human brain and differs from I and II by having six glutamic acid residues near the terminal part of the protein. Metallothionein III is thought to be a growth inhibitory factor, and its expression is not regulated by metals; however, it does bind zinc. Another proposed role for metallothionein III is participation in the use of zinc as a neuromodulator, as metallothionein III is present in the neurons that store zinc in their terminal vesicles. Metallothionein IV occurs during differentiation of stratified squamous epithelium, but it is known to have a role in the absorption or toxicity of cadmium. Metallothionein in the gastrointestinal mucosa plays a role in the gastrointestinal transport of cadmium. Its presence in cells of the placenta impairs the transport of cadmium from maternal to fetal blood and across blood–brain barriers, but only when the concentration of cadmium is low. Newborns are virtually cadmium free, whereas zinc and copper are readily supplied to the fetus. Rapid renal concentration occurs mainly during the early years of life. [9,11]

Cadmium bound to metallothionein in food does not appear to be absorbed or distributed in the same way as inorganic cadmium compounds. Low dietary concentration of calcium promotes absorption of cadmium from the intestinal tract of experimental animals. A low iron status in laboratory animals and humans has also been shown to result in greater absorption of cadmium. In particular, women with low body iron stores, as reflected by low serum ferritin concentrations, show an average gastrointestinal absorption rate that may be twice as high (ca. 10%) as that of a control group of women. High iron status results in decreasing total and fractional cadmium accumulation from diets, whereas low iron status in the organism promotes accumulation of cadmium. Studies in rats with reduced iron status showed that the inclusion of wheat bran (contains phylate, which hinders the absorption of iron, calcium and other minerals) in their diets increased the uptake of cadmium.[9,11] The LD_{50} value for rats and mice treated orally ranges from about 100–3000 mg/kg of body weight, after a single dose of cadmium chloride.

The high affinity of Cd for –SH groups and the ability of imparting moderate covalency in bonds result in increased lipid solubility, bioaccumulation, and toxicity. In humans, after normal levels of exposure, ca. 50% of the body burden is found in the kidneys, ca. 15% in the liver, and ca. 20% in the muscles. As in animals, the proportion of cadmium in the kidney decreases as the liver concentration increases. The lowest levels of cadmium are found in the brain, bone, and fat. Accumulation in the kidney continues until 50–60 years of age in humans and falls thereafter, possibly due to age-related changes in kidney integrity and function. In contrast, the cadmium level in the muscle continues to increase over the course of life. The average cadmium concentration in the renal cortex of nonoccupationally exposed persons aged 50 varies between 11 and 100 mg/kg in different regions.

Diet is the major route of human exposure to cadmium. Contamination of foods with cadmium results from its presence in soil and water. The concentrations of cadmium in foods range widely, and the highest average concentrations are found in mollusks, kidney, liver, cereals, cocoa, and leafy vegetables. A daily intake of ca. 60 μg would be required to reach a concentration of 50 mg/kg in the renal cortex of persons at the age of 50, assuming an absorption ratio of 5%. About 10% of the absorbed daily dose is rapidly excreted.[9,11]

5.7.6 Lead

The JECFA and other WHO committees have recognized that infants and children are the groups at highest risk of lead exposure from food and drinking water. Lead as an anthropogenic contaminant finds its way into the air, water, and surface soil. Lead-containing manufactured products also contribute to the lead body burden. The domestic environment in which infants and children spend the greater part of their time is of particular importance as the source of lead intake. In addition to exposure from general environmental sources, some infants and young children, as a result of normal, typical behavior, can receive high doses of lead through mouthing or swallowing of nonfood items. Pica, the habitual ingestion of nonfood substances, which occurs among many young children, has frequently been implicated in the etiology of lead toxicity. In the U.S., on average, 2-year old children may receive ca. 45% of their daily lead intake from dust, 40% from food, 15% from water and beverages, and 1% from inhaled air.[8] Lead absorption is heavily influenced by food intake, and much higher rates occur after fasting than when lead is ingested with a meal. This effect may be due mainly to competition from other ions, particularly iron and calcium, for intestinal transport pathways. Absorption is also affected by age, the typical absorption rates in adults and infants being 5–10% and ca. 50%, respectively. Children absorb lead from the diet with greater efficiency than adults.[10] After absorption and distribution in blood (where most lead is found in erythrocytes), lead is initially distributed to soft tissues throughout the body. Subsequently, lead is deposited in the bone, where it eventually accumulates. The half-life of lead in blood and other soft tissues is 28–36 days. Lead deposited in physiologically inactive cortical bone may persist for decades without substantially influencing its concentrations in blood and other tissues. On the other hand, lead accumulated early in life may be released later when bone resorption is increased, for instance, as a result of calcium deficiency or osteoporosis. Lead deposited in physiologically active trabecular bone is in equilibrium with the blood. The accumulation of high concentrations of lead in blood when exposure is reduced may be due to the ability of bone to store and release this element.

Dietary lead not absorbed in the gastrointestinal tract is excreted in the feces. Lead not distributed to other tissues is excreted through the kidney and, to lesser extent, by biliary clearance.[10] The biochemical basis of lead toxicity is its ability to bind to biologically important molecules, thereby interfering with their function by a number of mechanisms. At the subcellular level, the mitochondrion appears to be the main target organelle for the toxic effects of lead in many tissues. Lead has been shown to selectively accumulate in the mitochondria. There is evidence that it causes structural injury to these organelles and impairs basic cellular energetics and other mitochondrial functions. It is a cumulative poison, producing a continuum of effects primarily on the hematopoietic system, the nervous system, and the kidneys. At very low blood levels, lead may impair the normal metabolic pathways in children. At least three enzymes of the heme biosynthetic pathway are affected. Lead interferes with δ-aminolevulinic acid dehydratase at a level in blood amounting to ca. 10 μg/100 ml.[8] Alteration in the activity of the enzymes of the heme synthetic pathway leads to accumulation of the intermediates of the pathway. There is some evidence that accumulation of δ-aminolevulinic acid exerts toxic effects on neural tissues through interference with the activity of the neurotransmitter γ-aminobutyric acid. The reduction in heme production *per se* has also been reported to adversely affect nervous tissue by reducing the activity of tryptophan pyrollase, a heme-requiring enzyme. This results in increased metabolism of tryptophan via a second pathway that produces high blood and brain levels of the neurotransmitter serotonin.

Lead interferes with vitamin D metabolism, because it inhibits the hydroxylation of 25-hydroxy-vitamin D to produce the active form of vitamin D. The effect has been reported in children at blood lead levels as low as 10–15 μg/100 ml.[8] Measurements of the inhibitory effects of lead on heme synthesis is widely used in screening tests to determine whether medical treatment for lead toxicity is needed for children in high-risk populations who have not yet developed overt symptoms of lead poisoning.

5.8 SOME OTHER INTERACTIONS OF TOXIC ELEMENTS

Data on the toxicity of the five toxic minerals discussed so far are summarized in Table 5.5 and Table 5.6.

The general biochemistry of most elements is not entirely independent of one another. Elements of similar chemical properties tend to be taken up together. Sometimes one element has an inhibiting effect on another, or there can be a synergistic effect, e.g., enhancement of absorption of calcium in the presence of adequate amounts of phosphorus. Aluminum, for example, may exert negative effects on cell division and, after oral ingestion, can reduce the absorption and utilization of phosphorus, calcium, magnesium, iron, and manganese. Also, cadmium and lead may hinder calcium and iron absorption. Zinc–copper or molybdenum–copper antagonisms affect the ratios of Zn/Cu and Cu/Mo, leading to copper and molybdenum deficiencies, respectively. Metal ions such as Mg(II) and Zn(II) are involved in the transcription of the genetic message. Toxic metal ions may affect the activity of RNA polymerase and induce the expression of specific proteins that, in turn, change the susceptibility of cells to metals. Many other important interactions are described in a review by Beyersmann.[71]

We can expect that in the future the biological function of some minerals, now recognized only as potentially essential, may be confirmed as being essential. Possibly, their vital role in the metabolism of the living organism would be established. An increase in our knowledge in this field will make it possible to use some potentially essential elements in technological processes connected with the production of new forms of foods important for dietetic nutrition and for preventing some diseases such as obesity, bone weakening, and fractures, as well as for the active regulation of some enzymes and hormones. All these prospects are fascinating to scientists and should provide them ample new challenges in the future.

TABLE 5.5
Metals Toxicity for Human[7–11,29,60,63,65–70]

Metal	Toxic Effects	Daily Intake (mg/adult person)	Source of Exposure	Absorption (%)
Aluminum	Encephalopathy (dialysis dementia), osteodystrophy in patients exposed to high doses, elevated content in brain, cerebrospinal fluids, serum, and hair in patients with dialysis dementia. It is proposed that signs of neurotoxicity are manifested when the brain concentration exceeds 10–20 times the normal level (> 2 mg/kg dry weight of the grey matter). Aluminum concentration in human tissues probably reflects the geochemical environment of individuals and locally grown food products. There is no risk of toxicity to healthy people from typical dietary intakes	9–36	Grain, some bakery and dairy products made with aluminum additives acting as buffers, dough strengtheners, and leavening agents, acid-reacting ingredients in self-rising flour or cornmeal, emulsifying agents for processed cheeses, stabilizers, texturizes, and anticaking agents, thickeners and curing agents, and a few other intentional used additives. High levels are observed in seafood, black and herbal teas, as well as spices. Level of aluminum naturally present in water is generally low (0.001–1 mg/l), but when water is treated with aluminum sulphate salts (as coagulating agent), its content may rise threefold (to 0.6 mg/l) of the normal accepted level (0.2 mg/l)	< 1–7%
Arsenic	Inorganic compounds cause abnormal skin hyperpigmentation, hyperkeratosis, skin, and lung cancer. Organoarsenic compounds present in fish are less or nontoxic	0.0–0.29	Contaminated water, food containing residue of arsenic pesticides, and veterinary drugs, fish, and shellfish are the richest source of organic compounds, e.g., arsenobetaine and arsenocholine	> 90% organoarsenic compounds and high inorganic trivalent compounds

Cadmium	Accumulates mainly in liver and renal cortex, nephrotoxicity, tubular damage impairing the reabsorption of calcium and phosphate resulting in decalcification, osteoporosis, and osteomalacia. Disturbs vitamin D hydroxylation to its active form, Itai Itai disease, embryotoxic in early gestation. Impairs immune system, and calcium and iron absorption. Causes hypertension and cardiovascular disease; kidney is the critical organ	< 0.01–0.1	Oysters, cephalopods, crops growing on land fertilized with high contaminated phosphate and sewage sludge, cadmium leaching from enamel and pottery glazes, contaminated water	3–10%, cadmium bound to metallothionein is well absorbed
Lead	At blood levels greater than 40 μg/100 ml, exerts a significant effect on hemopoietic system, resulting in anemia; affects central nervous system	< 0.1–0.2	Food contaminated from leaching of glazes of ceramic foodware, as well as from motor vehicle exhausts, atmospheric deposits, canned foods, water supply from plumbing system	5–10% in adult person, 40–50% in children
Mercury	Methylmercury compounds easily pass the blood–brain and placental barriers, cause severe neurological damage, greater in young children, animals; also renal damage and anorexia	< 0.02–0.1	Fish and shellfish, meat from animal feed with mercury dressed grains	> 90% as methylmercury compounds, and 15% as inorganic mercuric compounds

TABLE 5.6
Provisional Tolerable Weekly Intake (PTWI) of Toxic Elements According to the Joint FAO/WHO Expert Committee on Food Additives for Man[8,9]

Element	PTWI	Comments
Aluminum mg/kg body weight	0–7.0	Includes intake of aluminum from food additive uses
Arsenic μg/kg body weight	15.0	For inorganic arsenic
Cadmium μg/kg body weight	7.0	
Lead μg/kg body weight	25.0	When blood lead levels in children exceed 25 μg/100 ml (in whole blood), investigations should be carried out to determine the major sources of exposure, and all possible steps should be taken to ensure that lead levels in food are as low as possible
Mercury μg/kg body weight	3.3 as methylmercury and 5.0 as total mercury	With the exception of pregnant and nursing women, who are at greater risk to adverse effects from methylmercury

REFERENCES

1. Anke, M. et al., Lithium, in *Handbook of Nutritionally Essential Mineral Elements*, O'Del, B.L. and Sunde, R.A., Eds., Marcel Dekker, New York, Basel, Hong Kong, 1997, p. 465.
2. Morohashi, T., Sano, T., and Yamada, S., Effects of strontium on calcium metabolism in rats: I. A distinction between the pharmacological and toxic doses, *Jpn. J. Pharmacol.*, 64, 155, 1994.
3. Morohashi, T. et al., Effects of strontium on calcium metabolism in rats. II. Strontium prevents the increased rate of bone turnover in ovariectomized rats, *Jpn. J. Pharmacol.*, 68, 153, 1995.
4. O'Dell, B.L. and Sunde, R.A., Eds., *Handbook of Nutritionally Essential Mineral Elements*, Marcel Dekker, New York, 1997.
5. Nabrzyski, M. and Gajewska, R., Content of strontium, lithium and calcium in selected milk products and in some marine smoked fish, *Nahrung/Food*, 46, 204, 2002.
6. Nielsen, F.H., Ultratrace elements in nutrition: Current knowledge and speculation, *J. Trace Elements Exp. Med.*, 11, 251, 1998.
7. WHO, *Trace Elements in Human Nutrition and Health,* WHO, Geneva, 1996, p. 222.
8. WHO, Food Additives Series, 21, Toxicological Evaluation of Certain Food Additives and Contaminants, 30th Meeting of the Joint FAO/WHO Expert Committee on Food Additives, Cambridge, U.K., 1986.
9. WHO, Food Additives Series, 24, Toxicological Evaluation of Certain Food Additives and Contaminants, 33rd Meeting of the Joint FAO/WHO Expert Committee on Food Additives, Cambridge, U.K., 1989.

10. WHO, Food Additives Series, 44, Safety Evaluation of Certain Food Additives and Contaminants, 53 Meeting of the Joint FAO/WHO Expert Committee on Food Additives, Geneva, 2000.
11. WHO, Food Additives Series, 46, Safety Evaluation of certain Food Additives and Contaminants, 55th Meeting of the Joint FAO/WHO Expert Committee on Food Additives. Geneva, 2001.
12. WHO, Food Additives Series, 17, Toxicological Evaluation of Certain Food Additives, 26 Report of Joint FAO/WHO Expert Committee on Food Additives, Geneva, WHO, Technical Report, Ser. 683, Rome, 1982.
13. WHO, Food Additives Series, 18, IPCS, Toxicological Evaluation of Certain Food Additives and Contaminants, 27 Report of Joint FAO/WHO Expert Committee on Food Additives, Technical Report, Ser. 696, Geneva, 1983.
14. *Recommended Dietary Allowances*, 10th ed., National Research Council, Food and Nutrition Board, National Academy Press, Washington, D.C., 1989.
15. Feltman, J., Ed., *Prevention's Giant Book of Health Facts*, Rodale Press, Emmaus, PA, 1990.
16. Johnson, J.L., Molybdenum, in *Handbook of Nutritionally Essential Mineral Elements*, O'Dell, B.L. and Sunde, R.A., Eds., Marcel Dekker, New York, 1997.
17. Bremner, I., Heavy metal toxicities, *Q. Rev. Biophys.*, 7, 75, 1974.
18. Rosenberg, J.H. and Solomons, N.W., Physiological and pathophysiological mechanism in mineral absorption, in *Absorption and Malabsorption of Minerals,* Vol. 12, Solomons, N.W. and Rosenberg, J.H., Eds., Alan R. Liss, Inc., New York, 1984.
19. Eschleman, M., Ed., *Introductory Nutrition and Diet Therapy*, Lippincot J.B. Co., London, 1984.
20. Gajewska, R. and Nabrzyski, M., Molybdenum content in children and adults full day's food and in some foodstuffs, *Bromat. Chem. Toksykol.*, 24, 185, 1991.
21. Hendler, S.S., Ed., *The Doctor's Vitamin and Mineral Encyclopaedia*, Simon and Schuster, New York, 1990.
22. Causeret, J., Fish as a source of mineral nutrition, in *Fish as Food,* Vol. 2, Borgstrom, G., Ed., New York, London, Academic Press, 1962, p. 205.
23. Çelik, U. and Oehlenschläger, J., Zinc and copper in marine fish samples collected from the Eastern Mediterranean Sea, *Eur. Food Res. Technol.*, 220, 37, 2005.
24. Gajewska, R. et al., Trace levels of metals in bee honey, *Bromat. Chem. Toksykol.*, 17, 259, 1984 (in Polish).
25. Gajek, O., Nabrzyski, M., and Gajewska, R., Metallic impurities in imported canned fruit and vegetables and bee honey. *Roczniki PZH*, 38, 14, 1987 (in Polish).
26. Lopez, A., Ward, D.R., and Williams, H.L., Essential elements in oysters (*Crassostrea virginica*) as affected by processing method, *J. Food Sci.*, 48, 1680, 1961, 1983.
27. Marzec, Z., Kunachowicz, H., Iwanow, K., and Rutkowska, U., Eds., *Tables of Trace Elements in Food Products*, National Food and Nutrition Institute, Warsaw, 1992 (in Polish).
28. Mateos, C.J. et al., Chromium content in breakfast cereals depending on cereal grains used in their manufacture, *Eur. Food Res. Technol.*, 220, 42, 2005.
29. Nabrzyski, M. and Gajewska, R., Aluminum and fluoride in hospital daily diets and teas, *Z. Lebensm. Unters. Forsch.*, 201, 307, 1995.
30. Pasławska, S. and Nabrzyski, M., Assay of iodine in powdered milk, *Bromat. Chem. Toxicol.*, 8, 73, 1975 (in Polish).
31. WHO, Environmental Health Criteria, in *Fluorine and Fluorides*, WHO, Geneva, 36, 1984.

32. Williams, D.M., Clinical significance of copper deficiency and toxicity in the world Population, in *Clinical Biochemical and Nutritional Aspects of Trace Elements*, Vol. 6, Prasad, A.S., Ed., Alan R. Liss, Inc., New York, 1982, p. 277.
33. Wilpinger, M., Schönsleben, I., and Pfanhauser, W., Chrom in Oesterreichischen Lebensmitteln, *Z. Lebensm. Unters. Forsch.* 201, 521, 1995 (in German).
34. Wojnowski, W., Składniki mineralne, in *Chemiczne i Funkcjonalne Właściwości Składników Żywności*, Sikorski, Z. E., Ed., Wydawnictwa Naukowo-Techniczne, Warszawa, 1994, p. 76 (in Polish).
35. FAO/WHO Ed., *Summary of Evaluations Performed by the Joint FAO/WHO Expert Committee on Food Additives 1956–1993*, International Life Science Inst. Press, Geneva, 1994.
36. Du, Z. and Bramlage, W.J., Superoxide dismutase activities in Senescin Aplle Fruit (*Malus domestica* borkh), *J. Food Sci.*, 59, 581, 1994.
37. Ramanathan, L., and Das, N. P., Effect of natural copper chelating compounds on the prooxidant activity of ascorbic acid in steam-cooked ground fish, *Int. J. Food Sci. Technol.,* 28, 279, 1993.
38. Hultin, H.O., Oxidation of Lipids in seafoods, in *Seafoods: Chermistry, Processing Technology and Quality*, Shahidi, F. and Botta, J.R., Eds., Chapman and Hall, London, 1994, p.49.
39. Wu, S.Y. and Brewer, M.S., Soy protein isolate antioxidant effect on lipid peroxidation of ground beef and microsomal lipids, *J. Food Sci.*, 59, 702, 1994.
40. Pearson, A.M., Love, J.D., and Shorland, F.B., "Warmed over" flavor in meat, poultry, and fish, *Adv. Food Chem.*, 23, 2, 1977.
41. Oelingrath, I.M. and Slinde, E., Sensory evaluation of rancidity and off-flavor in frozen stored meat loaves fortified with blood, *J. Food Sci.* 53, 967,1988.
42. Miller, D.K. et al., Lipid oxidation and warmed-over aroma in cooked ground pork from swine fed increasing levels of iron, *J. Food Sci.*, 59, 751, 1994.
43. Miller, D.K. et al., Dietary iron in swine affects nonheme iron and TBAR's in pork skeletal muscles, *J. Food Sci.*, 59, 747, 1994.
44. Castell, C.H., MacLeam, J., and Moore, B., Rancidity in lean fish muscle. IV Effect of sodium chloride and other salts, *J. Fish. Res. Board Can.*, 22, 929, 1965.
45. Castell, C.H. and Spears, D. M., Heavy metal ions and the development of rancidity in blended fish muscle, *J. Fish. Res. Board Can.*, 25, 639, 1968.
46. Kraniak, J.M. and Shelef, L. A., Effect of ethylenediaminetetraacetic acid (EDTA) and metal ions on growth of *Staphylococcus aureus* 196 E in culture media, *J. Food Sci.*, 53, 910, 1988.
47. Ha, Y.W. et al., Calcium binding of two microalgal polysaccharides and selected industrial hydrocolloids, *J. Food Sci.*, 54, 1336, 1989.
48. WHO, Food Additives Series, 30, Toxicological Evaluation of Certain Food Additives and Naturally Occurring Toxicants, 39 Meeting of the Joint FAO/WHO Expert Committee on Food Additives, Geneva, 1993.
49. WHO, Food Additives Series, 32, Toxicological Evaluation of Certain Food Additives and Contaminants, 41 Meeting of the Joint FAO/WHO Expert Committee on Food Additives, Geneva, 1993.
50. Samant, S.K. et al., Protein-polysaccharide interactions: a new approach in food formulation, *Int. J. Food Sci. Technol.*, 28, 547, 1993.
51. Barbut, S. and Mittal, G.S., Rheological and gelation properties of meat batters prepared with three chloride salts, *J. Food Sci.*, 53, 1296, 1985.

52. Asghar, A. and. Bhatti A. R., Endogenous proteolytic enzymes in skeletal muscle: their significance in muscle physiology and during postmortem aging events in carcasses, *Adv. Food Res.*, 31, 343, 1987.
53. True, R.H. et al., Changes in the nutrient composition of potatoes during home preparation, III, *Miner. Am. Potato J.*, 56, 339, 1979.
54. Trzebska-Jeske, I. et al., The effect of mechanical processing on nutritional value of groats produced in Poland, *Roczniki PZH.*, 24, 717, 1973 (in Polish).
55. Rutkowska, U., The effect of the grinding process on contents of copper, zinc, and manganese in rye and wheat flour, *Roczniki PZH.*, 26, 339, 1975 (in Polish).
56. Alfrey, A.C., Le Gendre, G.R., and Kaehney, W.D., The dialysis encephalopathy syndrome, *N. Engl. J. Med.*, 294, 184, 1976.
57. Alfrey, A.C., Hegg, A., and Craswell, P., Metabolism and toxicity of aluminum in renal failure, *Am. J. Clin. Nutr.*, 33, 1509, 1980.
58. Kraus, A.S. and Forbes, W.F., Aluminum, fluoride and the prevention of Alzheimer's disease, *Can. J. Publ. Health*, 83, 97, 1992.
59. Schenkel, H. and Klüber, J., Mögliche Auswirkungen einer erhöhten Aluminiumaufnahme auf Landwirtschaftliche Nutziere, *Obers. Tierernährung.*, 15, 273, 1987.
60. Pennington, J.A.T., Aluminum content of foods and diets, *Food Addit. Contam.*, 5, 161, 1987.
61. Fairweather-Tait, S.J. et al., Aluminum in the diet Human Nutrition, *Food Sci. Nutr.*, 41F, 183, 1987.
62. Müller, M., Anke M., and Illing-Günther, H., Aluminum in wild mushrooms and cultivated *Agaricus bisporus, Z. Lebensm. Unters. Forsch.*, 205, 242, 1997.
63. Ranau, R., Oelhlenschläger, J., and Steinhart, H., Aluminum content in edible parts of seafood, *Eur. Food Res. Technol.*, 212, 431, 2001.
64. Müller, J.P., Steinegger, A., and Schlatter, C., Contribution of aluminum from packaging materials and cooking utensils to the daily aluminum intake, *Z. Lebensm. Unters. Forsch.*, 197, 332, 1993.
65. Vaessen, H.A.M.G. and van Ooik, A., Speciation of arsenic in Dutch total diets: methodology and results, *Z. Lebensm. Unters. Forsch.*, 189, 232, 1989.
66. Peoples, S.A., The metabolism of arsenic in man and animals, in *Arsenic-Industrial, Biomedical, Environmental Perspectives*, Proceedings of the Arsenic Symposium, Lederer, W.H. and Fensterheim, R.J., Eds., Van Nostrand Reinhold Co., New York, 1983, p. 125.
67. Cappon, C.J. and Smith, J.C., Chemical form and distribution of mercury and selenium in edible sea food, *J. Anal. Toxicol.*, 6, 10, 1982.
68. Nabrzyski, M. and Gajewska, R., Determinations of mercury, cadmium and lead in food, *Roczniki PZH*, 35, 1, 1984 (in Polish).
69. Nabrzyski, M., Gajewska, R., and Lebiedzinska, A., Arsenic in daily food rations of adults and children, *Roczniki PZH*, 36, 113, 1985 (in Polish).
70. WHO, Environmental health criteria, in *Arsenic*, WHO, Geneva, 18, 1981.
71. Beyersmann, D., The significance of interactions in metal essentiality and toxicity, in *Metals and Their Compounds in the Environment*, Merian, E., Ed., VCH Verlagsgesellschaft Weinheim, NewYork., Basel, 1991, p. 491.

6 Mineral Components in Foods of Animal Origin and in Honey

Piotr Szefer and Małgorzata Grembecka

CONTENTS

6.1 INTRODUCTION

In general, variations in levels of mineral components in nonplant food are not as broad as those recorded for food of plant origin. Transfer of nutritional elements goes primarily from soil to humans by means of plants, although it can also go, except for vegetarians, via a secondary step from the bodies of farm animals.[1,2] This disparity can have serious implications for human health. For instance, iron has a great ability to concentrate in animals, especially in their livers, kidneys, and muscles, whereas this metal usually occurs in plant tissues at lower levels and in chemical form, not readily absorbed by the human gut. Zinc, like iron, occurs in animal tissues in greater amounts than in cereals. Moreover, meat does not contain phytic acid and other plant compounds responsible for inhibition of metal absorption in the digestive canal of humans. In general, meat is a poor source of trace elements, except for liver and kidney, which are enriched in some metals such as cadmium, copper, and zinc. Therefore, offal is generally a rich source of these metals. An example of accumulative abilities is the liver of sea mammals, which contains metalothionein (MT) as low-molecular-weight protein, which plays an important role in metal homeostasis and detoxification. MT exhibits a strong, competitive affinity for some trace elements (e.g., copper and zinc) and can constitute a reservoir for these metals.

Some common seafood, especially soft tissue of oysters and the whole body of squids, can contain extremely high levels of copper, zinc, and cadmium, respectively. Trace elements such as selenium, mercury, and arsenic can be biomagnified along the trophic chain in the marine ecosystems, thereby reaching higher concentrations at the top of the chain. For instance, significant bioaccumulation of methylmercury in seafood has resulted in a serious food safety problem. This organomercury compound is characterized by significantly higher toxicity as compared to inorganic species of mercury.[3] Seafood may be also contaminated by organotins, especially tributyltin (TBT). In the case of inorganic tin, few serious problems have been reported in food except when elevated levels are occasionally found in some canned products, usually acidic foods. As regards arsenic, its total concentration in fish products is relatively high, with a great percentage consisting of the nontoxic organoarsenic compound arsenobetaine. Therefore, consumers do not expose themselves to high risk by consuming significant amounts of arsenic in seafood.

Knowledge of the total level of trace elements in foodstuffs is very important for establishing dietary requirements; however, the levels of the forms of elements in foods are also needed to estimate their safety and nutritional quality. Although most regulations on metals and metalloids in food are based on their total concentrations, only a few regulations consider the specific forms of chemical elements. Some organizations, e.g., the World Health Organization (WHO) or the U.S. Food and Drug Administration (FDA), appear to pay great attention to the speciation of the elements, and it has been reported that they are just starting to recommend speciation analysis and regulate food, based on the species of mercury, arsenic, or tin.[4,5] With this requirement in mind, consideration has been given to individual species of mercury, arsenic, selenium, and tin in different groups of foods in Chapter 6 and Chapter 7 of this book. It should be stressed that there are various factors responsible for modifying metal toxicity, including interactions which occur among the metals themselves or between metals and other dietary components. Such processes can either decrease or enhance the metal toxicity.[6] For instance, selenium protects against acute toxicity of mercury, whereas zinc deficiency enhances lead toxicity, and iron deficiency favors gastrointestinal cadmium absorption. Metals such as arsenic, cadmium, chromium, nickel, and lead have been classified as posing a potential cancer risk to human beings.[6]

6.2 MACRONUTRIENTS IN FOOD

6.2.1 Calcium

The levels of calcium (Ca) in different groups of foods vary over a very wide range.[7,8] The low levels are observed for pork loin (ca. 40 mg/kg, see Table 6.1). Soft cheeses and especially hard cheeses contain up to 510 and 1120 mg Ca/100 g, respectively, and are always a major dietary source of this element. Calcium concentration in cow milk varies with typical mean value being around 120 mg/100 g (see Table 6.2). The concentrations of calcium in honey range from 0.30 to 24.8 mg/100 g wet weight (w.w.) with the highest level in thyme honey (see Table 6.3).

6.2.2 Magnesium

Magnesium (Mg) is commonly found in foods. Meat products contain this mineral from 13.0 to 42.6 mg Mg/100 g (see Table 6.1). As can be seen in Table 6.2, a similar range of concentrations is reported for dairy products (7.9 to 56.0 mg/100 g). Cow milk contains lower levels of Mg (10.9 mg/100 g) than raw goat milk (75.7 mg/100 g) and sheep milk (62.6 mg/100 g). The concentrations of magnesium in

TABLE 6.1
Macronutrients in Meat and Seafood (in mg/100 g w/w[a])

Name	Origin	n	Ca	K	Mg	Na	P	Ref.
Meat								
Pork loin	Spain	42	38.7 ± 6.55	3684 ± 300	—	405 ± 37.6	—	10
Carcass meat[b]	U.K.		9	—	—	—	—	7
Poultry[b]	U.K.		7.8	—	—	—	—	7
Poultry[b]	U.S.		22.4 ± 18.4	295 ± 42.8	26.9 ± 2.78	308 ± 283	221 ± 21.7	8
			8.0 – 54.6	246 – 350	25.2 – 30.7	59.5 – 663	197 – 254	
Meat products[b]	U.K.		35	—	—	—	—	7
Meat products[b]	U.S.		22.1 ± 19.5	318 ± 94.7	22.7 ± 5.96	620 ± 625	225 ± 57.2	8
			5.1 – 69.3	121 – 420	13.0 – 34.0	61.5 – 1950	133 – 354	
Offal[b]	U.K.		6.8	—	—	—	—	7
Offal[b]	U.S.		5	345	22.9	73.4	452	8
Morcilla de Burgos	Spain		—	—		0.07 ± 0.01	—	11
Blood sausage[c]						0.05 – 0.09		
Fish								
Fish[b]	U.K.		131	—	—	—	—	7
Fish[b]	U.S.		44.7 ± 51.8	235 ± 112	33.6 ± 6.27	286 ± 126.3	190 ± 18.2	8
			12.2 – 122	95.5 – 369	28.9 – 42.6	136 – 445	179 – 217	
Seafood								
Shrimp	Polish market		29.9 ± 14.6	40.1 ± 45.3	21.7 ± 13.7	536 ± 226	—	14
Shrimp	Polish market		12.3	30.3	14.5	667	—	14
Crab claws meat	Turkey		149	256	—	266	—	12
Crab claws meat	Turkey		151	308	—	354	—	12
Crab body meat	Turkey		64.9	244	—	327	—	12
Crab body meat	Turkey		87.6	303	—	319	—	12
Crab claws meat boiled	Polish market		379	221	72.1	255	—	14

Lobster claws meat boiled	Polish market	162	73.7	38.3	369	—	14
Mussel	Greenland	—	—	—	272 – 802	—	13
Mussel boiled	Polish market	7.13 ± 6.83	117 ± 159	19.7 ± 3.36	184 ± 197	—	14
Mussel raw	Polish market	14.0	97.5	53.2	194	—	14
Octopus raw tentacles	Polish market	6.12 ± 2.85	155 ± 117	36.3 ± 12.8	411 ± 404	—	14
Octopus raw small	Polish market	9.3	62.8	13.3	534	—	14

[a] w/w = wet weight.

[b] Total diet studies, mean of all the products from the group.

[c] Recalculated from dry to wet weight.

TABLE 6.2
Macronutrients in Milk and Dairy Products (in mg/100 g w/w)

Name	Origin	n	Ca	K	Mg	Na	P	Ref.
Milk								
Dairy milk	Sweden		114 ± 3	160 ± 4	12 ± 0	40 ± 2	90 ± 2	15
			110 – 121	160 – 170	12.0 – 12.0	37 – 45	88 – 93	
Low-fat (2%) milk, fluid	U.S.		106	153	11.1	40.4	89.2	8
Milk[b]	U.S.		104 ± 6.92	149 ± 8.14	10.8 ± 0.74	39.5 ± 2.29	86.6 ± 5.30	8
			95.8 – 109	140 – 155	10.0 – 11.4	36.9 – 41.2	80.5 – 90.1	
Milk[b]	U.K.		83.5	—	—	—	—	7
Raw cow milk[c]	Canary Islands	151	161 ± 20.1	11.1 ± 1.81	11.1 ± 1.81	51.8 ± 10.6	—	16
			116 – 217	6.94 – 15.4	6.94 – 15.4	27.0 – 84.5		
Raw milk		7	—	—	62.8 ± 9.29	—	—	17
					51.7 – 78.8			
Skimmed cow milk[c]		5	89.9 ± 2.72	—	9.22 ± 0.68	48.3 ± 3.79	—	18
Whole milk, fluid	U.S.		95.8	140	10	36.9	80.5	8
Cow whole milk[c]		5	115 ± 3.50	—	10.9 ± 0.87	49.2 ± 2.91	—	18
Sterilized cow milk[c]	Canary Islands	18	127 ± 6.02	11.8 ± 1.37	11.8 ± 1.37	53.3 ± 11.6	—	16
			120 – 135	10.7 – 14.4	10.7 – 14.4	41.8 – 74.5		
UHT cow whole milk[c]		5	90.4 ± 2.14	—	8.64 ± 0.58	49.6 ± 2.52	—	18
Skimmed UHT cow milk[c]		5	63.2 ± 1.85	—	7.77 ± 0.39	46.1 ± 4.0	—	18
Raw goat milk		2	—	—	75.5 ± 0.25	—	—	19
Raw sheep milk		4	—	—	62.6 ± 2.52	—	—	19
					58.8 – 65.3			
Human milk[c]		60	29.8 ± 1.14	—	2.23 ± 0.05	—	—	20
Human milk[c]	Italy	6	29.8 ± 17.0	—	2.23 ± 2.12	—	—	20
			9.38 – 63.9		0.27 – 5.19			
Human whole milk[c]		5	24.4 ± 1.46	—	2.33 ± 0.39	15.9 ± 1.75	—	18

Skimmed human milk[c]		5	15.8 ± 0.97	—	1.84 ± 0.29	15.3 ± 0.97	—	18
Skim milk powder		2	1250 ± 10	—	150 ± 2.0	—	—	21
			1220 – 1270		143 – 158			
Goat milk infant formula		5	457 ± 12.8	746 ± 17.0	53.6 ± 0.99	231 ± 5.16	334 ± 6.05	22
			435 – 468	720 – 769	51.7 – 54.4	221 – 235	324 – 342	
Formula milk[c]		5	110 ± 3.01	—	7.57 ± 0.49	50.1 ± 3.01	—	18
Formula milk whey[c]		5	0.06 ± 0.001	—	5.73 ± 0.19	45.3 ± 0.97	—	23
Milk based liquid first and follow-on formulas[c]	U.K.	18	64.3 ± 14.7	—	5.82 ± 1.29	32.2 ± 3.12	—	24
Milk-based follow-on partially oil-filled		5	851 ± 15.6	1051 ± 12.2	75.6 ± 1.98	274 ± 4.58	596 ± 13.7	22
			838 – 882	1035 – 1068	71.8 – 77	268 – 277	581 – 620	
Milk-based infant formula, oil-filled		5	503 ± 13.2	486 ± 17.5	46.6 ± 1.08	219 ± 5.7	361 ± 11.7	22
			485 – 522	466 – 512	45.1 – 47.8	212 – 228	352 – 372	
Skimmed formula milk[c]		5	76.7 ± 2.23	—	6.89 ± 0.39	47.6 ± 2.72	—	18
Milk-based powder formulas[c]	Nigeria	6	37.4 ± 3.38	—	2.54 ± 0.16	16.4 ± 1.66	—	24
Milk-based powder formulas[c]	U.K.	12	33.4 ± 5.15	—	4.10 ± 0.37	17.9 ± 4.33	—	24
Milk-based powder formulas[c]	U.S.	9	38.6 ± 6.90	—	3.53 ± 0.96	18.6 ± 4.46	—	24
Soy-based infant formula, oil-filled		5	377 ± 7.91	530 ± 16.2	39.5 ± 1.01	172 ± 13.5	220 ± 5.95	22
			374 – 391	512 – 564	31.5 – 40.2	158 – 197	214 – 231	
Soy-based powder formula[c]	U.S.	6	50 ± 5.01	—	3.81 ± 1.07	22.5 ± 4.53	—	24
Whey-based formula, partially oil-filled		5	599 ± 8.97	722 ± 30.5	54.4 ± 1.52	220 ± 6.18	387 ± 16.4	22
			584 – 608	674 – 764	52.6 – 56.5	212 – 231	357 – 407	
Whey-based infant formula, oil-filled		5	389 ± 20.0	544 ± 36.2	38.6 ± 22.6	203 ± 11.1	275 ± 14.8	22
			370 – 427	499 – 596	36.6 – 42.9	187 – 211	249 – 295	
Whey-based formula, partially oil-filled		5	599 ± 8.97	722 ± 30.3	54.4 ± 1.52	220 ± 6.18	387 ± 16.4	22
			584 – 608	674 – 764	52.6 – 56.5	212 – 231	357 – 407	
Cow milk whey[c]		5	0.07 ± 0.001	—	8.15 ± 0.29	41.7 ± 0.87	—	23
Milk whey coming from cow milk[c]		5	46.1 ± 3.40	—	7.09 ± 0.582	46.31 ± 3.5	—	18
Milk whey coming from UHT cow milk[c]		5	31.7 ± 2.82	—	5.05 ± 0.49	44.7 ± 3.20	—	18
UHT cow milk whey[c]		5	0.04 ± 0.001	—	5.82 ± 0.19	42.8 ± 0.87	—	23

TABLE 6.2 (CONTINUED)
Macronutrients in Milk and Dairy Products (in mg/100 g w/w)

Name	Origin	n	Ca	K	Mg	Na	P	Ref.
Human milk whey[R]		5	0.013 ± 0.001	—	2.04 ± 0.19	12.4 ± 0.39	—	23
Milk whey coming from human milk[c]		5	13.9 ± 0.87	—	1.65 ± 0.29	14.2 ± 1.07	—	18
Milk whey coming from formula milk[c]		5	34.0 ± 1.75	—	5.34 ± 0.29	45.9 ± 3.98	—	18
Dairy Products								
Commercial yoghurts[c]		16	140 ± 24.8	193 ± 30.0	12.5 ± 1.63	65.1 ± 8.44	—	25
			106 – 199	150 – 255	9.80 – 17.2	46.2 – 75.4		
Dairy products[b]	U.K.		232	—	—	—	—	7
Milk products[b]	U.S.		305 ± 289	153 ± 74.1	19.2 ± 10.2	349 ± 453	254 ± 207	8
			71.9 – 792	74.7 – 303	7.9 – 37.8	51.7 – 1460	84.3 – 567	
Plain yoghurt, low-fat	U.S.		155	214	15.7	59.7	123	8
Fruit-added yoghurts		7	98.5 ± 7.1	118 ± 15.2	9.4 ± 0.8	36.6 ± 3.5	—	26
Yoghurt, low-fat, fruit	U.S.		112	174	13	51.7	96.5	8
Mayonnaise	U.S.		7.8	10.2	< 1.4	504	24.8	8
Butter (regular, salted)	U.S.		23.4	23.7	1.7	576	22.9	8
Swiss cheese	U.S.		792	77.2	37.8	192	567	8
American cheese	U.S.		551	171	27	1460	548	8
Cheddar cheese	U.S.		688	74.7	28.4	589	458	8
Emmental cheese		6	0.96 ± 0.05	89.4 ± 6.39	36.4 ± 1.91	181 ± 51.9	—	27
			0.93 – 1.05	80 – 97.7	33.6 – 39	132 – 256		
Hard cheeses			969	48	35	876	—	28
			766 – 1120	20 – 60	30 – 45	180 – 2270		
Semi-hard cheese	Canary Islands		1020 ± 310	200 ± 80	56 ± 18	1090 ± 450	—	29
Cream cheese	U.S.		72	110	7.9	297	93.4	8
Cottage cheese, 4%	U.S.		71.9	97	8.1	341	142	8

Commercial goat cheese		2	—	—	36.5 ± 8.05	—	—	19
Soft cheeses			294 93 – 510	86 50 – 115	18 10 – 44	234 25 – 645	—	28
Commercial sheep cheese		4	—	—	32.2 ± 9.38 21.3 – 44.3	—	—	19
Quark cheeses			90 73 – 109	127 107 – 156	12 9 – 14	163 27 – 397	—	28
Fresh cheese	Canary Islands		980 ± 170	170 ± 20	47 ± 8	600 ± 280	—	29
Market white brine cheese	Turkey	10	309 ± 18	90 ± 10	18 ± 4	999 ± 225	—	30
Soft white cheese	Turkey	20	300 ± 58	79 ± 20	23 ± 6	1583 ± 650	—	30
Stirred curd cheeses			646 337 – 800	45 8 – 80	23 10 – 40	557 140 – 966	—	28
Soft white cheese melted	Turkey	20	347 ± 13	53 ± 14	17 ± 4	1431 ± 190	—	30
Eggs								
Eggs[b]	U.K.		52.8	—	—	—	—	7
Egg products[b]	U.S.		59.3 ± 4.19 54.5 – 61.8	130 ± 6.81 122 – 135	12.1 ± 0.2 11.9 – 12.3	163 ± 60.8 123 – 233	190 ± 16.0 172 – 201	8

TABLE 6.3
Macronutrients in Honey (in mg/100 g w/w)

Name	Origin	n	Ca	K	Mg	Na	P	Ref.
Acacia honey	France	150	2.29 0.30 – 10.9	—	0.87 0.14 – 11.0	—	7.35 3.21 – 39.8	31
Apiaceae honey	Northwest Morocco	7	—	31.2 ± 7.0 26.5 – 58.7	1.87 ± 0.4 0.75 – 3.48	—	—	32
Citrus honey	Northwest Morocco	10	—	8.54 ± 3.2 1.5 – 16.2	2.10 ± 0.68 0.48 – 6.48	—	—	32
Clover honey	Egypt		—	17.3 ± 1.38	24.4 ± 4.15	38.9 ± 2.44	—	33
Eucalyptus honey	Northwest Morocco	12	—	20.5 ± 7.4 3.02 – 77.7	2.60 ± 0.15 1.36 – 3.32	 —	—	32
Honey	Turkey	30	5.1 ± 4.26	29.6 ± 15.3	3.3 ± 1.43	11.8 ± 5.44	—	34
Honey	U.S.		4.5	29.2	1.3	< 3	< 4	8
Honeydew	Northwest Morocco	3	—	188 ± 8.7 210 – 239	15.5 ± 2.7 10.1 – 22.0	—	—	32
Honey	France	86	5.41 0.89 – 13.1	—	1.92 0.36 – 6.88	—	12.9 8.44 – 35.5	31
Honey	Italy		8.4 ± 3.70 4.7 – 13.2	—	5.50 ± 7.91 1.6 – 17.3	—	 —	35
Honey	U.S.		—	17.5 ± 6.41 9.1 – 23.0	—	2.15 ± 0.25 2.01 – 2.53	—	36
Indian honey	India	30	5.88 ± 1.95 3.37 – 8.46	73.6 ± 15.1 49.0 – 93.3	—	19.4 ± 6.55 9.79 – 24.7	—	37
Industrial Galician honey	Spain	20	—	61.8 ± 52.2	10.5 ± 5.1	5.42 ± 1.62	—	38
Lazio honey	Italy	84	4.77 ± 0.66	47.2 ± 3.33	3.7 ± 0.42	9.6 ± 0.57		39
Lythrum honey	Northwest Morocco	7	—	10.8 ± 3.0 3.8 – 22.4	1.57 ± 0.4 0.68 – 3.88	—	—	32

Natural Galician honey	Spain	22	—	134.5 ± 66.5	7.7 ± 4.34	11.5 ± 5.28	—	38
Orange honey	Egypt			43.0 ± 3.09	18.3 ± 0.73	33.5 ± 6.51	—	33
Pine tree honey	Slovenia	25	—	314 ± 13	—	—	—	40
Saudi Arabia honey	Saudi Arabia		—	28.4 ± 42.6	—	3.07 ± 1.92	3.86 ± 4.95	41
				0.93 – 53.8		1.5 – 7.75	1.28 – 17	
Sesame honey	Egypt		—	122 ± 28.5	8.30 ± 0.16	30.8 ± 1.63	—	33
Syrup-feed honey	Egypt		—	1266 ± 21.2	108 ± 5.37	208 ± 18.6	—	33
Thyme honey	Spain	25	18.1 ± 3.5	67.9 ± 28.7	7.7 ± 2.48	38.9 ± 6.95	5.1 ± 1.79	32
			11.0 – 24.8	26.1 – 138	3.7 – 13.9	25.6 – 50.1	2.6 – 9.6	

honey show a wide range in values, from 0.14 to 30.0 mg/100 g; the highest level observed in syrup-feed honey was 133 mg Mg/100 g (see Table 6.3).

6.2.3 Potassium

Potassium (K), like sodium, plays an important role in animal physiology and is distributed abundantly in human diets with relatively uniform concentrations in most animal foods. Pork loin is characterized by higher levels of potassium, i.e., 3680 mg/100 g and seafood contained an order of magnitude lower its concentration (244 to 308 mg/100 g) (see Table 6.1). The potassium concentrations in cow milk ranged from 6.94 to 15.4 mg/100 g (see Table 6.2) and were smaller than those in dairy products, e.g., cheeses (20 to 200 mg/100 g). Honey contained variable levels of potassium, ranging from 1.5 to 239 mg/100 g in honeydew and 150 mg/100 g in sesame honey. However, the highest levels of this macroelement occurred in syrup-feed honey, i.e., 1555 mg/100 g (see Table 6.3).

6.2.4 Sodium

Sodium chloride is widely used as a preservative and hence is found in most animal foods. Its levels are generally in the range of 25.0 and 277 mg/100 g, although higher levels have been reported in some meat products (405 mg/100 g in pork loin) and hard cheeses (2270 mg/100 g) (see Table 6.1 and Table 6.2). Honey from the U.S. contains from 2.01 to 2.53 mg Na/100 g; however, the highest levels of the macroelement were noted for thyme honey (38.9 mg/100 g) and sesame honey (37.8 mg/100 g). Syrup-feed honey was characterized by an order of magnitude greater sodium concentration of about 255 mg/100 g (see Table 6.3).

6.2.5 Phosphorus

As can be seen in Table 6.2, foods such as milk and dairy products, especially many types of cheeses such as Swiss cheese, are good sources of phosphorus (P) (up to 567 mg/100 g). Offal and eggs are also rich in this element containing 452 and up to 201 mg P/100 g, respectively (see Table 6.1 and Table 6.2). Phosphorus in foodstuff is not exclusively natural in origin as polyphosphates are common additives in many meat products, e.g., ready-sliced ham, luncheon meat, and prepared frozen poultry.[9] Concentrations of this macroelement in honey range from 1.28 to 39.8 mg/100 g (see Table 6.3).

6.3 MICRONUTRIENTS IN FOOD

6.3.1 Chromium

Chromium (Cr) is an essential trace element for humans although it should be kept in mind that Cr^{+6}, unlike Cr^{+3}, has toxic properties. Therefore, differentiation between these two species of chromium is very important for assessing food safety.[42] Available data on chromium concentrations in meat and meat products are summarized in Table 6.4. The reported levels of this metal in milk and products are usually very low and found at or below the detection limits of most analytical methods, generally < 0.0001 mg/100 g. Somewhat higher concentrations of this element (up to 0.010 and 0.020 mg/100 g) are detected in some meat products and dairy products, especially cheeses (see Table 6.5). Concentrations of Cr in honey range from 0.0001 to 0.070 mg/100 g with the highest values noted for pine tree honey (see Table 6.6).

6.3.2 Cobalt

Whether cobalt (Co) is an essential trace element remains equivocal. Adequate information on the distribution of cobalt in food is not available. As can be deduced from Table 6.4, the concentrations of this element in meat are very low (< 0.013 mg/100 g) with its somewhat higher levels reported for offal. Cobalt is a constituent of vitamin B_{12} or cobalamin, and is a good source for yeast extract.[1] Human milk contains very small amount of cobalt of < 0.00001 to 0.002 mg/100 g whereas the levels in raw goat and sheep milk range from 0.001 to 0.003 mg/100 g (see Table 6.5). Significantly higher mean level is reported for soft white cheese, i.e., 0.15 mg Co/100 g (see Table 6.5). The metal occurs in honey in wide range of levels, from < 0.005 to 0.18 mg/100 g with the lowest concentration found in acacia and Galician honey and the highest in orange and sesame honey (see Table 6.6).

6.3.3 Copper

As an essential element, copper (Cu) is widely distributed in different kinds of animal foods, occurring at levels of between 0.00001 mg/100 g (veal) and 5.37 mg/100 g fresh weight (crab meat). Among the foodstuffs, mammalian liver is exceptionally rich in copper, with concentrations of up to 8.0 mg/100 g.[1] Offal, like bovine muscle, can contain considerable amounts of this element (up to 4 mg/100 g, see Chapter 10 of this book) but somewhat less than in mammalian liver. The kidney and liver of hens contain 0.28 and 0.58 mg Cu/100 g, respectively, whereas copper levels in geese muscle meat (5.4 mg/100 g) are among the highest values reported (see Table 6.4). Crab meat also has high levels of copper, up to 5.37 mg/100 g (see Table 6.4). The lowest values are observed in human and cow milk (0.01 to 0.14 mg/100 g). Goat and sheep milk contain from 0.05 to 0.08 mg Cu/100 g. The richest dietary sources of the metal are infant milk formula and emmenthal cheese which can contain up to 0.64 and 1.31 mg Cu/100 g, respectively. Concentrations of copper in honey range from 0.003 to 0.29 mg/100 g, depending on geographical origin of the product and species of tree or plant from which it is obtained (see Table 6.6). For instance, pine tree honey contains one order of magnitude higher levels of copper (0.15 mg/100 g) than Lythrum honey (0.03 mg/100 g).

6.3.4 Iron

Iron (Fe) is an essential element that is present in all food products, and its concentration can vary significantly depending on the kind of food and its technological processing.[1] It should be emphasized that iron in animal food is considerably more bioavailable to man than that found in plant food. In meat and fish skeletal muscles 15 to 25% of iron is bound as heme iron, which is easily adsorbed and hence

TABLE 6.4
Micronutrients in Meat and Its Products (in mg/100 g w/w)

Name	Origin	n	Co	Cr	Cu	Fe	Mn	Mo	Ni	Se	Zn	Ref.
Meat												
Carcass meat[b]	U.K.		0.0004	0.02	0.14	2.1	0.01	0.001	0.004	0.01	5.1	7
Pork meat	Poland	10	—	—	0.07 ± 0.01 0.06 – 0.08	1.28 ± 0.10 1.20 – 1.35	0.02 ± 0.001 0.02 – 0.02	—	—	—	2.57 ± 0.13 2.48 – 2.66	51
Pork loin	Spain	42	—	—	—	13.3 ± 4.67	—	—	—	—	22.0 ± 5.56	10
Pork shoulder[c]	Switzerland	25	—	—	—	1.89 ± 0.49	—	—	—	—	7.84 ± 1.44	52
Pork[c]	Switzerland	25	—	—	—	0.63 ± 0.07	—	—	—	—	2.15 ± 0.30	52
Swine muscle meat	Poland	334	—	—	n = 658 0.11 ± 0.06 0.06 – 0.15	n = 658 1.28 ± 0.46 0.95 – 1.6	n = 658 0.01 ± 0.003 0.01 – 0.01	—	—	—	n = 658 2.55 ± 0.07 2.5 – 2.6	53
Beef	Galicia	56	—	—	0.0002 ± 0.000003 0.0001 – 0.0002	—	—	—	—	—	0.005 ± 0.0001 0.004 – 0.01	54
Beef shoulder[c]	Switzerland	25	—	—	—	2.10 ± 0.48	—	—	—	—	6.67 ± 1.02	52
Beef[c]	Switzerland	25	—	—	—	1.99 ± 0.49	—	—	—	—	5.31 ± 0.88	52
Bovine muscle	Egypt	54	—	—	0.30 ± 0.02 0.28 – 0.31	7.14 ± 0.71 6.63 – 7.64	0.13 ± 0.02 0.11 – 0.14	—	—	—	4.95 ± 2.02 3.48 – 6.33	55
Bovine muscle[c]	Morocco		—	—	0.11 ± 0.03 0.07 – 0.16	—	—	—	—	—	3.19 ± 0.55 2.30 – 3.99	56
Buffalo muscle	Egypt	54	—	—	0.18 ± 0.02 0.16 – 0.19	6.13 ± 0.40 5.84 – 6.41	0.09	—	—	—	5.09 ± 2.76 3.13 – 7.04	55
Cattle muscle meat	Poland	93	—	—	n = 147 0.12 ± 0.05 0.01 – 0.84	—	—	—	—	—	—	57
Beef meat	Poland	10	—	—	0.07 ± 0.001 0.07 – 0.07	3.34 ± 0.63 2.89 – 3.78	0.02 ± 0.001 0.01 – 0.01	—	—	—	3.94 ± 0.85 3.34 – 4.54	51
Veal	Galicia	438	—	—	0.0001 ± 0.000001 0.00001 – 0.0001	—	—	—	—	—	0.005 ± 0.0001 0.003 – 0.01	54

Veal[c]	Switzerland	25	—	—	—	0.41 ± 0.14	—	—	—	—	2.54 ± 0.38	52
Goat muscle	Egypt	54	—	—	0.12 ± 0.01 0.11 – 0.12	4.56 ± 0.78 4.01 – 5.11	0.08 ± 0.02 0.06 – 0.09	—	—	—	4.05 ± 0.13 3.96 – 4.14	55
Sheep muscle meat	Egypt	54	—	—	0.13 ± 0.014 0.12 – 0.14	4.30 ± 0.54 3.91 – 4.68	0.1 ± 0.03 0.08 – 0.12	—	—	—	3.89 ± 1.14 3.08 – 4.69	55
Sheep muscle meat	Poland	5	—	—	0.07 0.05 – 0.09	1.8 0.51 – 3.0	0.01 0.01 – 0.02	—	—	—	4.0 1.2 – 5.4	58
Rabbits muscle meat	Poland	30	—	—	0.06 0.03 – 0.10	1.0 0.51 – 1.8	0.01 0.003 – 0.02	—	—	—	1.7 0.89 – 3.8	58
Elk muscle	Egypt	54	—	—	0.16 ± 0.03 0.14 – 0.18	5.81 ± 0.88 5.19 – 6.44	0.14	—	—	—	5.14 ± 1.45 4.11 – 6.16	55
Cattle liver	Galicia	437	—	—	0.01 ± 0.0002 0.0003 – 0.03	—	—	—	—	—	0.005 ± 0.0001 0.002 – 0.02	54
Offal[b]	U.K.		0.006	0.01	4	6.9	0.28	0.12	0.01	0.04	4.3	7
Offals[b]	U.S.		—	—	11.1	6.16	0.37	—	ND	0.06	5.4	8
Pork liver pastes	Spain	80	0.01 ± 0.001 0.01 – 0.01	0.01 ± 0.003 0.01 – 0.01	0.05 ± 0.03 0.01 – 0.08	0.16 ± 0.01 0.14 – 0.17	0.03 ± 0.02 0.01 – 0.05	—	0.02 ± 0.002 0.02 – 0.02	—	0.05 ± 0.005 0.04 – 0.06	59
Polish sausages	Poland	140	—	—	0.04 – 0.13	0.61 – 16.2	0.01 – 0.07	—	—	—	0.83 – 4.5	51
Morcilla de Burgos blood sausage[c]	Spain		—	—	—	9.18 ± 2.89 3.78 – 17.8	—	—	—	—	—	11
Meat products[b]	U.K.		0.001	0.02	0.15	2.3	0.14	0.01	0.01	0.01	2.5	7
Meat products[b]	U.S.		—	—	0.11 ± 0.02 < 0.09 – 0.14	1.52 ± 0.73 0.7 – 2.77	< 0.2	—	< 0.02	0.03 ± 0.01 0.01 – 0.05	3.57 ± 1.92 1.42 – 7.96	8
Poultry												
Poultry[b]	U.K.		0.0003	0.02	0.07	0.71	0.02	0.004	0.004	0.015	1.6	7
Poultry[b]	U.S.		—	—	< 0.09	0.79 ± 0.32 0.47 – 1.22	0.12 ± 0.03 < 0.1 – 0.14	—	< 0.01 – 0.12	0.03 ± 0.01 0.02 – 0.04	1.26 ± 0.41 0.86 – 1.87	8
Chicken breast[c]	Switzerland	25	—	—	—	0.44 ± 0.06	—	—	—	—	0.80 ± 0.06	52

TABLE 6.4 (CONTINUED)
Micronutrients in Meat and Its Products (in mg/100 g w/w)

Name	Origin	n	Co	Cr	Cu	Fe	Mn	Mo	Ni	Se	Zn	Ref.
Chicken thigh[c]	Switzerland	25	—	—	—	0.88 ± 0.09	—	—	—	—	2.20 ± 0.18	52
Hen's muscle meat	Poland	1	—	—	0.05	1.1	0.01	—	—	—	0.57	58
Hen's muscle meat	Poland	10	—	—	0.07 ± 0.01 0.06 – 0.09	1.02 ± 0.26 0.72 – 1.41	0.05 ± 0.004 0.04 – 0.05	—	—	—	1.55 ± 0.11 0.44 – 1.70	51
Chicken breasts	Poland	10	—	—	0.05 ± 0.01 0.04 – 0.06	0.44 ± 0.05 0.37 – 0.50	0.03 ± 0.005 0.03 – 0.04	—	—	—	0.58 ± 0.04 0.55 – 0.64	51
Chicken thighs	Poland	10	—	—	0.08 ± 0.02 0.06 – 0.10	0.77 ± 0.10 0.69 – 0.93	0.04 ± 0.01 0.04 – 0.05	—	—	—	1.41 ± 0.21 0.04 – 1.59	51
Smoked chicken	Poland	5	—	—	0.05 ± 0.01 0.04 – 0.05	0.70 ± 0.16 0.55 – 0.96	0.04 ± 0.01 0.04 – 0.05	—	—	—	0.95 ± 0.21 0.68 – 1.20	51
Smoked chicken muscles	Poland	4	—	—	0.03 ± 0.01	1.0 ± 0.5	0.02 ± 0.01	—	—	—	1.0 ± 0.4	60
Geese muscle meat	Poland	32	—	—	0.73 0.15 – 5.4	3.5 1.5 – 5.6	0.03 0.02 – 0.04	—	—	—	1.4 0.6 – 2.5	58
Turkey's breasts	Poland	10	—	—	0.05 ± 0.02 0.03 – 0.09	0.50 ± 0.09 0.41 – 0.60	0.04 ± 0.01 0.03 – 0.05	—	—	—	1.02 ± 0.22 0.76 – 1.29	51
Turkey's thighs	Poland	10	—	—	0.09 ± 0.01 0.07 – 0.10	1.24 ± 0.37	0.04 ± 0.01 0.03 – 0.05	—	—	—	2.76 ± 0.23 0.53 – 3.1	51
Ducks muscle meat	Poland	57	—	—	0.6 0.22 – 1.7	3.0 1.6 – 8.1	0.02 0.01 – 0.04	—	—	—	1.3 0.56 – 5.2	58
Duck meat	Polska	10	—	—	0.31 ± 0.03 0.28 – 0.35	2.81 ± 0.27 0.46 – 3.10	0.03 ± 0.003 0.03 – 0.04	—	—	—	1.82 ± 0.22 0.61 – 2.13	51
Hen kidney	Poland	1	—	—	0.28	18	0.19	—	—	—	2.1	58
Hen liver	Poland	13	—	—	0.38 0.17 – 0.58	7.1 1.8 – 33.0	0.16 0.10 – 0.23	—	—	—	2.9 1.8 – 5.9	58

Poultry Polish sausages	Poland	25	—	—	0.03 – 0.10	0.37 – 1.46	0.04 – 0.08	—	—	—	0.48 – 2.54	51
Poultry sausage	Poland	11	—	—	0.1 ± 0.09	1.6 ± 0.1	0.02 ± 0.001	—	—	—	1.7 ± 0.2	60
Fish												
Fish[b]	U.K.		0.001	0.02	0.11	1.6	0.11	0.003	0.008	0.039	0.91	7
Fish[b]	U.S.		—	—	< 0.1 – 0.23	1.09 ± 0.30 < 0.3 – 1.43	< 0.1 – 0.21	—	< 0.01 – 0.01	0.04 ± 0.02 0.02 – 0.07	0.79 ± 0.56 0.43 – 1.62	8
Atlantic salmon	Southern Baltic	2	0.001	—	0.01 – 0.06	0.3 – 0.49	0.01 – 0.03	—	0.01	—	1.12 – 1.9	96
Cod	Southern Baltic	70	< 0.0005 – 0.001	—	0.02 ± 0.005 0.001 – 0.11	0.4 ± 0.06 0.11 – 1.29	0.02 ± 0.005 0.005 – 0.08	—	0.03 ± 0.01 0.002 – 0.10	—	0.3 ± 0.05 0.09 – 0.88	96
Eel	Southern Baltic	2	0.001	—	0.20 – 0.29	0.45	0.03	—	0.02	—	1.36 – 1.7	96
Flounder	Southern Baltic	2	0.001 0.0005 – 0.001	—	0.02 0.02 – 0.03	0.68 0.43 – 0.92	0.02 0.01 – .0.3	—	0.01 0.002 – 0.02	—	0.4 0.25 – 0.55	96
Garfish	Southern Baltic	4	0.001 < 0.0005 – 0.002	—	0.02 0.02 – 0.03	0.69 0.52 – 0.86	0.01 0.01 – 0.02	—	0.07 0.04 – 0.1	—	1.62 1.18 – 2.27	96
Greater sand eel	Southern Baltic	3	< 0.0003 – 0.001	—	0.01 0.003 – 0.03	4.5 2.2 – 5.8	0.35 0.33 – 0.37	—	0.01 0.01 – 0.02	—	1.84 1.7 – 1.93	96
Herring	Southern Baltic	34	0.001 < 0.0005 – 0.002	—	0.04 ± 0.01 0.01 – 0.07	1.02 0.67 – 1.53	0.03 0.01 – 0.07	—	0.04 0.004 – 0.11	—	0.85 0.39 – 1.41	96
Perch	Southern Baltic		—	—	0.01 – 0.04	—	—	—	—	—	0.37 – 0.76	96
Sprat	Southern Baltic	5	0.001 ± 0.0001 < 0.001 – 0.001	—	0.04 ± 0.01 0.015 – 0.05	1.35 0.95 – 1.9	0.04 ± 0.005 0.02 – 0.05	—	0.03 ± 0.01 0.01 – 0.07	—	1.26 ± 0.06 1.14 – 1.38	96
Whiting	Southern Baltic	2	< 0.0005	—	0.01 – 0.02	0.32 – 0.38	0.03	—	0.04	—	0.13 – 0.21	96
Seafood												
Shrimp	Taiwan		—	—	0.24 – 0.30	—	—	—	—	—	1.02 – 1.25	61
Shrimp	Spain		—	—	—	—	—	—	—	0.03	—	62
Shrimp	Spain		—	—	0.67 – 1.10	0.22 – 0.51	0.21 – 0.39	—	—	—	—	63

TABLE 6.4 (CONTINUED)
Micronutrients in Meat and Its Products (in mg/100 g w/w)

Name	Origin	n	Co	Cr	Cu	Fe	Mn	Mo	Ni	Se	Zn	Ref.
Shrimp	Spain		—	—	1.92 – 2.72	0.90 – 3.10	0.15 – 0.60	—	—	—	—	63
Shrimp	Mexico		—	—	0.25 – 0.59	0.51 – 6.36	0.03 – 0.21	—	—	—	0.13 – 1.12	64
Shrimp	Mexico		—	—	0.29	0.56	0.05	—	—	—	0.84	65
Shrimp	Mexico		—	—	0.28	0.68	0.04	—	—	—	0.68	65
Shrimp	Mexico		—	—	0.29	0.66	0.04	—	—	—	0.90	65
Shrimp	Mexico		—	—	0.28	0.80	0.05	—	—	—	0.60	65
Shrimp	Greenland		—	—	—	—	—	—	—	0.15	—	66
Shrimp boiled	Polish market		—	—	0.16 ± 0.15	0.50 ± 0.77	0.03 ± 0.02	—	—	0.02 ± 0.01	0.78 ± 0.24	14
Shrimp raw	Polish market		—	—	0.15	0.06	0.01	—	—	0.02	0.90	14
Crab	Taiwan		—	—	0.32 – 0.90	—	—	—	—	—	0.41 – 4.02	61
Crab	Spain		—	—	—	—	—	—	—	0.08	—	62
Crab	Spain		—	—	2.98 – 5.37	3.75 – 42.1	1.46 – 8.19	—	—	—	—	67
Crab claw meat	Turkey		—	—	2.53	1.04	0.39	—	—	—	6.99	12
Crab claw meat	Turkey		—	—	2.08	0.45	0.06	—	—	—	4.68	12
Crab body meat	Turkey		—	—	3.13	1.13	0.37	—	—	—	4.7	12
Crab body meat	Turkey		—	—	1.49	0.68	0.16	—	—	—	3.72	12
Crab claws	Australia		—	—	1.8	—	—	—	—	—	9.9	67
Crab	Texas		—	—	1.33 – 4.65	—	—	—	—	—	0.30 – 5.22	67
Crab meat from claws boiled	Polish market		—	—	0.74	1.28	0.44	—	—	0.09	9.71	14
Lobster	Mexico		—	—	0.84 – 1.73	0.38 – 2.04	0.01 – 0.04	—	—	—	2.04 – 3.74	68

Lobster	Mexico	—	—	0.91 – 2.06	0.72 – 1.66	0.04 – 0.07	—	—	—	2.16 – 4.46	68
Lobster meat from claws, boiled	Polish market	—	—	1.84	0.26	0.26	—	—	0.07	4.27	14
Clam	Taiwan	—	—	0.02 – 0.48	—	—	—	—	—	0.38 – 1.11	61
Clam	Spain	—	—	—	—	0.01	—	—	—	—	69
Clam	Spain	—	—	0.29 – 0.44	15.9 – 45.8	0.82 – 1.76	—	—	—	—	63
Mussel	Taiwan	—	—	0.20 – 0.48	—	—	—	—	0.07 – 0.13	—	61
Mussel	Malaysia	—	—	0.16 – 0.40	—	—	—	—	—	1.50 – 2.58	70
Mussel	Spain	—	—	—	—	0.05 – 0.32	—	—	—	—	69
Mussel	Spain	—	—	—	—	—	—	—	0.04	—	62
Mussel	Greenland	—	—	0.15 – 0.21	2.64 – 6.72	—	—	—	0.06 – 0.11	1.33 – 2.34	13
Mussel	Hong Kong	—	—	0.12 – 0.16	—	—	—	—	—	1.80 – 2.70	71
Mussel	Greenland	—	—	—	—	—	—	—	0.09	—	66
Mussel	Texas	—	—	0.11 – 0.32	10.7 – 19.7	0.14 – 0.44	—	—	—	0.90 – 1.22	72
Mussel boiled	Polish market	—	—	0.12 ± 0.07	3.54 ± 3.10	2.49 ± 2.13	—	—	0.04 ± 0.01	2.99 ± 1.93	14
Mussel raw	Polish market	—	—	0.08	9.47	0.23	—	—	0.06	1.80	14
Octopus	Polish market	—	—	0.24 ± 0.00	0.091 ± 0.001	0.02 ± 0.01	—	—	0.015 ± 0.0	1.44 ± 0.36	14
Octopus	Polish market	—	—	0.17	0.40	0.06	—	—	0.026	1.19	14

TABLE 6.5
Micronutrients in Milk, Dairy Products, and Eggs (in mg/100 g w/w)

Name	Origin	n	Co	Cr	Cu	Fe	Mn	Mo	Ni	Se	Zn	Ref.
Milk												
Bottled cow milk[c]		3	—	ND	—	—	0.004 ± 0.0002	0.003 ± 0.001	—	—	—	73
Cow milk fat content 3.8%[c]			—	—	0.004 ± 0.0001	—	—	—	—	—	0.31 ± 0.0001	74
Cow milk		22	—	—	—	—	—	—	—	0.002 ± 0.0005 0.001 – 0.003	—	75
Dairy milk	Sweden		—	< 0.0004	0.01 ± 0.01 < 0.01 – 0.3	0.04 ± 0.01 0.01 – 0.06	<0.01	0.003 ± 0.0002 0.003 – 0.003	—	0.002 ± 0.0003 0.001 – 0.002	0.44 ± 0.06 0.38 – 0.54	15
Low-fat (2%) milk, fluid	U.S.		—	—	—	—	—	—	—	< 0.004	0.4	8
Low-fat milk[c]	Brazil	3	—	—	—	0.09 ± 0.002	—	—	—	0.005 ± 0.001	—	76
Milk[b]	U.S.		—	—	ND	<0.3	ND	—	ND	< 0.004	0.38 ± 0.02 0.36 – 0.4	8
Milk[b]	U.K.		0.0002	0.03	0.005	0.41	0.003	0.003	< 0.002	0.002	0.35	7
High-fat milk	Poland	19	—	—	0.02 ± 0.01 0.001 – 0.06	0.07 ± 0.06 0.02 – 0.24	0.005 ± 0.002 0.002 – 0.01	—	—	—	0.41 ± 0.11 0.17 – 0.56	77
Milk	Poland	8	—	—	0.05 ± 0.02 0.03 – 0.08	0.61 ± 0.43 0.31 – 1.6	0.004 ± 0.005 0.003 – 0.004	—	—	—	2.8 ± 0.4 2.2 – 2.9	77
Nonfat milk[c]		10	—	—	—	0.03 ± 0.01 0.01 – 0.04	—	—	—	—	0.39 ± 0.07 0.26 – 0.48	78
Raw cow milk[c]	Canary Islands Tenerife	151	—	—	0.01 ± 0.003 0.003 – 0.02	0.05 ± 0.02 0.02 – 0.10	—	—	—	0.002 ± 0.0004 0.001 – 0.003	0.43 ± 0.07 0.22 – 0.64	16
Skimmed bottled cow milk[c]		3	—	ND	—	—	0.005 ± 0.0004	0.005 ± 0.0005	—	—	—	73

Standard milk fat content 6%[c]			—	—	0.01 ± 0.0002	—	—	—	—	—	0.35 ± 0.0001	74
Whole milk >6% fat content[c]			—	—	0.01 ± 0.0002	—	—	—	—	—	0.41 ± 0.0001	74
Whole raw bovine milk	Calabria, Italy	40	—	0.0002 < 0.0001 – 0.01	0.0002 < 0.00001 – 0.07	—	—	—	—	0.001 < 0.0001 – 0.01	0.20 < 0.003 – 0.50	79
Raw milk		7	0.004 ± 0.002 0.001 – 0.01	0.03 ± 0.003 0.02 – 0.03	0.41 ± 0.41 0.09 – 1.21	0.73 ± 0.41 0.22 – 1.42	0.02 ± 0.01 0.001 – 0.02	—	0.05 ± 0.02 0.02 – 0.07	—	3.09 ± 0.59 2.16 – 4.15	17
Buffalo milk fat content 7.5%[c]			—	—	0.01 ± 0.0001	—	—	—	—	—	0.36 ± 0.0001	74
Skimmed cow milk[c]		5	—	—	0.005 ± 0.0003	0.02 ± 0.001	0.005 ± 0.0004	—	—	0.001 ± 0.0001	0.24 ± 0.02	18
Skimmed milk[c]	U.S., Mexico, Brazil, Uruguay	8	—	—	—	0.08 ± 0.02 0.06 – 0.11	—	—	—	0.005 ± 0.002 0.002 – 0.01	—	76
Whole milk, fluid	U.S.		—	—	—	< 0.3	—	—	—	< 0.004	0.36	8
Whole milk[c]	Brazil, Uruguay, Chile	4	—	—	—	0.07 ± 0.004	—	—	—	0.01 ± 0.002 0.005 – 0.01	—	76
Whole milk[c]		10	—	—	—	0.03 ± 0.01 0.02 – 0.04	—	—	—	—	0.52 ± 0.12 0.42 – 0.78	78
Cow whole milk[c]		5	—	0.0001 ± 0.00004	0.01 ± 0.0005	0.02 ± 0.002	0.005 ± 0.0004	—	0.001 ± 0.0001	0.001 ± 0.0001	0.37 ± 0.03	18
Sterilized cow milk[c]	Canary Islands	18	—	—	0.01 ± 0.004 0.01 – 0.02	0.02 ± 0.002 0.01 – 0.02	—	—	—	0.001 ± 0.0003 0.001 – 0.002	0.30 ± 0.01 0.27 – 0.31	16
UHT cow whole milk[c]		5	—	0.0001 ± 0.00003	0.002 ± 0.0003	0.02 ± 0.001	0.002 ± 0.0001	—	0.001 ± 0.0002	0.001 ± 0.0002	0.30 ± 0.02	18
Skimmed UHT cow milk[c]		5	—	0.0001 ± 0.00002	0.002 ± 0.0002	0.02 ± 0.001	0.002 ± 0.0001	—	—	0.001 ± 0.0001	0.22 ± 0.02	18
Nonfat milk powder		3	—	ND	—	—	0.03 ± 0.01	0.03 ± 0.003	—	—	—	73

TABLE 6.5 (CONTINUED)
Micronutrients in Milk, Dairy Products, and Eggs (in mg/100 g w/w)

Name	Origin	n	Co	Cr	Cu	Fe	Mn	Mo	Ni	Se	Zn	Ref.
Powdered cow milk		3	—	0.004 ± 0.0005	—	—	0.04 ± 0.002	0.03 ± 0.01	—	—	—	73
Goat's milk		10	—	—	—	—	—	—	—	0.002 ± 0.0004 0.001 – 0.003	—	75
Raw goat milk		2	0.002 ± 0.0	0.003 ± 0.0004	0.06 ± 0.001	0.27 ± 0.03	0.01 ± 0.002	—	0.002 ± 0.0004	—	1.81 ± 0.13	19
Raw sheep milk		4	0.002 ± 0.001 0.001 – 0.003	0.002 ± 0.0004 0.001 – 0.003	0.06 ± 0.01 0.05 – 0.08	0.38 ± 0.06 0.27 – 0.44	0.03 ± 0.003 0.03 – 0.03	—	0.01 ± 0.001 0.004 – 0.01	—	2.40 ± 0.21 2.12 – 2.6	19
Human milk			< 0.00001 – 0.0002	< 0.0001 – 0.02	0.01 – 0.14	0.01 – 0.16	0.0002 – 0.003	—	< 0.00001 – 0.001	—	—	80
Human milk[c]			—	—	0.02 ± 0.0001	—	—	—	—	—	0.17 ± 0.0001	74
Human milk[c]		58	—	—	—	—	—	—	—	0.002 ± 0.0004 0.001 – 0.002	—	81
Human milk[c]		3	—	ND	—	—	0.01 ± 0.001	ND	—	—	—	73
Human milk[c]		60	—	—	0.04 ± 0.003	0.06 ± 0.004	0.003 ± 0.0002	—	—	—	0.26 ± 0.01	20
Human milk[c]	Italy	6	—	—	0.03 ± 0.02	0.06 ± 0.05 0.02 – 0.18	0.003 ± 0.002 0.001 – 0.01	—	—	—	0.26 ± 0.30 0.02 – 0.88	20
Human whole milk[c]		5	—	0.0001 ± 0.00003	0.03 ± 0.002	0.04 ± 0.003	0.001 ± 0.0001	—	0.001 ± 0.0002	0.001 ± 0.0002	0.37 ± 0.02	18
Skimmed human milk[c]		5	—	—	0.02 ± 0.001	0.02 ± 0.002	0.001 ± 0.0001	—	—	0.001 ± 0.0001	0.20 ± 0.01	18
Vegetal-based milk	Vizcaya, Spain	2	—	—	—	—	—	—	—	0.001 ± 0.00003 0.001 – 0.001	—	82
Powder milk	Poland	9	—	—	0.05 ± 0.04 0.03 – 0.13	0.53 ± 0.6 0.29 – 1.5	0.04 ± 0.01 0.03 – 0.05	—	—	—	2.7 ± 0.1 0.78 – 4.3	77
Skim milk powder		2	—	—	0.02 ± 0.001 0.02 – 0.03	0.23 ± 0.02 0.21 – 0.24	0.05 ± 0.002 0.05 – 0.05	—	—	—	4.71 ± 0.04 4.47 – 4.88	21

Whole milk powder		3	—	0.01 ± 0.002	—	—	0.02 ± 0.002	0.03 ± 0.003	—	—	—	73
Goat milk infant formula		5	—	—	0.40 ± 0.02 0.37 – 0.94	6.12 ± 0.11 5.96 – 6.27	0.09 ± 0.004 0.08 – 0.09	—	—	—	3.49 ± 0.07 3.37 – 3.53	22
Formula milk[c]		5	—	0.001 ± 0.0002	0.04 ± 0.001	0.11 ± 0.01	0.02 ± 0.001	—	0.002 ± 0.0002	0.002 ± 0.0002	0.41 ± 0.01	18
Formula milk whey[c]		5	—	0.0001 ± 0.00001	0.005 ± 0.0004	0.03 ± 0.002	0.01 ± 0.0001	—	0.0003 ± 0.00002	0.001 ± 0.00005	0.18 ± 0.01	23
Infant formula		3	—	0.004 ± 0.0004	—	—	0.13 ± 0.02	0.02 ± 0.003	—	—	—	73
Milk-based liquid first and follow-on formulas[c]	London, U.K.	18	—	0.001 ± 0.0004	0.05 ± 0.02	1.10 ± 0.22	0.01 ± 0.003	0.003 ± 0.001	0.00002 ± 0.0001	—	0.54 ± 0.12	24
Milk-based follow-on partially oil-filled		5	—	—	0.44 ± 0.03 0.41 – 0.48	7.67 ± 0.2 7.33 – 7.89	0.11 ± 0.01 0.11 – 0.12	—	—	—	3.98 ± 0.12 3.89 – 4.20	22
Milk-based infant formula, oil-filled		5	—	—	0.35 ± 0.11 0.33 – 0.36	6.41 ± 0.11 6.27 – 6.53	0.08 ± 0.002 0.07 – 0.08	—	—	—	3.69 ± 0.05 3.62 – 3.76	22
Milk-infant formula adapted		3	—	—	0.34 ± 0.06 0.27 – 0.41	6.9 ± 1.23 5.28 – 8.26	—	—	—	—	3.24 ± 0.33 2.8 – 3.58	83
Milk-infant formula follow-up		3		—	0.24 ± 0.08 0.13 – 0.3	7.91 ± 0.19 7.67 ± 8.13	—	—	—	—	3.71 ± 0.28 3.42 – 4.09	83
Milk-infant formula preterm		3	—	—	0.50 ± 0.09 0.38 – 0.59	8.19 ± 1.63 6.35 – 10.3	—	—	—	—	3.29 ± 0.34 2.93 – 3.75	83
Ready-to-drink infant formulas			0.0001 – 0.0001	< 0.0001 – 0.002	0.09 – 0.26	0.14 – 1.25	0.003 – 0.01	—	0.001 – 0.002	—	—	80

TABLE 6.5 (CONTINUED)
Micronutrients in Milk, Dairy Products, and Eggs (in mg/100 g w/w)

Name	Origin	n	Co	Cr	Cu	Fe	Mn	Mo	Ni	Se	Zn	Ref.
Milk-infant formula without lactose		3	—	—	0.37 ± 0.06 0.31 – 0.46	6.88 ± 0.89 5.62 – 7.55	—	—	—	—	3.37 ± 0.36 3.07 – 3.87	83
Milk-infant formula hydrolysate		3	—	—	0.43 ± 0.07 0.36 – 0.53	7.77 ± 0.19 7.59 – 8.04	—	—	—	—	3.75 ± 0.31 3.32 – 4.02	83
Skimmed formula milk[c]		5	—	0.001 ± 0.00004	0.03 ± 0.001	0.08 ± 0.01	0.01 ± 0.001	—	—	—	0.27 ± 0.01	18
Milk-based powder formulas[c]	Lagos, Nigeria,	6	—	0.001 ± 0.0003	0.04 ± 0.01	0.82 ± 0.12	0.01 ± 0.002	0.002 ± 0.0005	—	—	0.34 ± 0.02	24
Milk-based powder formulas[c]	London, U.K.	12	—	0.0005 ± 0.0005	0.04 ± 0.01	0.61 ± 0.23	0.01 ± 0.002	0.002 ± 0.001	—	—	0.31 ± 0.09	24
Milk-based powder formulas[c]	U.S.	9	—	0.001 ± 0.001	0.05 ± 0.01	0.90 ± 0.05	0.01 ± 0.004	0.001 ± 0.0003	0.000002 ± 0.00001	—	0.36 ± 0.08	24
Milk-infant formula soya-based		3	—	—	0.62 ± 0.022 0.59 – 0.64	8.65 ± 2.28 5.93 – 11.5	—	—	—	—	4.18 ± 0.25 3.94 – 4.53	83
Soy-based infant formula, oil-filled		5	—	—	0.21 ± 0.02 0.18 – 0.25	2.53 ± 0.14 2.40 – 2.79	0.40 ± 0.01 0.38 – 0.41	—	—	—	1.46 ± 0.08 1.37 – 1.56	22
Soy-based powder formula[c]	U.S.	6	—	0.001 ± 0.001	0.07 ± 0.011	0.89 ± 0.03	0.02 ± 0.004	0.003 ± 0.001	0.0002 ± 0.0004	—	0.50 ± 0.11	24
Powdered infant formula		5	—	—	—	—	—	—	—	0.01 ± 0.003 0.002 – 0.01	—	81

Whey-based formula, partially oil-filled	5	—	—	0.47 ± 0.02 0.45 – 0.49	7.83 ± 0.22 7.57 – 8.09	0.11 ± 0.06 0.10 – 0.11	—	—	—	4.3 ± 0.12 4.14 – 4.49	22
Whey-based infant formula, oil-filled	5	—	—	0.39 ± 0.01 0.38 – 0.41	6.81 ± 0.13 6.62 – 7.02	0.08 ± 0.003 0.08 – 0.09	—	—	—	3.92 ± 0.13 3.76 – 4.15	22
Whey-based formula, partially oil-filled	5	—	—	0.47 ± 0.02 0.45 – 0.49	7.83 ± 0.22 7.57 – 8.09	0.11 ± 0.06 0.10 – 0.11	—	—	—	4.3 ± 0.12 4.14 – 4.49	22
Cow milk whey[c]	5	—	—	0.001 ± 0.0001	0.01 ± 0.0003	0.002 ± 0.0001	—	—	0.0004 ± 0.00002	0.13 ± 0.01	23
Milk whey coming from cow milk[c]	5	—	—	0.001 ± 0.0002	0.01 ± 0.001	0.002 ± 0.0003	—	—	—	0.10 ± 0.01	18
Milk whey coming from UHT cow milk[c]	5	—	—	0.001 ± 0.0001	0.01 ± 0.001	0.001 ± 0.0001	—	—	—	0.09 ± 0.01	18
UHT cow milk whey[c]	5	—	—	0.001 ± 0.00005	0.004 ± 0.0002	0.0005 ± 0.00004	—	—	0.0004 ± 0.00002	0.08 ± 0.01	23
Human milk whey[c]	5	—	—	0.01 ± 0.001	0.02 ± 0.001	0.001 ± 0.00004	—	—	0.001 ± 0.0005	0.12 ± 0.01	23
Milk whey coming from human milk[c]	5	—	—	0.02 ± 0.001	0.019 ± 0.0016	0.001 ± 0.0001	—	—	—	0.17 ± 0.02	18
Whey milk[c]	10	—	—	—	0.01 ± 0.004 0.001 – 0.01	—	—	—	—	0.04 ± 0.01 0.03 – 0.08	78
Milk whey coming from formula milk[c]	5	—	0.0001 ± 0.00003	0.01 ± 0.0004	0.03 ± 0.002	0.005 ± 0.0004	—	—	—	0.16 ± 0.02	18

TABLE 6.5 (CONTINUED)
Micronutrients in Milk, Dairy Products, and Eggs (in mg/100 g w/w)

Name	Origin	n	Co	Cr	Cu	Fe	Mn	Mo	Ni	Se	Zn	Ref.
Dairy Products												
Commercial yoghurts[c]		16	—	—	—	—	—	—	—	—	0.5 ± 0.08 0.39 – 0.71	25
Dairy products[b]	U.K.		0.0004	0.09	0.05	1.2	0.03	0.01	0.002	0.005	1.4	7
Milk products[b]	U.S.		—	—	< 0.09 – < 0.12	< 0.3	< 0.1 – < 0.2	—	< 0.01	0.02 ± 0.01 < 0.004 – 0.02	1.53 ± 1.59 0.38 – 4.36	8
Plain yoghurt, lowfat	U.S.		—	—	< 0.09	< 0.3	—	—	—	< 0.004	0.56	8
Fruit-added yoghurts		7	—	—	0.02 ± 0.01	0.14 ± 0.14	0.07 ± 0.06	—	—	—	0.31 ± 0.03	26
Yoghurt, lowfat, fruit	U.S.		—	—	—	< 0.3	< 0.1	—	< 0.01	< 0.004	0.44	8
Mayonnaise	U.S.		—	—	—	< 0.3	—	—	—	< 0.005	< 0.2	8
Butter	Poland	3	—	—	0.01 ± 0.003 0.004 – 0.01	0.11 ± 0.07 0.04 – 0.17	0.01 ± 0.002 0.004 – 0.01	—	—	—	0.09 ± 0.02 0.06 – 0.11	77
Butter (regular, salted)	U.S.		—	—	—	< 0.3	—	—	—	—	< 0.1	8
Swiss cheese	U.S.		—	—	< 0.12	< 0.3	< 0.2	—	—	0.02	4.36	8
American cheese	U.S.		—	—	< 0.12	< 0.3	< 0.2	—	—	0.02	2.8	8
Cheddar cheese	U.S.		—	—	< 0.12	< 0.3	< 0.2	—	—	0.02	3.55	8
Emmenthal cheese		6	—	—	0.73 ± 0.39 < 0.1 – 1.31	—	0.03 ± 0.01 0.02 – 0.04	0.01 ± 0.003 0.001 – 0.02	—	—	4.44 ± 0.24 4.13 – 4.78	27
Hard cheeses			0.0003 0.0003 – 0.0005	0.005 0.001 – 0.01	—	0.17 0.13 – 0.21	—	—	—	0.01 0.005 – 0.01	3.88 3.37 – 4.50	46

Swiss Cheese	Poland	13	—	—	0.03 ± 0.01 0.01 – 0.04	0.55 ± 0.51 0.16 – 2.1	0.07 ± 0.02 0.04 – 0.12	—	—	—	1.6 ± 1.3 0.44 – 5.3	77
Cream cheese	U.S.		—	—	—	< 0.3	—	—	—	< 0.005	0.51	8
Cream cheese	Poland	4	—	—	0.02 ± 0.005 0.02 – 0.03	0.43 ± 0.12 0.25 – 0.51	0.07 ± 0.04 0.03 – 0.12	—	—	—	2.0 ± 0.2 1.7 – 2.2	77
Cottage cheese, 4%	U.S.		—	—	—	< 0.3	—	—	< 0.01	0.01	0.38	8
Commercial goat cheese		2	0.001 ± 0.00005	0.01 ± 0.0005	0.11 ± 0.03	0.24 ± 0.04	0.02 ± 0.0001	—	0.001 ± 0.0004	—	1.81 ± 0.11	19
Soft cheeses			0.0002 0.0002 – 0.0003	0.01 0.004 – 0.03	—	0.15 0.04 – 0.24	—	—	—	0.004 0.001 – 0.01	1.06 0.36 – 2.81	46
Commercial sheep cheese		4	0.002 ± 0.0003 0.002 – 0.002	0.004 ± 0.001 0.003 – 0.005	0.10 ± 0.04 0.05 – 0.16	0.33 ± 0.13 0.23 – 0.56	0.04 ± 0.01 0.03 – 0.05	—	0.002 ± 0.0004 0.001 – 0.002	—	1.94 ± 0.33 1.55 – 2.29	19
Quark cheeses			0.0005 0.0003 – 0.001	0.003 0.001 – 0.01	—	0.13 0.08 – 0.21	—	—	—	0.003 0.002 – 0.005	4.41 3.37 – 5.18	46
Semihard cheese	Canary Islands		—	—	0.09 ± 0.03	0.21 ± 0.05	—	—	—	0.02 ± 0.005	0.47 ± 0.18	29
Cream partially skimmed	Spain	6	—	—	—	—	—	—	—	0.002 ± 0.0005 0.001 – 0.003	—	84
Cream partially skimmed	Spain	6	—	—	—	—	—	—	—	0.002 ± 0.0005 0.001 – 0.003	—	82
Cream skimmed	Spain	8	—	—	—	—	—	—	—	0.002 ± 0.0005 0.001 – 0.002	—	84
Cream skimmed	Spain	6	—	—	—	—	—	—	—	0.002 ± 0.0005 0.001 – 0.002	—	82
Full cream	Spain	4	—	—	—	—	—	—	—	0.001 ± 0.0002 0.001 – 0.002	—	84
Full cream	Spain	4	—	—	—	—	—	—	—	0.001 ± 0.0002 0.001 – 0.002	—	82
Fresh cheese	Canary Islands		—	—	0.08 ± 0.03	0.22 ± 0.05	—	—	—	0.01 ± 0.0002	0.65 ± 0.25	29
Market white brine cheese	Turkey	10	0.02 ± 0.005	0.01 ± 0.0003	0.04 ± 0.01	0.20 ± 0.03	0.005 ± 0.003	0.03 ± 0.01	0.11 ± 0.01	—	1.26 ± 0.08	30

TABLE 6.5 (CONTINUED)
Micronutrients in Milk, Dairy Products, and Eggs (in mg/100 g w/w)

Name	Origin	n	Co	Cr	Cu	Fe	Mn	Mo	Ni	Se	Zn	Ref.
Soft white cheese	Turkey	20	0.15 ± 0.16	0.02 ± 0.02	0.05 ± 0.04	0.54 ± 0.85	0.01 ± 0.004	0.03 ± 0.03	0.12 ± 0.16	—	1.77 ± 0.42	30
Stirred curd cheeses			0.0001 0.0002 – 0.001	0.01 0.01 – 0.01	—	0.17 0.10 – 0.21	—	—	—	0.01 0.01 – 0.01	3.47	46
Soft cheeses			0.0002 0.0002 – 0.0003	0.01 0.004 – 0.03	—	0.15 0.04 – 0.24	—	—	—	0.004 0.001 – 0.01	1.06 0.36 – 2.81	46
Commercial sheep cheese		4	0.002 ± 0.0003 0.002 – 0.002	0.004 ± 0.001 0.003 – 0.005	0.10 ± 0.04 0.05 – 0.16	0.33 ± 0.13 0.23 – 0.56	0.04 ± 0.01 0.03 – 0.05	—	0.002 ± 0.0004 0.001 – 0.002	—	1.94 ± 0.33 1.55 – 2.29	19
Quark cheeses			0.0005 0.0003 – 0.001	0.003 0.001 – 0.01	—	0.13 0.08 – 0.21	—	—	—	0.003 0.002 – 0.005	4.41 3.37 – 5.18	46
Semi-hard cheese	Canary Islands		—	—	0.09 ± 0.03	0.21 ± 0.05	—	—	—	0.02 ± 0.005	0.47 ± 0.18	29
Creampartially skimmed	Spain	6	—	—	—	—	—	—	—	0.002 ± 0.0005 0.001 – 0.003	—	84
Creampartially skimmed	Spain	6	—	—	—	—	—	—	—	0.002 ± 0.0005 0.001 – 0.003	—	82
Cream skimmed	Spain	8	—	—	—	—	—	—	—	0.002 ± 0.0005 0.001 – 0.002	—	84
Cream skimmed	Spain	6	—	—	—	—	—	—	—	0.002 ± 0.0005 0.001 – 0.002	—	82
Full cream	Spain	4	—	—	—	—	—	—	—	0.001 ± 0.0002 0.001 – 0.002	—	84
Full cream	Spain	4	—	—	—	—	—	—	—	0.001 ± 0.0002 0.001 – 0.002	—	82
Fresh cheese	Canary Islands		—	—	0.08 ± 0.03	0.22 ± 0.05	—	—	—	0.01 ± 0.0002	0.65 ± 0.25	29
Market white brine cheese	Turkey	10	0.02 ± 0.005	0.01 ± 0.0003	0.04 ± 0.01	0.20 ± 0.03	0.005 ± 0.003	0.03 ± 0.01	0.11 ± 0.01	—	1.26 ± 0.08	30
Soft white cheese	Turkey	20	0.15 ± 0.16	0.02 ± 0.02	0.05 ± 0.04	0.54 ± 0.85	0.01 ± 0.004	0.03 ± 0.03	0.12 ± 0.16	—	1.77 ± 0.42	30
Stirred curd cheeses			0.0001 0.0002 – 0.001	0.01 0.01 – 0.01	—	0.17 0.10 – 0.21	—	—	—	0.01 0.01 – 0.01	3.47	46

Soft white cheese melted	Turkey	20	0.08 ± 0.03	0.02 ± 0.004	0.05 ± 0.01	0.70 ± 0.03	0.01 ± 0.01	0.02 ± 0.002	0.09 ± 0.04	—	2.99	30
Fat White cheese		5	—	—	0.03 ± 0.01 0.01 – 0.04	0.28 ± 0.22 0.12 – 0.62	0.03 ± 0.005 0.02 – 0.04	—	—	—	0.77 ± 0.35 0.53 – 1.4	77
Eggs	Nigeria	151	0.001 ± 0.0 0.001 – 0.001	—	0.08 ± 0.002 0.07 – 0.08	2.32 ± 0.08 2.18 – 2.41	—	—	0.003 ± 0.001 0.002 – 0.003	—	1.38 ± 0.02 1.29 – 1.45	85
Eggs[b]	U.K.		0.0002	0.02	0.06	2	0.031	0.01	0.003	0.02	1.1	7
Egg products[b]	U.S.		—	—	< 0.09	1.70 ± 0.21 1.46 – 1.85	< 0.1	—	< 0.01	0.03 ± 0.004 0.02 – 0.03	1.23 ± 0.125 1.09 – 1.32	8
Eggs	Poland	20	—	—	0.06 ± 0.003 0.06 – 0.06	1.8 ± 0.14 1.7 – 1.9	0.03 ± 0.003 0.02 – 0.03	—	—	—	1.2 ± 0.14 1.1 – 1.3	86
Eggs		20	—	—	0.06 ± 0.01 0.05 – 0.07	1.8 ± 0.3 1.4 – 2.3	0.02 ± 0.01 0.02 – 0.05	—	—	—	1.2 ± 0.2 0.86 – 1.6	77

bioavailable.[9] Therefore, the richest source of this element are animal liver, offal, and meat products, which can contain up to 6.9 mg Fe/100 g. Concentrations of iron range from 0.14 to 33.0 mg/100 g in meat (see Table 6.4) with the highest values for chicken kidney (18.0 mg/100 g) and chicken liver (up to 33.0 mg/100 g). Elevated levels of iron have been reported in crab meat (up to 42.1 mg/100 g). On the other hand, low levels of iron are observed in cow milk (0.018 to 0.097 mg/100 g) containing the nonheme iron protein. This protein, named *lactoferrin*, transports iron in the bloodstream.[1] Honey contains iron in a wide range of values, from 0.01 to 20.2 mg/100 g (see Table 6.6); sesame honey contains two orders of magnitude higher levels of iron (20.2 mg/100 g) compared to acacia honey (0.12 mg/100 g).

Some attention should be paid to controlling the concentration of iron in canned food. According to Reilly,[1] the age of the can as well as the duration of storage are important factors which can affect the iron release from the container. An example of extreme contamination with iron is a can of anchovies stored for 4 years that contained 580 mg Fe/100 g.

6.3.5 Manganese

As can be seen in Table 6.4, animal foods contain small amounts of manganese (Mn), in the range of 0.003 to 0.14 mg/100 g in meat with higher levels in bovine and elk muscles. Crab meat has elevated levels of manganese, up to 8.19 mg/100 g. Human and cow milk and their products contain very low levels, from 0.0002 to 0.02 mg Mn/100 g. The concentration of manganese in infant formula can become elevated (reaching 0.12 mg/100 g), whereas soy-based infant milk formula contains up to 11.5 mg Mn/100 g (see Table 6.5). As for geographic variations in Mn concentrations, it seems that there are insignificant differences in reported food levels for several countries.[1] Honey contain from 0.003 to 4.28 mg Mn/100 g; higher levels are noted for eucalyptus and acacia tree honey whereas lower values are associated with citrus and Apiaceae honey (see Table 6.6).

6.3.6 Molybdenum

Molybdenum (Mo) concentrations in animal foods range from 0.001 to 0.12 mg/100 g. The lowest values are reported for carcass meat, and the highest for offal. Milk and dairy products contain from 0.003 to 0.005 mg Mo/100 g and from 0.001 to 0.03 mg Mo/100 g, respectively (see Table 6.5).

6.3.7 Nickel

Nickel (Ni) is probably an essential trace element. A number of reports[1,43] suggest that Ni can migrate from stainless steel containers to dairy products and other foods. Nickel occurs in animal foods in small concentrations, ranging between < 0.00001 and 0.12 mg/100 g in milk and dairy products. Meat and its products contain on the average 0.020 mg Ni/100 g (see Table 6.5). Concentrations of nickel in honey vary over a broad range, from 0.0002–0.41 mg/100 g. The highest levels are observed for clover honey (see Table 6.6). According to Ysart et al.,[7] canned foods contain higher levels of Ni, the result of contamination from cooking equipment and containers.[1]

TABLE 6.6
Micronutrients in Honey (in mg/100 g w/w)

Name	Origin	n	Co	Cr	Cu	Fe	Mn	Ni	Se	Zn	Ref.
Acacia honey	France	150	0.01 0.003 – 0.03	0.02 0.005 – 0.05	0.02 0.003 – 0.23	0.12 0.01 – 1.0	0.08 0.01 – 1.03	—	—	0.08 0.004 – 0.60	31
Acacia tree honey	Slovenia	25	—	—	0.14 ± 0.01	0.73 ± 0.06	0.66 ± 0.04	—	—	0.51 ± 0.02	40
Apiaceae honey	Northwest Morocco	7	—	—	0.06 ± 0.03 0.03 – 0.11	0.81 ± 0.17 0.54 – 1.85	0.04 ± 0.01 0.003 – 0.10	—	—	0.09 ± 0.02 0.04 – 0.17	32
Citrus honey	Northwest Morocco	10	—	—	0.06 ± 0.02 0.005 – 0.21	0.683 ± 0.16 0.17 – 1.592	0.05 ± 0.01 0.01 – 0.10	—	—	0.19 ± 0.09 0.05 – 1.01	32
Clover honey	Egypt		0.20 ± 0.06	—	0.15 ± 0.03	4.72 ± 0.33	0.04 ± 0.08	0.33 ± 0.03	—	0.76 ± 0.04	33
Eucalyptus honey	Northwest Morocco	12	—	—	0.10 ± 0.03 0.04 – 0.05	0.98 ± 0.13 0.38 – 2.01	0.60 ± 0.07 0.09 – 1.15	—	—	0.27 ± 0.11 0.04 – 1.25	32
Heuchalipt honey	Italy	10	—	—	0.16 ± 0.20	—	—	—	—	0.42 ± 0.56	87
Honey	Turkey	30	0.1 ± 0.06	—	0.18 ± 0.17	0.66 ± 0.32	0.1 ± 0.07	—	—	0.27 ± 0.25	34
Honey			—	—	—	0.42	<0.1	< 0.01	—	0.11	8
Honeydew	Northwest Morocco	3	—	—	0.23 ± 0.01 0.17 – 0.27	3.13 ± 0.27 2.60 – 3.50	0.28 ± 0.05 0.20 – 0.37	—	—	0.30 ± 0.06 0.18 – 0.39	32
Honey	Italy		—	0.0003 ± 0.00003 0.0001 – 0.0004	0.01 ± 0.001 0.01 – 0.02	0.23 ± 0.02 0.02 – 0.07	—	0.001 ± 0.00005 0.0002 – 0.005	—	—	88
Honey	France	86	0.02 0.01 – 0.02	0.02 0.01 – 0.04	0.03 0.01 – 0.17	1.10 0.06 – 8.68	0.37 0.01 – 4.28	0.02 0.01 – 0.03	—	0.13 0.02 – 0.64	31
Honey	Italy		—	—	—	—	—	—	—	0.16 ± 0.05 0.12 – 0.23	35
Honey	India	10	< 0.03 – 0.03	—	0.17 ± 0.08 0.12 – 0.20	1.6 ± 1.22 0.36 – 2.84	—	0.04 ± 0.03 < 0.005 – 0.07	—	1.15 ± 1.24 0.47 – 2.9	89
Honey	U.S.		0.01 ± 0.01 < 0.005 – 0.02	0.005 ± 0.001 0.004 – 0.01	—	1.20 ± 0.54 <0.4 – 1.59	—	0.06 ± 0.02 0.04 – 0.08	0.001 ± 0.001 < 0.002 – 0.002	0.38 ± 0.06 0.34 – 0.46	36

TABLE 6.6 (CONTINUED)
Micronutrients in Honey (in mg/100 g w/w)

Name	Origin	n	Co	Cr	Cu	Fe	Mn	Ni	Se	Zn	Ref.
Honey	Italy	10	< 0.03 – 0.03	—	0.09 ± 0.002 < 0.005 – 0.09	0.12 ± 0.06 0.07 – 0.18	—	0.04 ± 0.001 < 0.005 – 0.04	—	0.29 0.26 – 0.32	89
Indian honey	India	30	—	—	0.21 ± 0.04 0.17 – 0.29	1.08 ± 0.15 0.89 – 1.33	—	—	—	0.82 ± 0.55 0.26 – 1.68	37
Industrial Galician honey	Spain	20	< 0.005	—	0.01 ± 0.01	0.38 ± 0.3	0.11 ± 0.11	< 0.002	—	0.15 ± 0.07	38
Lazio honey	Italy	84	—	—	0.03 ± 0.005	0.45 ± 0.04	0.3 ± 0.04	—	—	0.31 ± 0.03	39
Lythrum honey	Northwest Morocco	7	—	—	0.03 ± 0.01 0.002 – 0.06	0.83 ± 0.24 0.19 – 1.87	0.03 ± 0.005 0.008 – 0.04	—	—	0.26 ± 0.11 0.04 – 0.83	32
Orange honey	Egypt		0.15 ± 0.03	—	0.08 ± 0.0	6.51 ± 1.38	0.04 ± 0.08	0.14 ± 0.02	—	0.41 ± 0.08	33
Natural Galician honey	Spain	22	< 0.005	—	0.09 ± 0.07	0.37 ± 0.17	0.52 ± 0.31	< 0.002	—	0.2 ± 0.13	38
Pine tree honey	Slovenia	25	—	0.07 ± 0.01	0.15 ± 0.01	0.7 ± 0.04	0.5 ± 0.1	0.08 ± 0.02	—	0.54 ± 0.03	40
Saudi Arabia honey	Saudi Arabia		—	—	0.05 ± 0.01 0.04 – 0.06	0.13 ± 0.03 0.10 – 0.19	0.01 ± 0.005 0.003 – 0.02	—	—	0.21 ± 0.07 0.15 – 0.30	41
Sesame honey	Egypt		0.15 ± 0.03	—	0.14 ± 0.03	16.4 ± 1.79	0.14 ± 0.02	0.11 ± 0.03	—	0.59 ± 0.03	33
Syrup-feed honey	Egypt		0.26 ± 0.03	—	0.08 ± 0.0	300 ± 6.43	0.46 ± 0.03	0.24 ± 0.0	—	0.57 ± 0.08	33

6.3.8 Selenium

The range between toxic and safe doses of selenium (Se) (an essential element) is very narrow. Selenium is present in foods in levels ranging from 0.0007 to 0.020 mg/kg in milk and dairy products and 0.04 to 0.15 mg/100 g in seafood and offal (see Table 6.5). The protective role of selenium against certain toxic metals has been reported and discussed extensively by Reilly[1,2]. For instance, no sign of mercury intoxication was observed in the study of an Inuit fishing community with elevated blood mercury levels that often exceeded 200 μg/l. The reduced risk of mercury poisoning was attributed to the protective presence of high amounts of selenium in fish and seal meat, which constituted the major component of the diet of the local people.[1,44,45] Concentration of selenium in foods of animal origin is determined by its levels in plant foods eaten by the animals. Because toxic properties of this metalloid are strongly dependent on its chemical speciation, it is important to assess the chemical forms of Se in certain foods. Van Deal[46] has quantified the species of selenium in foodstuffs, including cooked cod and cow, goat, and sheep milk. Speciation analysis of selenium in food is discussed elsewhere (see Chapter 2 of this book).

6.3.9 Zinc

Zinc (Zn) is an essential element found at elevated levels in foods of animal origin. In many African and Asian regions, zinc intakes are low, especially where the customary diets of children contain insufficient amounts of animal products.[1,47] For instance, poor communities in developing countries of the Middle East, e.g., Iran and Egypt, as well as in Bangladesh, suffer from marginal zinc deficiency as their diet consists mainly of cereals rich in phytate and fiber in which the zinc is unavailable for absorption.[1,47–49] In order to improve zinc nutrition especially in the developing countries, some effort is underway to provide zinc supplements to local populations or fortify staple foods with this element.[1,50]

Concentrations of zinc in meat and its products range from 0.004 to 22.0 mg/100 g with lower values for beef and the highest for pork loin (see Table 6.4). Mammalian liver as well as carcass meat and offal may be important sources of zinc with some seafoods also being highly enriched in this element.[1] Milk and milk products are characterized by lower levels of zinc. Concentrations of this element in human and cow milk range between 0.17 and 0.88 mg/100 g. Higher levels are reported for infant formula (4.53 mg/100 g) and some cheeses (5.30 mg/100 g) (see Table 6.5). Concentrations of zinc in honey range from 0.004 mg/100 g (acacia tree honey) to 2.9 mg/100 g (Indian honey). High levels are also reported for clover honey, up to 0.93 mg/100 g (see Table 6.6).

6.4 "POSSIBLY ESSENTIAL" MICRONUTRIENTS IN FOOD

6.4.1 Arsenic

Arsenic (As) is a nonessential metalloid present in inorganic and organic forms in foodstuffs. It occurs in animal foods generally at low levels, except for fish and seafoods (see Chapter 10 of this book). The lowest values are observed in poultry and meat products (0.0003-0.005 mg/100 g), but much higher levels have been reported in fish meat (up to 0.53 mg/100 g, see Table 6.7). As can be deduced from Table 6.8, very low levels of As are reported for milk (< 0.0001 to 0.004 mg/100 g) with 2 to 3 times higher values in dairy products (0.0003 to 0.004 mg/100 g). Honey contain from 0.00005 to 0.003 mg As/100 g (see Table 6.9).

It should be kept in mind that arsenic in seafoods (edible mollusks, shrimp, lobster, and algae) occurs mainly in nontoxic compounds such as arsenobetaine and arsenocholine. Therefore, food safety assessment requires data not only on total arsenic concentration but also on its speciation in foods. In consequence, the consumers may not be exposing themselves to much risk by eating food of marine origin that is rich in arsenobetaine. On the other hand, it is still impossible to assess the health risk for arseno-sugars because of the lack of toxicity data for this organoarsenic compound.[3] Environmental contamination of foods with arsenic is discussed in detail in Chapter 10 of this book.

6.4.2 Boron

Dietary boron (B) is characterized by a low degree of toxicity, because it is rapidly excreted. Levels of this element in animal foods are lower than those in plant foods, with reported values ranging between < 0.04 and 0.05 mg/100 g (see Table 6.7 and Table 6.8).

6.4.3 Vanadium

Vanadium (V) is probably an essential trace element. There is limited information on levels of vanadium in animal foods. Data listed in Table 6.7 to Table 6.9 show that levels of vanadium are generally within the values of < 0.00001 and 0.02 mg/100 g. According to data in Table 6.8, milk contains significantly lower levels (< 0.00001 to 0.0004 mg/100 g) of vanadium compared to dairy products (0.012 to 0.020 mg/100 g). Seafood represent the richest source of vanadium.[1] Concentration of the element in honey is small and range between 0.0001 and 0.0002 mg/100 g (see Table 6.9).

6.5 NONTOXIC, NONESSENTIAL METALS

6.5.1 Aluminum

Concentrations of aluminum (Al) in animal foods are frequently comparable with those in plant products, with range of values from 0.002 to 0.08 mg/100 g in cow milk and 0.11 to 1.12 mg/100 g in dairy products (see Table 6.8). Infant formulas designated for low-birth-weight infants can contain higher than normal levels of aluminum.[90–92] A reduction in the levels of aluminum in soy-base infant formulas has recently been observed.[93] The levels of aluminum in fish can be up to 0.55 mg Al/100 g (see Chapter 10 of this book). Meat offal may contain up to 5.30 mg Al/100 g (see Chapter 10 of this book). Aluminum occurs in honey in a broad range of values, from 0.005 to 0.97 mg/100 g (see Table 6.9). Occurrence of this element in processed food is discussed in Chapter 10 of this book.

TABLE 6.7
"Possibly Essential" Micronutrients and Nontoxic, Nonessential Metals in Meat, Fish, and Poultry (in mg/100 g w/w)

Name	Origin	n	Al	As	B	Sn	Ref.
Meat							
Carcass meat[b]	U.K.		0.05	0.0004	< 0.04	0.002	7
Beef	Galicia	56	—	n = 12 0.0005 ± 0.00005 0.001 – 0.002	—	—	54
Veal	Galicia	438	—	n = 48 0.0004 ± 0.00001 0.001 – 0.002	—	—	54
Cattle liver	Galicia	437	—	n = 134 0.004 ± 0.0004 0.001 – 0.04	—	—	54
Offal[b]	U.K.		0.035	0.0004	< 0.04	0.002	7
Offals[b]	U.S.		—	< 0.004	—	—	8
Meat products[b]	U.K.		0.32	0.0004	0.04	0.03	7
Meat products[b]	U.S.		—	< 0.004 – < 0.005	—	—	8
Poultry							
Poultry[b]	U.K.		0.03	0.0003	< 0.04	< 0.002	7
Poultry[b]	U.S.		—	< 0.004	—	—	8
Fish							
Fish[b]	U.K.		0.55	0.43	0.05	0.04	7
Fish[b]	U.S.		—	0.20 ± 0.22 0.08 – 0.53	—	—	8

6.5.2 Tin

Tin (Sn) does not constitute any serious problems in most foods except when there is incidental contamination in canned food products that are acidic.[1] The levels of tin range from < 0.002 mg/100 g in poultry and eggs to 0.044 mg/100 g in fish (see Chapter 10 of this book). Milk-based powder formulas contain from 0.002 to 0.01 mg Sn/100 g (see Table 6.8). Lower levels are reported for honey with values in the range of < 0.0004 to 0.002 mg Sn/100 g (see Table 6.9). Meat and offal contain, on the average, 0.002 mg Sn/100 g (see Chapter 10 of this book). Uptake of tin by foodstuffs is dependent on the kind of the food, the age of the cans or whether cans are lacquered or not.[1] It has been reported that tin in canned food occurs in the inorganic form; however, certain food, especially fish, may also contain organic tin species that are more toxic than the inorganic compounds. Among the organotin compounds such as tributyltin (TBT), dibutyltin (DBT), and monobutyltin (MBT), the TBT is known to be the most highly bioaccumulated in fish and other edible marine organisms such as mollusks and crustaceans.[1,3] A major source of butyltins in fish and seafood is TBT-based antifouling paint agents used extensively on ships.[3]

TABLE 6.8
"Possibly Essential" Micronutrients and Nontoxic, Nonessential Metals in Milk, Dairy Products, and Eggs (in mg/100 g w/w)

Name	Origin	n	Al	As	B	Sn	V	Ref.
Milk								
Bottled cow milk[c]		3	0.08 ± 0.005	—		—	—	73
Milk[b]	U.S.		—	< 0.004	—	—	—	8
Milk[b]	U.K.		<0.03	0.0002	< 0.04	< 0.002	—	7
Skimmed bottled cow milk[c]		3	0.04 ± 0.002	—	—	—	—	73
Whole raw bovine milk	Calabria, Italy	40	—	0.004 < 0.00002 – 0.07	—	—	—	79
Skimmed cow milk[c]		5	0.002 ± 0.0003	—	—	—	—	18
Cow whole milk[c]		5	0.002 ± 0.0003	—	—	—	—	18
UHT cow whole milk[c]		5	0.003 ± 0.0004	—	—	—	—	18
Skimmed UHT cow milk[c]		5	0.002 ± 0.0004	—	—	—	—	18
Nonfat milk powder		3	0.21 ± 0.02	—	—	—	—	73
Powdered cow milk		3	0.93 ± 0.03	—	—	—	—	73
Goat milk		3	—	0.001 ± 0.00004	—	—	—	82
Raw goat milk		2	0.11 ± 0.03 0.09 – 0.14	—	—	—	—	19
Raw sheep milk		4	0.29 ± 0.09 0.15 – 0.39	—	—	—	—	19
Human milk			< 0.00001 – 0.04	< 0.0001 – 0.003	—	—	< 0.00001 – 0.0004	80
Human milk[c]		3	0.01 ± 0.001	—	—	—	—	73
Human whole milk[c]		5	0.002 ± 0.0003	—	—	—	—	18
Skimmed human milk[c]		5	ND	—	—	—	—	18
Vegetal based milk	Spain	2	—	0.0003 ± 0.000005 0.0003 – 0.0004	—	—	—	82

Raw milk		7	0.34 ± 0.17 0.14 – 0.59	—	—	—	—	17
Skim milk powder		2	0.05 ± 0.002 < 0.002 – 0.11	—	—	—	—	21
Whole milk powder		3	0.1 ± 0.01	—	—	—	—	73
Vegetal based milk	Spain	2	—	0.0003 ± 0.000005 0.0003 – 0.0004	—	—	—	82
Formula milk[c]		5	0.03 ± 0.001	—	—	—	—	18
Formula milk whey[c]		5	0.01 ± 0.0005	—	—	—	—	23
Infant formula		3	0.41 ± 0.10	—	—	—	—	73
Milk based liquid first and follow-on formulas[c]	London, U.K.	18	0.01 ± 0.02	—	—	0.003 ± 0.004	0.0001 ± 0.0001	24
Ready to drink infant formulas			0.005 – 0.05	—	—	—	0.00003 – 0.0001	80
Skimmed formula milk[c]		5	0.02 ± 0.001	—	—	—	—	18
Milk-based powder formulas[c]	Lagos, Nigeria	6	0.01 ± 0.002	—	—	0.002 ± 0.003	—	24
Milk-based powder formulas[c]	London, U.K.	12	0.01 ± 0.01	—	—	0.002 ± 0.003	—	24
Milk-based powder formulas[c]	U.S.	9	0.01 ± 0.01	—	—	0.01 ± 0.004	0.02 ± 0.03[d]	24
Soy-based powder formula[c]	U.S.	6	0.05 ± 0.02	—	—	0.01 ± 0.005	0.3 ± 0.3[d]	24
Milk whey coming from cow milk[c]		5	ND	—	—	—	—	18
Milk whey coming from UHT cow milk[c]		5	ND	—	—	—	—	18
Milk whey coming from human milk[c]		5	ND	—	—	—	—	18
Milk whey coming from formula milk[c]		5	0.01 ± 0.001	—	—	—	—	18

TABLE 6.8 (CONTINUED)
"Possibly Essential" Micronutrients and Nontoxic, Nonessential Metals in Milk, Dairy Products, and Eggs (in mg/100 g w/w)

Name	Origin	n	Al	As	B	Sn	V	Ref.
Dairy Products								
Dairy products[b]	U.K.		0.06	0.3[d]	0.04	0.03	—	7
Milk products[b]	U.S.		—	< 4 – < 5[d]	—	—	—	8
Yoghurt, low-fat, fruit	U.S.		—	< 4[d]	—	—	—	8
Butter (regular, salted)	U.S.		—	< 4[d]	—	—	—	8
Swiss cheese	U.S.		—	< 5[d]	—	—	—	8
Commercial goat cheese		2	0.14 ± 0.04 0.11 – 0.18	—	—	—	—	19
Commercial sheep cheese		4	0.47 ± 0.24 0.26 – 0.88	—	—	—	—	19
Cream cheese	U.S.		—	< 0.005	—	—	—	8
Skimmed cream		4	—	0.0005 ± 0.0001 0.0003 – 0.001	—	—	—	82
Cream partially skimmed	Spain	6	—	0.001 ± 0.0002 0.001 – 0.001	—	—	—	84
Cream partially skimmed	Spain	6	—	0.001 ± 0.0002 0.001 – 0.001	—	—	—	82
Cream partially skimmed		3	—	0.001 ± 0.0002 0.001 – 0.001	—	—	—	82
Cream skimmed	Spain	8	—	0.0005 ± 0.0001 0.0003 – 0.001	—	—	—	84
Cream skimmed	Spain	6	—	0.0005 ± 0.0001 0.0003 – 0.001	—	—	—	82
Full cream	Spain	4	—	0.001 ± 0.0003 0.0004 – 0.001	—	—	—	84

Full cream	Spain	4	—	0.001 ± 0.0003 0.0004 – 0.001	—	—	—	82
Full cream		2	—	0.003 ± 0.002 0.001 – 0.004	—	—	—	82
Market white brine cheese	Turkey	10	0.24 ± 0.01	—	—	—	0.01 ± 0.0002	30
Soft white cheese	Turkey	20	0.53 ± 0.21	—	—	—	0.02 ± 0.02	30
Soft white cheese melted	Turkey	20	1.12 ± 0.04	—	—	—	0.02 ± 0.01	30
Eggs								
Eggs[b]	U.K.		0.03	0.0002	< 0.04	< 0.002	—	7
Egg products[b]	U.S.		—	< 0.004	—	—	—	8

d = µg/100 g w/w

TABLE 6.9
"Possibly Essential" Micronutrients and Nontoxic, Nonessential Metals in Honey (in mg/100 g w/w)

Name	Origin	n	Al	As	Sn	V	Ref.
Acacia honey	France	150	0.04 0.005 – 0.14	—	—	—	31
Honey	Italy		—	0.0004 ± 0.00003 < 0.00005 – 0.0001	0.001 ± 0.0001 < 0.0004 – 0.003	0.0004 ± 0.00003 0.0001 – 0.0002	88
Honey	France	86	0.23 0.02 – 0.97	—	—	—	31
Honey	U.S.		—	0.003	—	—	36
Saudi Arabia honey	Saudi Arabia		0.01 ± 0.01 0.005 – 0.02	—	—	—	41

The occurrence of inorganic tin and butyltins as food contaminants is discussed in detail in Chapter 10 of this book.

6.6 TOXIC METALS

6.6.1 Beryllium

There is a lack of data on beryllium (Be), in contrast to those on Cd, Hg, and Pb (see Table 6.10) in meat and dairy products in spite of the great interest in its behavior from ecotoxicological point of view.[1] Milk-based formulas contain from 0.00001 to 0.00010 mg Be/100 g (see Table 6.11).

6.6.2 Cadmium

The concentration of cadmium (Cd) in foodstuffs is usually very low. Shellfish may be rich in this element. Australian beef and sheep offal have been reported to contain high levels of cadmium.[1] Animal foods contain, on the average, from < 0.00001 to 0.02 mg Cd/100 g in milk and dairy products (see Table 6.11) and from 0.0001 to 0.03 mg Cd/100 g in meat and its products (see Table 6.10). Cattle liver and chicken kidneys contain up to 0.80 and 0.51 mg Cd/100 g, respectively. Concentrations in honey range from < 0.00005 to 0.06 mg Cd/100 g. Further discussion of food contamination with cadmium is provided in Chapter 10 of this book.

6.6.3 Lead

Organic species of lead (Pb) are generally characterized by higher toxicity than inorganic lead compounds. The concentrations of lead in animal foods range from 0.00001 to 0.02 mg/100 g (see Table 6.11). Meat and offal can contain high levels of lead. Milk and milk products contain, on the average, 0.0002 and 0.001 mg Pb/100 g, respectively (see Table 6.11). High levels of lead have been reported in chicken

TABLE 6.10
Toxic Metals in Meat and Its Products (in mg/100 g w/w)

Name	Origin	n	Cd	Hg	Pb	Ref.
Meat						
Carcass meat[b]	U.K.		0.0001	0.0003	0.001	7
Swine muscle meat	Poland	334	n = 658 0.001 ± 0.0001 0.0005 – 0.001	n = 1181 0.0002 ± 0.0001 0.0001 – 0.0003	0.002 ± 0.004 <0.001 – 0.01	53
Beef	Galicia	56	n = 27 0.0001 ± 0.00002 0.0001 – 0.001	—	n = 40 0.002 ± 0.0002 0.001 – 0.005	54
Bovine muscle	Egypt	54	0.002 ± 0.001 0.001 – 0.003	—	0.01 ± 0.002 0.01 – 0.01	55
Bovine muscle[c]	Morocco		0.02 ± 0.005 0.01 – 0.03	—	—	56
Buffalo muscle	Egypt	54	0.001 ± 0.0003 0.001 – 0.001	—	0.01 ± 0.002 0.005 – 0.01	55
Cattle muscle meat	Poland	93	n = 92 0.001 ± 0.0005 < 0.0005 – 0.002	n = 291 0.0001 ± 0.00004 0.0001 – 0.0002	0.004 ± 0.003 < 0.001 – 0.01	57
Muscle cows	Galicia	56	—	0.00004 0.00004 – 0.00005	—	99
Meat calves	North Spain		0.0002	—	—	98
Muscle calves	Galicia	184	—	0.00004 0.00004 – 0.00005	—	99
Muscle calves	Asturia	100		0.00004 0.00004 – 0.00004		99
Veal	Galicia	438	n = 173 0.0001 ± 0.00001 0.0001 – 0.003	—	n = 135 0.001 ± 0.00004 0.001 – 0.005	54

TABLE 6.10 (CONTINUED)
Toxic Metals in Meat and Its Products (in mg/100 g w/w)

Name	Origin	n	Cd	Hg	Pb	Ref.
Goat muscle	Egypt	54	0.003 ± 0.002 0.001 – 0.004	—	0.005 ± 0.005 0.001 – 0.01	55
Sheep muscle	Egypt	54	0.002 ± 0.001 0.001 – 0.002	—	0.005 ± 0.005 0.001 – 0.01	55
Sheep muscle meat	Poland	5	<0.0005 < 0.0005	< 0.003 < 0.002	< 0.003 < 0.002	58
Rabbits muscle meat	Poland	30	0.0005 < 0.0005 – 0.002	< 0.0001 < 0.0001 – 0.0002	0.002 < 0.001 – 0.01	58
Elk muscle	Egypt	54	0.002 ± 0.001 0.001 – 0.002	—	0.01 ± 0.01 0.004 – 0.02	55
Cattle liver	Galicia	437	n = 381 0.003 ± 0.002 0.0002 – 0.80	—	n = 289 0.005 ± 0.0003 0.002 – 0.05	54
Kidney calves	North Spain		0.02	—	—	98
Liver calves	North Spain		0.003	—	—	98
Offal[b]	U.K.		0.01	0.001	0.01	7
Offal[b]	U.S.		0.01	< 0.004	< 0.005	8
Meat Products						
Meat products[b]	U.K.		0.001	0.0003	0.001	7
Meat products[b]	U.S.		< 0.001 – < 0.001	NA	< 0.005	8
Poultry						
Poultry[b]	U.K.		0.0002	0.0004	< 0.001	7
Poultry[b]	U.S.		< 0.0005 – 0.001	ND	< 0.004	8
Hen muscle meat	Poland	1	0.003	< 0.003	< 0.003	58

Smoked chicken	Poland	4	0.001 ± 0.001	0.0002	0.01 ± 0.01	60
Geese muscle meat	Poland	32	0.0005 < 0.0005 – 0.001	< 0.001 < 0.001 – 0.003	< 0.001 < 0.001 – 0.003	58
Turkey muscle meat	Poland	30	—	0.001 0.0003 – 0.001	—	58
Duck muscle meat	Poland	57	0.0005 < 0.0005 – 0.001	0.002 < 0.001 – 0.01	0.002 < 0.001 – 0.01	58
Chicken liver	Kuwait	33	0.01 0.004 – 0.02	—	0.01 0.002 – 0.04	100
Hens kidney	Poland	1	0.51	0.001	0.13	58
Hens liver	Poland	13	0.01 0.002 – 0.04	0.0005 0.0002 – 0.001	0.02 0.001 – 0.01	58
Poultry sausage	Poland	11	< 0.0005	0.0002 ± 0.0001	0.005 ± 0.002	60
Fish						
Fish[b]	U.K.		0.002	0.005	0.002	7
Fish[b]	U.S.		0.002 ± 0.0004 < 0.001 – 0.002	0.01 ± 0.01 < 0.004 – 0.02	< 0.004	8
Atlantic salmon	Southern Baltic	2	0.0003 – 0.001	—	0.001 – 0.01	96
Cod	Southern Baltic	70	0.0003 ± 0.0001 < 0.0001 – 0.005	—	0.01 < 0.001 – 0.03	96
Eel	Southern Baltic	2	0.0004 – 0.001	—	0.005 – 0.01	96
Flounder	Southern Baltic	2	0.0003 0.0001 – 0.0004	—	0.01 0.01 – 0.01	96
Garfish	Southern Baltic	4	0.001 0.0004 – 0.003	—	0.005 0.001 – 0.01	96
Greater sand eel	Southern Baltic	3	0.004 0.001 – 0.01	—	0.02 0.02 – 0.03	96
Herring	Southern Baltic	34	0.001 ± 0.0003 0.0003 – 0.005	—	0.01 0.001 – 0.07	96
Perch	Southern Baltic		0.0002 – 0.004	0.003 – 0.01	0.001 – 0.003	96

TABLE 6.10 (CONTINUED)
Toxic Metals in Meat and Its Products (in mg/100 g w/w)

Name	Origin	n	Cd	Hg	Pb	Ref.
Sprat	Southern Baltic	5	0.002 ± 0.0005 0.001 – 0.004	—	0.01 ± 0.002 0.003 – 0.05	96
Stickleback	Southern Baltic	1	0.004	—	0.09	96
Whiting	Southern Baltic	2	0.0005	0.0005	0.005 – 0.02	96
Seafood						
Shrimp	Taiwan		0.0005 – 0.001	0.004 – 0.02	< 0.001	61
Shrimp	Spain		0.00 – 0.001	—	0.005 – 0.01	63
Shrimp			0.002 – 0.005	—	0.002 – 0.01	63
Shrimp	Mexico		0.003 – 0.02	—	—	64
Shrimp	Greenland		0.002	0.01	0.004	66
Shrimp boiled	Polish market		0.01 ± 0.03	0.003 ± 0.003	0.001 ± 0.00002	14
Shrimp raw	Polish market		0.001	0.002	0.0004	14
Crab	Taiwan		0.0003 – 0.01	0.001 – 0.03	0.002 – 0.08	61
Crab	Spain		0.002 – 0.03	—	0.00 – 0.06	67
Crab claws	Australia		0.001	0.04	—	67
Crab	Texas		0.0003 – 0.02	—	0.05 – 0.77	67
Crab meat from claws, boiled	Polish market		0.01	0.02	0.01	14
Lobster	Mexico		0.004 – 0.01	—	—	68
Lobster	Mexico		0.005 – 0.01	—	—	68
Lobster meat, from claws boiled	Polish market		0.02	0.01	< 0.00005	14
Clam	Taiwan		0.002 – 0.06	0.001 – 0.004	0.001 – 0.66	61
Clam	Spain		0.02 – 0.04	—	0.18 – 0.31	63
Mussel	Taiwan		0.01 – 0.02	0.001 – 0.004	0.001 – 0.002	61
Mussel	Malaysia		0.01 – 0.03	—	0.05 – 0.18	70
Mussel	Greenland		0.02 – 0.05	0.001 – 0.002	0.02 – 0.05	13

Mussel	Japan (Minamata Bay)	—	0.001 – 0.012	—	101
Mussel	Japan (Kagoshima Bay)	—	0.0002 – 0.003	—	101
Mussel	Hong Kong	0.01 – 0.03	—	0.04 – 0.09	71
Mussel	Greenland	0.07	0.002	0.01	66
Mussel	Texas	0.03 – 0.04	—	0.04 – 0.10	72
Mussel, boiled	Polish market	0.02 ± 0.01	0.001 ± 0.001	0.01 ± 0.01	14
Mussel raw	Polish market	0.01	0.003	0.005	14
Octopus	Portugal	0.006	0.005	0.005	102
Octopus raw tentacles	Polish market	0.001 ± 0.001	0.003 ± 0.0005	0.002	14
Octopus small	Polish market	0.09	0.0005	0.004	14

TABLE 6.11
Toxic Metals in Milk, Dairy Products, and Eggs (in mg/100 g w/w)

Name	Origin	n	Be	Cd	Hg	Pb	Ref.
Milk							
Cow milk fat content 3.8%[c]			—	0.00001 ± 0.0002	—	0.0002 ± 0.0001	74
Lowfat (2%) milk, fluid	U.S.		—	—	—	<0.003	8
Milk[b]	U.S.		—	<0.0004	ND	<0.002 – <0.003	8
Milk[b]	U.K.		—	0.0001	0.00007	<0.001	7
Milk		6	—	—	0.0004	—	103
fat content 2%					0.0001 – 0.001		
Milk		4	—	—	0.0001	—	103
fat content 3.2%					0.0001 – 0.0001		
Standard milk fat content 6%[c]			—	0.00001 ± 0.0003	—	0.0002 ± 0.0002	74
Whole milk >6% fat content[c]			—	0.00001 ± 0.0002	—	0.0003 ± 0.0001	74
Whole raw bovine milk	Calabria, Italy	40	—	0.000002	—	0.0001	79
				<0.00001 – 0.002		0.00001 – 0.001	
Buffalo milk fat content 7.5%[c]			—	0.00001 ± 0.0003	—	0.0003 ± 0.0001	74
Skimmed cow milk[c]		5	—	—	—	0.0001 ± 0.00002	18
Whole milk, fluid	U.S.		—	—	—	<0.003	8
Whole milk[c]	U.S.	8	—	0.00003 ± 0.00001	—	0.0001 ± 0.00003	104
				0.00002 – 0.0001		0.0001 – 0.0002	
Cow whole milk[c]		5	—	0.00004 ± 0.00001	—	0.0002 ± 0.00002	18
Milk UHT 0.5% fat		4	—	—	0.001	—	103
					0.0004 – 0.001		
Milk UHT 1.5% fat		8	—	—	0.0004	—	103
					0.0003 – 0.001		
Milk UHT 3.2% fat		23	—	—	0.0004	—	103
					0.00004 – 0.001		

Milk UHT 0% fat		10	—	—	0.001 0.00005 – 0.003	—	103
UHT cow whole milk[c]		5	—	0.00005 ± 0.00002	—	0.0002 ± 0.00003	18
Skimmed UHT cow milk[c]		5	—	—	—	0.0002 ± 0.00002	18
Raw goat milk		2	—	0.02 ± 0.002	—	0.005 ± 0.00005	19
Raw sheep milk		4	—	0.02 ± 0.002 0.01 – 0.02	—	0.01 ± 0.002 0.01 – 0.01	19
Human milk[c]			—	0.00001 ± 0.0003	—	0.0002 ± 0.0002	74
Human whole milk[c]		5	—	0.00005 ± 0.00002	—	0.0001 ± 0.00004	18
Raw milk		7	—	0.01 ± 0.002 0.01 – 0.01	—	0.01 ± 0.002 0.01 – 0.02	17
Milk condensed in cans		10	—	—	0.001 0.0003 – 0.002	—	103
Formula milk[c]		5	—	0.0001 ± 0.00002	—	0.0005 ± 0.00004	18
Formula milk whey[c]		5	—	0.00003 ± 0.00001	—	0.0001 ± 0.00002	23
Milk based liquid first and follow-on formulas[c]	U.K.	18	0.0001 ± 0.0001	0.00003 ± 0.0001	—	0.0001 ± 0.00004	24
Skimmed formula milk[c]		5	—	0.00005 ± 0.00002	—	0.0004 ± 0.00003	18
Milk-based powder formulas[c]	Nigeria	6	—	—	—	0.00004 ± 0.0001	24
Milk-based powder formulas[c]	U.K.	12	—	—	—	0.0001 ± 0.0002	24
Milk-based powder formulas[c]	U.S.	9	0.00001 ± 0.00002	—	—	—	24
Soy-based powder formula[c]	U.S.	6	0.00001 ± 0.00001	—	—	—	24
Cow milk whey[c]		5	—	—	—	0.0001 ± 0.00003	23
Human milk whey[c]		5	—	0.00001 ± 0.000005	—	—	23
Dairy Products							
Bioyoghurt natural		10	—	—	0.0004 0.0001 – 0.001	—	103
Dairy products[b]	U.K.		—	0.0002	0.0002	0.001	7
Milk products[b]	U.S.		—	<0.0005 – <0.001	NA	<0.003 – <0.005	8
Plain yoghurt, lowfat	U.S.		—	<0.0005	—	<0.003	8

TABLE 6.11 (CONTINUED)
Toxic Metals in Milk, Dairy Products, and Eggs (in mg/100 g w/w)

Name	Origin	n	Be	Cd	Hg	Pb	Ref.
Bioyoghurt with fruits		18	—	—	0.0003 0.0003 – 0.002	—	103
Fruit yoghurts		14	—	—	0.001 0.001 – 0.002	—	103
Yoghurt, lowfat, fruit	U.S.		—	<0.0005	—	<0.003	8
Mayonnaise	U.S.		—	<0.004	—	<0.01	8
Butter (regular, salted)	U.S.		—	<0.002	—	—	8
Swiss cheese	U.S.		—	<0.001	—	<0.004	8
American cheese	U.S.		—	<0.001	—	—	8
Cheddar cheese	U.S.		—	—	—	<0.005	8
Hard cheeses			—	—	0.0003 0.0001 – 0.001	—	46
Cream cheese			—	—	—	<0.004	8
Cottage cheese, 4%	U.S.		—	<0.0005	—	<0.003	8
Commercial goat cheese		2	—	0.01 ± 0.002	—	0.01 ± 0.002	19
Soft cheeses			—	—	0.0002 0.0001 – 0.0005	—	46
Commercial sheep cheese		4	—	0.01 ± 0.004 0.01 – 0.02	—	0.01 ± 0.001 0.01 – 0.01	19
Quark cheeses			—	—	0.0005 0.0001 – 0.001	—	46
Stirred curd cheeses			—	—	0.001 0.0001 – 0.002	—	46
Turkish cheese	Turkey	45				0.06 ± 0.04 0.01 – 0.12	105

Chicken egg	Kuwait	190	—	0.001 ND-0.002	—	0.01 ND-0.02	100
Eggs	Nigeria	151	—	0.01 ± 0.0 0.01 – 0.01	—	0.06 ± 0.003 0.05 – 0.06	85
Eggs	Poland	20	—	<0.0005	0.0002 ± 0.0001 0.0001 – 0.0002	0.005 ± 0.001 0.004 – 0.005	60
Eggs[b]	U.K.		—	0.0001	0.0004	0.001	7
Egg products[b]	U.S.		—	<0.0005	ND	<0.003	8

TABLE 6.12
Toxic Metals in Honey (in mg/100 g w/w)

Name	Origin	n	Cd	Hg	Pb	Ref.
Clover honey	Egypt		0.001 ± 0.01	—	0.34 ± 0.03	33
Heuchalipt honey	Italy	10	0.01 ± 0.02	—	0.08 ± 0.10	87
Honey	U.S.		< 0.001	—	< 0.004	8
Honey	Italy		0.0001 ± 0.000005 < 0.00005 – 0.0001	—	0.002 ± 0.0002 0.0003 – 0.02	88
Honey	France	86	0.02 0.01 – 0.03	—	0.08 0.03 – 0.11	31
Honey	India	10	0.06 ± 0.001 < 0.005 – 0.06	—	0.06 ± 0.04 < 0.005 – 0.08	89
Honey	U.S.		—	0.0001	—	36
Honey	Italy	10	0.03 < 0.005 – 0.03	—	0.06 < 0.005 – 0.06	89
Orange honey	Egypt		0.001 ± 0.01	—	0.46 ± 0.02	33
Pine tree honey	Slovenia	25	—	—	0.06 ± 0.02	40
Saudi Arabia honey	Saudi Arabia		—	—	0.01 ± 0.01 0.003 – 0.02	41
Sesame honey	Egypt		0.04 ± 0.0	—	0.51 ± 0.08	33
Syrup-feed honey	Egypt		0.04 ± 0.0	—	0.76 ± 0.08	33

kidney (0.13 mg/100 g) and in offal (up to 0.011 mg/100 g) (see Table 6.10). Concentrations of lead in honey range from 0.0003 mg/100 g (Saudi Arabia honey) to 0.63 mg/100 g (sesame honey). Syrup-feed honey contains 0.93 mg Pb/100 g (see Table 6.12). More detailed information on occurrence of lead as a contaminant in animal foods is presented in Chapter 10 of this book.

6.6.4 Mercury

The levels of mercury (Hg) in animal foods are generally low, ranging from 0.00005 to 0.0020 mg/100 g in milk and dairy products (see Table 6.11) and 0.00004 to 0.0002 mg/100 g in meat and its products (see Table 6.10). Seafoods may bioaccumulate higher levels of mercury, up to 0.07 mg/100 g (see Table 6.10). Organ meat also tends to contain higher levels of the element, derived mostly from industrial sources (see Chapter 10 of this book). Manifestations of such contamination include the elevated renal levels of 0.0001 to 0.014 mg Hg/100 g in cattle from the Netherlands and the concentrations of up to 0.16 mg/100 g Hg in fish from the Alexandria coast.[94,95] Concentration of the metal in honey can be up to 0.0001 mg/100 g (see Table 6.12).

Analysis of foodstuffs for particular species of mercury is very important from both the hygienic and toxicological points of view. Organic compounds of mercury are characterized by high toxicity, greater than for its inorganic species. A spectacular example of marine contamination by mercury is the Minamata Bay in Japan where widespread contamination of edible fish and seafoods with methylmercury (CH_3Hg) resulted in many deaths from mercurialism.[96] South America tuna contain elevated levels, up to 0.69 mg Hg/100 g, of which only 6.2% is in the organic form.[97] Significant bioaccumulation of CH_3Hg in fish muscle constitutes a safety problem, hence the FDA has established a health advisory for organomercury compounds in seafood.[46] Contamination of food by mercury is further discussed in Chapter 10 of this book.

6.7 OTHER ELEMENTS

6.7.1 Antimony

Animal foods generally contain very low levels of antimony (Sb), often below the detection limit of the methods used. Meat products and offal contain about 0.0004 and 0.0001 mg Sb/100 g, respectively (see Table 6.13). As seen in Table 6.14, concentrations of antimony range from < 0.00001 mg/100 g in milk to 0.001 mg/100 g in vegetal- and milk-based formulas.

6.7.2 Barium

It is questionable whether barium (Ba) is an essential trace element for human beings. Barium is detected in animal foods in levels from 0.002 mg/100 g in poultry to ca. 0.03 mg/100 g in meat and dairy products (see Table 6.13 and Table 6.14). Cow, goat, and sheep milk contain from 0.11 to 0.39 mg Ba/100 g; lower levels are observed in milk-based powder formulas. The concentrations of this metal in cheese range widely, from 0.03 to 0.50 mg/100 g (see Table 6.14). Toxicity, as well as absorption of this element from food, depends on its chemical speciation. Most of the barium present in foods seems to be relatively insoluble, and hence a large proportion of the barium ingested is promptly excreted.[1] Other chemical forms of barium are effectively dissolved in digestive fluids. Chronic barium poisoning due to consumption of contaminated table salt in a Chinese community was connected with increased cardiovascular disease.[1,106]

6.7.3 Bismuth

Levels of bismuth (Bi) in meat and meat products are usually very low, typically in the range of 0.00001 to 0.00009 mg/100 g (see Table 6.13). Milk and eggs contain low amount of the element (0.00001 mg/100 g). Among dairy products, cream skimmed contain from 0.001 to 0.003 mg Bi/100 g (see Table 6.14).

6.7.4 Cerium and Other Rare Earth Elements

Data on concentrations of rare earth elements (REEs) in animal foods is extremely limited.

TABLE 6.13
Other Metals in Meat, Fish, and Poultry (in μg/100 g w/w)

Name	Origin	Au	Ba	Bi	Ge	Ir	Li	Pd	Pt	Rh	Ru	Sb	Sr	Th	Tl	U	Ref.
Carcass meat[b]	U.K.	0.1	5	0.02	0.2	0.1	2	0.04	0.01	< 0.01	< 0.2	0.2	10	—	0.3	—	7
Emu meat		—	ND	—	—	—	—	—	—	—	—	—	10	—	—	—	112
Meat products[b]	U.K.	0.1	30	0.04	0.2	< 0.1	1	0.1	< 0.01	< 0.01	< 0.2	0.4	60	—	0.1	—	7
Poultry[b]	U.K.	0.1	2	0.01	< 0.2	< 0.1	1	< 0.03	0.01	< 0.01	< 0.2	0.1	10	—	0.2	—	7
Offal[b]	U.K.	0.1	10	0.03	0.2	0.1	3	0.2	< 0.01	0.05	< 0.2	0.1	10	—	0.3	—	7
Fish[b]	U.K.	0.1	20	0.1	0.2	< 0.1	10	0.2	< 0.01	0.02	< 0.2	0.3	360	—	0.1	—	7
Fish	Poland	—	—	—	—	—	—	—	—	—	—	—	—	< 0.03 – 0.2		< 0.02 – 0.1	113

TABLE 6.14
Other Metals in Milk, Dairy Products, and Eggs (in mg/100 g w/w)

Name	Origin	n	Ag	Au	Ba	Bi	Ge	Ref.
Milk								
Milk[b]	U.K.		—	0.00004	0.01	0.00001	<0.0002	7
Raw goat milk		2	—	—	0.19 ± 0.02 0.17 – 0.21	—	—	19
Raw sheep milk		4	—	—	0.31 ± 0.05 0.26 – 0.39	—	—	19
Human milk			< 0.00001 – 0.004	0.00001 – 0.0002	—	—	—	80
Raw milk		7	—	—	0.21 ± 0.08 0.11 – 0.37	—	—	17
Milk based liquid first and follow-on formulas[c]	London, U.K.	18	—	—	0.005 ± 0.004	—	—	24
Ready to drink infant formulas			< 0.00001 – 0.0001	< 0.000005 – 0.00002	—	—	—	80
Milk-based powder formulas[c]	Lagos, Nigeria	6	—	—	0.004 ± 0.002	—	—	24
Milk-based powder formulas[c]	London, U.K.	12	—	—	0.002 ± 0.001	—	—	24
Milk-based powder formulas[c]	U.S.	9	—	—	0.002 ± 0.001	—	—	24
Soy-based powder formula[c]	U.S.	6	—	—	0.005 ± 0.001	—	—	24
Dairy products								
Commercial goat cheese		2	—	—	0.17 ± 0.04	—	—	19
Commercial sheep cheese		4	—	—	0.26 ± 0.14 0.14 – 0.50	—	—	19

TABLE 6.14 (CONTINUED)
Other Metals in Milk, Dairy Products, and Eggs (in mg/100 g w/w)

Name	Origin	n	Ag	Au	Ba	Bi	Ge	Ref.
Cream partially skimmed	Spain	6	—	—	—	0.002 ± 0.0003 0.001 – 0.002	—	84
Cream, skimmed	Spain	8	—	—	—	0.002 ± 0.001 0.001 – 0.003	—	84
Full cream	Spain	4	—	—	—	0.002 ± 0.0002 0.002 – 0.002	—	84
Market white brine cheese	Turkey	10	—	—	0.03 ± 0.01	—	—	30
Soft white cheese	Turkey	20	—	—	0.08 ± 0.03	—	—	30
Soft white cheese melted	Turkey	20	—	—	0.08 ± 0.02	—	—	30
Dairy products[b]	U.K.		—	0.00004	0.031	0.0001	0.0002	7
Eggs[b]	U.K.			0.00004	0.05	0.00001	< 0.0002	7

Name	Origin	n	Ir	Li	Pd	Pt	Rb	Rh	Ru	Ref.
Milk[b]	U.K.		< 0.0002	0.0003	< 0.00003	< 0.00001	—	0.00003	0.0002	7
Raw goat milk		2	—	—	—	0.05 ± 0.00005	—	—	—	19
Raw sheep milk		4	—	—	—	0.04 ± 0.002 0.04 – 0.05	—	—	—	19
Human milk			—	—	—	< 0.000001 – 0.000004	—	—	—	80
Raw milk		7	—	—	—	< 0.00001	—	—	—	17
Ready to drink infant formulas			—	—	—	< 0.000001	—	—	—	80
Dairy Products										
Commercial goat cheese		2	—	—	—	0.03 ± 0.002	—	—	—	19

Soft cheeses			—	—	—	—	0.15 0.07 – 0.35	—	—	28
Commercial sheep cheese		4	—	—	—	0.03 ± 0.004 0.02 – 0.03	—	—	—	19
Hard cheeses			—	—	—	—	0.08 0.07 – 0.11	—	—	28
Quark cheeses			—	—	—	—	0.11 0.1 – 0.16	—	—	28
Stirred curd cheeses			—	—	—	—	0.06 0.01 – 0.11	—	—	28
Dairy products[b]	U.K.		0.0002	0.0005	0.00004	0.00001	—	0.00001	0.0002	7
Eggs[b]	U.K.		< 0.0002	0.001	0.00004	< 0.00001	—	< 0.00001	< 0.0002	7

Name	Origin	n	Sb	Sc	Sr	Te	Ti	Tl	Ref.
Milk									
Cow milk partially skimmed[c]		3	—	—	—	0.001 ± 0.0001 0.001 – 0.001	—	—	111
Cow milk skimmed[c]		3	—	—	—	0.001 ± 0.0004 0.00001 – 0.001	—	—	111
Skimmed cow milk[c]		5	—	—	0.05 ± 0.002	—	—	—	18
Cow whole milk[c]		5	—	—	0.06 ± 0.002	—	—	—	18
UHT cow whole milk[c]		5	—	—	0.05 ± 0.003	—	—	—	18
Skimmed UHT cow milk[c]		5	—	—	0.05 ± 0.002	—	—	—	18
Cow milk vegetable base[c]		3	—	—	—	0.0003 ± 0.0001	—	—	111
Milk[b]	U.K.		< 0.0001	—	0.03	—	—	< 0.0001	7
Goat milk		3	0.001 ± 0.00005	—	—	—	—	—	82
Raw goat milk		2	—	—	0.47 ± 0.01	—	—	—	19
Raw sheep milk		4	—	—	0.56 ± 0.06 0.54 – 0.62	—	—	—	19
Human milk			—	< 0.00001 – 0.0001	—	—	< 0.0003 – 0.01	—	80

TABLE 6.14 (CONTINUED)
Other Metals in Milk, Dairy Products, and Eggs (in mg/100 g w/w)

Name	Origin	n	Sb	Sc	Sr	Te	Ti	Tl	Ref.
Human whole milk[c]		5	—	—	0.01 ± 0.001	—	—	—	18
Skimmed human milk[c]		5	—	—	0.01 ± 0.0005	—	—	—	18
Vegetal-based milk	Spain	2	0.0003 – 0.00003	—	—	—	—	—	82
Vegetal-based milk	Vizcaya, Spain	2	—	—	—	0.0003 ± 0.00002 0.0002 – 0.0003	—	—	82
Raw milk		7	—	—	0.42 ± 0.31 0.10 – 1.01	—	—	—	17
Vegetal-based milk	Spain	2	0.0003 – 0.00003	—	—	—	—	—	82
Vegetal-based milk	Vizcaya, Spain	2	—	—	—	0.0003 ± 0.00002 0.0002 – 0.0003	—	—	82
Formula milk[c]		5	—	—	0.07 ± 0.003	—	—	—	18
Formula milk whey[c]		5	—	—	0.03 ± 0.001	—	—	—	23
Milk-based liquid first and follow-on formulas[c]	London, U.K.	18	0.0004 ± 0.001	—	0.02 ± 0.005	—	0.01 ± 0.005	0.001 ± 0.002	24
Ready-to-drink infant formulas			—	< 0.00001	—	—	< 0.0003 – 0.0005	—	80
Skimmed formula milk[c]		5	—	—	0.06 ± 0.003	—	—	—	18
Milk-based powder formulas[c]	Lagos, Nigeria	6	0.0003 ± 0.0003	—	0.01 ± 0.001	—	0.001 ± 0.0003	0.003 ± 0.001	24
Milk-based powder formulas[c]	London, U.K.	12	0.001 ± 0.001	—	0.01 ± 0.003	—	0.001 ± 0.0004	0.001 ± 0.001	24
Milk-based powder formulas[c]	U.S.	9	0.001 ± 0.002	—	0.02 ± 0.01	—	0.002 ± 0.001	0.004 ± 0.002	24
Soy-based powder formula[c]	U.S.	6	0.001 ± 0.001	—	0.02 ± 0.01	—	0.003 ± 0.001	0.002 ± 0.001	24

Cow milk whey[c]		5	—	—	0.03 ± 0.001	—	—	—	23
Milk whey coming from cow milk[c]		5	—	—	0.03 ± 0.001	—	—	—	18
Milk whey coming from UHT cow milk[c]		5	—	—	0.02 ± 0.001	—	—	—	18
UHT cow milk whey[c]		5	—	—	0.03 ± 0.002	—	—	—	23
Human milk whey[c]		5	—	—	0.01 ± 0.0005	—	—	—	23
Milk whey coming from human milk[c]		5	—	—	0.01 ± 0.0005	—	—	—	18
Milk whey coming from formula milk[c]		5	—	—	0.02 ± 0.001	—	—	—	18
Dairy products									
Commercial goat cheese		2	—	—	0.38 ± 0.09	—	—	—	19
Commercial sheep cheese		4	—	—	0.35 ± 0.07 0.27 – 0.45	—	—	—	19
Goat full cream[c]		3	—	—	—	0.001 ± 0.00002	—	—	111
Skimmed cream		4	0.001 ± 0.0002 0.0003 – 0.001	—	—	—	—	—	82
Cream partially skimmed	Spain	6	0.001 ± 0.0003 0.0005 – 0.001	—	—	0.001 ± 0.0001 0.001 – 0.001	—	—	84
Cream partially skimmed	Spain	6	0.001 ± 0.0003 0.0005 – 0.001	—	—	—	—	—	82
Cream partially skimmed	Spain	6	—	—	—	0.001 ± 0.0001 0.001 – 0.001	—	—	82
Cream partially skimmed		3	0.001 ± 0.0003 0.0005 – 0.001	—	—	—	—	—	82
Cream skimmed	Spain	8	0.001 ± 0.0002 0.0003 – 0.001	—	—	0.0005 ± 0.0004 0.0001 – 0.001	—	—	84
Cream skimmed	Spain	6	0.001 ± 0.0001 0.001 – 0.001	—	—	—	—	—	82

TABLE 6.14 (CONTINUED)
Other Metals in Milk, Dairy Products, and Eggs (in mg/100 g w/w)

Name	Origin	n	Sb	Sc	Sr	Te	Ti	Tl	Ref.
Cream skimmed	Spain	6	—	—	—	0.001 ± 0.0004 0.0001 – 0.001	—	—	82
Cow full cream[c]	Spain	2	—	—	—	0.001 ± 0.0001 0.001 – 0.001	—	—	111
Full cream	Spain	4	0.001 ± 0.0001 0.001 – 0.001	—	—	0.001 ± 0.0001 0.001 – 0.001	—	—	84
Full cream	Spain	4	0.001 ± 0.0001 0.001 – 0.001	—	—	—	—	—	82
Full cream	Spain	4	—	—	—	0.001 ± 0.0001 0.001 – 0.001	—	—	82
Full cream		2	0.001 ± 0.0001 0.001 – 0.001	—	—	—	—	—	82
Dairy products[b]	U.K.		0.0001	—	0.1	—	—	0.0001	7
Eggs[b]	U.K.		0.0001	—	0.05	—	—	0.0001	7

6.7.5 Caesium and Germanium

Levels of caesium (Cs) in food products can be very low, mostly below the detection limit, e.g., < 0.00005 mg/100 g, except when accidental contamination is introduced.[1] Most of the studies on occurrence of radiocaesium ($^{134/137}Cs$) in food are recent and were performed just after the Chernobyl accident[96] (see Chapter 11 of this book).

Concentrations of germanium (Ge) in animal foods (offal, poultry, eggs, fish, milk, and dairy products) are low and range from < 0.0002 to 0.0002 mg/100 g (see Table 6.13 and Table 6.14).

6.7.6 Gold

Concentrations of gold (Au) in animal foods range from 0.00004 to 0.00010 mg/100 g. Milk, dairy products, and eggs contain 0.00001 to 0.00004 mg Au/100 g, but the levels in offal, fish, and meat products are somewhat higher, being up to 0.0001 mg/100 g (see Table 6.14). Lower levels are noted for ready-to-drink instant formulas (< 0.000005 to 0.000020 mg/100 g).

6.7.7 Lithium

Different groups of animal food contain low levels of lithium (Li), ranging from 0.0003 to 0.006 mg/kg. The lowest values are registered for milk and dairy products, whereas the highest levels are found in fish (see Table 6.13 and Table 6.14). Concentrations of this element in honey broadly range from 0.0003 to 0.005 mg/100 g (see Table 6.15).

6.7.8 Platinum, Palladium, and other Platinum Metals

Platinum (Pt) and palladium (Pd) (well-known catalytic metals), as well as ruthenium (Ru), iridium (Ir), and rhodium (Rh), are present in animal foods in extremely low and varying amounts, ranging from < 0.000001 to 0.045 mg/100 g (see Table 6.13 and Table 6.14). The lowest levels are found in human milk, but goat and sheep milk shows high values, from 0.04 to 0.05 mg Pt/100 g. As shown in Table 6.14, commercial goat and sheep cheeses are also characterized by higher levels of platinum (0.02 to 0.03 mg/100 g). The levels of palladium are generally several times higher compared to those of platinum. Higher values for different platinum metals can be attributable to their emission by cars equipped with catalytic converters for automobile exhaust purification.[107,108] It has been reported that some species of sea fish can accumulate platinum metals. According to Sures et al.,[109] Pd emitted in this way is bioavailable for European eels (*Anguilla anguilla*).

6.7.9 Rubidium

The concentrations of rubidium (Rb) in dairy products range from 0.01 to 0.35 mg /100 g (see Table 6.14). Honey contains similar levels of the metal, in the range of 0.01 to 0.15 mg /100 g (see Table 6.15).

6.7.10 Scandium

Concentrations of scandium (Sc) in animal foods are low and range from < 0.00001 to 0.0001 mg/100 g in human milk and infant formulas (see Table 6.14) and < 000003 to 0.02 mg/100 g in honey (see Table 6.15).

6.7.11 Silver

Silver (Ag) occurs naturally at low levels in animal foods, mostly at or below detection limits of many analytical instruments. The intake of silver may be increased where silver or silver-plated utensils are used in preparation and storage of food.[1] Concentration of this metal in milk and ready to drink formula range from < 0.00001 to 0.004 mg/100 g (see Table 6.14). Honey contains from 0.01 to 0.02 mg Ag/100 g (see Table 6.15).

6.7.12 Strontium

Meat products typically contain 0.06 mg Sr/100 g, whereas lower levels of this metal have been reported in carcass meat and offal (0.01 mg/100 g), and higher in fish meat (0.36 mg/100 g) (see Table 6.13). The concentrations of strontium (Sr) in cow milk range from 0.03 to 0.06 mg/100 g. Sheep and goat milk shows higher its levels, from 0.47 to 0.62 mg/100 g (see Table 6.14). Similar values for strontium are reported in dairy products, e.g., cheese (0.27-0.45 mg/100 g). Honey contains from 0.05 to 0.32 mg Sr/100 g (see Table 6.15). Most of studies on occurrence of radiostrontium

TABLE 6.15
Other Metals in Honey (in mg/100 g w/w)

Name	Origin	n	Ag	Li	Rb	Sc	Sr	Ref.
Clover honey	Egypt		—	—	—	—	0.26 ± 0.02	33
Honey	France	86	0.01 0.01 – 0.02	0.004 0.003 – 0.005	—	—	—	31
Honey	U.S.		—	—	0.02 ± 0.01 0.01 – 0.03	0.01 ± 0.01 < 0.00003 – 0.02	< 0.2	36
Industrial Galician honey	Spain	20	—	0.0003 ± 0.0001	0.03 ± 0.07	—	—	38
Natural Galician honey	Spain	22		0.001 ± 0.001	0.15 ± 0.12	—	—	38
Orange honey	Egypt		—	—	—	—	0.04 ± 0.01	33
Sesame honey	Egypt		—	—	—	—	0.04 ± 0.0	33
Syrup-feed honey	Egypt		—	—	—	—	0.04 ± 0.1	33

(^{90}Sr) in food, especially fish and other seafood, were undertaken after the Chernobyl disaster[96] (see Chapter 11 of this book).

6.7.13 Tellurium

Levels of tellurium (Te) in animal products are very low, generally below 0.0005 mg/100 g.[1] Among dairy products, skimmed cow milk contains from 0.00001 to 0.001 mg Te/100 g whereas cheese (parmesan) and cream may contain up to 0.001 mg Te/100 g (see Table 6.14). Concentration of this element in vegetal-based milk range from 0.0002 to 0.0003 mg/100 g.

6.7.14 Thallium

Thallium (Tl) is totally absorbed from food and highly toxic. Its levels in animal food are somewhat lower than in plant products and are almost near or below the analytical detection, i.e., 0.0001 mg/100 g. The concentrations of this element in milk-based formulas and soy-based powder formula range from < 0.0001 to 0.004 mg/100 g (see Table 6.14). Meats, meat products, offal, and poultry contain from 0.0001 to 0.0003 mg Tl/100 g (see Table 6.13).

6.7.15 Titanium

There is little or no information on levels of titanium (Ti) in animal foods. It is believed that most of the reported concentrations (low) can be attributed to external contamination.[1] Among the animal foods, somewhat higher levels are noted for dairy products; these data need to be validated using reliable certified materials (CRMs), however.[1,110] Recent studies report concentration of titanium in milk and milk-based infant formula to be in the range of < 0.0003 to 0.01 mg/100 g (see Table 6.14).

6.7.16 Uranium and Thorium

Information on concentrations of uranium (U) and thorium (Th) in edible parts of marine fish is very limited. Szefer et al.[113] have reported a wide range of uranium levels in fish muscle, from < 0.00001 to 0.004 mg/100 g. Fish muscle contains similar thorium levels, in the range of < 0.00001 to 0.01 mg/100 g (see Table 6.13).

REFERENCES

1. Reilly, C., *Metal Contamination of Food: Its Significance for Food Quality and Human Health*, Blackwell Science, Oxford, 2002.
2. Reilly, C., *The Nutritional Trace Metals*, Blackwell Publishing, Oxford, 2004.
3. Capar, S.G. and Szefer P., Determination and speciation of trace elements in foods, in *Methods of Analysis of Food Components and Additives*, Otles, S., Ed., CRC Press, Boca Raton, FL, 2005, chap. 6.

4. Sutton, K.L. and Heitkemper, D.T., Speciation analysis of biological, clinical and nutritional samples using plasma spectrometry, in *Elemental Speciation. New Approaches for Trace Element Analysis,* Caruso, J.A., Sutton, K.L., and Ackley, K.L., Eds., Elsevier Science, Amsterdam, 2000, p. 501.
5. González, E.B. and Sanz-Medel, A., Liquid chromatographic techniques for trace element speciation analysis, in *Elemental Speciation: New Approaches for Trace Element Analysis,* Caruso, J.A., Sutton, K.L., and Ackley, K.L., Eds., Elsevier Science, Amsterdam, 2000, p. 81.
6. Rojas, E. et al., Are metals dietary carcinogens?, *Mutation Res.*, 443, 157, 1999.
7. Ysart, G. et al., Dietary exposure estimates of 30 elements from the U.K. Total Diet Study, *Food Addit. Contam.*, 16, 391, 1999.
8. Capar, G.S. and Cunningham, W.C., Element and radionuclide concentrations in food: FDA total diet study 1991–1996, *J. AOAC Int.*, 83, 157, 2000.
9. Coultate, T.P., *Food: The Chemistry of Its Components*, Royal Society of Chemistry, Cambridge, MA, 2002.
10. González-Martín, I. et al., Mineral analysis (Fe, Zn, Ca, Na, K) of fresh Iberian pork loin by near infrared reflectance spectrometry: determination of Fe, Na, and K with remote fiber-optic reflectance probe, *Anal. Chim. Acta*, 468, 293, 2002.
11. Santos, E.M. et al., Physicochemical and sensory characterisation of Morcilla de Burgos, a traditional Spanish blood sausage, *Meat Sci.*, 65, 893, 2003.
12. Gökoǯlu, N. and Yerlikaya, P., Determination of proximate composition and mineral contents of blue crab *(Callincetes sapidus)* and swim crab *(Portunus pelagicus)* caught off the Gulf of Antalya, *Food Chem.*, 80, 495, 2003.
13. Riget, F., Johansen, P., and Asmund, G., Influence of length of elements concentrations in Blue Mussels (*Mytilus edulis*), *Mar. Pollut. Bull.*, 32, 745, 1996.
14. Kwoczek, M. et al., Essential and toxic elements in seafood available in Poland from different geographical regions, *J. Agric. Food Chem.*, 54, 3015, 2006.
15. Lindmark-Månsson, H., Fondén, R., and Pettersson, H.-E., Composition of Swedish dairy milk, *Int. Dairy J.,* 13, 409, 2003.
16. Rodríguez Rodríguez, E.M., Sanz Alaejos, M., and Romero, C.D., Mineral concentrations in cow's milk from the Canary Island, *J. Food Compos. Anal.*, 14, 419, 2001.
17. Coni, E. et al., Preliminary evaluation of the factors influencing the trace element content of milk and dairy products, *Food Chem.*, 52, 123, 1995.
18. Rivero Martino, F.A., Fernández Sánchez, M.L., and Sanz-Medel, A., The potential of double focusing-ICP-MS for studying elemental distribution patterns in whole milk, skimmed milk and milk whey of different milks, *Anal. Chim. Acta*, 442, 191, 2001.
19. Coni, E. et al., Minor and trace element content in sheep and goat milk and dairy products, *Food Chem.*, 57, 253, 1996.
20. Bocca, B. et al., Determination of the total content and binding pattern of elements in human milk by high performance liquid chromatography-inductively coupled plasma atomic emission spectrometry, *Talanta*, 53, 295, 2000.
21. Booth, C.K., Reilly, C., and Farmakalidis, E., Mineral composition of Australian ready-to-eat breakfast cereals, *J. Food Compos. Anal.*, 9, 135, 1996.
22. Hua, K.M., Kay, M., and Indyk, H.E., Nutritional element analysis in infant formulas by direct dispersion and inductively coupled plasma-optical emission spectrometry, *Food Chem.*, 68, 463, 2000.
23. Rivero Martino, F.A., Fernández, M.L., and Sanz Medel, A., Total determination of essential and toxic elements in milk whey by double focusing ICP-MS, *J. Anal. At. Spectrom.*, 15, 163, 2000.

24. Ikem, A. et al., Levels of 26 elements in infant formula from USA, U.K., and Nigeria by microwave digestion and ICP–OES, *Food Chem.*, 77, 439, 2002.
25. De La Fuente, M.A. et al., Total and soluble contents of calcium, magnesium, phosphorus and zinc in yoghurts, *Food Chem.*, 80, 573, 2003.
26. Sánchez-Segarra, P.J. et al., Influence of the addition of fruit on the mineral content of yoghurts: nutritional assessment, *Food Chem.*, 70, 85, 2000.
27. Pillonel, L. et al., Stable isotope ratios, major, trace and radioactive elements in emmental cheeses of different origins, *Lebensm.-Wiss. u-Technol.*, 36, 615, 2003.
28. Gambelli, L. et al., Minerals and trace elements in some Italian dairy products, *J. Food Compos. Anal.*, 12, 27, 1999.
29. Peláez Puerto, P. et al., Chemometric studies of fresh and semi-hard goats' cheeses produced in Tenerife (Canary Islands), *Food Chem.*, 88, 361, 2004.
30. Merdivan, M. et al., Basic nutrients and element contents of white cheese of diyarbakir in Turkey, *Food Chem.*, 87, 163, 2004.
31. Devillers, J. et al., Chemometrical analysis of 18 metallic and nonmetallic elements found in honeys sold in France, *J. Agric. Food Chem.*, 50, 5998, 2002.
32. Terrab, A. et al., Mineral content and electrical conductivity of the honeys produced in Northwest Morocco and their contribution to the characterization of unifloral honeys, *J. Sci. Food Agric.*, 83, 637, 2003.
33. Rashed, M.N. and Soltan, M.E., Major and trace elements in different types of Egyptian mono-floral and non-floral bee honeys, *J. Food Compos. Anal.*, 17, 725, 2004.
34. Yilmaz, H. and Yavuz, O., Content of some trace metals in honey from south-eastern Anatolia, *Food Chem.*, 65, 475, 1999.
35. López-García, I. et al., Fast determination of calcium, magnesium and zinc in honey using continuous flow flame atomic absorption spectrometry, *Talanta*, 49, 597, 1999.
36. Iskander, F.Y., Trace and minor elements in four commercial honey brands, *J. Radioanal. Nucl. Chem. Lett.*, 201, 401, 1995.
37. Nanda, V. et al., Physico-chemical properties and estimation of mineral content in honey produced from different plants in Northern India, *J. Food Compos. Anal.*, 16, 613, 2003.
38. Latorre, M.J. et al., Chemometric classification of honeys according to their type. II. Metal content data, *Food Chem.*, 66, 263, 1999.
39. Conti, M.E., Lazio region (central Italy) honeys: a survey of mineral content and typical quality parameters, *Food Contr.*, 11, 459, 2000.
40. Kump, P., Nečemer, M., and Šnajder, J., Determination of trace elements in bee honey, pollen and tissue by total reflection and radioisotope x-ray fluorescence spectrometry, *Spectrochim. Acta* Part B, 51, 499, 1996.
41. Al-Khalifa, A.S. and Al-Arify, I.A., Physicochemical characteristics and pollen spectrum of some Saudi honeys, *Food Chem.*, 67, 21, 1999.
42. Darrie, G., The importance of chromium in occupational health, in *Trace Element Speciation for Environment, Food and Health*, Ebdon, L. et al., Eds., Royal Society of Chemistry, Cambridge, MA, 2001, p. 315.
43. Koops, J., Klomp, H., and Westerbeek, D., Spectroscopic determination of nickel and furildioxime with special reference to milk and milk products and to the release of nickel from stainless steel by acidic dairy products and by acid cleaning, *Neth. Milk Dairy J.*, 36, 333, 1982.
44. Margolin, S., Mercury in marine seafood: the scientific medical margin of safety as a guide to the potential risk to public health, *World Rev. Nutr. Diet*, 34, 182, 1980.

45. Hansen, J.C., Kromann, N., and Wulf H.C., Selenium and its interrelation with mercury in wholeblood and hair in an East Greenland population, *Sci. Total Environ.*, 38, 33, 1984.
46. Van Deal, P., Trace element speciation in food: a tool to assure food safety and nutritional quality, in *Trace Element Speciation for Environment, Food and Health*, Ebdon, L. et al., Eds., Royal Society of Chemistry, Cambridge, MA, 2001, p. 232.
47. Osendarp, S.J.M. et al., Zinc supplementation during pregnancy and effects on growth and morbidity in low birthweight infants: a randomized placebo controlled trial, *Lancet*, 357, 1080, 2001.
48. Prasad, A.S. et al., Zinc metabolism in patients with the syndrome of iron deficiency anemia, hepatosplenomegaly, dwarfism and hypogonadism, *J. Lab. Clin. Med.*, 61, 537, 1963.
49. Walsh, C.T. et al., Zinc health effects and research priorities for the 1990s, *Environ Health Perspect.*, 102, 5, 1994.
50. Gibson, R.S. and Ferguson, E.L., Nutrition interventions to combat zinc deficiencies in developing countries, *Nutr. Res. Rev.*, 11, 115, 1998.
51. Kot, A., Zaręba, S., and Wyszogrodzka-Koma, L., Copper, zinc, manganese and iron in Polish poultry, pork, bovine meat and meat products, *Bromat. Chem. Toksykol.*, 35, 39, 2002.
52. Leonhardt, M. and Wenk, C., Variability of selected vitamins and trace elements of different meat cuts, *J. Food Compos. Anal.*, 10, 218, 1997.
53. Falandysz, J., Some toxic and essential trace metals in swine from Northern Poland, *Sci. Total Environ.*, 136, 193, 1993.
54. López Alonso, M. et al., Contribution of cattle products to dietary intake of trace and toxic elements in Galicia, Spain, *Food Addit. Contam.*, 19, 533, 2002.
55. Abou-Arab, A.A.K., Heavy metal contents in Egyptian meat and the role of detergent washing on their levels, *Food Chem. Toxicol.*, 39, 593, 2001.
56. Sedki, A. et al., Toxic and essential trace metals in muscle, liver and kidney of bovines from a polluted area of Morocco, *Sci. Total Environ.*, 317, 201, 2003.
57. Falandysz, J., Some toxic and essential trace metals in cattle from the northern part of Poland, *Sci. Total Environ.*, 136, 177, 1993.
58. Falandysz, J., Manganese, copper, zinc, iron, cadmium, mercury and lead in muscle meat, liver and kidneys of poultry, rabbit and sheep slaughtered in the northern part of Poland, 1987, *Food Addit. Contam.*, 8, 71, 1991.
59. Brito, G. et al., Differentiation of heat- treated pork liver pastes according to their metal content using multivariate data analysis, *Eur. Food Res. Technol.,* 218, 584, 2004.
60. Falandysz, J. and Kotecka, W., Stężenia metali w wybranych produktach spożywczych Trójmiasta, *Bromat. Chem. Toksykol*, 26, 143, 1993.
61. Jeng, M.-S. et al., Mussel Watch: a review of Cu and other metals in various marine organisms in Taiwan, 1991–1998, *Environ. Pollut.*, 110, 207, 2000.
62. Mendez, H. et al., Ultrasonic extraction combined with fast furnace analysis as an improved methodology for total selenium determination in seafood by electrothermal-atomic absorption spectrometry, *Anal. Chim. Acta.*, 452, 217, 2002.
63. Blasco, J., Arias, A.M., and Saenz, V., Heavy metals in organisms of the River Guadalquivir estuary: possible incidence of the Aznalcollar disaster, *Sci. Total Environ.*, 242, 249, 1999.
64. Páez-Osuna, F. and Ruiz-Fernández, C., Trace metals in the Mexican shrimp *Penaeus vannamei* from estuarine and marine environments, *Environ. Pollut.*, 87, 243, 1995.

65. Páez-Osuna, F. and Tron-Mayen, L., Concentration and distribution of heavy metals in tissues of wild and farmed shrimp *Panaeus vannamei* from the north-west coast of Mexico, *Environ. Int.*, 22, 442, 1996.
66. Johansen, P., Pars T., and Bjerregaard, P., Lead, cadmium, mercury and selenium intake by Greenlanders from local marine food, *Sci. Total Environ.*, 245, 187, 2000.
67. Turoczy, N.J. et al., Cadmium, copper, mercury, and zinc concentrations in tissues of the King Crab (*Pseudocarcinus gigas*) from Southeast Australian waters, *Environ. Int.*, 27, 327, 2001.
68. Páez-Osuna, F. et al., Trace metal concentrations and their distribution in the lobster *Panulirus inflatus* (Bouvuer, 1895) from the Mexican Pacific coast, *Environ. Pollut.*, 90, 163, 1995.
69. Yerba, M.C. and Moreno-Cid, A., On-line determination of manganese in solid seafood samples by flame atomic absorption spectrometry, *Anal. Chim. Acta*, 477, 149, 2003.
70. Yap, C.K., Ismail, A., and Tan, S.G., Heavy metal (Cd, Cu, Pb and Zn) concentrations in the green-lipped mussel *Perna viridis* (Linnaeus) collected from some wild and aquacultural sites in the west coast of Peninsular Malaysia, *Food Chem.*, 84, 569, 2004.
71. Wong, C.K.C., Cheung, R.Y.H., and Wong, M.H., Heavy metal concentrations in green-lipped mussels collected from Tolo Harbor and markets in Hong Kong and Shenzhen, *Environ. Pollut.*, 109, 165, 2000.
72. Park, J. and Presley, B.J., Trace metal contamination of sediments and organisms from the Swan Lake area of Galveston Bay, *Environ. Pollut.*, 98, 209,1997.
73. Viñas, P. et al., Electrothermal atomic absorption spectrometric determination of molybdenum, aluminum, chromium and manganese in milk, *Anal. Chim. Acta*, 356, 267, 1997.
74. Tripathi, R.M. et al., Daily intake of heavy metals by infants through milk and milk products, *Sci. Total Environ.*, 227, 229, 1999.
75. Rodríguez Rodríguez, E.M., Alaejos, M.S., and Romero, C.D., Comparison of mineralization methods for fluorimetric determination of selenium in milks, *Z. Lebensm. Unters. Forsch. A*, 204, 425, 1997.
76. Aleixo, P.C. and Nóbrega, J.A., Direct determination of iron and selenium in bovine milk by graphite furnace atomic absorption spectrometry, *Food Chem.*, 83, 457, 2003.
77. Falandysz, J. and Kotecka, W., Zawartość manganu, miedzi, cynku i żelaza w produktach nabiałowych, odżywkach dla dzieci i słodyczach, *Bromat. Chem. Toksykol.*, XXVII, 77, 1994.
78. Bermejo, P., Dominguez, R., and Bermejo, A., Direct determination of Fe and Zn in different components of cow milk by FAAS with a high performance nebulizer, *Talanta*, 45, 325, 1997.
79. Licata, P. et al., Levels of "toxic" and "essential" metals in samples of bovine milk from various dairy farms in Calabria, Italy, *Environ. Int.*, 30, 1, 2004.
80. Prohaska, T. et al., Determination of trace elements in human milk by inductively coupled plasma sector field mass spectrometry (ICP-SFMS), *J. Anal. At. Spectrom.*, 15, 335, 2000.
81. Rodríguez Rodríguez, E.M., Alaejos, M.S., and Romero, C.D., Concentrations of selenium in human milk, *Z. Lebensm. Unters. Forsch. A*, 207, 174, 1998.
82. Cava-Montesinos, P. et al., Determination of arsenic and antimony in milk by hydride generation atomic fluorescence spectrometry, *Talanta*, 60, 787, 2003.
83. López, J.C. et al., Mathematic predictive models for calculating copper, iron and zinc dialysability in infant formulas, *Eur. Food. Res. Technol.*, 212, 608, 2001.

84. Cava-Montesinos, P. et al., Determination of As, Sb, Se, Te, and Bi in milk by slurry sampling hydride generation atomic fluorescence spectrometry, *Talanta*, 62, 175, 2004.
85. Fakayode, S.O. and Olu-Owolabi, I.B., Trace metal content and estimated daily human intake from chicken eggs in Ibadan, Nigeria, *Arch. Environ. Health*, 58, 245, 2003.
86. Falandysz, J. and Kotecka, W., Stężenia metali w wybranych produktach spożywczych Trójmiasta, *Bromat. Chem. Toksykol.*, 26, 143, 1993.
87. Sanna, G. et al., Determination of heavy metals in honey by anodic stripping voltammetry at microelectrodes, *Anal. Chim. Acta*, 415, 165, 2000.
88. Caroli, S. et al., Determination of essential and potentially toxic trace elements in honey by inductively coupled plasma-based techniques, *Talanta*, 50, 327, 1999.
89. Buldini, P.L. et al., Ion chromatographic and voltammetric determination of heavy and transition metals in honey, *Food Chem.*, 73, 487, 2001.
90. Koo, W.W.K., Kaplan, L.A., and Krug-Wispe, S.K., Aluminum contamination of infant formulas, *J. Parenter. Enteral. Nutr.*, 12, 170, 1988.
91. Bouglé, D. et al., Concentrations en aluminum des formules pour prématurés, *Arch. Pédiatr.*, 46, 768, 1989.
92. Baxter, M.J., Burrell, J.A., and Massey, R.C., The aluminum content of infant formula and tea, *Food Addit. Contam.*, 7, 101, 1990.
93. Coni, E., Bellomonte, G., and Caroli, S., Aluminum content in of infant formulas, *J. Trace Elem. Electrol. Health Dis.*, 7, 83, 1993.
94. Vos, G., Hovens, J.P.C., and Delft, W.V., Arsenic, cadmium, lead and mercury in meat, livers and kidneys of cattle slaughtered in the Netherlands during 1980–1985, *Food Addit. Contam.*, 4, 73, 1987.
95. Moharram, Y.G. et al., Mercury content of some marine fish from the Alexandria coast, *Nahrung*, 31, 899, 1987.
96. Szefer, P., *Metals, Metalloids and Radionuclides in the Baltic Sea Ecosystem*, Elsevier Science, Amsterdam, 764 pp, 2002.
97. Capon, C.J. and Smith, J.C., Chemical form and distribution of mercury and selenium in edible seafood, *J. Anal. Toxicol.*, 6, 10, 1982.
98. Miranda, M. et al., Cadmium levels in liver, kidney and meat in calves from Asturias (North Spain), *Eur. Food Res. Technol.*, 212, 426, 2001.
99. López Alonso, M. et al., Mercury concentration in cattle from NW Spain, *Sci. Total Environ.*, 302, 93, 2003.
100. Husain, A. et al., Toxic metals in food products originating from locally reared animals in Kuwait, *Bull. Environ. Contam. Toxicol.*, 57, 549, 1996.
101. Haraguchi, K. et al., Detection of localized methylmercury contamination by use of the mussel adductor muscle in Minamata Bay and Kagoshima Bay, Japan, *Sci. Total Environ.*, 261, 75, 2000.
102. Vaz-Pires, P. and Barbosa, A., Sensory, microbiological, physical and nutritional properties of iced whole common octopus (*Octopus vulgaris*), *Lebensm.-Wiss. u-Technol.*, 37, 105, 2004.
103. Gajewska, R., Nabrzyski, M., and Kania, P., Zawartość rtęci w mleku i jego przetworach, *Bromat. Chem. Toksykol.*, XXVII, 29, 1994.
104. Schaum, J. et al., A national survey of persistent, bioaccumulative and toxic (PBT) pollutants in the U.S. milk supply, *J. Expo. Anal. Environ. Epidemiol.*, 13, 177, 2003.
105. Mendil, D., Mineral and trace metal levels in some cheese collected from Turkey, *Food Chem.*, 96, 532, 2006.

106. Rossa, O. and Berman, L.B., Barium poisoning due to consumption of contaminated salt, *J. Pharmacol. Exp. Ther.*, 177, 433, 1971.
107. Hodge, V.F. and Stallard, M.O., Platinum and palladium in roadside dust, *Environ. Sci. Technol.*, 20, 1058, 1986.
108. Helmers, E. and Kummerer, K., Platinum group elements in the environment — anthropogenic impact: platinum fluxes: quantification of sources and sinks, and outlook, *Environ. Sci. Pollut. Res.*, 6, 29, 1999.
109. Sures, B. et al., First report on the uptake of automobile catalyst emitted palladium by European eels (*Anguilla anguilla*) following experimental exposure to road dust, *Environ. Pollut.*, 113, 341, 2001.
110. Iyengar, V. and Wottiez, J., Trace elements in human clinical specimens evaluation of literature data to identify reference values, *Clin. Chem.*, 34, 474, 1988.
111. Ródenas-Torralba, E., Morales-Rubio, A., and de la Guardia, M., Multicommutation hydride generation atomic fluorescence determination of inorganic tellurium species in milk, *Food Chem.*, 91, 181, 2005.
112. Pegg, R.B., Amarowicz, R., and Code W.E., Nutritional characteristics of emu (*Dromaius novaehollandiae*) meat and its value-added products, *Food Chem.*, 97, 193, 2006.
113. Szefer, P., Szefer, K., and Falandysz J., Uranium and thorium in muscle tissue of fish taken from the southern Baltic, *Helgol. Meeresunters.*, 44, 31, 1990.

7 Mineral Components in Food Crops, Beverages, Luxury Food, Spices, and Dietary Food

Piotr Szefer and Małgorzata Grembecka

CONTENTS

7.1 INTRODUCTION

The levels of essential and nonessential elements in food crops are dependent on many factors, including genetic properties of the plants, nature of soil on which the plants grow, climatic conditions, and the degree of maturity of the plants during harvesting. There are marked regional variations in metal concentrations even for the same group of plants and vegetables. The spatial variability can be explained by different geochemical structures of soils constituting the substratum for cultivated plants. For instance, the concentration of selenium in rice grown in a selenium-rich region of China is much higher than that in rice produced from New Zealand.[1] Although soils can be a rich source of essential trace elements for plants requiring mineral nutrients to grow, nonessential and toxic elements, e.g., mercury, cadmium, and lead, can also be taken up by plants. An example of such accumulation is the contamination of potatoes and cereals by cadmium in some areas of Australia as a result of the use of cadmium-containing phosphate fertilizers.[1,2] The concentration of a given element in plants is dependent not only on the geochemical composition of the soil material but also on the availability of the element in the soil. Some plants themselves can assist in the uptake by modifying the geochemical composition of the associated soil solution by releasing hydrogen ions and organic chelating agents. The distribution of mineral components within the plant depends on whether or not the absorbed metal can be translocated upwards to the leaves and fruits, seeds, and other storage organs. Many metals can be retained in the roots. Some plants are characterized by the unusual ability to take up several metals from the soil substratum and accumulate them in specific tissues to abnormally high levels. Among plants known to accumulate metals are the tea plant, *Camellia sinensis*.

Environmental contamination may sometimes be a serious source of toxic elements such as lead and cadmium in food. The use of cast iron cooking pots may result in the significant increase of iron concentration in the diet. Stainless steel used in food processing and as storage equipment can be an important source of chromium and nickel, whereas brass and copper utensils are responsible for copper release into foods.[1,3,4]

Fortification of food by the addition of several nutrients, including mineral components, is commonly practiced and can have a serious effect on the nutritional status of the food supply. For instance, most of the cereals are poor dietary sources of trace elements and hence are often fortified with iron, copper, manganese, and zinc.[1] On the other hand, spinach and cauliflower appear to be good sources of cadmium, whereas Brussels sprout and Chinese beets are rich in lead content.[5] Besides interactions occurring among metals and with other major food components, any risk assessment must also consider metal pharmacokinetics and differences in the dietary habits of local populations and in the geographical distribution of metals.[5]

7.2 MACRONUTRIENTS IN FOOD

7.2.1 Calcium

Concentrations of calcium (Ca) in miscellaneous cereals range from 0.94 to 370 mg/100 g, with the lowest values in maize products (flour, corn flour, and grains) and the highest levels in soya flour (Table 7.1). White flours fortified with additional calcium carbonate are a rich source of calcium compared to wholemeal flour, which contains ca. 35.0 mg Ca/100 g.[6] The concentrations of calcium in food crops cover a wide range with lower values in apples, green pepper, and potatoes (< 8.7 mg/100 g) and higher values in broccoli (ca. 100 mg/100 g) and spinach (600 mg/100 g) (Table 7.2). Tea leaves and green coffee contain 138–723 mg Ca/100 g and 89.8–156 mg Ca/100 g, respectively (Table 7.3). Nuts (260 mg/100 g), cocoa powder (200 mg/100 g), and especially Egyptian sugar (1000 mg/100 g) are also a rich source of calcium (Table 7.4). Fruit juices generally contain from 3.5 to 18.5 mg Ca/100 g, whereas maple syrup and cactus pear juice contain the highest levels of this macroelement, namely, 77.5 and 59.0 mg/100 g, respectively. The concentrations of calcium in beer and wine are comparable and range from 1.99 to 10.8 mg/100 g (Table 7.5).

7.2.2 Magnesium

As can be seen in Table 7.1 and Table 7.2, grain products contain somewhat more magnesium (Mg, 7.57–156 mg/100 g) than vegetables and fruits (5.5–67.5 mg/100 g). Higher levels of this element are reported for tea leaves (56.3–266 mg/100 g), nuts (118–191 mg/100 g), and especially for instant coffee (212–415 mg/100 g)

TABLE 7.1
Macroelements in Miscellaneous Cereals[a]

Name	Origin	n	Ca	K	Mg	Na	P	Ref.
Bread								
Bread[b]	U.K.		146	—	—	—	—	7
Bread[b]	U.S.		91.5 ± 14.4	163 ± 39.1	39.8 ± 22.3	519 ± 65.3	140 ± 37.8	9
			71.8 – 108	121 – 223	18.6 – 73.7	441 – 622	91 – 193	
Bread			23	—	27	—	—	8
Bagel plain	U.S.		76.9	102	25.5	467	94.1	9
White bread	U.S.		103	121	22.3	518	91	9
Whole white bread	U.S.		89.1	223	73.7	508	193	9
Cornbread	U.S.		108	136	18.6	441	150	9
Rye bread	U.S.		71.8	164	36.4	622	121	9
Bread, white flour[c]			13.8	—	8.64	—	47.5	10
Bread, wholemeal flour[c]			2.06	—	38.2	—	116	10
Cereals								
Corn flakes	U.S.		2.8	90	8.5	1010	39	9
Oat flakes			50 ± 4	397 ± 1	154 ± 1	—	—	11
Rolled oats		6	47.1 ± 1.9	—	174 ± 12	—	—	1
			44 – 61		151 – 262			
Raisin bran cereal	U.S.		46.1	599	143	489	372	9
Fruit- flavored cereal	U.S.		27.9	112	28.3	489	100	9
Rice								
Rice	India, Pakistan	28	—	—	2.58	—	—	12
					7.57 – 46.9			

Rice	Europe	25	—	—	81.1	—	—	12
					21.4 – 138			
Rice	Different		7.65	141	24.4	—	—	13
Rice[c]			20.0 – 60.8	130–313	130 – 136	5.21 – 52.1	269 – 282	14
White rice		6	4.94 ± 0.06	—	32.5 ± 0.6	—	—	1
			4.66–5.36		28.6 – 36.7			
Brown rice		7	10.6 ± 0.2	—	124 ± 1.3	—	—	1
			9.8–11.7		118 – 134			
Grits				—		—	—	
Grits		10	0.82 ± 0.06	—	16.7 ± 1.6	—	—	1
			0.54 – 1.05		8.4 – 31.3			
				—		—	—	
Flour								
Flour			23	—	27	—	—	8
Maize flour		4	1.08 ± 0.03	—	24.8 ± 2.0	—	—	1
			0.94 – 1.19		18.1 – 28.1			
Refined oat flour		1	14.3 ± 0.2	—	33 ± 6.5	—	—	1
			14.0 – 15.0		10.5 – 42.3			
Wholemeal oat flour		2	51.2 ± 2.3	—	153 ± 0.9	—	—	1
			42.4 – 56.2		138 – 156			
White wheat flour		4	19.2 ± 0.3	—	30.8 ± 1.5	—	—	1
			17.6 – 21.1		28 – 33.6			
Soya flour		4	343 ± 7.7	—	401 ± 6.3	—	—	1
			326 – 37		374 – 429			
Miscellaneous Cereals								
Miscellaneous cereals[b]	U.K.		73.1	—	—	—	—	7
Miscellaneous cereals[b]	U.S.		62.0 ± 54.4	158 ± 140	36.6 ± 39.8	507 ± 366	169 ± 130	9
			< 1.0 – 191	22.7 ± 599	60 – 143	< 3.0 – 1090	17.5 – 447	
Oat bran			81 ± 0.8	630 ± 12.0	231 ± 8.3	—	—	11

TABLE 7.1 (CONTINUED)
Macroelements in Miscellaneous Cereals[a]

Name	Origin	n	Ca	K	Mg	Na	P	Ref.
Stabilized wheatgerm		3	84 ± 13.6	—	385 ± 78	—	—	1
			63 – 93.2		245 – 457			
Quinoa[c]			43.7 – 140	812–1894	175	3.49 – 83.8	236 – 437	14
Corn grains[c]			13.1 – 35	289 – 398	105 – 123	7.88 – 26.3	224 – 298	14
Soft whole wheat		4	36.3 ± 0.4	—	127 ± 1.2	—	—	1
			35.3 – 36		113 – 132			
Spelt wheat		9	26 ± 4	411 ± 26	133 ± 8	0.6 ± 0.1	421 ± 27	15
Raw maize		6	3.22 ± 0.2	—	108 ± 3.0	—	—	1
			2.04 – 4.98		99 – 122			
Oat grains[c]			60.9 – 91.4	341 – 383	114 – 122	15.7 – 69.6	290 – 331	14
Kibbled wheat		6	40.3 ± 0.4	—	129 ± 1.5	—	—	1
			39.6 – 40.5		127 – 140			
Wheat grains[c]			33.1 – 61.0	332 – 410	112 – 140	6.98 – 43.6	297 – 366	14
Decorticated rye		4	38.9 ± 0.4	—	136 ± 2.0	—	—	1
			36.8 – 41.6		127 – 143			
White roll	U.S.		73.9	128	25.6	518	112	9
Gluten		4	62.2 ± 0.7	—	26.5 ± 2.4	—	—	1
			33 – 90		14 – 37			
Dough			23	—	27	—	—	8

[a] in mg/100 g w/w
[b] Total diet studies, mean of all the products from the group
[b] recalculated from dry weight to wet weight

TABLE 7.2
Macronutrients in Vegetables, Mushrooms, and Fruits (in mg/100 g w/w)

Name	Origin	n	Ca	K	Mg	Na	P	Ref.
Vegetables								
Butterhead lettuce			47 ± 14	318 ± 92	18 ± 5	5 ± 2	—	16
Cabbage			44 ± 6	266 ± 87	14 ± 2	3 ± 1	—	16
Cabbages	Saudi Arabia	5	93.8	25.9	19.1	29.6	—	17
Canned vegetables[b]	U.K.		27.8	—	—	—	—	7
Canned vegetables[b]	U.S.		28.0 ± 25.0	117 ± 53.9	10.6 ± 6.08	340 ± 36.1	39.2 ± 22.6	9
			3.2 – 63.8	21.5 – 245	4.3 – 31.3	224 – 577	13.9 – 91	
Carrots	Saudi Arabia	5	66.5	23.4	20.48	77.2	—	17
Celery	U.S.		32.4	260	9.2	73.1	23	9
Chicory			50 ± 6	638 ± 593	17 ± 4	19 ± 9	—	16
Chinese cabbage			47 ± 6	228 ± 112	13 ± 3	5 ± 2	—	16
Cucumber	U.S.		12.7	145	10.9	< 3	21	9
Cucumber	Saudi Arabia	5	16.9	8.24	7.45	8.8	—	17
Eggplant	Saudi Arabia	5	20.3	15.7	14.4	16.4	—	17
Green pepper	Tenerife Island	60	17.6 ± 2.76	143 ± 21.2	13.6 ± 2.76	6.73 ± 1.00	13.6 ± 2.76	18
			10.8 – 23.4	101 – 199	10.1 – 17.7	2.28 – 9.85	16.2 – 50.3	
Green pepper	U.S.		8.7	162	9.6	< 3	19.5	9
Green pepper	Saudi Arabia	5	14.2	16.0	17.1	14.4	—	17
Green vegetables[b]	U.K.		37.9	—	—	—	—	7
Green vegetables[b]	U.S.		32.2 ± 23.2	218 ± 117	23.5 ± 16.3	22.3 ± 17.3	62.9 ± 51.0	9
			8.1 – 97.7	104 – 437	7.8 – 49.6	< 3.0 – 53.7	17.9 – 147	
Iceberg lettuce	U.S.		17	161	7.8	11.6	22.7	9
Kale			286 ± 43	712 ± 517	51 ± 4	12 ± 4	—	16
Leek	Saudi Arabia	5	283	29.3	25.6	67.0	—	17

TABLE 7.2 (CONTINUED)
Macronutrients in Vegetables, Mushrooms, and Fruits (in mg/100 g w/w)

Name	Origin	n	Ca	K	Mg	Na	P	Ref.
Lentils canned			17 ± 3	—	14.3 ± 2.2	—	58 ± 6	19
Lettuce	Saudi Arabia	5	31.8	10.6	9.7	0.11	—	17
Onion	Saudi Arabia	5	84.9	26.9	22.8	3.19	—	17
Onion	U.S.		19	143	9.2	3.3	30.1	9
Onion conventional	Denmark		19.7 ± 8.41	164 ± 93.5	10.2 ± 1.69	21.4 ± 11.1	43.9 ± 7.47	20
			8.75 – 40.7	20 – 368	6.88 – 16.6	7.81 – 49.6	23.9 – 66	
Onion organic	Denmark		14.2 ± 5.56	183 ± 118	10.9 ± 1.82	15.9 ± 5.44	42.1 ± 8.04	20
			6.77 – 40	69.1 – 453	6.84 – 16.4	8.36 – 37.6	28.7 – 68.5	
Other vegetables[b]	U.K.		20.8	—	—	—	—	7
Other vegetables[b]	U.S.		28.0 ± 34.0	191 ± 83.2	15.3 ± 9.10	261 ± 352	41.9 ± 32.5	9
			2.9 – 138	61.1 – 356	3.2 – 35.9	< 3 – 1330	11.3 – 145	
Parsley	Saudi Arabia	5	396	41.5	34.1	84.7	—	17
Peas conventional	Denmark		23.1 ± 9.6	—	—	—	129 ± 24.5	20
			1.52 – 64.5				80.1 – 179	
Peas organic	Denmark		21.9 ± 5.79	—	—	—	182 ± 63.9	20
			14 – 53.5				110 – 329	
Potatoes	Spain		8.5 ± 1.13	427 ± 68.6	24.4 ± 3.75	9.8 ± 7.35	—	21
			7.7 – 9.3	378 – 475	21.7 – 27	4.6 – 15.0		
Potatoes	Saudi Arabia	5	8.52	37.7	37.0	60.4	—	17
Potatoes[c]	Idaho	342	11.8 ± 3.64	464 ± 74.8	26.7 ± 4.22	—	55.6 ± 15.3	22
			4.38 – 26.0	228 – 728	17.0 – 44.7		26.0 – 114	
Potatoes[c]	non-Idaho	266	7.94 ± 4.09	472 ± 63.6	25.9 ± 4.26	—	57.4 ± 15.1	22
			2.22 – 25.8	302 – 694	16.3 – 41.3		27.8 – 98.2	
Potatoes[b]	U.K.		10.9	—	—	—	—	7
Potatoes[b]	U.S.		21.5 ± 16.1	393 ± 127	22.0 ± 6.69	109 ± 101	71.1 ± 31.0	9
			5.6 – 50.8	225 – 596	14.6 – 34.4	< 2.0 – 261	43 – 129	

Radish	U.S.		19.4	207	8.2	28.6	16.5	9
Red pepper	Tenerife	60	18.6 ± 3.86	195 ± 31.7	16.0 ± 2.19	6.61 ± 11.8	30.5 ± 3.91	18
	Island		13.0 – 33.2	120 – 250	12.3 – 23.1	4.13 – 8.60	24.4 – 40	
Rucola			98 ± 12	363 ± 92	30 ± 9	4 ± 2	—	16
Salq	Saudi Arabia	5	284	23.1	24.4	534	—	17
Spinach	Saudi Arabia	5	146	26.4	20.1	46.6	—	17
Spinach substitute			64 ± 21	537 ± 130	55 ± 16	94 ± 49	—	16
Tomato	Saudi Arabia	5	19.6	1.01	11.2	14.2	—	17
Tomato fruits, rockwool high EC	Denmark		5.1	153	8.7	0.78	31	23
Tomato fruits, rockwool norm EC	Denmark		7.1	151	8.5	0.73	32	23
Tomato fruits, soil norm EC	Denmark		11	164	9	0.55	33	23
Tomato red	U.S.		8.3	221	10.3	< 3	24.9	9
Marrow (zucchini)	Saudi Arabia	5	44.7	12.6	14.41	10.5	—	17
Watercress			127 ± 24	271 ± 54	30 ± 5	14 ± 4	—	16
Watercress	Saudi Arabia	5	199	12.6	14.1	29.9	—	17
White asparagus			11.8 ± 1.8	—	5.5 ± 1.0	—	25.7 ± 2.4	19
Whole peeled tomatoes, canned			13 ± 7	—	10.2 ± 1.4	—	19.3 ± 2.1	19
Mushrooms								
Bay bolete[c]	Poland	166	1.26 ± 0.67	279 ± 39.0	3.88 ± 2.02	8.69 ± 6.16	—	24
			0.14 – 4.43	129 – 518	0.58 – 7.56	1.23 – 44.8		
Mushrooms	U.S.		2.1	270	7.7	3.6	68.6	9
Mushrooms, canned			19 ± 7	—	6.2 ± 0.4	—	43 ± 6	19
Fruits								
Apple red	U.S.		4.6	104	5	< 3	9.9	9
Dried apple		6	17.9 ± 0.8	—	23.6 ± 0.9	—	—	1
			13.0 – 23.0		18 – 28			
Apricot	U.S.		14.9	261	10.3	<3	22.7	9
Avocado	U.S.		12.9	498	29.7	4.9	50.1	9
Banana	U.S.		5.1	351	29.5	< 2	22.8	9

TABLE 7.2 (CONTINUED)
Macronutrients in Vegetables, Mushrooms, and Fruits (in mg/100 g w/w)

Name	Origin	n	Ca	K	Mg	Na	P	Ref.
Banana pulp	Tenerife		18.9 ± 1.41	509 ± 36.1	37.4 ± 5.37	11.7 ± 0.42	59.1 ± 6.15	25
			10.1 – 30.8	59 – 733	21.2 – 67.5	9.08 – 23.9	42.4 – 94.3	
Cantaloupe	U.S.		9.1	261	11.1	17.4	13.8	9
Currants		6	90.8 ± 4.6	—	52.9 ± 1.1	—	—	1
			64.5 – 115		48.3 – 58.2			
Fresh fruit[b]	U.K.		11.3	—	—	—	—	7
Fresh fruit[b]	U.S.		11.8 ± 7.55	208 ± 109	12.1 ± 7.73	11.2 ± 8.84	20.1 ± 9.97	9
			4.6 – 32.2	104 – 498	< 5.0 – 29.7	< 2.0 – 17.4	9.9 – 50.1	
Fruit products[b]	U.K.		12.7	—	—	—	—	7
Fruit products[b]	U.S.		14.1 ± 14.5	211 ± 238	12.7 ± 11.7	3.95 ± 2.74	21.9 ± 29.3	9
			2.9 – 49	34.2 – 744	2.9 – 41.9	< 2.0 – 10.5	5.6 – 101	
Grapefruit	U.S.		20.6	138	8.9	< 2	16.4	9
Grapes, seedless	U.S.		9.8	182	7	< 2.0	19.3	9
Orange	U.S.		32.2	158	10.7	< 2	18.7	9
Peach	U.S.		5.3	189	8.7	< 2.0	20.2	9
Pear	U.S.		9.1	120	6.8	< 2.0	10.8	9
Plums	U.S.		5.1	160	6.9	—	16.4	9
Strawberries	U.S.		14.9	151	12.3	< 2.0	25.2	9
Strawberry	Finland	12	18.9	209	16.1	—	—	26
			17.1 – 22.3	164 – 253	11.2 – 22.3			
Sultanas		6	53.6 ± 2.5	—	43.5 ± 0.8	—	—	1
			38.7 – 64.6		40.5 – 48.7			
Sweet cherries	U.S.		13.7	238	13.1	< 3.0	23.9	9
Watermelon	U.S.		7.2	104	9.9	< 2.0	11.3	9

TABLE 7.3
Macronutrients in Tea, Coffee, and Confectionary Products (in mg/100 g w/w)

Name	Origin	n	Ca	K	Mg	Na	P	Ref.
Tea								
Tea	Africa	18	358 ± 84.7	—	157 ± 23.4	—	—	27
Tea	Asia	36	351 ± 57.2	—	56.3 ± 31.6	—	—	27
Tea	China	13	323 ± 85.6	—	165 ± 26.6	—	—	27
Tea	India, Sri Lanka	13	366 ± 58.2	—	157 ± 14.1	—	—	27
Tea	U.S.		< 0.4	17.6	1.1	1	< 2	9
(from tea bag)								
US tea brands	U.S.	7	—	1810 ± 350	—	33.8 ± 28.6	—	28
				1310 – 2370		11.4 – 79.6		
Tea[c]	China	39	314 ± 80.3	1829 ± 296	—	—	429 ± 105	29
			139 – 554	1385 – 2836			259 – 665	
Beiqishen tea	China		646 ± 37.2	1923 ± 239	225 ± 5.7	28.8 ± 4.51	247 ± 4.4	30
Chinese tea	China	5	396 ± 2.76	—	266 ± 7.73	—	—	31
Indian tea brands	India	15	—	2110 ± 200	—	5.35 ± 2.74	—	28
				1770 – 2400		2.14 – 11.8		
Red tea		9	534 ± 108	—	216 ± 12.9	—	—	32
			400 – 723		187 – 224			
Instant tea[d] (< 2% tea extract)		3	0.24	21.3	0.3	0.14	—	33
			0.18 – 0.26	20.4 – 23	0.3 – 0.5	0.12 – 0.16		
Instant tea[d] (100% extract)		2	0.17	25.5	2.02	0.03	—	33
			0.14 – 0.21	25.1 – 26	2.0 – 2.04	0.02 – 0.04		
Beiqishen tea extract[d]	China		1.74 ± 0.05	15.9 ± 0.57	1.46 ± 0.005	0.27 ± 0.02	1.45 ± 0.06	30
Black tea beverages[d]		23	0.95 ± 0.07	27 ± 2.83	0.8 – 2.2	0 – 0.6	—	33
			0.6 – 1.8	17.9 – 36.4				

TABLE 7.3 (CONTINUED)
Macronutrients in Tea, Coffee, and Confectionary Products (in mg/100 g w/w)

Name	Origin	n	Ca	K	Mg	Na	P	Ref.
Green tea beverages[d]		20	0.74 ± 0.05	18 ± 1.41	0.5 – 1.4	0.04 – 0.33	—	33
			0.57 – 0.99	9.4 – 25.9				
Tea soft drinks[d]		8	1.7	7.8	1.4	14.6	—	33
			0.1 – 5.7	1.0 – 17.6	0.5 – 5.0	2.6 – 33		
Coffee								
Decaffeinated instant coffee	U.S.		2.5	46.2	4.8	1.5	4.2	9
Ground coffee	U.S.		1.5	36.7	2.6	< 2	< 3	9
Soluble coffee	Brazil	21	139 ± 20	3810 ± 530	340 ± 50	111 ± 148	350 ± 50	34
			106 – 189	3250 – 5170	212 – 415	27.4 – 667	223 – 410	
Roasted coffee beans	Costa Rica	20	104 ± 14.5	1793 ± 115	213 ± 9.7	4 ± 0.67	185 ± 13.5	35
Roasted coffee beans	Colombia	20	109 ± 14.5	1851 ± 133	219 ± 13.5	3.87 ± 1.06	191 ± 19.3	35
Roasted coffee beans	Guatemala	20	119 ± 28	1836 ± 854	233 ± 13.5	2.79 ± 0.57	189 ± 17.4	35
Roasted coffee beans	Panama	20	96.3 ± 16.4	1804 ± 122	210 ± 10.6	0.93 ± 0.22	168 ± 8.7	35
Roasted coffee beans	Ethiopia	20	97.8 ± 13.5	1862 ± 133	199 ± 12.5	1.94 ± 0.45	180 ± 14.5	35
Roasted coffee beans	Kenya	20	94.2 ± 16.4	1690 ± 99.5	208 ± 15.4	3.6 ± 0.97	165 ± 13.5	35
Roasted coffee beans	Sulawesi	20	90.2 ± 11.6	1850 ± 105	227 ± 15.4	142 ± 0.51	204 ± 15.4	35
Roasted coffee beans	Sumatra	20	110 ± 18.3	1893 ± 135	203 ± 10.6	1.02 ± 0.94	187 ± 20.3	35
Roasted coffee[c]	Mixture	18	101 ± 9.7	1371 ± 57.9	168 ± 9.66	1.8 ± 0.75	150 ± 19.3	36
			84 – 119	1229 – 1488	141 – 187	0.64 – 3.3	120 – 189	
Green coffee[c]	Mixture	41	110 ± 19.3	1540 ± 160	178 ± 9.3	5.16 ± 2.03	160 ± 19.3	36
			89.8 – 156	1170 – 1831	155 – 199	1.76 – 11.4	136 – 212	

Confectionary Products								
Sugar and preserves[b]	U.K.		61.2	—	—	—	—	7
Sugar and preserves[b]	U.S.		47.1 ± 46.6	140.2 ± 88.4	17.1 ± 15.0	232 ± 136	89.5 ± 61.2	9
			1.1 – 188	< 3.0 – 403	< 1.0 – 65.8	9.8 – 410	< 4.0 – 232	
Sugar	Egypt	2	145 ± 7.07	3.45 ± 0.21	0.10 ± 0.01	0.16 ± 0.01	—	37
Sugar	Egypt		1000 ± 500	100 ± 100	0.0 ± 0.0	8 ± 13	0.001 ± 0.004	38
White sugar, granulated	U.S.		< 2.0	< 3	—	< 3	—	9
Sugar cane plant	Egypt	3	490 ± 271	118 ± 19.7	2.14 ± 0.05	0.38 ± 0.77	—	37
Cocoa powder		4	191 ± 4.4	—	727 ± 11	—	—	1
			182 – 202		691 – 774			
Molasses	Egypt		7100 ± 4500	2200 ± 700	100 ± 100	130 ± 60	2.6 ± 1.9	38
Sugar cane juice	Egypt	3	393 ± 344	66.0 ± 13.6	1.24 ± 0.30	1.92 ± 0.99	—	37
Sugar juice	Egypt		2200 ± 1300	1900 ± 500	200 ± 100	80 ± 30	7.2 ± 6.3	38

[d] recalculated as follows; 100 ml = 100 g

TABLE 7.4
Macronutrients in Nuts, Seeds and Oils (in mg/100 g w/w)

Name	Origin	n	Ca	K	Mg	Na	P	Ref.
Nuts and Seeds								
Almond			262 ± 4	—	240 ± 0.4	—	—	39
Sliced almonds		6	302 ± 6.0	—	273 ± 5.0	—	—	1
			287–337		242 – 295			
Hazelnut	Turkey	11	83.5 ± 5.14	637 ± 105	144 ± 14.9	0.70 ± 0.10	—	40
			78.8–92.5	527 – 895	118 – 144	0.61 – 0.86		
Hazelnut	Turkey		237 ± 20.5	619 ± 50.3	173 ± 11.4	42 ± 6.06	309 ± 27.8	41
Hazelnut			131 ± 2	—	170 ± 0.2	—	—	39
Hazelnut[c]	Turkey		173 ± 44.4	439 ± 35.6	191 ± 21.8	2.24 ± 0.71	—	42
Mixed nuts (no peanuts)	U.S.		105	578	238	283	448	9
Nuts[b]	U.K.		86.1	—	—	—	—	7
Nuts[b]	U.S.		64.6 ± 35.0	633 ± 56.0	199 ± 34.4	386 ± 95.7	392 ± 49.5	9
			43.2 – 105	578 – 690	174 – 238	283 – 472	354 – 448	
Peanut			65 ± 10	—	160 ± 50	—	—	39
Peanut butter	U.S.		43.2	632	174	404	354	9
Peanut greens[c]	U.S.	2	1172 ± 225	1703 ± 722	641 ± 435	139 ± 57.1	—	43
			986 – 1422	882 – 2237	322 – 1138	85.3 – 199		
Walnut			76 ± 5	—	160 ± 30	—	—	39
Oils								
Oils and fats[b]	U.K.		8.7	—	—	—	—	7
Oils and fats[b]	U.S.		14.4 ± 12.7	22.9 ± 1.13	1.55 ± 0.21	657 ± 115	16.3 ± 9.40	9
			< 2.0 – 23.4	< 3 – 23.7	< 1.2 – 1.7	< 3.0 – 738	< 4.0 – 22.9	
Olive or sunflower oil	U.S.		< 2	< 3	< 1.2	< 3	< 4	9

TABLE 7.5
Macronutrients in Beverages (in mg/100 g w/w)

Name	Origin	n	Ca	K	Mg	Na	P	Ref.
Beverages[b]	U.K.		1.8	—	—	—	—	7
Beverages[b]	U.S.		4.99 ± 3.87	27.0 ± 23.5	3.84 ± 2.52	3.56 ± 1.82	9.41 ± 4.64	9
			< 0.4 – 13.1	< 2.0 – 79.9	< 0.4 – 8.7	< 1.0 – 6.9	< 2.0 – 14.6	
Juices								
Cactus pear juice			59	—	98.4	—	—	44
Grapefruit juice			5.72	137	8.66	1.18	18.2	45
Lemon juice			11.6	145	11.4	1.82	19.1	45
Lime juice			5.82	137	6.71	0.42	15	45
Mandarin juice			7.93	229	12.2	1	18.1	45
Maple syrup[d]		80	77.5 ± 27.9	202.6 ± 37.5	16.7 ± 7.2	—	—	46
Orange juice	Australia	290	8.16 ± 0.18	169 ± 1.41	11.0 ± 0.53	0.90 ± 0.05	18.1 ± 0.28	45
			3.5 – 16.0	77.7 – 235	5.51 – 17	0.06 – 7.1	7.8 – 27.0	
Orange juice[d]		8	3.6 – 13.5	198 – 257	10.2 – 20.8	0.05 – 0.13	17.6 – 22.1	47
Orange juice concentrate	Australia	83	7.9 ± 1.67	164.5 ± 25.4	9.78 ± 1.69	1.77 ± 1.47	16.5 ± 3.37	45
			3.78 – 12.5	106.1 – 225	4.11 – 15.3	0.35 – 9.47	9.81 – 24.2	
Orange juice concentrate	Brazil	42	6.78 ± 1.01	162.6 ± 14.4	9.62 ± 0.89	0.28 ± 0.3	14.9 ± 1.37	45
			5.17 – 10.0	135 – 203	8.4 – 12.0	0.04 – 1.34	11.9 – 19.0	
Pineapple juice[d]			14.6 ± 3.48	96.7 ± 35.2	13.4 ± 3.69	5.11 ± 3.35	3.14 ± 0.83	48
			9.64 – 18.5	40.2 – 132	6.96 – 17.3	1.52 – 9.66	1.87 – 4.01	
Pineapple nectar[d]			9.75 ± 3.23	33.1 ± 7.49	6.39 ± 2.59	9.34 ± 10.2	1.14 ± 0.52	48
			5.40 – 14.8	21.3 – 40.9	4.25 – 10.8	1.72 – 32.0	0.54 – 1.86	
Pummelo juice			10.4	156	7.5	0.42	18.5	45
Seville orange juice			10.8	147	8.71	1.04	13.3	45
Tangello juice			12.9	167	10.4	1.84	17.3	45

TABLE 7.5 (CONTINUED)
Macronutrients in Beverages (in mg/100 g w/w)

Name	Origin	n	Ca	K	Mg	Na	P	Ref.
Tomato juice	U.S.		10.5	223	11.2	277	18	9
Alcoholic Beverages								
Beer	U.S.		4.2	24.5	5.8	3.1	14.4	9
Beer[d]		32	5.71 ± 2.34	40.6 ± 12.7	8.74 ± 5.09	3.90 ± 2.60	21.6 ± 7.22	49
			1.99 – 10.8	20.0 – 83.3	4.31 – 26.7	0.84 – 10.3	10.9 – 38.1	
Martini	U.S.		< 0.4	2.8	< 0.4	< 1	< 2	9
Whiskey	U.S.		< 0.5	< 2	< 0.5	< 1	< 2	9
Whiskey[d]	Scotland	35	0.11 ± 0.08	—	0.06 ± 0.08	1.19 ± 0.47	—	50
			0.04 – 0.43		0.002 – 0.39	0.31 – 2.26		
Dry table wine	U.S.		7.9	79.9	8.7	4	12.2	9
Red wines[d]	Brazil		5.76 – 9.28	—	—	—	3.15 – 8.76	51
Red wines[d]	Portugal		6.57 – 9.95	—	—	—	6.89 – 15	51
Red wines[d]	Chile		7.21	—	—	—	12.3	51
White wines[d]	Brazil		6.26 – 10.3	—	—	—	2.23 – 5.33	51
White wines[d]	Chile		5.51	—	—	—	8.33	51
Wine[d]	Spain	125	6.43 ± 1.59	68.8 ± 27.2	8.38 ± 1.31	8.8 ± 3.82	—	52
			2.9 – 10.6	25.5 – 160	5.5 – 11.8	2.6 – 22.7		
Wine[d]	Italy		3.0 – 9.0	75 – 146	7.0 – 11.5	0.34 – 20.0	28.0 – 63.0	53

(Table 7.3 and Table 7.4). Data in Table 7.3 suggests that cocoa powder is the richest source of magnesium (691–774 mg/100 g). Fruit juices contain from 4.11 to 20.8 mg Mg/100 g except cactus pear juice concentrate, which has 98.4 mg Mg/100 g. The concentration of magnesium in alcoholic beverages such as beer and wine range from 4.31 to 26.7 mg/100 g (Table 7.5).

7.2.3 Potassium

Grain products, fruits, and vegetables show levels of potassium (K) in the range of 20 to 730 mg/100 g (Table 7.1). Concentrations of this macroelement in some plant products are relatively high, with values >700 mg/100 g in Idaho potatoes, banana pulp, and avocado (Table 7.2). Seeds and nuts are rich in potassium, with values of up to 2240 mg/100 g (Table 7.4). Tea leaves (2840 mg K/100 g) and instant coffee (5170 mg K/100 g) are very important sources of potassium (Table 7.3). Fruit juices contain from 21.3 to 257 mg K/100 g (Table 7.5), similar to typical values for beer and wine (20.0–160 mg/100 g).

7.2.4 Sodium

Concentrations of sodium (Na) in cereals range from 5.21 to 1010 mg/100 g (Table 7.1). Raw vegetables and fruit juices contain relatively low levels of sodium in the range of 2.28 to 94.0 mg/100 g and from 0.04 to 277 mg/100 g (Table 7.2). Tea leaves (2.14–79.6 mg/100 g) and grain coffee (0.64–11.39 mg/100 g) are also poor sources of sodium (Table 7.3), as are fruit juices and alcoholic beverages (Table 7.5).

7.2.5 Phosphorus

Concentrations of phosphorus (P) in miscellaneous cereals and vegetables are similar and range from 16.2 to 437 mg/100 g (Table 7.1 and Table 7.2). Table 7.3 and Table 7.4 show that tea leaves (247–665 mg/100 g) and nuts (309–448 mg/100 g) have somewhat higher levels of this macroelement. Fruits contain smaller amounts of phosphorus, typically in the range of 9.9 to 94.3 mg/100 g (Table 7.2).

7.3 MICRONUTRIENTS IN FOOD

7.3.1 Chromium

Concentrations of chromium (Cr) are generally ranging from 0.003 to 0.022 mg/100 g in miscellaneous cereals (Table 7.6) and from 0.00004 to 0.006 mg/100 g in vegetables and 0.005 to 0.018 mg/100 g in fruits (Table 7.7). Higher levels are found in grain products (< 0.0014–0.10 mg/100 g), with the highest levels in macaroni and soya (Table 7.6). Tea leaves contain from 0.11 to 0.33 mg Cr/100 g (Table 7.8), whereas nuts show values up to 0.042 mg/100 g (Table 7.9). Concentration of chromium in beer and wine range from 0.0004 to 0.004 mg/100 g and from 0.002 to 0.011 mg/100 g, respectively (Table 7.10).

With concentrations that can reach up to 0.5 mg/100 g, brewer's yeast is a good source of chromium.[1] Canned and other processed foods can be contaminated by

TABLE 7.6
Micronutrients in Miscellaneous Cereals (in mg/100 g w/w)

Name	Origin	n	Co	Cr	Cu	Fe	Mn	Mo	Ni	Se	Zn	Ref.
Bread												
Bread[b]	U.K.		0.002	0.01	0.16	2.1	0.8	0.02	0.01	0.004	0.9	7
Bread[b]	U.S.		—	—	0.19 ± 0.04 < 0.09 – 0.23	2.66 ± 0.65 1.55 – 3.18	0.85 ± 0.64 0.2 – 1.82	—	0.02 ± 0.003 < 0.01 – 0.02	0.02 ± 0.01 0.01 – 0.03	1.0 ± 0.43 0.6 – 1.63	9
Bread			—	—	—	1	—	—	—	—	1	8
Bagel plain	U.S.		—	—	0.13	3.03	0.44	—	< 0.012	0.03	0.77	9
Bread	Poland	26	—	—	0.10 – 0.21	1.4 – 3.0	0.82 – 2.3	—	—	—	0.88 – 1.7	65
White bread	U.S.		—	—	< 0.12	3.03	0.42	—	0.02	0.02	0.62	9
Whole white bread	U.S.		—	—	0.23	2.83	1.82	—	0.02	0.03	1.63	9
Cornbread	U.S.		—	—	< 0.09	1.55	0.2	—	0.02	0.01	0.6	9
Rye bread	U.S.		—	—	0.15	2.73	0.7	—	< 0.01	0.03	0.96	9
Bread, white flour[c]			—	—	0.31	0.61	0.16	—	—	—	0.35	10
Bread, wholemeal flour[c]			—	—	1.22	1.25	0.86	—	—	—	1.05	10
Cereals												
Cereals		27	—	—	—	—	—	—	—	0.003 0.0002 – 0.01	—	66
Corn flakes			—	—	< 0.09	19.6	0.12	—	< 0.01	0.005	0.81	9
Oat flakes			—	—	—	5.5 ± 0.2	5.3 ± 0.4	—	—	—	3.9 ± 0.1	11
Rolled oats		6	—	—	0.53 ± 0.06 0.43 – 0.61	5.39 ± 0.29 4.51 – 7.49	7.14 ± 0.95 5.0 – 13.9	—	—	—	3.66 ± 0.48 2.74 – 7.26	56

Raisin bran cereal	U.S.		—	—	0.45	34	2.88	—	0.04	<0.004	10.3	9
Breakfast cereals	Spain	36	—	0.02 ± 0.01 0.01–0.05	—	—	—	—	—	—	—	67
Fruit- flavored cereal	U.S.		—	—	0.1	18	0.73	—	0.04	0.01	14	9
Rice												
Rice	India, Pakistan	28	—	—	—	—	—	—	—	0.01 0.0005 – 0.02	—	12
Rice	Europe	25	—	—	—	—	—	—	—	0.005 0.001 – 0.02	—	12
Rice	Different		0.002	0.06	0.37	—	0.64	—	0.10	0.03	1.05	13
Rice[c]			0.004	—	0.22 – 0.26	2.26 – 4.95	0.96 – 1.74	—	—	0.02	1.04 – 1.48	14
White rice		6	—	—	0.14 ± 0.00 0.14 – 0.14	0.38 ± 0.01 0.31 – 0.47	2.14 ± 0.06 1.81 – 2.51	—	—	—	1.21 ± 0.02 1.11 – 1.31	56
Brown rice		7	—	—	0.22 ± 0.003 0.22 – 0.23	1.23 ± 0.01 1.15 – 1.3	5.3 ± 0.07 5.03 – 5.74	—	—	—	1.73 ± 0.02 1.66 – 1.95	56
Wild rice		26	—	—	0.87 0.26 – 1.31	3.39 0.65 – 43.7	—	—	—	—	2.32 0.66 – 3.92	68
Unpolished rice[c]			—	—	—	4.91	—	—	—	—	2.9	69
Grits												
Grits		10	—	—	0.06 ± 0.02 0.04 – 0.09	0.39 ± 0.03 0.22 – 0.59	0.14 ± 0.01 0.08 – 0.27	—	—	—	0.46 ± 0.11 0.23 – 0.53	56
Flour												
Chickpea flour		1	—	0.01 ± 0.001	—	—	—	—	0.05 ± 0.003	—	—	70
Flour			—	—	—	1.63	—	—	—	—	0.95	8

TABLE 7.6 (CONTINUED)
Micronutrients in Miscellaneous Cereals (in mg/100 g w/w)

Name	Origin	n	Co	Cr	Cu	Fe	Mn	Mo	Ni	Se	Zn	Ref.
Common buckwheat flour[c]	Slovenia		0.01 ± 0.0002	0.01 ± 0.001	—	7.10 ± 0.07	—	—	0.09 ± 0.01	0.003 ± 0.0005	1.73 ± 0.02	71
Tartary buckwheat fine flour[c]	Slovenia		0.002 ± 0.0001	0.01 ± 0.001	—	3.20 ± 0.03	—	—	0.1 ± 0.02	<0.001	1.23 ± 0.01	71
Tartary buckwheat flour[c]	Slovenia		0.01 ± 0.0001	0.03 ± 0.001	—	12.8 ± 0.04	—	—	0.16 ± 0.02	0.001 ± 0.001	2.26 ± 0.02	71
Corn flour		1	—	0.003 ± 0.00	—	—	—	—	0.02 ± 0.001	—	—	70
Maize flour		4	—	—	0.06 ± 0.002 0.06 – 0.07	0.55 ± 0.02 0.49 – 0.62	0.17 ± 0.01 0.13 – 0.21	—	—	—	0.46 ± 0.02 0.37 – 0.56	56
Refined oat flour		1	—	—	0.32 ± 0.003 0.31 – 0.32	1.29 ± 0.01 1.27 – 1.33	2.76 ± 0.03 2.68 – 2.78	—	—	—	1.46 ± 0.01 1.42 – 1.48	56
Wholemeal oat flour		2	—	—	0.37 ± 0.002 0.37 – 0.37	5.33 ± 0.07 4.95 – 5.33	5.48 ± 0.09 5.24 – 5.68	—	—	—	2.79 ± 0.01 2.66 – 2.8	56
Wheat flour		7	0.002 ± 0.0001	0.004 ± 0.001	—	—	—	—	0.02 ± 0.004 0.02 – 0.05	—	—	70
Wheat flour[c]	Sweden		n = 19 < 0.0003 – 0.0005	n = 56 < 0.001 – 0.001	n = 105 0.15 ± 0.04 0.1 – 0.32	n = 21 1.03 ± 0.24 0.67 – 1.72	n = 105 0.53 ± 0.32 0.23-1.98	—	n = 61 0 – 0.02	—	n = 105 0.65 ± 0.27 0.39 – 1.81	72
White wheat flour		4	—	—	0.16 ± 0.001 0.16 – 0.17	1.24 ± 0.02 1.13 – 1.31	1.29 ± 0.02 1.25 – 1.37	—	—	—	0.92 ± 0.01 0.83 – 0.96	56
Soya flour		4	—	—	1.53 ± 0.04 1.37 – 1.71	9.26 ± 0.29 7.87 – 10.1	5.17 ± 0.1 4.99 – 5.43	—	—	—	3.12 ± 0.05 3.0 – 3.26	56

Rye flour[c]	Sweden		n = 32 < 0.001 – 0.004	n = 41 < 0.002 – 0.005	n = 90 0.33 ± 0.05 0.25 – 0.46	n = 60 0.01 ± 0.005 0.002 – 0.04	n = 90 2.57 ± 0.74 1.37 – 5.23	—	n = 60 0.01 ± 0.005 0.002 – 0.04	—	n = 90 2.40 ± 0.34 1.8 – 3.17	72
Tartary buckwheat leaf flour[c]	Slovenia		0.36 ± 0.004	< 0.5 ± 0.005	—	138 ± 1.38	—	—	0.31 ± 0.05	0.04 ± 0.001	2.51 ± 0.02	71
Miscellaneous Cereals												
Common buckwheat bran[c]	Slovenia		0.01 ± 0.0002	0.01 ± 0.0002	—	8.02 ± 0.08	—	—	0.17 ± 0.03	0.004 ± 0.001	2.67 ± 0.03	71
Miscellaneous cereals[b]	U.K.		0.001	0.01	0.18	3.2	0.68	0.02	0.02	0.003	0.86	7
Miscellaneous cereals[b]	U.S.		—	—	0.20 ± 0.12 < 0.09 – 0.45	7.39 ± 9.83 0.81 – 34	0.90 ± 0.93 < 0.1 – 3.38	—	0.05 ± 0.06 < 0.01 – 0.23	0.014 ± 0.01 < 0.004 – 0.031	2.62 ± 4.28 0.12 – 14	9
Tartary buckwheat bran[c]	Slovenia		0.05 ± 0.0005	0.02 ± 0.001	—	13.0 ± 0.13	—	—	0.35 ± 0.01	0.005 ± 0.0005	6.97 ± 0.07	71
Oat bran			—	—	—	7.5 ± 1.0	5.5 ± 1.9	—	—	—	4.6 ± 0.2	11
Wheat bran[c]	Sweden		n = 13 < 0.001 – 0.003	n = 19 < 0.0035 – 0.003	n = 30 1.24 ± 0.20 0.72 – 1.50	n = 30 13.4 ± 2.12 9.11 – 17.8	n = 30 11.1 ± 1.68 8.32—15.0	—	n = 22 0.06 ± 0.05 0.02 – 0.19	—	n = 30 8.23 ± 1.42 6.11 – 11.5	72
Stabilized wheatgerm		3	—	—	1.47 ± 0.08 1.23 – 1.55	10.5 ± 0.4 6.8 – 12.3	19.4 ± 1.1 13.3 – 25.2	—	—	—	9.2 ± 0.05 6.65 – 11.8	56
Quinoa[c]			0.003	—	0.48	5.06 – 16.6	0.86	—	—	0.01	2.36	14
Corn grains[c]			0.003 – 0.004	—	0.29 – 0.35	1.66 – 2.62	0.42 – 0.44	—	—	0.003 – 0.01	1.31 – 2.19	14
Soft whole wheat		4	—	—	0.33 ± 0.003 0.33 – 0.34	3.54 ± 0.07 3 – 4	6.72 ± 0.12 6.1 – 7.7	—	—	—	2.1 ± 0.04 1.9 – 2.2	56
Spelt wheat		9	—	—	0.4 ± 0.1	4.4 ± 0.6	4.2 ± 0.7	—	—	—	3.7 ± 0.4	15
Rye grains[c]	Northeast Poland		—	—	0.30 ± 0.003	—	2.52 ± 0.03	—	—	0.02 ± 0.001	2.09 ± 0.03	73

TABLE 7.6 (CONTINUED)
Micronutrients in Miscellaneous Cereals (in mg/100 g w/w)

Name	Origin	n	Co	Cr	Cu	Fe	Mn	Mo	Ni	Se	Zn	Ref.
Common buckwheat grain[c]	Slovenia		0.01 ± 0.0002	0.01 ± 0.001	—	5.28 ± 0.053	—	—	0.1 ± 0.01	0.002 ± 0.0002	2.27 ± 0.02	71
Tartary buckwheat grain[c]	Slovenia		0.05 ± 0.0004	0.10 ± 0.01	—	40.3 ± 0.40	—	—	0.27 ± 0.05	0.003 ± 0.002	3.05 ± 0.03	71
Barley grains[c]	Northeast Poland		—	—	0.22 ± 0.002	—	1.19 ± 0.01	—	—	0.02 ± 0.001	2.19 ± 0.03	73
Raw maize		6	—	—	0.26 ± 0.08 0.19 – 0.41	2.54 ± 0.07 2.28 – 2.98	0.78 ± 0.02 0.66 – 0.89	—	—	—	2.45 ± 0.07 1.96 – 2.86	56
Oat grains[c]			0.005 – 0.01	—	0.29 – 0.61	4.52 – 7.39	3.31 – 3.65	—	—	0.002 – 0.003	1.57 – 3.57	14
Oat grains[c]	Northeast Poland		—	—	0.221 ± 0.002	—	4.17 ± 0.02	—	—	0.015 ± 0.001	2.54 ± 0.10	73
Kibbled wheat		6	—	—	0.34 ± 0.003 0.33 – 0.36	3.29 ± 0.04 3.24 – 3.59	6.34 ± 0.11 5.51 – 6.74	—	—	—	1.93 ± 0.03 1.87 – 2.06	56
Wheat grains[c]			0.01 – 0.01	—	0.23 – 0.61	2.62 – 5.32	2.27 – 3.66	—	—	0.003 – 0.005	2.01 – 4.36	14
Wheat grains[c]	Northeast Poland		—	—	0.28 ± 0.02	—	3.34 ± 0.09	—	—	0.02 ± 0.002	2.33 ± 0.08	73
Decorticated rye		4	—	—	0.38 ± 0.01 0.34 – 0.41	3.73 ± 0.15 3.19 – 4.6	3.75 ± 0.15 2.47 – 3.59	—	—	—	2.37 ± 0.07 2.21 – 2.58	56
Buckwheat groats	Polish market		0.01 ± 0.004	0.01 ± 0.006	0.50 ± 0.02	3.03 ± 0.16	3.44 ± 0.06	—	0.15 ± 0.02	—	3.17 ± 0.07	75
Macaroni, graham	Polish market		0.01 ± 0.003	0.02 ± 0.002	0.51 ± 0.02	6.76 ± 0.83	9.92 ± 0.21	—	0.01 ± 0.01	—	4.84 ± 0.10	75

Macaroni, soya	Polish market		0.03 ± 0.001	0.22 ± 0.002	0.08 ± 0.01	2.50 ± 0.14	0.28 ± 0.08	—	0.08 ± 0.03	—	7.75 ± 0.92	75
Wheat germ	Polish market		0.01 ± 0.003	0.01 ± 0.001	0.79 ± 0.03	10.4 ± 0.51	38.1 ± 1.75	—	0.12 ± 0.04	—	18.3 ± 1.2	75
White roll	U.S.		—	—	0.13	3.19	0.46	—	< 0.01	0.03	0.81	9
Gluten		4	—	—	0.55 ± 0.03 0.41 – 0.68	5.19 ± 0.1 4.82 – 5.6	1.73 ± 0.1 1.32 – 2.3	—	—	—	2.96 ± 0.22 3.95 – 1.81	56
Dough			—	—	—	1.18	—	—	—	—	1	8

TABLE 7.7
Micronutrients in Vegetables, Mushrooms, and Fruits (in mg/100 g w/w and in μg/100 g w/w[a])

Name	Origin	n	Co	Cr	Cu	Fe	Mn	Mo	Ni	Se	Zn	Ref.
Vegetables												
Artichoke	Spain		—	—	—	—	—	—	—	—	—	76
Butterhead lettuce			—	—	0.04 ± 0.02	0.5 ± 0.2	0.4 ± 0.3	—	—	—	0.33 ± 0.04	16
Cabbage			—	—	0.05 ± 0.05	0.14 ± 0.03	0.2 ± 0.1	—	—	—	0.2 ± 0.1	16
Cabbages	Saudi Arabia	5	0.01	—	0.004	0.78	0.22	—	0.003	—	0.15	17
Canned vegetables[b]	U.K.		0.001	0.01	0.15	1.2	0.18	0.03	0.03	0.002	0.39	7
Canned vegetables[b]	U.S.		—	—	0.15 ± 0.04 < 0.08 – 0.18	0.57 ± 0.35 < 0.3 – 1.47	0.24 ± 0.02 < 0.1 – 0.34	—	0.02 ± 0.01 < 0.01 – 0.04	< 0.003	0.28 ± 0.13 0.1 – 0.76	9
Carrot	Greece		0.001	0.01	0.07	0.46	0.08	—	0.01	0.0005	0.14	77
Carrots	Saudi Arabia	5	0.02	—	0.01	0.35	0.07	—	0.21	—	0.11	17
Celery	U.S.		—	—	< 0.09	< 0.3	< 0.1	—	<0.01	< 0.004	< 0.1	9
Celery leaf	Greece		0.001	0.03	0.18	1.7	0.43	—	0.04	< 0.0001	0.44	77
Celery root	Greece		0.003	0.09	0.22	1.67	0.49	—	0.03	< 0.0001	0.49	77
Chicory			—	—	0.06 ± 0.03	0.5 ± 0.2	0.3 ± 0.1	—	—	—	0.24 ± 0.03	16
Chinese cabbage			—	—	0.04 ± 0.02	0.20 ± 0.03	0.05 ± 0.14	—	—	—	0.23 ± 0.04	16
Cucumber	U.S.		—	—	< 0.09	< 0.3	< 0.1	—	< 0.01	—	0.11	9
Cucumber	Saudi Arabia	5	0.00	—	0.01	0.33	0.03	—	0.04	—	0.13	17
Eggplant	Saudi Arabia	5	0.01	—	0.02	3.19	0.16	—	0.09	—	0.37	17
Green pepper	Tenerife Island	60	—	—	0.05 ± 0.02 0.02 – 0.08	0.27 ± 0.07 0.14 – 0.41	0.05 ± 0.01 0.03 – 0.07	—	—	—	0.13 ± 0.04 0.08 – 0.23	18
Green pepper	U.S.		—	—	< 0.09	< 0.3	0.11	—	0.0103	—	0.12	9

Green pepper	Saudi Arabia	5	0.01	—	0.04	0.44	0.05	—	0.12	—	0.08	17
Green vegetables[b]	U.K.		0.001	0.02	0.08	1.1	0.2	0.02	0.01	0.001	0.34	7
Green vegetables[b]	U.S.			—	0.18 ± 0.08 < 0.07 – 0.26	1.21 ± 0.76 < 0.2 – 2.19	0.28 ± 0.17 0.11 – 0.55	—	0.04 ± 0.03 < 0.01 – 0.07	<0.003 – 0.01	0.49 ± 0.38 < 0.1 – 1.04	9
Iceberg lettuce	U.S.		—	—	< 0.07	0.32	0.12	—	0.01	< 0.003	0.13	9
Kale			—	—	0.04 ± 0.02	0.4 ± 0.2	0.3 ± 0.1	—	—	—	0.29 ± 0.05	16
Leek	Saudi Arabia	5	0.02	—	0.02	1.43	0.17	—	0.001	—	0.21	17
Leek leaf	Greece		—	—	—	—	0.21	—	0.04	< 0.0001	0.24	77
Leek root	Greece		< 0.001	0.003	0.17	0.35	0.08	—	0.02	< 0.0001	0.24	77
Legumes		7	—	—	—	—	—	—	—	0.01 0.002 – 0.03	—	66
Legumes		40	—	0.005 – 0.06	0.15 – 0.5	—	—	—	0.002 – 0.03	—	3.26 – 7.02	78
Lentils canned			—	—	—	1.24 ± 0.15	—	—	—	—	—	19
Lettuce	Greece		< 0.001	0.004	0.02	0.40	0.09	—	0.005	< 0.0001	0.10	77
Lettuce	Saudi Arabia	5	0.003	—	0.005	1.68	0.11	—	0.004	—	0.22	17
Maize			—	—	—	—	0.27 0.25 – 0.28	—	—	—	1.71 1.63 – 1.78	79
Onion	Saudi Arabia	5	0.01	—	0.03	1.07	0.04	—	0.21	—	0.20	17
Onion	U.S.		—	—	< 0.09	< 0.3	0.12	—	< 0.01	< 0.004	0.15	9
Onion conventional	Denmark		0.2 ± 0.1[a] 0.01 – 0.5	0.001 ± 0.001 0.0002 – 0.005	0.05 ± 0.01 0.02 – 0.10	0.27 ± 0.08 0.13 – 0.52	0.16 ± 0.06 0.05 – 0.36	0.001 ± 0.001 0.0 – 0.003	—	0.01 ± 0.002 0.003 – 0.01	0.34 ± 0.14 0.14 – 1.0	20
Onion leaf	Greece		0.001	0.03	0.09	0.72	0.11	—	0.01	<0.0001	0.28	77
Onion organic	Denmark		0.1 ± 0.1[a] 0.01 – 0.5	0.001 ± 0.0004 0.0003 – 0.003	0.06 ± 0.02 0.025 – 0.11	0.32 ± 0.07 0.19 – 0.53	0.15 ± 0.05 0.07 – 0.38	0.002 ± 0.004 0.0 – 0.02	—	0.005 ± 0.002 0.002 – 0.01	0.35 ± 0.12 0.17 – 0.83	20
Onion root	Greece		0.002	0.01	0.19	0.73	0.22	—	0.03	0.0005	0.68	77
Other vegetables[b]	U.K.		0.001	0.01	0.09	0.75	0.16	0.01	0.01	0.002	0.26	7
Other vegetables[b]	U.S.		—	—	0.15 ± 0.05 < 0.07 – 0.22	0.66 ± 0.40 < 0.3 – 1.56	0.23 ± 0.17 < 0.1 – 0.59	—	0.02 ± 0.02 < 0.01 – 0.05	0.01 ± 0.004 < 0.003 – 0.01	0.29 ± 0.15 < 0.1 – 0.62	9
Parsley	Greece		0.002	0.01	0.03	0.72	0.15	—	0.03	< 0.0001	0.09	77

TABLE 7.7 (CONTINUED)
Micronutrients in Vegetables, Mushrooms, and Fruits (in mg/100 g w/w and in μg/100 g w/w[a])

Name	Origin	n	Co	Cr	Cu	Fe	Mn	Mo	Ni	Se	Zn	Ref.
Parsley	Saudi Arabia	5	0.03	—	0.05	2.94	0.21	—	0.01	—	0.34	17
Peas, conventional	Denmark		0.5 ± 0.4[a] 0.1 – 2.0	0.3 ± 0.1[a] 0.04 – 1.0	0.14 ± 0.03 0.08 – 0.21	1.02 ± 0.29 0.51 – 2.08	0.22 ± 0.06 0.13 – 0.42	0.02 ± 0.03 0.003 – 0.15	—	0.01 ± 0.003 0.002 – 0.02	0.79 ± 0.25 0.42 – 1.63	20
Peas, organic	Denmark		0.5 ± 0.3[a] 0.1 – 1.0	0.4 ± 0.2[a] 0.1 – 1.0	0.12 ± 0.03 0.07 – 0.22	1.15 ± 0.20 0.60 – 1.51	0.24 ± 0.08 0.14 – 0.56	0.03 ± 0.02 0.004 – 0.11	—	0.01 ± 0.002 0.004 – 0.01	1.0 ± 0.31 0.56 – 1.94	20
Potatoes	Spain		—	—	0.14 ± 0.03 0.12 – 0.16	0.81 ± 0.30 0.60 – 1.03	0.13 ± 0.01 0.13 – 0.14	—	—	—	0.40 ± 0.01 0.40 – 0.41	21
Potatoes	Saudi Arabia	5	0.03	—	0.02	1.07	0.13	—	0.24	—	0.1	17
Potatoes[c]	Idaho	342	0.01 ± 0.02 0 – 0.07	0.01 ± 0.01 0 – 0.06	0.1 ± 0.03 0 – 0.19	0.78 ± 0.30 0.29 – 2.02	0.15 ± 0.04 0.04 – 0.41	0.01 ± 0.02 0 – 0.08	0.02 ± 0.02 0 – 0.10	—	0.28 ± 0.07 0.09 – 0.52	22
Potatoes[c]	Non-Idaho	266	0.01 ± 0.02 0 – 0.07	0.02 ± 0.01 0 – 0.05	0.12 ± 0.06 0 – 0.40	0.90 ± 0.39 0.26 – 2.91	0.23 ± 0.11 0.03 – 0.78	0.01 ± 0.02 0 – 0.11	0.02 ± 0.03 0 – 0.11	—	0.39 ± 0.11 0.14 – 1.29	22
Potatoes[b]	U.K.		0.002	0.01	0.13	0.81	0.19	0.01	0.01	0.001	0.45	7
Potatoes[b]	U.S.		—	—	0.13 ± 0.02 < 0.08 – 0.14	0.58 ± 0.28 0.29 – 0.91	0.21 ± 0.11 0.11 – 0.4	—	0.01 ± 0.01 < 0.01 – 0.02	< 0.003	0.31 ± 0.09 0.21 – 0.46	9
Radish	U.S.		—	—	< 0.09	< 0.3	< 0.1	—	—	—	0.11	9
Red pepper	Tenerife Island	60	—	—	0.07 ± 0.03 0.03 – 0.13	0.31 ± 0.07 0.22 – 0.46	0.07 ± 0.04 0.03 – 0.08	—	—	—	0.17 ± 0.03 0.10 – 0.23	18
Rucola			—	—	0.1 ± 0.2	1.1 ± 0.6	0.3 ± 0.3	—	—	—	0.4 ± 0.1	16
Salq	Saudi Arabia	5	0.03	—	0.02	1.49	0.31	—	0.21	—	0.29	17
Spinach	Greece		0.003	0.01	0.24	2.15	0.44	—	0.05	< 0.0001	0.30	77
Spinach	Saudi Arabia	5	0.01	—	0.02	1.41	0.08	—	0.15	—	0.08	17
Spinach substitute			—	—	0.05 ± 0.03	1 ± 0.8	1 ± 0.8	—	—	—	0.3 ± 0.1	16
Tomato	Saudi Arabia	5	0.01	—	0.03	0.35	0.04	—	0.09	—	0.08	17

Tomato fruits rockwool high EC	Denmark		—	0.0001	0.04	0.28	0.09	0.01	0.0005	—	0.11	23
Tomato fruits rockwool norm EC	Denmark		—	0.0001	0.03	0.24	0.09	0.01	0.0005	—	0.10	23
Tomato fruits soil norm EC	Denmark		—	0.0001	0.03	0.26	0.07	0.003	0.001	—	0.12	23
Tomato red	U.S.		—	—	< 0.09	< 0.3	0.1	—	< 0.01	—	0.13	9
Marrow (zucchini)	Saudi Arabia	5	0.01	—	0.04	0.91	0.09	—	0.06	—	0.23	17
Watercress				—	0.2 ± 0.1	0.9 ± 0.2	0.2 ± 0.1	—	—	—	0.8 ± 0.4	16
Watercress	Saudi Arabia	5	0.02	—	0.01	1.62	0.12	—	0.28	—	0.68	17
White asparagus			—	—	—	0.44 ± 0.01	—	—	—	—	—	19
Whole peeled tomatoes, canned			—	—	—	0.38 ± 0.09	—	—	—	—	—	19
Mushrooms												
Bay bolete[c]	Poland	166	0.004 ± 0.001 0.0003 – 0.02	0.005 ± 0.002 0.0002 – 0.08	0.33 ± 0.14 0.01 – 0.84	0.98 ± 0.52 0.09 – 6.10	0.19 ± 0.1 0.02 – 1.91	—	0.01 ± 0.004 0.0002 – 0.03	—	1.61 ± 0.42 0.66 – 4.80	24
Lactarius deliceus[c]	Turkey		0.005 ± 0.0002[a]	0.01 ± 0.0005[a]	0.31 ± 0.03	2.88 ± 0.21	0.4 ± 0.02[a]	—	0.02 ± 0.001[a]	—	0.82 ± 0.07	80
Morchella esculenta[c]	Turkey			0.01 ± 0.001[a]	0.43 ± 0.03	1.46 ± 0.12	0.3 ± 0.02[a]	—	0.01 ± 0.001[a]	—	0.45 ± 0.04	80
Mushrooms	U.S.		—	—	0.22	< 0.3	< 0.1	—	—	0.01	0.36	9
Mushrooms canned			—	—	—	0.37 ± 0.05	—	—	—	—	—	19
Russula delica[c]	Turkey		0.005 ± 0.0004[a]	0.01 ± 0.001[a]	0.25 ± 0.01	4.85 ± 0.33	0.2 ± 0.02[a]	—	0.03 ± 0.001[a]	—	0.78 ± 0.05	80
Fruits												
Apple red	U.S.		—	—	< 0.09	< 0.3	< 0.1	—	< 0.009	< 0.004	< 0.1	9
Apricot	U.S.		—	—	< 0.09	0.41	< 0.1	—	0.01	—	0.18	9
Avocado	U.S.			—	0.22	0.48	< 0.2	—	0.02	< 0.005	0.61	9

TABLE 7.7 (CONTINUED)
Micronutrients in Vegetables, Mushrooms, and Fruits (in mg/100 g w/w and in μg/100 g w/w[a])

Name	Origin	n	Co	Cr	Cu	Fe	Mn	Mo	Ni	Se	Zn	Ref.
Banana	U.S.		—	—	0.11	0.28	0.27	—	< 0.01	< 0.003	0.16	9
Banana pulp	Tenerife Island		—	—	0.12 ± 0.01 0.09 – 0.20	0.32 ± 0.057 0.16 – 0.54	0.07 ± 0.0 0.06 – 0.12	—	—	—	0.17 ± 0.014 0.10 – 0.24	25
Cantaloupe	U.S.		—	—	< 0.07	< 0.2	< 0.1	—	0.01	< 0.003	0.14	9
Currants		6	—	—	1.98 ± 0.13 1.49 – 2.7	1.98 ± 0.09 1.62 – 2.54	0.49 ± 0.03 0.33 – 0.66	—	—	—	1.1 ± 0.5 0.23 – 12.7	56
Dried apple		6	—	—	0.20 ± 0.01 0.16 – 0.24	1.17 ± 0.04 0.95 – 1.34	0.19 ± 0.01 0.13 – 0.27	—	—	—	0.19 ± 0.01 0.13 – 0.27	56
Dry fruits		9	—	—	—	—	—	—	—	0.03 0.01 – 0.06	—	66
Fresh fruit[b]	U.K.		0.0004	< 0.01	0.09	0.27	0.2	0.001	0.003	0.0004	0.08	7
Fresh fruit[b]	U.S.		—	—	0.13 ± 0.06 < 0.07 – 0.22	0.34 ± 0.10 < 0.2 – 0.48	0.32 ± 0.06 < 0.1 – 0.36	—	0.01 ± 0.01 < 0.01 – 0.03	< 0.003	0.23 ± 0.19 < 0.1 – 0.61	9
Fruit products[b]	U.K.		0.0005	< 0.01	0.07	0.32	0.22	0.001	0.01	0.0005	0.07	7
Fruit products[b]	U.S.		—	—	0.32 ± 0.04 < 0.07 – 0.34	0.85 ± 0.59 < 0.2 – 1.85	0.42 ± 0.39 < 0.1 – 1.0	—	0.03 ± 0.02 0.005 – 0.07	< 0.003	0.28 ± 0.13 < 0.1 – 0.43	9
Grapefruit	U.S.		—	—	< 0.07	< 0.2	< 0.1	—	< 0.01	< 0.003	< 0.1	9
Grapes, seedless	U.S.		—	—	0.11	0.31	< 0.1	—	< 0.01	—	< 0.1	9
Orange	U.S.		—	—	< 0.07	< 0.2	< 0.1	—	< 0.01	—	< 0.1	9
Peach	U.S.		—	—	< 0.07	0.22	< 0.1	—	0.01	—	0.17	9
Pear	U.S.		—	—	0.08	< 0.2	< 0.1	—	0.01	—	< 0.1	9
Plums	U.S.		—	—	< 0.07	< 0.2	< 0.1	—	< 0.01	—	< 0.1	9
Strawberries	U.S.		—	—	< 0.07	0.41	0.36	—	0.01	< 0.003	0.1	9
Strawberry	Finland	12	—	—	0.06 0.04 – 0.1	0.32 0.21 – 0.45	0.32 0.23 – 0.44	—	—	—	0.12 0.08 – 0.15	26
Sultanas		6	—	—	0.51 ± 0.02 0.40 – 0.64	2.05 ± 0.13 1.72 – 3.01	0.42 ± 0.02 0.29 – 0.60	—	—	—	0.42 ± 0.04 0.30 – 0.67	56

Sweet cherries	U.S.		—	—	< 0.09	< 0.3	< 0.1	—	—	—	< 0.1	9
Watermelon	U.S.		—	—	< 0.07	0.25	< 0.1	—	0.01	< 0.003	< 0.1	9
Frozen Fruits												
Gooseberry	Polish market	5	—	0.01 ± 0.004 0.01 – 0.02	0.06 ± 0.01 0.05 – 0.07	—	—	—	0.01 ± 0.004 0.004 – 0.01	—	0.15 ± 0.03 0.12 – 0.2	81
Black chokeberry	Polish market	5		0.03 ± 0.01 0.02 – 0.05	0.10 ± 0.01 0.08 – 0.11	—	—	—	0.03 ± 0.01 0.03 – 0.05	—	0.19 ± 0.02 0.16 – 0.22	81
Blackberry	Polish market	5	—	0.02 ± 0.01 0.01 – 0.03	0.14 ± 0.03 0.11 – 0.18	—	—	—	0.01 ± 0.004 0.01 – 0.02	—	0.22 ± 0.03 0.19 – 0.26	81
Raspberry	Polish market	5	—	0.01 ± 0.003 0.01 – 0.02	0.10 ± 0.01 0.08 – 0.12	—	—	—	0.03 ± 0.01 0.02 – 0.05	—	0.31 ± 0.02 0.28 – 0.33	81
Fruits	Polish market	71	—	0.02 0.005 – 0.06	0.09 0.03 – 0.18	—	—	—	0.02 0.004 – 0.05	—	0.17 0.06 – 0.33	81
Black currant	Polish market	5		0.02 ± 0.01 0.01 – 0.03	0.08 ± 0.01 0.08 – 0.09	—	—	—	0.01 ± 0.005 0.005 – 0.02	—	0.21 ± 0.04 0.16 – 0.26	81
Red currant	Polish market	5		0.01 ± 0.003 0.01 – 0.02	0.08 ± 0.02 0.06 – 0.11	—	—	—	0.01 ± 0.01 0.01 – 0.02	—	0.15 ± 0.03 0.12 – 0.19	81
Plum	Polish market	5	—	0.01 ± 0.004 0.01 – 0.02	0.05 ± 0.02 0.03 – 0.07	—	—	—	0.01 ± 0.003 0.005 – 0.01	—	0.12 ± 0.05 0.08 – 0.20	81
Strawberry	Polish market	6	—	0.01 ± 0.004 0.005 – 0.02	0.06 ± 0.01 0.05 – 0.08	—	—	—	0.01 ± 0.005 0.01 – 0.02	—	0.15 ± 0.02 0.12 – 0.18	81
Cherry	Polish market	7	—	0.01 ± 0.005 0.01 – 0.02	0.09 ± 0.01 0.07 – 0.1	—	—	—	0.01 ± 0.01 0.01 – 0.03	—	0.10 ± 0.03 0.06 – 0.14	81

[a] μg/100 g w/w

TABLE 7.8
Micronutrients in Tea, Coffee, and Confectionary Products (in mg/100 g w/w)

Name	Origin	n	Co	Cr	Cu	Fe	Mn	Mo	Ni	Se	Zn	Ref.
Tea												
Tea	Africa	18	0.02 ± 0.02	0.2 ± 0.12	1.91 ± 0.88	17.9 ± 14.8	56.3 ± 31.6	—	0.48 ± 0.26	—	2.51 ± 0.98	27
Tea	Asia	36	0.02 ± 0.02	0.22 ± 0.1	1.23 ± 0.41	15.3 ± 4.89	84.4 ± 41.5	—	0.42 ± 0.16	—	2.63 ± 1.13	27
Tea	China	13	0.03 ± 0.01	0.12 ± 0.07	1.91 ± 0.96	28.1 ± 20.3	72.0 ± 31.4	—	0.49 ± 0.19	—	2.78 ± 1.24	27
Tea	India, Sri Lanka	13	0.02 ± 0.01	0.11 ± 0.08	2.12 ± 0.88	10.2 ± 1.63	36.7 ± 13.8	—	0.41 ± 0.13	—	2.44 ± 0.74	27
Tea (from tea bag)			—	—	—	—	0.29	—	< 0.005	< 0.001	< 0.04	9
U.S. tea brands	U.S.	7	—	—	1.23 ± 0.48 0.44 – 1.73	—	32.9 ± 23.1 7.9 – 76.8	—	—	—	—	28
Tea[c]	China	39	—	—	1.71 ± 0.71 0.74 – 3.69	—	64.6 ± 31.5 14.8 – 139	—	0.78 ± 0.39 0.18 – 2.12	0.02 0.003 – 0.69	3.64 ± 0.85 1.85 – 5.54	29
Beiqishen tea	China		0.08 ± 0.01	0.33 ± 0.06	2.13 ± 0.03	39.5 ± 7.94	112.9 ± 17.2	< 0.005	0.68 ± 0.06	—	3.72 ± 0.44	30
Chinese tea	China	5	—	—	—	—	11.0 ± 0.27	—	—	—	—	31
Chinese tea	China	5	—	—	1.70 ± 0.48	—	—	—	—	—	2.78 ± 0.65 2.23 – 3.73	82
Red tea		9	0.16 ± 0.04 0.11 – 0.24	0.20 ± 0.07 0.13 – 0.34	1.88 ± 0.25 1.65 – 2.36	46.4 ± 22.0 28.7 – 97.2	71.0 ± 15.2 56.2 – 99.5	—	0.53 ± 0.1 0.39 – 0.66	—	3.86 ± 0.54 2.90 – 4.67	32
Indian tea	India	3	—	—	2.36 ± 0.57 1.70 – 2.71	—	—	—	—	—	2.48 ± 0.19 2.31 – 2.69	82
Indian tea brands	India	15	—	—	1.48 ± 0.82 0.16 – 3.5	—	57.5 ± 9.6 37.1 – 75.8	—	—	—	—	28
Instant tea[d] (<2% tea extract)		3	—	—	0.001 0.001 – 0.001	0.01 0.005 – 0.01	0.31 0.25 – 0.4	—	—	—	0.005 0.004 – 0.01	33
Instant tea[d] (100% extract)		2	—	—	0.005 0.003 – 0.01	0.01 0.01 – 0.01	0.68 0.63 – 0.73	—	—	—	0.02 0.02 – 0.02	33

Beiqishen tea extract[d]	China		0.0003 ± 0.0001	0.001 ± 0.0003	0.01 ± 0.001	0.01 ± 0.001	0.44 ± 0.01	< 0.00001	0.01 ± 0.0004	—	0.01 ± 0.0003	30
Chinese tea infusions[d]	China	5	—	—	0.01 ± 0.003	—	—	—	—	—	0.02 ± 0.005 0.01 – 0.03	82
Black tea beverages[d]		23	—	—	0.005 ± 0.01 0.003 – 0.02	0.01 ± 0.001 0.003 – 0.02	0.1 – 0.6	—	—	—	0.01 – 0.05	33
Indian tea infusions	India	3	—	—	0.01 ± 0.005 0.001 – 0.01	—	—	—	—	—	0.01 ± 0.004 0.01 – 0.02	82
Green tea beverages[d]		20	—	—	0.01 ± 0.001 0.001 – 0.01	0.01 ± 0.0 0.002 – 0.02	0.13 – 0.38	—	—	—	0.01 – 0.03	33
Tea soft drinks[d]		8	—	—	0.0005 0.0003 – 0.001	0.01 0 – 0.03	0.07 0.0 – 0.11	—	—	—	0.004 0.002 – 0.01	33
Coffee												
Decaffeinated instant coffee	U.S.		—	—	—	< 0.1	< 0.04	—	< 0.005	< 0.001	< 0.04	9
Ground coffee	U.S.		—	—	—	—	<0.1	—	—	—	<0.1	9
Ground coffee	Polish market		0.04 0.01 – 0.09	0.02 0.01 – 0.05	1.63 1.36 – 1.99	3.98 2.03 – 6.78	2.30 1.65 – 4.06	—	0.16 0.05 – 0.42	—	0.47 0.37 – 0.64	83
Ground decaffeinated coffee	Polish market		0.05 0.02 – 0.08	0.03 0.02 – 0.04	1.80 1.59 – 2.0	6.39 3.87 – 8.26	2.01 1.69 – 2.33	—	0.25 0.0.8 – 0.43	—	0.49 0.47 – 0.51	83
Soluble coffee	Polish market		0.06 0.03 – 0.08	0.03 0.02 – 0.04	0.08 0.05 – 0.12	2.75 1.93 – 4.34	1.59 0.93 – 2.27	—	0.18 0.09 – 0.34	—	0.23 0.18 – 0.29	83
Soluble coffee	Brazil	21	—	<1.0 – 0.052 <1.0 – 0.052	0.13 ± 0.06 0.05 – 0.23	8.42 ± 1.13 1.4 – 45.1	1.51 ± 0.82 0.36 – 3.88	—	—	—	0.72 ± 0.35 0.32 – 1.51	34
Roasted coffee beans	Costa Rica	20	—	—	1.75 ± 0.25	1.45 ± 0.29	2.22 ± 0.39	—	—	—	0.77 ± 0.13	35
Roasted coffee beans	Colombia	20	—	—	1.66 ± 0.16	1.64 ± 0.29	3.67 ± 0.97	—	—	—	0.77 ± 0.17	35

TABLE 7.8 (CONTINUED)
Micronutrients in Tea, Coffee, and Confectionary Products (in mg/100 g w/w)

Name	Origin	n	Co	Cr	Cu	Fe	Mn	Mo	Ni	Se	Zn	Ref.
Roasted coffee beans	Guatemala	20	—	—	1.35 ± 0.22	1.25 ± 0.19	2.41 ± 0.48	—	—	—	0.77 ± 0.13	35
Roasted coffee beans	Panama	20	—	—	1.62 ± 0.17	1.93 ± 0.48	2.51 ± 0.39	—	—	—	0.68 ± 0.17	35
Roasted coffee beans	Ethiopia	20	—	—	1.33 ± 0.09	1.16 ± 0.29	2.03 ± 0.39	—	—	—	0.75 ± 0.18	35
Roasted coffee beans	Kenya	20	—	—	1.72 ± 0.21	1.45 ± 0.48	3.77 ± 1.06	—	—	—	0.69 ± 0.18	35
Roasted coffee beans	Sulawesi	20	—	—	1.21 ± 0.33	2.03 ± 0.58	2.8 ± 0.67	—	—	—	0.76 ± 0.16	35
Roasted coffee beans	Sumatra	20	—	—	1.27 ± 0.19	2.99 ± 0.68	1.83 ± 0.48	—	—	—	0.63 ± 0.11	35
Roasted coffee[c]	Mixture	18	—	—	1.39 ± 0.21 1.08 – 1.72	5.23 ± 0.82 3.89 – 7.09	1.96 ± 0.94 1.18 – 4.31	—	—	—	1.41 ± 1.03 0.48 – 3.56	36
Green coffee[c]	Mixture	41	—	—	1.90 ± 0.94 1.38 – 7.43	3.78 ± 1.40 2.39 – 9.01	2.64 ± 0.97 1.4 – 4.83	—	—	—	1.01 ± 0.95 0.35 – 5.92	84
Confectionary Products												
Sugar and preserves[b]	U.K.		0.003	0.02	0.15	0.94	0.15	0.003	0.03	0.001	0.38	7
Sugar and preserves[b]	U.S.		—	—	0.24 ± 0.09 < 0.08 – 0.41	1.45 ± 1.05 < 0.3 – 3.85	0.31 ± 0.10 < 0.1 – 0.49	—	0.04 ± 0.03 < 0.01 – 0.09	0.008 ± 0.004 < 0.003 – 0.01	0.43 ± 0.25 < 0.1 – 1.34	9
Sugar	Egypt	2	0.001 ± 0.0001	0.001 ± 0.001	0.002 ± 0.001	0.02 ± 0.01	0.001 ± 0.0004	—	0.001 ± 0.0004	—	0.0004 ± 0.0	37
Sugar	Egypt		0.0001 ± 0.0	0.0001 ± 0.0	0.001 ± 0.0004	2.1 ± 0.4	0.04 ± 0.02	—	—	—	0.002 ± 0.001	38

White sugar, granulated	U.S.		—	—	—	< 0.3	—	—	—	—	< 0.1	9
Sugarcane plant	Egypt	3	0.003 ± 0.0002	0.02 ± 0.005	0.01 ± 0.003	0.44 ± 0.15	0.05 ± 0.02	—	0.01 ± 0.002	—	0.04 ± 0.05	37
Cocoa powder		4	—	—	5.55 ± 0.04 5.48 – 5.62	21.9 ± 1.8 16 – 25.8	6.34 ± 0.12 6.38 – 6.52	—	—	—	8.96 ± 0.14 8.5 – 9.43	56
Molasses	Egypt		0.002 ± 0.001	0.03 ± 0.02	0.005 ± 0.0025	17.4 ± 16.3	1.02 ± 0.62	—	—	—	0.01 ± 0.003	38
Sugarcane juice	Egypt	3	0.003 ± 0.0005	0.021 ± 0.001	0.01 ± 0.004	0.31 ± 0.18	0.01 ± 0.001	—	0.005 ± 0.001	—	0.005 ± 0.002	37
Sugar juice	Egypt		0.002 ± 0.001	0.02 ± 0.02	0.01 ± 0.01	42.4 ± 7.4	1.79 ± 1.17	—	—	—	0.01 ± 0.01	38

TABLE 7.9
Micronutrients in Nuts, Seeds, and Oils (in mg/100 g w/w)

Name	Origin	n	Co	Cr	Cu	Fe	Mn	Mo	Ni	Se	Zn	Ref.
Nuts, Seeds												
Almond	Spain	6	—	0.04	1.11	4.5	—	—	0.03	—	3.88	78
				0.04 – 0.05	1.05 – 1.26	4.03 – 4.87			0.02 – 0.05		2.79 – 5.0	
Almond			—	—	2 ± 0.0	14 ± 1	—	—	—	—	6 ± 0.3	39
Sliced almonds		6	—	—	1.04 ± 0.04	2.74 ± 0.09	1.48 ± 0.05	—	—	—	2.75 ± 0.08	56
					0.91 – 1.23	2.06 – 3.09	1.25 – 1.56				2.08 – 2.95	
Cashew	Spain	3	—	0.03	0.78	1.42	—	—	0.02	—	4.0	78
				0.03 – 0.03	0.66 – 0.85	1.03 – 1.56			0.02 – 0.02		3.75 – 4.96	
Chestnut	Spain	6	—	0.03	0.48	0.9	—	—	0.02	—	3.39	78
				0.03 – 0.03	0.4 – 0.56	0.73 – 1.02			0.01 – 0.02		3.06 – 3.77	
Different types of nuts	Brazil	3	—	—	—	—	—	—	—	0.74 ± 1.39	—	85
										0.01 – 3.51		
Poppy seeds	Polish market		0.02 ± 0.01	0.05 ± 0.01	1.39 ± 0.12	12.11 ± 1.17	7.18 ± 1.43	—	0.07 ± 0.06	—	7.02 ± 1.66	86
			0.01 – 0.02	0.05 – 0.06	1.28 – 1.60	10.6 – 13.3	5.72 – 8.49		0.01 – 0.12		5.43 – 8.60	
Sesame	Polish market		0.03 ± 0.01	0.02 ± 0.01	1.32 ± 0.10	5.58 ± 0.29	1.55 ± 0.04	—	0.06 ± 0.01	—	5.24 ± 0.23	86
			0.01 – 0.04	0.02 – 0.03	1.21 – 1.50	5.23 – 5.93	1.51 – 1.62		0.05 – 0.07		4.95 – 5.55	
Hazelnut	Spain	5	—	0.03	1.66	1.73	—	—	0.04	—	4.25	78
				0.04 – 0.04	1.4 – 1.72	1.33 – 1.99			0.03 – 0.05		3.8 – 4.56	
Hazelnut	Turkey	11	—	—	0.65 ± 0.29	2.32 ± 0.21	6.09 ± 4.17	—	—	—	1.95 ± 0.25	40
					0.10 – 0.90	2.05 – 2.56	1.77 – 15.8				1.71 – 2.44	
Hazelnut	Turkey		—	—	3.16 ± 0.28	3.63 ± 0.92	8.37 ± 3.22	—	—	—	4.11 ± 0.95	41
Hazelnut	Spain	24	—	—	1.60 ± 0.42	3.49 ± 0.61	2.35 ± 1.46	—	—	—	—	87
					1.0 – 2.52	2.5 – 4.5	0.73 – 5.8					

Hazelnut			—	—	2 ± 0.0	8 ± 2	—	—	—	—	3 ± 1	39
Hazelnut[c]	Turkey		—	—	1.72 ± 0.36	3.79 ± 0.44	1.95 ± 0.50	—	—	—	2.48 ± 0.2	42
Mixed nuts (no peanuts)	U.S.		—	—	1.55	4.16	2.32	—	0.30	0.06	4.1	9
Nuts[b]	U.K.		0.01	0.07	0.85	3.4	1.5	0.10	0.25	0.03	3.1	7
Nuts[b]	U.S.		—	—	0.88 ± 0.58	2.55 ± 1.40	1.99 ± 0.34	—	0.16 ± 0.13	0.03 ± 0.04	3.25 ± 0.74	9
					0.52 – 1.55	1.71 – 4.16	1.65 – 2.32		0.07 – 0.304	< 0.005 – 0.06	2.71 – 4.1	
Peanut	Spain	5	—	0.04	0.69	2.28	—	—	0.03	—	3.19	78
				0.04 – 0.04	0.56 – 1.03	2.0 – 2.66			0.02 – 0.04		2.99 – 3.78	
Peanut	Spain		—	—	1.01 ± 0.01	3.94 ± 0.02	—	—	—	—	—	88
					0.99 – 1.02	3.91 – 3.96						
Peanut			—	—	1 ± 0.1	6 ± 0.1	—	—	—	—	4 ± 0.2	39
Peanut butter			—	—	0.52	1.78	1.65	—	0.07	0.01	2.71	9
Peanut greens[c]	U.S.	2	—	—	—	15.2 ± 1.64	—	—	—	—	5.05 ± 2.89	43
						13.3 – 16.1					1.89 – 7.58	
Pine nut kernel	Spain	7	—	0.03	2.04	2.23	—	—	0.01	—	5.02	78
				0.03 – 0.04	1.97 – 2.33	1.96 – 2.44			0.01 – 0.01		4.07 – 6.03	
Pistachio	Spain	8	—	0.04	0.92	7.35	—	—	0.02	—	3.34	78
				0.04 – 0.04	0.56 – 1.03	7.0 – 7.56			0.01 – 0.03		3.02 – 3.67	
Pumpkin seeds	Slovenia		—	—	—	—	—	—	—	0.003 ± 0.001	—	89
Roasted salted corn	Spain	6	—	0.04	0.66	5.03	—	—	0.04	—	4.05	78
				0.03 – 0.04	0.5 – 0.93	4.28 – 5.54			0.02 – 0.06		3.79 – 4.52	
Sunflower seed	Spain	5	—	0.07	1.36	4.09	—	—	0.02	—	5.88	78
				0.04 – 0.1	1.0 – 1.57	3.88 – 4.33			0.02 – 0.03		5.06 – 6.9	
Walnut			—	—	2 ± 0.2	7 ± 0.3	—	—	—	—	4 ± 0.4	39
Walnuts	Spain	5	—	0.04	2.2	2.25	—	—	0.02	—	2.85	78
				0.03 – 0.04	2.01 – 2.56	2.0 – 2.44			0.02 – 0.02		2.56 – 3.97	

TABLE 7.9 (CONTINUED)
Micronutrients in Nuts, Seeds, and Oils (in mg/100 g w/w)

Name	Origin	n	Co	Cr	Cu	Fe	Mn	Mo	Ni	Se	Zn	Ref.
Oils												
Grape seeds oil		3	—	—	0.01 ± 0.001	—	—	—	—	—	0.03 ± 0.01	90
Maize oil		5	—	—	0.005 ± 0.002	—	—	—	—	—	0.005 ± 0.002	90
Nuts oil		3	—	—	0.02 ± 0.004	—	—	—	—	—	0.06 ± 0.01	90
Oils and fats[b]	U.K.		0.0003	0.04	0.005	0.05	0.002	0.002	0.003	0.001	0.06	7
Oils and fats[b]	U.S.		—	—	ND	< 0.3	ND	—	< 0.04	ND	< 0.1	9
Olive or sunflower oil	U.S.		—	—	—	< 0.3	—	—	—	—	< 0.1	9
Peanuts oil		6	—	—	0.01 ± 0.002	—	—	—	—	—	0.03 ± 0.02	90
Pumpkin oil			—	—	—	—	—	—	—	< 0.0001	—	89
Rice oil		5	—	—	0.01 ± 0.003	—	—	—	—	—	0.02 ± 0.005	90
Soy oil		5	—	—	0.07 ± 0.02	—	—	—	—	—	0.004 ± 0.0005	90
Soy oil	Poland	1	—	—	0.02	0.16	< 0.03	—	—	—	0.07	65
Sunflower oil		5	—	—	0.01 ± 0.004	—	—	—	—	—	0.02 ± 0.01	90

TABLE 7.10
Micronutrients in Beverages (in mg/100 g w/w)

Name	Origin	n	Co	Cr	Cu	Fe	Mn	Mo	Ni	Se	Zn	Ref.
Beverages[b]	U.K.		0.0002	< 0.01	0.01	0.04	0.27	0.0003	0.003	0.0003	0.03	7
Beverages[b]	U.S.		—	—	< 0.05	<0.1 – 0.28	0.2 ± 0.13 < 0.04 – 0.29	—	<0.005	< 0.003	< 0.1	9
Juices												
Grapefruit juice			0.0001	—	0.04	0.07	0.02	0.001	0.002	—	0.05	45
Lemon juice			< 0.0001	—	0.02	0.10	0.02	< 0.0001	0.001	—	0.05	45
Lime juice			< 0.0001	—	0.03	0.05	0.03	0.001	0.003	—	0.06	45
Mandarin juice			< 0.0001	—	0.14	0.1	0.03	< 0.0001	0.01	—	0.05	45
Orange juice	Australia	290	0.0003 ± 0.0 < 0.0001 – 0.002	—	0.03 ± 0.001 0.004 – 0.1	0.07 ± 0.01 0.002 – 0.18	0.020 ± 0.002 0.01 – 0.05	0.0003 ± 0.0 < 0.0001 – 0.003	0.003 ± 0.0 < 0.0001 – 0.02	—	0.03 ± 0.001 0.01 – 0.07	45
Orange juice[b]		8	—	—	0.01 – 0.02	0.09 – 0.36	0.04 – 0.06	—	—	—	0.03 – 0.03	47
Orange juice concentrate	Australia	83	0.0002 ± 0.0001 < 0.0001 – 0.0004	—	0.03 ± 0.005 0.01 – 0.05	0.07 ± 0.02 0.03 – 0.15	0.02 ± 0.003 0.01 – 0.03	0.0003 ± 0.0003 < 0.0001 – 0.001	0.004 ± 0.002 0.001 – 0.003	—	0.02 ± 0.01 0.01 – 0.04	45
Orange juice concentrate	Brazil	42	0.0001 ± 0.0001 < 0.0001 – 0.001	—	0.02 ± 0.004 0.02 – 0.03	0.06 ± 0.01 0.03 – 0.09	0.03 ± 0.01 0.01 – 0.05	0.0001 ± 0.0002 <0.0001 – 0.001	0.001 ± 0.001 0.0003 – 0.003	—	0.03 ± 0.01 0.01 – 0.04	45
Pineapple juice[b]			—	—	0.05 ± 0.01 0.03 – 0.06	0.23 ± 0.07 0.13 – 0.32	0.84 ± 0.36 0.29 – 1.26	—	—	—	0.07 ± 0.03 0.04 – 0.11	48
Pineapple nectar[b]			—	—	0.03 ± 0.004 0.02 – 0.03	0.25 ± 0.14 0.09 – 0.44	0.38 ± 0.19 0.14 – 0.76	—	—	—	0.03 ± 0.01 0.02 – 0.06	48
Pummelo juice			< 0.0001	—	0.02	0.08	0.01	0.001	0.001	—	0.05	45
Seville orange juice			< 0.0001	—	0.03	0.07	0.01	< 0.0001	0.002	—	0.08	45

TABLE 7.10 (CONTINUED)
Micronutrients in Beverages (in mg/100 g w/w)

Name	Origin	n	Co	Cr	Cu	Fe	Mn	Mo	Ni	Se	Zn	Ref.
Tangello juice			< 0.0001	—	0.02	0.06	0.02	< 0.0001	0.003	—	0.04	45
Tomato juice			—	—	< 0.09	0.36	< 0.1	—	0.02	< 0.004	0.11	9
Alcoholic Beverages												
Beer			—	—	—	—	< 0.1	—	—	< 0.003	< 0.1	9
Beer[d]		32	—	—	—	0.02 ± 0.02 0.01 – 0.11	0.01 ± 0.01 0.003 – 0.03	—	—	—	0.01 ± 0.02 0.0001 – 0.1	49
Beer[d]	Poland	35	0.00002 0.00001 – 0.0001	0.001 0.0004 – 0.004	0.01 0.003 – 0.01	0.02 0.004 – 0.05	0.02 0.005 – 0.05	—	0.001 0.0004 – 0.02	—	0.002 0.0004 – 0.01	91
Martini			—	—	< 0.04	< 0.1	< 0.04	—	—	—	<0.04	9
Whiskey			—	—	< 0.04	< 0.1	—	—	—	—	—	9
Whiskey[d]	Scotland	35	—	—	0.05 ± 0.03 0.01 – 0.17	0.20 ± 0.63 0.002 – 2.77	—	—	0.01 ± 0.02 <0.00001 – 0.06	—	0.27 ± 0.61 0.001 – 2.06	50
Dry table wine			—	—	< 0.05	0.28	0.11	—	< 0.005	< 0.002	< 0.05	9
Red wines[d]	Brazil		—	0.01	0.001 – 0.04	0.17 – 0.52	0.08 – 0.25	—	0.002 – 0.03	—	0.02 – 0.13	51
Red wines[d]	Portugal		—	0.01	0.001 – 0.02	0.33 – 0.84	0.05 – 0.26	—	0.004 – 0.01	—	0.04 – 0.11	51
Red wines[d]	Chile		—	0.01	0.01	0.24	0.08	—	0.002	—	0.08	51
White wines[d]	Brazil		—	0.01	0.001 – 0.01	0.08 – 0.24	0.11 – 0.17	—	0.01	—	0.03 – 0.06	51
White wines[d]	Chile		—	—	0.001	0.09	0.07	—	0.01	—	0.03	51
Wine[d]	Spain	125	—	—	0.02 ± 0.02 0.003 – 0.12	0.22 ± 0.09 0.09 – 0.51	0.11 ± 0.04 0.04 – 0.26	—	—	—	0.07 ± 0.02 0.02 – 0.19	52
Wine[d]	Italy		0.000004 – 0.00002	0.002 – 0.005	0.0001 – 0.13	0.13 – 2.78	0.07 – 0.25	0.000001 – 0.00001	0.001 – 0.02	—	0.01 – 0.48	53

chromium compounds used to increase the lacquer adherence and resistance to oxidation of tinplate. Foods may also contain incidentally introduced chromium that originates from cooking utensils, storage vessels, as well as stainless steel commonly used in food processing equipment. For instance, the concentration of chromium in red cabbage pickled in vinegar and preserved in stainless steel has been found to be 0.007 mg Cr/100 g, which is about seven times higher than the levels in fresh vegetables.[1,54] According to Saner,[55] some spices such as black pepper can contain high levels of chromium.

7.3.2 Cobalt

Published levels of cobalt (Co) in food crops and animal food products are very scarce in the scientific literature. The reported levels are generally low (often $\leq$ 0.001 mg/100 g), with the lowest levels observed in vegetables, i.e., 0.00001–0.02 mg/100 g (Table 7.7) and the highest values in some cereal products such as common buckwheat bran, grain, and flour (0.05 mg/kg), tartary buckwheat grain (0.05 mg/100 g), and tartary buckwheat leaf flour (0.356 mg/100 g; see Table 7.6). A good source of cobalt is Marmite or Vegemite yeast extracts sold in the U.K. and Australia.[1] The concentrations of cobalt in tea leaves range from 0.015 to 0.08 mg/100 g (Table 7.8). Fruit juices and alcoholic beverages contain very low levels of the metal, generally below 0.0001 mg/100 g (Table 7.10). The use of cobalt salts as food additives, for example in beer to improve its foaming efficiency, has been well known.[1]

7.3.3 Copper

Grain products contain from < 0.04 to 1.71 mg Cu/100 g, with the lowest values noted for grits and maize flour (0.04–0.09 mg/100 g) and the highest for soya flour and stabilized wheatgerm (Table 7.6). Cereals, especially in wholegrain, contain up to 10 mg Cu/kg in ready-to-eat Australian breakfast cereals that are based on wheat bran.[1,56] Among other miscellaneous cereals, stabilized wheatgerm (1.55 mg/100 g), wheat bran (1.50 mg/100 g), and wild rice (1.31 mg/100 g) have high levels of this metal (Table 7.6). Low levels of copper occur in vegetables (Table 7.7), ranging from 0.004 to 0.24 mg/100 g, except legumes (which can be up to 0.5 mg/100 g). Fruits contain small amounts of Cu (0.01–0.24 mg/100 g), whereas currants bioconcentrate copper to 2.7 mg/100 g (Table 7.7). Tea leaves and coffee beans contain generally from 0.16 to 3.69 mg Cu/100 g and from 1.08 to 7.43 mg/100 g, respectively (Table 7.8). Copper contents of nuts range widely (0.10–3.16 mg/100 g), with the highest level in Turkish hazelnut — up to 3.16 mg/100 g (Table 7.9).

7.3.4 Iron

Vegetables and fruits contain low levels of iron (Fe), from 0.13 to 3.01 mg/100 g (Table 7.7). The iron in food crop is mostly present in the form of insoluble complexes of Fe^{3+} with phytic acid, phosphates, oxalates, and carbonates. Less than 8% of Fe in food crops is found to be present in bioavailable form. Food crops, therefore, tend to be a poor source of iron. Eggplant from Saudi Arabia, however, is an exception in having relatively high levels of iron (3.2 mg/100 g). Among cereal products, rye flour contains from 0.002 to 0.04 mg Fe/100 g, whereas other miscellaneous products can be a significant source of iron (0.22–19.6 mg/100 g). Wild rice and especially tartary buckwheat leaf flour contain as much as 43.7 and 138 mg Fe/100 g (Table 7.6). Breakfast cereals fortified voluntarily, and flour with the addition of iron, show relatively high levels of this metal.[1] As seen in Table 7.8, the concentration of iron in tea leaves (up to 97.2 mg/100 g) is significantly higher than that of grain coffee (up to 9.01 mg/100 g). Nuts and cocoa powder may be good sources of iron, containing up to 16.1 and 25.8 mg Fe/100 g (Table 7.8 and Table 7.9). Among the spices, curry powder can have as much as 58 mg Fe/100 g.[1] Fruit juices and alcoholic beverages are poor sources of iron (Table 7.10).

It has been reported[1,57] that canned foods can have particularly high levels of iron — up to 130 mg/100 g in canned blackcurrants.

7.3.5 Manganese

In general, cereals and cereal products are relatively rich in manganese (Mn), whereas vegetables and fruits are characterized by low Mn contents. For instance, the concentrations of the element in wheat bran and stabilized wheatgerm range from 8.32 to 15.0 and 13.3 to 25.2 mg/100 g, respectively (Table 7.6). Hazelnuts contain high levels of manganese, up to 15.8 mg/100 g (Table 7.9). Vegetables (0.01–0.0.78 mg/100) and fruits (0.01–0.66 mg/100) are a minor source of the element (Table 7.7).

Although foods of plant origin are estimated to be relatively low sources of manganese, it is interesting to note that tea can be the richest source with concentrations of up to 139 mg per 100 g in tea leaves (Table 7.8). According to Reilly[1] and Wenlock et al.,[58] British dry tea leaves may contain up to 90 mg Mn/100 g, and the tea beverage from 0.71 to 3.8 mg/100 g. By comparison, concentrations of Mn in ground coffee and instant coffee are 1.18–4.83 and 0.36–3.88 mg/100 g, respectively (Table 7.8). Nuts may be rich in manganese, with high levels (up to 15.9 mg/100 g) found in hazelnut (Table 7.9). Fruit juices and alcoholic beverages are poor source of the metal (Table 7.10).

7.3.6 Molybdenum

In spite of large-scale variations in levels of individual food samples, molybdenum (Mo) occurs in food crops generally at low levels. Some food crops may bioconcentrate extremely high levels of the metal. In certain parts of India and Armenia, inhabitants have been exposed to abnormally high levels of molybdenum as a result of consuming locally produced foods cultivated on molybdenum-rich soil.[1,59,60] Anthropogenic activity such as treatment of soils with sewage sludge and fertilizers may be also responsible for Mo contamination of plants grown in some areas.

Concentrations of molybdenum in foods range widely, from values of 0.000001 (wine) to 0.15 mg/100 g (peas) (Table 7.7 and Table 7.10). Ranges in concentrations of molybdenum in U.S. peas cultivated conventionally and organically were found to be 0.003–0.148 and 0.004–0.109 mg/100 g, respectively (Table 7.7). Canned vegetables can contain up to 0.03 mg Mo/100 g.[1]

7.3.7 Nickel

Concentrations of nickel (Ni) in most foodstuffs are generally low, often being ≤ 0.001 mg/100 g. Reported levels in vegetables vary from 0.0005 to 0.28 mg/100 g, and in fruits between < 0.004 and 0.05 mg/100 g (Table 7.7). Low concentrations are likewise found in the sugarcane plant (0.01 mg/100 g) and sugar (0.001 mg/100 g) (Table 7.8). As can be seen in Table 7.7, grain products contain from 0.002 to 0.35 mg Ni/100 g. The values (up to 2.12 mg/100 g) are significantly higher for leaves of tea (Table 7.8). According to Reilly,[1] and Smart and Sherlock,[61] British instant tea can contain up to 1.7 mg Ni/100 g. The cacao bean, used to produce cocoa and chocolate, is also rich in nickel, with concentrations that can reach 1.0 mg/100 g. Some herbs and spices are also characterized by high levels of the element.[1] Table 7.9 shows that good sources of nickel also include nuts (up to 0.30 mg/100 g).

Concentration of nickel in processed foods can be increased because of leaching of Ni from cooking equipment and containers. Canned vegetables show higher nickel levels than other vegetables.[1]

7.3.8 Selenium

Concentrations of this metalloid (Se) range from < 0.0001 to 0.06 mg/100 g in grain products, vegetables, and fruit products (Table 7.6 and Table 7.7). The highest levels reported include 0.69 mg/100 g in tea leaves to 0.74 mg/100 g in nuts with one extraordinarily high value of 3.51 mg/100 g found in an individual nut (Table 7.8 and Table 7.9). Selenium deficiency has significant negative health effects manifested as diseases such as Keshan disease, which is common in Jilin Province of China.[62] This endemic cardiovascular syndrome occurs especially in children and young women and is common in parts of China where levels of selenium in soils are very low.[1] It has been reported by Capar and Szefer[63] that intestinal absorption of selenium is higher for its organic compounds. In view of this, food crops such as soybean, wheat-enriched, garlic, onion, broccoli, and selenium-rich yeast have been analyzed for selected chemical species of selenium. The following species have been quantified: Se^{4+}, Se^{6+}, selenomethionine, selenocysteine, and methylselenocysteine. According to Reilly,[1] the dominant chemical form of selenium in cereals is the amino acid selenomethionine, although in some plants significant amounts of the selenate occur. It has also been reported that in leaves of cabbage, beets, and garlic, up to 50% of the total selenium can be present as inorganic species.[1] Selenocystathionine has been identified as a toxic compound in South American nut.[64] It is important to note that selenate, selenite, and selenomethionine are the main components of nutritional supplements and enriched yeast preparations.[1,60]

7.3.9 Zinc

Concentrations of zinc (Zn) in foods of plant origin are generally small and vary from 0.05 to 11.8 mg/100 g fresh weight. Important sources of this element are cereals such as wheat bran and stabilized wheatgerm, which may contain up to 11.5 and 11.8 mg Zn/100 g, respectively (Table 7.6). Significant sources of zinc include nuts shown to contain from 1.71 to 6.03 mg Zn/100 g (Table 7.9). Seeds with elevated levels of zinc include poppy and sunflower seeds, which contain up to 8.6 mg Zn/100 g (Table 7.9). The lowest levels are generally found in fresh fruits and their products (0.02–0.61 mg/100 g), with the exception of dried apple, sultanas and, especially, currants, which contain up to 0.27, 0.67 and 12.7 mg Zn/100 mg (Table 7.7). Cocoa powder contain from 8.5 to 9.43 mg Zn/100 g (Table 7.8). Fruit juices and alcoholic beverages are characterized by lower levels of zinc (Table 7.10).

7.4 POSSIBLY ESSENTIAL MICRONUTRIENTS IN FOOD

7.4.1 Arsenic

Concentrations of arsenic (As) in food crops are generally very low, ranging from 0.00001 to 0.046 mg/100 g with the highest levels in miscellaneous cereals such as wheat flour (0.002–0.003 mg/100 g). Some Indian vegetables contain up to 0.05 mg As/100 g (Table 7.11 and Table 7.12). Beiqishen tea contains, on the average, 1.02 ± 0.28 mg As/100 g (Table 7.13). Nuts are characterized by low levels of the metalloid, < 0.005 mg/100 g (Table 7.14). The concentration of arsenic in beer (0.0002–0.001 mg/100 g) is two orders of magnitude higher than the level in wine, 0.000004–0.0001 mg As/100 g (Table 7.15). High levels of As (up to 20.0 mg/100 g) have been reported in edible wild mushroom (*Laccaria amethystina*). However, consuming these mushrooms may not pose a significant to risk to human health because the dominant arsenic species is methylated arsenic acid, which has low mammalian toxicity.[1,92] Cooking oil and cotton seed by-product, often used in human food, can introduce significant amounts of arsenic into meals.[1,93]

7.4.2 Boron

Concentrations of boron (B) in food crops are generally higher than those in animal food products.[1] Grain products contain boron in range of 0.01 to 0.15 mg/100 g, whereas values in vegetables are from 0.01 to 2.64 mg/100 g (Table 7.11 and Table 7.12). Among the vegetables, pea is characterized by the highest levels of boron. High levels of boron are reported in raisins (2.2 mg/100 g), peanuts (1.7 mg/100 g), peanut butter (1.45 mg/100 g), nuts (1.4 mg/100 g)[1,94] and in Beiqishen tea (1.29 mg/100 g) (Table 7.13). Banana pulp and fruit juices contained from 0.08 to 0.32 mg B/100 g and from 0.03 to 0.29 mg B/100 g, respectively (Table 7.15).

Among alcoholic beverages, wine and beer contain from 0.21 to 1.21 mg B/100g and 0.01 to 0.05 mg B/100 g (Table 7.15). According to Murray[95] and Reilly,[1] concentrations of boron in dry table wine and brewed leaf tea vary from 0.09 to 6.1 mg/100 g.

7.4.3 Vanadium

Vanadium (V) is present in vegetables in a wide range of concentrations, from 0.000005 to 0.16 mg/100 g (Table 7.12), with the highest values in potatoes. The concentration in cereals and cereal products average about 2.3 mg V/100 g.[1] According to Schroeder and Nason[96] and Badmaev et al.,[97] vegetable oils, some cereals, and vegetables generally contain elevated levels of vanadium. The richest food sources of this element include skimmed milk, seafoods, especially lobster, and mushrooms. Some dietary supplements can contain up to 1.3 mg V/100 g.[1,98] Concentrations of vanadium in alcoholic beverages vary significantly (Table 7.15), with the levels in beer (0.001–0.005 mg/100 g) being higher than those in wine (0.000001–0.0001 mg/100 g). Beer is a good source of vanadium, especially in Germany where beer drinking it is the principal exposure route for men.[1,99] Processed foods always contain less vanadium than unprocessed ones.[1,100]

7.5 NONTOXIC AND NONESSENTIAL METALS

7.5.1 Aluminum

Levels of aluminum (Al) in food crops generally range from 0.003 to 0.12 mg/100 g in vegetables, and from 0.019 to 1.89 mg/100 g in fresh fruit. Higher levels occur in legumes (up to 4.58 mg/100 g) and sultanas (up to 1.89 mg/100 g) (Table 7.12). Miscellaneous cereals contain from < 0.002 to 1.29 mg Al/100 g (Table 7.11). As shown in Table 7.15, wines contain higher amounts of aluminum (0.02–0.18 mg/100 g) compared to beers (0.004–0.06 mg/100 g). The contents of the metal in cocoa powder tend to be high, in the range of 6.1 to 9.03 mg/100 g (Table 7.13). Tea leaves are characterized by high concentrations of aluminum, typically 26.9 to 171 mg/100g (Table 7.13). Levels as high as 3000 mg Al/100 g have been reported in dried tea leaves (Chapter 10). Higher-than-normal levels of Al have also been reported[1,101,102] in infant formulas, especially those meant for low-birth-weight infants.[1,103] The occurrence of elevated levels of Al in some vegetables, herbs, beverages, and water is considered in more detail in Chapter 10.

7.5.2 Tin

The concentrations of tin (Sn) in food crops are usually very low, typically in the range of < 0.00016 to 0.001 mg/100 in vegetables (Table 7.12), and < 0.001 to 0.38 mg in grain products (Table 7.11). Tartary buckwheat leaf flour can be a rich source of tin. Canned vegetables can contain as much as 4.4 mg Sn/100 g. Some fruit products seem to have elevated levels of this element, up to 1.7 mg/100 g. Normal levels of tin in foods are low, except for canned products (see Chapter 10) such as canned vegetables where the Sn content can be as high as 4.4 mg/100 g. According to Biégo et al.,[104] fresh tomatoes contain, on average, 0.005 ± 0.001 mg Sn/100 g, whereas the average levels of tin in tomatoes preserved in unlacquered and lacquered cans may increase significantly — up to 0.60 ± 0.39 and 8.43 ± 6.21 mg/100 g, respectively. Concentration of this element in tins of Italian peaches reach levels of

TABLE 7.11
Possibly Essential Micronutrients and Nontoxic, Nonessential Metals in Miscellaneous Cereals (in mg/100 g w/w)

Name	Origin	n	Al	As	B	Sn	Ref.
Bread							
Bread[b]	U.K.		0.37	0.001	0.05	0.003	7
Bread[b]	U.S.		—	< 0.004	—	—	9
Whole white bread	U.S.		—	< 0.004	—	—	9
Cornbread	U.S.		—	< 0.004	—	—	9
Rye bread	U.S.		—	< 0.004	—	—	9
Cereals							
Rolled oats		6	0.2 ± 0.04 0.09 – 0.51	—	—	—	56
Raisin bran cereal	U.S.		—	< 0.004	—	—	9
Fruit-flavored cereal	U.S.		—	< 0.004	—	—	9
Rice							
Rice	India, Pakistan	28	—	—	0.05 0.04 – 0.09	—	12
Rice	Europe	25	—	—	0.06 0.01 – 0.15	—	12
Rice	Different		—	0.02	—	—	13
White rice		6	< 0.002	—	—	—	56
Brown rice		7	0.20 ± 0.02 0.14 – 0.3	—	—	—	56
Wild rice		26	—	0.01 0.001 – 0.01	—	—	68
Grits							
Grits		10	< 0.002	—	—	—	56
Flour							
Common buckwheat flour[c]	Slovenia		—	—	—	< 0.001	71
Tartary buckwheat fine flour[c]	Slovenia		—	—	—	< 0.001	71
Tartary buckwheat flour[c]	Slovenia		—	—	—	0.09 ± 0.05	71
Maize flour		4	0.03 ± 0.01 0.03 – 0.04	—	—	—	56
Refined oat flour		1	0.04 ± 0.02 < 0.002 – 0.1	—	—	—	56
Wholemeal oat flour		2	0.33 ± 0.014 < 0.002 – 0.37	—	—	—	56
Wheat flour	Spain		—	0.002 ± 0.0002	—	—	107
Wheat flour	U.S.		—	0.002 ± 0.0001	—	—	106
Wheat flour	Hungary		—	0.002 ± 0.0002	—	—	106
Wheat flour	France		—	0.002 ± 0.0001	—	—	106

TABLE 7.11 (CONTINUED)
Possibly Essential Micronutrients and Nontoxic, Nonessential Metals in Miscellaneous Cereals (in mg/100 g w/w)

Name	Origin	n	Al	As	B	Sn	Ref.
Wheat flour	England		—	0.002 ± 0.0002	—	—	106
Wheat flour	Italy		—	0.002 ± 0.0001	—	—	106
Wheat flour	Canada		—	0.001 ± 0.0001	—	—	106
Wheat flour	Australia		—	0.002 ± 0.0002	—	—	106
Wheat flour	Norway		—	0.003 ± 0.0002	—	—	106
Wheat flour	Sweden		—	0.002 ± 0.0001	—	—	106
White wheat flour		4	0.03 ± 0.01 0.002 – 0.12	—	—	—	56
Soya flour		4	0.59 ± 0.022 0.45 – 0.75	—	—	—	56
Tartary buckwheat leaf flour[c]	Slovenia		—	—	—	0.38 ± 0.08	71
Miscellaneous Cereals							
Common buckwheat bran[c]	Slovenia		—	—	—	< 0.001	71
Miscellaneous cereals[b]	U.K.		7.8	0.001	0.09	0.002	7
Miscellaneous cereals[b]	U.S.		—	0.01 ± 0.004 < 0.004 – 0.01	—	—	9
Tartary buckwheat bran[c]	Slovenia		—	—	—	0.12 ± 0.06	71
Stabilized wheatgerm		3	0.41 ± 0.05 0.4 – 0.71	—	—	—	56
Soft whole wheat		4	0.03 ± 0.01 0.01 – 0.1	—	—	—	56
Common buckwheat grain[c]	Slovenia		—	—	—	< 0.001	71
Tartary buckwheat grain[c]	Slovenia		—	—	—	0.20 ± 0.1	71
Raw maize		6	0.16 ± 0.03 0.03 – 0.4	—	—	—	56
Kibbled wheat		6	0.14 ± 0.01 0.06 – 0.17	—	—	—	56
Decorticated Rye		4	< 0.002	—	—	—	56
Gluten		4	0.75 ± 0.13 0.49 – 1.29	—	—	—	56

TABLE 7.12
Possibly Essential Micronutrients and Nontoxic, Nonessential Metals in Vegetables (in mg/100 g w/w)

Name	Origin	n	Al	As	B	Sn	V	Ref
Vegetables								
Bottle ground leaf			—	0.03	—	—	—	107
Canned vegetables[b]	U.K.		0.11	0.0003	0.12	4.4	—	7
Canned vegetables[b]	U.S.		—	< 0.004 – 0.01	—	—	—	9
Carrot	Greece		—	< 0.0001	—	—	0.01	77
Celery leaf	Greece		—	0.001	—	—	0.02	77
Celery root	Greece		—	< 0.0001	—	—	0.02	77
Cucumber	U.S.		—	< 0.004	—	—	—	9
Elephant foot			—	0.03	—	—	—	107
Ghotkol	India		—	0.05	—	—	—	107
Green pepper	Tenerife Island	60		—	0.05 ± 0.03 0.01 – 0.12	—	—	18
Green vegetables[b]	U.K.		0.18	0.0003	0.2	0.002	—	7
Green vegetables[b]	U.S.		—	< 0.004	—	—	—	9
Iceberg lettuce	U.S.		—	< 0.003	—	—	—	9
Leek leaf	Greece		—	< 0.0001	—	—	0.01	77
Leek root	Greece		—	< 0.0001	—	—	0.01	77
Legumes		40	0.27 – 4.58	—	—	—	—	78
Lettuce	Greece		—	0.001	—	—	0.01	77
Onion	U.S.		—	< 0.004	—	—	—	9
Onion conventional	Denmark		0.02 ± 0.02 0.005 – 0.12	—	0.16 ± 0.08 0.03 – 0.38	0.0003 ± 0.0002 0.0001 – 0.001	0.00004 ± 0.00001 0.00001 – 0.0001	20
Onion leaf	Greece		—	< 0.0001	—	—	0.01	77

Onion organic	Denmark		0.01 ± 0.01	—	0.09 ± 0.03	0.0004 ± 0.0002	0.00005 ± 0.00003	20
			0.005 – 0.07		0.01 – 0.17	0.0001 – 0.001	0.00001 – 0.0002	
Onion root	Greece		—	< 0.0001	—	—	0.004	77
Other vegetables[b]	U.K.		0.32	0.001	0.14	0.002		7
Other vegetables[b]	U.S.		—	< 0.004 – 0.01	—	—	—	9
Parsley	Greece		—	0.0004	—		0.004	77
Peas conventional	Denmark		0.01 ± 0.005	—	1.22 ± 0.43	—	0.00003 ± 0.00001	20
			0.003 – 0.03		0.11 – 2.64		0.00001 – 0.0001	
Peas, organic	Denmark		0.01 ± 0.005	—	1.29 ± 0.34	—	0.00003 ± 0.00001	20
			0.003 – 0.03		0.53 – 2.45		0.000005 – 0.0001	
Potatoes[c]	Idaho	342	—	—	—	—	0.03 ± 0.03	22
							0 – 0.1	
Potatoes[c]	Non-Idaho	266	—	—	—	—	0.03 ± 0.04	22
							0 – 0.16	
Potatoes[b]	U.K.		0.22	0.0005	0.14	< 0.002	—	7
Potatoes[b]	U.S		—	< 0.004	—	—	—	9
Radish	U.S.		—	< 0.004	—	—	—	9
Red pepper	Tenerife Island	60	—	—	0.07 ± 0.04	—	—	18
					0.01 – 0.21			
Snake gourd	India		—	0.05	—	—	—	108
Spinach	Greece		—	0.001	—	—	0.01	77
Taro	India		—	0.04	—	—	—	108
Tomato fruits, rockwool high EC	Denmark		—	—	—	0.0001	0.00001	23
Tomato fruits, rockwool norm EC	Denmark		—	—	—	0.0001	0.00001	23
Tomato fruits, soil norm EC	Denmark		—	—	—	0.0001	0.00001	23

TABLE 7.12 (CONTINUED)
Possibly Essential Micronutrients and Nontoxic, Nonessential Metals in Vegetables (in mg/100 g w/w)

Name	Origin	n	Al	As	B	Sn	V	Ref
Mushrooms								
Mushrooms	U.S		—	0.01	—	—	—	9
Fruits								
Apple red	U.S		—	< 0.004	—	—	—	9
Avocado	U.S.		—	< 0.005	—	—	—	9
Banana pulp	Tenerife Island[d]		—	—	0.20 ± 0.02 0.08 – 0.32	—	—	25
Cantaloupe	U.S		—	< 0.003	—	—	—	9
Currants		6	1.1 ± 0.08 0.84 – 1.71	—	—	—	—	56
Dried apple		6	0.11 ± 0.02 0.02 – 0.21	—	—	—	—	56
Fresh fruit[b]	U.K.		0.06	0.0002	0.34	0.003	—	7
Fresh fruit[b]	U.S.		—	< 0.003 – < 0.005	—	—	—	9
Fruit products[b]	U.K.		0.1	0.0003	0.24	1.7	—	7
Fruit products[b]	U.S.		—	< 0.003 – < 0.005	—	—	—	9
Green papaya			—	0.04	—	—	—	108
Peach	U.S.		—	< 0.003	—	—	—	9
Strawberries	U.S		—	< 0.003	—	—	—	9
Sultanas		6	1.2 ± 0.1 0.91 – 1.89	—	—	—	—	56
Sweet cherries	U.S		—	< 0.004	—	—	—	9

TABLE 7.13
Possibly Essential Micronutrients and Nontoxic, Nonessential Metals in Tea, Coffee, and Confectionary Products (in mg/100 g w/w)

Name	Origin	n	Al	As	B	Sn	V	Ref.
Tea								
Tea	Africa	18	79.0 ± 40.6	—	—	—	0.03 ± 0.02	27
Tea	Asia	36	75.4 ± 26.1	—	—	—	0.02 ± 0.01	27
Tea	China	13	87.1 ± 49.8	—	—	—	0.04 ± 0.03	27
Tea	India, Sri Lanka	13	58.2 ± 13.9	—	—	—	0.01 ± 0.01	27
Beiqishen tea	China		171 ± 19.9	1.02 ± 0.28	1.29 ± 0.07	—	< 0.03	30
Chinese tea	China	5	76.8 ± 1.12	—	—	—	—	31
Black tea[c]	Ceylon, India, Kenya	36	83.0 ± 27.0	—	—	—	—	108
Flavored black tea[c]		30	70.5 ± 18.0	—	—	—	—	108
Peppermint tea[c]		15	44.1 ± 4.8	—	—	—	—	108
Fruit tea[c]		27	27.0 ± 14.5	—	—	—	—	108
Instant tea[d] (< 2% tea extract)		3	0.37	—	—	—	—	33
			0.29 – 0.43					
Instant tea[d] (100% extract)		2	0.35	—	—	—	—	33
			0.35 – 0.35					
Mountain herbal tea[c]		15	38.7 ± 14.7	—	—	—	—	108
Beiqishen tea extract[d]	China		0.56 ± 0.02	< 0.0003	0.01 ± 0.001	—	< 0.001	30
Black tea beverages[d]		23	0.5 ± 0.14	—	—	—	—	33
			0.1 – 1.0					
Black tea infusion[d]	Ceylon, India, Kenya	12	0.42	—	—	—	—	108
Peppermint tea infusion[d]		6	0.01	—	—	—	—	108
Fruit tea infusions[d]		6	0.04	—	—	—	—	108
Green tea beverages[d]		20	0.3 ± 0.0	—	—	—	—	33
			0.2 – 0.7					

TABLE 7.13 (CONTINUED)
Possibly Essential Micronutrients and Nontoxic, Nonessential Metals in Tea, Coffee, and Confectionary Products (in mg/100 g w/w)

Name	Origin	n	Al	As	B	Sn	V	Ref.
Tea soft drinks[d]		8	0.11 0.02 ± 0.25	—	—	—	—	33
Coffee								
Ground coffee[c]	Mixture	48	1.83 ± 0.97	—	—	—	—	108
Soluble coffee		21	8.75 ± 8.10 < 0.5 – 23.3	—	—	0.86 ± 0.12 0.64 – 1.16	—	34
Roasted coffee beans	Costa Rica	20	1.25 ± 0.19	—	—	—	—	35
Roasted coffee beans	Colombia	20	1.83 ± 0.39	—	—	—	—	35
Roasted coffee beans	Guatemala	20	0.77 ± 0.01	—	—	—	—	35
Roasted coffee beans	Panama	20	0.29 ± 0.19	—	—	—	—	35
Roasted coffee beans	Ethiopia	20	0.67 ± 0.19	—	—	—	—	35
Roasted coffee beans	Kenya	20	0.42 ± 0.29	—	—	—	—	35
Roasted coffee beans	Sulawesi	20	1.25 ± 0.66	—	—	—	—	35
Roasted coffee beans	Sumatra	20	3.48 ± 1.25	—	—	—	—	35
Ground coffee infusions[d]			0.01	—	—	—	—	108
Confectionary Products								
Sugar and preserves[b]	U.K.		0.36	0.001	0.08	0.002	—	7
Sugar and preserves[b]	U.S.		—	< 0.004 – 0.01	—	—	—	9
Sugar	Egypt		0.0005 ± 0.0002	0.00001 ± 0.00001	—	0.0001 ± 0.0002	0.001 ± 0.001	38
Cocoa powder		4	7.42 ± 0.34 6.1 – 9.03	—	—	—	—	56
Molasses	Egypt		0.004 ± 0.01	0.0001 ± 0.00001	—	0.0003 ± 0.0003	0.005 ± 0.002	38
Sugar juice	Egypt		0.004 ± 0.002	0.00001 ± 0.0	—	0.0002 ± 0.0002	0.003 ± 0.001	38

TABLE 7.14
Possibly Essential Micronutrients and Nontoxic, Nonessential Metals in Nuts, Seeds, and Oils (in mg/100 g w/w)

Name	Origin	n	Al	As	B	Sn	Ref.
Nuts and Seeds							
Hazelnut	Turkey	6	—	—	1.66 ± 0.21 1.39 – 2.14	—	109
Nuts[b]	U.K.		1.1	0.001	1.4	0.003	7
Nuts[b]	U.S.		—	< 0.005	—	—	9
Sliced almonds		6	0.1 ± 0.02 0.044 – 0.203	—	—	—	56
Mixed nuts (no peanuts)	U.S.		—	< 0.005	—	—	9
Peanut butter	U.S.		—	< 0.005	—	—	9
Oils							
Olive or sunflower oil	U.S.		—	< 0.004	—	—	9
Oils and fats[b]	U.K.		0.12	0.001	0.04	0.002	7
Oils and fats[b]	U.S.		—	< 0.004	—	—	9

up to 40.0 mg/100 g.[1] It should be noted that a high level of corrosion of the tin plate may result in high levels of nitrate in the water used to process fruits.[1,105] Contamination of foods by tin is discussed further in Chapter 10.

7.6 TOXIC METALS

7.6.1 Beryllium

Levels of beryllium (Be) in food crops tend to be very low, typically in the µg/kg range. For instance, onions generally contain from 0.000003 to 0.00001 mg Be/100 g (Table 7.16 and Table 7.17), whereas levels in wines commonly vary from 0.000001 to 0.0001 mg Be/l00 g (Table 7.20).

7.6.2 Cadmium

The levels of cadmium (Cd) in food crops range from ca. < 0.00003 to 0.004 mg/100 g in fruits, and from < 0.0004 to 0.03 mg/100 g in vegetables (Table 7.17). The high levels of this element in some vegetables can be attributable, in many instances, to the application of fertilizers and contaminated sewage sludge to agricultural soils.[1] Contamination of rice paddies with cadmium was the cause of the outbreak of itai-itai syndrome in the Jintsu River basin in Japan, which killed about 200 people. In Zhejiang province in China, levels of cadmium in rice have been reported to be about 50 times higher than those in the control area.[62] Green leaf vegetables, some of which are "cadmium accumulators" grown on soil in brownfield urban sites of England, contain Cd from 0.31 to 1.00 mg/100 g (dry weight).[1,110] Further discussion on the contamination foods and alcoholic beverages with cadmium can be found in Chapter 10.

TABLE 7.15
Possibly Essential Micronutrients and Nontoxic, Nonessential Metals in Beverages (in mg/100 g w/w)

Name	Origin	n	Al	As	B	Sn	V	Ref.
Beverages[b]	U.K.		0.17	0.0002	0.04	0.002	—	7
Beverages[b]	U.S.		—	< 0.002 – < 0.003	—	—	—	9
Juices								
Grapefruit juic			0.002	—	0.07	< 0.0001	< 0.0001	45
Lemon juice			0.001	—	0.17	< 0.0001	< 0.0001	45
Lime juice			0.005	—	0.05	< 0.0001	< 0.0001	45
Mandarin juice			0.04	—	0.03	< 0.0001	< 0.0001	45
Orange juice	Australia	290	0.01 ± 0.004	—	0.14 ± 0.01	0.0002 ± 0.0	0.0001 ± 0.0	45
			0.0001 – 0.1		0.06 – 0.29	< 0.0001 – 0.01	< 0.0001 – 0.001	
Orange juice[d]		8	0.02 – 0.06	—	—	—	—	47
Orange juice concentrate	Australia	83	0.01 ± 0.01	—	0.11 ± 0.02	0.0002 ± 0.0001	< 0.0001	45
			0.001 – 0.05		0.03 – 0.17	< 0.0001	< 0.0001	
Orange juice concentrate	Brazil	42	0.01 ± 0.01	—	0.06 ± 0.02	0.0004 ± 0.0003	< 0.0001	45
			0.003 – 0.04		0.03 – 0.1	< 0.0001 – 0.004	< 0.0001	
Pummelo juice			0.01	—	0.07	< 0.0001	< 0.0001	45
Seville orange juice			0.03	—	0.11	< 0.0001	< 0.0001	45
Tangello juice			0.003	—	0.13	< 0.0001	< 0.0001	45
Alcoholic Beverages								
Beer[d]		32	0.02 ± 0.01	—	0.02 ± 0.01	—	—	49
			0.004 – 0.06		0.01 – 0.04			
Beer[d]	Poland	35	—	0.001	—	0.0001	0.002	91
				0.0002 – 0.001		0.00004 – 0.0003	0.001 – 0.005	
Dry table wine	U.S.		—	< 0.002	—	—	—	9
Wine[d]	Italy		0.02 – 0.18	0.000004 – 0.0001	0.21 – 1.21	0.002 – 0.1	0.000001 – 0.0001	53

7.6.3 Lead

Lead (Pb) is present in all foods and beverages, primarily as a natural component as well as an accidental additive picked up during food processing. The levels of this metal in food crops, except those affected by environmental contamination, can range between 0.00004 and 0.08 mg/100 g. Lead concentrations in cereals and vegetables typically are 0.0004–0.06 and 0.00004–0.08 mg/100 g, respectively (Table 7.16 and Table 7.17). Low levels of lead have also been reported for fruits and oils (0.001–0.023 mg/100 g), whereas elevated concentrations (up to 0.77 mg/100 g) have been observed in tea (Table 7.17, Table 7.18, and Table 7.19). Foods in lead-soldered cans generally contain higher levels of lead as compared to fresh and frozen foods.[1] For instance, extremely high levels of lead, up to 2.0 mg/100 g, have been reported in Italian canned vegetables.[1,111] Contamination of alcoholic beverages and vegetables by lead, with special attention paid to contamination of wines, is discussed in Chapter 9 and Chapter 10. Wines preserved in bottles with lead seals represent a potential health hazard. According to Reilly[2] and Shaper et al.,[112] concentrations of lead in the blood of middle-aged British men were positively correlated with alcohol consumption. Serious exposure to lead can also occur among children and pregnant and lactating women who consume calcium supplements made from dolomite and bonemeal; these dietary supplements can have unexpectedly high levels of contaminant lead.[1,113]

7.6.4 Mercury

The levels of mercury (Hg) in food crops range from 0.00006 mg/100 g in fruit products to 0.0004 mg/100 g in canned vegetables. Beer contain very small amounts of this element, in the range of 0.00001 to 0.0001 mg/100 g (Table 7.20). Very high levels of mercury have been reported in vegetables grown on mercury-contaminated soil — for example, up to 0.004 mg Hg/100 g in lettuce cultivated on sludge-amended soil.[1,114] Although fish and seafoods are well known to be a major source of methylmercury in the diet (see Chapter 10), contamination may also occur from adventitious sources. The best-known incidence occurred in northern Iraq where a large number of the local population was poisoned from eating homemade bread prepared from grains that had been dressed with alkylmercury compounds as antifungal agents.[115] It is estimated that 100,000-500,000 inhabitants had suffered permanent disabilities from this inadvertent exposure.[62,116,117]

7.7 OTHER ELEMENTS

7.7.1 Antimony

Food crops contain very low levels of antimony (Sb), for instance, wines, which contain from 0.000001 to 0.0001 mg/100 g (Table 7.25), whereas concentrations in grain products and vegetables ranged from <0.0004 to 0.0012 mg/100 g and from 0.00002 to 0.0005 mg/100 g (Table 7.21 and Table 7.22). High levels of this element have been reported in beverages in contact with enameled surfaces. Low-pH fruit drinks are capable of dissolving up to 100 mg Sb/l from enamel.[1]

TABLE 7.16
Toxic Metals in Miscellaneous Cereals (in mg/100 g w/w)

Name	Origin	n	Cd	Hg	Pb	Ref.
Bread						
Bread[b]	U.K.		0.003	0.0004	0.002	7
Bread[b]	U.S.		0.002 ± 0.001 0.001 – 0.002	—	< 0.004	9
Bread[e]		10	0.002	—	—	118
Bagel plain	U.S.		0.002	—	< 0.004	9
Bread	Poland	26	0.001 – 0.004	—	0.005 – 0.01	65
White bread	U.S.		0.002	—	< 0.004	9
Whole white bread	U.S.		0.002	—	< 0.004	9
Cornbread	U.S.		0.001	—	< 0.004	9
Rye bread	U.S.		0.002	—	< 0.004	9
				—		
Cereals						
Corn flakes	U.S.		< 0.001	—	< 0.004	9
Raisin bran cereal	U.S.		0.005	—	< 0.004	9
Fruit-flavored cereal	U.S.		0.001	—	< 0.004	9
Rice						
Raw rice[e]		10	0.005	—	—	118
Rice	Different		—	—	0.01	13
Rice nonglutinous	NE China	17	0.001 ± 0.001 0.0004 – 0.003	—	0.002 ± 0.002 0.0004 – 0.01	74
Wild rice		26	0.005 0.001 – 0.01	—	0.004 0.0005 – 0.01	68
Rice glutinous	NE China	6	0.004 ± 0.004 0.0005 – 0.01	—	0.01 ± 0.02 0.001 – 0.04	74
Grits						
Foxtail millet	NE China	8	0.001 ± 0.0003 0.001 – 0.002	—	0.01 ± 0.004 0.003 – 0.01	74
Noodle[e]		10	0.0004	—	—	118
Flour						
Flour[e]		10	0.002	—	—	118
Common buckwheat flour[c]	Slovenia		—	< 0.001	—	71
Tartary buckwheat fine flour[c]	Slovenia		—	< 0.001	—	71
Tartary buckwheat flour[c]	Slovenia		—	< 0.001	—	71
Maize flour	NE China	7	0.001 ± 0.0003 0.0002 – 0.001	—	0.004 ± 0.002 0.001 – 0.01	74
Wheat flour	NE China	9	0.002 ± 0.001 0.001 – 0.003	—	0.004 ± 0.003 0.0015 – 0.01	74

TABLE 7.16 (CONTINUED)
Toxic Metals in Miscellaneous Cereals (in mg/100 g w/w)

Name	Origin	n	Cd	Hg	Pb	Ref.
Wheat flour[c]	Sweden		n = 105 0.003 ± 0.001 0.001 – 0.01	—	n = 60 < 0.001 – 0.003	72
Rye flour[c]	Sweden		n = 90 0.002 ± 0.001 0.0004 – 0.003	—	n = 61 < 0.001 – 0.02	72
Tartary buckwheat leaf flour[c]	Slovenia		—	< 0.00	—	71
Miscellaneous Cereals						
Common buckwheat bran[c]	Slovenia		—	0.001 ± 0.0004	—	71
Miscellaneous cereals[b]	U.K.		0.002	0.0004	0.002	7
Miscellaneous cereals[b]	U.S.		0.002 ± 0.001 < 0.0005 – 0.01	< 0.004	< 0.005	9
Tartary buckwheat bran[c]	Slovenia		—	< 0.001	—	71
Wheat bran[c]	Sweden		n = 30 0.01 ± 0.005 0.005 – 0.02	—	n = 26 < 0.002 – 0.01	72
Adzuki bean	NE China	3	0.01 ± 0.01 0.002 – 0.03	—	0.003 ± 0.001 0.002 – 0.004	74
Fava bean (broad bean)	NE China	1	0.01	—	0.002	74
Kidney bean	NE China	11	0.003 ± 0.003 0.001 – 0.01	—	0.003 ± 0.001 0.001 – 0.01	74
Green gram (mung bean)	NE China	5	0.01 ± 0.002 0.005 – 0.01	—	0.003 ± 0.002 0.001 – 0.01	74
Cowpea (blackeyed pea)	NE China	1	0.04	—	0.01	74
Soybean	NE China	11	0.01 ± 0.01 0.002 – 0.02	—	0.004 ± 0.004 0.001 – 0.01	74
Sorghum	NE China	3	0.001 ± 0.001 0.001 – 0.002	—	0.01 ± 0.005 0.002 – 0.01	74
Common buckwheat grain[c]	Slovenia		—	< 0.001	—	71
Tartary buckwheat grain[c]	Slovenia		—	< 0.001	—	71
Maize grain	NE China	9	0.0004 ± 0.0001 0.0002 – 0.001	—	0.01 ± 0.02 0.001 – 0.06	74
White roll	U.S.		0.002	—	< 0.004	9

[e] geometrical mean

TABLE 7.17
Toxic Metals in Vegetables, Mushrooms, and Fruits (in mg/100 g w/w)

Name	Origin	n	Be	Cd	Hg	Pb	Ref.
Vegetables							
Artichoke	Spain		—	—	—	0.01 ± 0.01	76
Cabbage	Spain		—	—	—	0.004 ± 0.004	76
Cabbage leaves		5	—	—	—	0.08 ± 0.001	119
Cabbages	Saudi Arabia	5	—	0.01	—	—	17
Canned vegetables[b]	U.K.			0.001	0.0004	0.002	7
Canned vegetables[b]	U.S.			0.001 ± 0.0003 < 0.0005 – 0.002	NA	< 0.003	9
Carrot	Greece		—	0.001	—	0.004	77
Carrot	Spain		—	—	—	0.003 ± 0.001	76
Carrots	Saudi Arabia	5	—	0.01	—	0.09	17
Cauliflower	Spain		—	—	—	0.003 ± 0.002	76
Celery	Spain		—	—		0.01 ± 0.01	76
Celery	U.S.		—	0.004	—	< 0.002	9
Celery leaf	Greece		—	0.002	—	0.01	77
Celery root	Greece		—	0.002	—	0.01	77
Chard	Spain		—	—	—	0.02 ± 0.01	76
Cucumber	U.S.		—	< 0.0004	—	< 0.002	9
Cucumber	Saudi Arabia	5	—	0.002	—	0.02	17
Eggplant	Saudi Arabia	5	—	0.01	—	0.03	17
Garlic	Spain		—	—	—	0.002 ± 0.001	76
Green pepper	U.S.		—	0.002	—	< 0.003	9
Green pepper	Saudi Arabia	5	—	0.01	—	0.02	17
Green vegetables[b]	U.K.		—	0.001	0.0002	0.001	7

Green vegetables[b]	U.S.		—	0.003 ± 0.004	< 0.004	< 0.003	9
				< 0.0005 – 0.01			
Iceberg lettuce	U.S.		—	0.01	—	—	9
Leek	Spain		—	—	—	0.01 ± 0.01	76
Leek	Saudi Arabia	5	—	0.01	—	—	17
Leek leaf	Greece		—	0.001	—	0.01	77
Leek root	Greece		—	0.001	—	0.002	77
Legumes		40	—	0 – 0.002	—	0.03 – 0.07	78
Lettuce	Greece		—	0.004	—	0.001	77
Lettuce	Spain		—	—	—	0.01 ± 0.01	76
Lettuce	Saudi Arabia	5	—	0.005	—	0.02	17
Lettuce leaves		5	—	—	—	0.06 ± 0.001	119
Onion	Spain		—	—	—	0.002 ± 0.002	76
Onion	Saudi Arabia	5	—	0.01	—	0.12	17
Onion	U.S.		—	0.002	—	< 0.003	9
Onion, conventional	Denmark		0.000003 ± 0.000003	0.002 ± 0.001	—	0.001 ± 0.0004	20
			0.0 – 0.00001	0.001 – 0.01		0.0001 – 0.002	
Onion leaf	Greece		—	0.002	—	0.01	77
Onion, organic	Denmark		0.00001 ± 0.00001	0.002 ± 0.001	—	0.001 ± 0.0003	20
			0 – 0.00004	0.0004 ± 0.005		0.0001 – 0.002	
Onion root	Greece		—	0.001	—	0.01	77
Other vegetables[b]	U.K.		—	0.001	0.0002	0.002	7
Other vegetables[b]	U.S.		—	0.002 ± 0.001	NA	< 0.002 – < 0.012	9
				< 0.0005 – 0.004			
Parsley	Greece		—	0.0002	—	0.003	77
Parsley	Spain		—	—	—	0.01 ± 0.01	76
Parsley	Saudi Arabia	5	—	—	—	0.05	17
Peas, conventional	Denmark		0.00001 ± 0.00002	0.001 ± 0.0005	—	0.001 ± 0.0003	20
			0.0 – 0.0001	0.0005 – 0.003		0.00004 – 0.002	

TABLE 7.17 (CONTINUED)
Toxic Metals in Vegetables, Mushrooms, and Fruits (in mg/100 g w/w)

Name	Origin	n	Be	Cd	Hg	Pb	Ref.
Peas, organic	Denmark		0.00001 ± 0.00001 0.0 – 0.0001	0.001 ± 0.0004 0.0003 – 0.002	—	0.0004 ± 0.0003 0.0001 – 0.002	20
Potatoes[b]	U.K.		—	0.003	0.0003	0.001	7
Potatoes[b]	U.S.		—	0.003 ± 0.002 0.001 – 0.005	NA	< 0.005	9
Potatoes	Saudi Arabia	5	—	0.02	—	0.06	17
Potatoes[c]	Idaho	342	—	0.01 ± 0.01 0 – 0.03	—	0.05 ± 0.06 0 – 0.23	22
Potatoes[c]	non-Idaho	266	—	0.01 ± 0.01 0 – 0.03	—	0.03 ± 0.06 0 – 0.29	22
Radish			—	0.001	—	—	9
Radish root	Spain		—	—	—	0.003 ± 0.002	76
Salq	Saudi Arabia	5	—	0.02	—	0.47	17
Spinach	Greece		—	0.004	—	0.01	77
Spinach	Spain		—	—	—	0.01 ± 0.01	76
Spinach	Saudi Arabia	5	—	0.01	—	0.08	17
Spinach leaves		5	—	—	—	0.08 ± 0.0004	119
Tomato	Saudi Arabia	5	—	0.005	—	0.02	17
Tomato fruits, rockwool high EC	Denmark		—	0.0001	—	0.0002	23
Tomato fruits, rockwool norm EC	Denmark		—	0.0001	—	0.0002	23
Tomato fruits, soil norm EC	Denmark		—	0.002	—	0.0001	23
Tomato leaves		5	—	—	—	0.05 ± 0.0003	119
Tomato red	U.S.		—	0.001	—	—	9
Veg. Marrow	Saudi Arabia	5	—	0.01	—	0.04	17
Watercress	Saudi Arabia	5	—	0.01	—	0.09	17

Mushrooms							
Agaricus maleolens[c]	Czech Republic	177	—	0.15 ± 0.22	0.13 ± 0.08	0.14 ± 0.11	120
				0.01 – 1.38	0.03 – 0.41	0.01 – 0.71	
Agaricus arvensis[c]	Czech Republic	125	—	0.26 ± 0.32	0.10 ± 0.08	0.11 ± 0.10	120
				0.01 – 1.33	0.003 – 0.51	0.01 – 0.58	
Bay bolete[c]	Poland	166	—	0.01 ± 0.005	—	0.02 ± 0.02	24
				0.0002 – 0.05		0.001 – 0.23	
Boletus edulis[c]	Slovakia	56	—	0.01 – 0.09	0.12 – 0.33	0.01 – 0.02	121
Boletus reticulatus[c]	Slovakia	56	—	0.01 – 0.09	0.12 – 0.21	0.004 – 0.01	121
Leccinum scabrum[c]	Slovakia	56	—	0.01 – 0.03	0.01 – 0.02	0.004 – 0.02	121
Leccinum griseum[c]	Slovakia	56	—	0.01	0.01	0.01	121
Lactarius deliceus[c]	Turkey		—	0.00003 ± 0.000002	—	0.00002 ± 0.000002	80
Morchella esculenta[c]	Turkey		—	0.00001 ± 0.000001	—	0.00001 ± 0.000001	80
Mushrooms	U.S.		—	0.001	—	<0.003	9
Russula delica[c]	Turkey		—	0.00004 ± 0.000003	—	0.00002 ± 0.000001	80
Xercomus badius[c]	Czech Republic		—	0.08 – 0.12	0.01 – 0.01	0.04 – 0.09	122
Fruits							
Apple, red	U.S.		—	< 0.001	—	< 0.003	9
Apricot	U.S.		—	< 0.001	—	< 0.003	9
Avocado	U.S.		—	0.003	—	< 0.01	9
Cantaloupe	U.S.		—	0.001	—	< 0.002	9
Fresh fruit[b]	U.K.		—	0.0002	0.0003	< 0.001	7
Fresh fruit[b]	U.S.		—	0.002 ± 0.001	NA	< 0.002 – < 0.005	9
				< 0.0004 – 0.003			
Fruit products[b]	U.K.		—	0.0001	0.0002	0.001	7
Fruit products[b]	U.S.		—	< 0.0003 – 0.001	ND	0.003 ± 0.001	9
						< 0.002 – 0.004	

TABLE 7.17 (CONTINUED)
Toxic Metals in Vegetables, Mushrooms, and Fruits (in mg/100 g w/w)

Name	Origin	n	Be	Cd	Hg	Pb	Ref.
Grapefruit	U.S.		—	<0.0004	—	< 0.002	9
Grapes, seedless	U.S.		—	<0.0004	—	< 0.002	9
Orange	U.S.		—	—	—	< 0.003	9
Peach	U.S.		—	< 0.001	—	< 0.003	9
Pear	U.S.		—	< 0.001	—	< 0.003	9
Plums	U.S.		—	< 0.0004	—	< 0.002	9
Strawberries	U.S.		—	0.002	—	< 0.002	9
Strawberry	Finland	12	—	< 0.00003 – 0.002	—	< 0.001	26
Sweet cherries	U.S.		—	—	—	< 0.002	9
Watermelon	U.S.		—	< 0.0004	—	—	9
Frozen Fruits							
Gooseberry		5	—	0.001 ± 0.0004 0.001 – 0.002	—	0.01 ± 0.01 0.01 – 0.02	81
Black chokeberry		5	—	0.003 ± 0.001 0.002 – 0.004	—	0.02 ± 0.005 0.01 – 0.02	81
Blackberry		5	—	0.002 ± 0.0003 0.001 – 0.002	—	0.02 ± 0.003 0.012 – 0.02	81
Raspberry		5	—	0.002 ± 0.0003 0.001 – 0.002	—	0.01 ± 0.005 0.005 – 0.02	81
Fruits		71	—	0.002 0.0004 – 0.004	—	0.01 0.004 – 0.03	81

Blackcurrant	5	—	0.001 ± 0.001 0.0004 – 0.002	—	0.01 ± 0.004 0.01 – 0.02	81
Red currant	5	—	0.001 ± 0.0004 0.001 – 0.002	—	0.01 ± 0.01 0.004 – 0.02	81
Plum	5	—	0.001 ± 0.0003 0.001 – 0.001	—	0.01 ± 0.01 0.01 – 0.02	81
Strawberry	6	—	0.002 ± 0.001 0.0004 – 0.004	—	0.01 ± 0.004 0.005 – 0.02	81
Cherry	7	—	0.001 – 0.0003 0.0005 – 0.001	—	0.01 ± 0.003 0.005 – 0.01	81

TABLE 7.18
Toxic Metals in Tea, Coffee, and Confectionary Products (in mg/100 g w/w)

Name	Origin	n	Cd	Hg	Pb	Ref.
Tea						
Tea	Africa	18	—	—	0.11 ± 0.01	27
Tea	Asia	36	—	—	0.02 ± 0.01	27
Tea	China	13	—	—	0.21 ± 0.10	27
Tea	India, Sri Lanka	13	—	—	0.05 ± 0.02	27
Tea[c]	China	39	—	—	0.14 ± 0.12	29
					0.03 ± 0.77	
Beiqishen tea	China		< 0.01	< 0.05	< 0.13	30
Chinese tea	China	5	0.003 ± 0.001	—	0.02 ± 0.02	82
					0.005 – 0.05	
Indian tea	India	3	0.001 ± 0.0001	—	0.01 ± 0.004	82
			0.001 – 0.001		0.005 – 0.01	
Beiqishen tea extract[d]	China		< 0.0003	< 0.00013	< 0.002	30
Coffee						
Decaffeinated instant coffee	U.S.		—	—	< 0.002	9
Ground coffee	U.S.		< 0.0003	—	< 0.002	9
						7
Confectionary Products						
Sugar and preserves[b]	U.K.		0.001	0.0004	0.001	
Sugar and preserves[b]	U.S.		0.002 ± 0.001	NA	< 0.005	9
			< 0.0004 – 0.003			
Sugar	Egypt	2	—	—	0.001 ± 0.001	37
Sugar	Egypt		—	—	0.0001 ± 0.0001	38
White sugar, granulated	U.S.		< 0.001	—	< 0.01	9
Sugar cane plant	Egypt	3	—	—	0.002 ± 0.002	37
Sugar cane root[c]	Brazil		0.02 ± 0.01	0.003 ± 0.002	0.80 ± 0.42	123
			0.01 – 0.04	0.001 – 0.01	0.29 – 2.23	
Sugar cane stem[c]	Brazil		0.02 ± 0.01	0.002 ± 0.001	0.67 ± 0.24	123
			0.01 – 0.03	0.001 – 0.01	0.35 – 1.38	
Sugar cane leaf[c]	Brazil		0.01 ± 0.06	0.002 ± 0.001	0.21 ± 0.12	123
			0.004 – 0.03	0.001 – 0.01	0.05 – 0.55	
Cocoa powder	Malaysia	7	0.06 ± 0.01	—	0.02 ± 0.01	125
Cocoa powder	Venezuela	3	0.18 ± 0.002	—	0.02 ± 0.001	125
Cocoa powder	Ecuador	1	0.07 ± 0.005	—	0.02 ± 0	125
Cocoa powder	El Oro, Ecuador	1	0.05 ± 0.002	—	0.001 ± 0.0003	125
Cocoa powder	Bahia, Brazil	1	0.01 ± 0.001	—	0.02 ± 0.001	125
Cocoa powder	Habuna, Brazil	1	0.02 ± 0.002	—	0.08 ± 0.015	125

TABLE 7.18 (CONTINUED)
Toxic Metals in Tea, Coffee, and Confectionary Products (in mg/100 g w/w)

Name	Origin	n	Cd	Hg	Pb	Ref.
Cocoa powder	Ivory Coast	3	0.01 ± 0.001	—	0.02 ± 0.004	125
Cocoa powder	Ghana	3	0.01 ± 0.002	—	0.005 ± 0.001	125
Cocoa liquor	El Oro, Ecuador	3	0.03 ± 0.0004	—	0.001 ± 0.0	125
Cocoa liquor	Ecuador	3	0.05 ± 0.001	—	0.015 ± 0.001	125
Cocoa liquor	Venezuela	3	0.11 ± 0.0005	—	0.01 ± 0.001	125
Cocoa butter	El Oro, Ecuador	3	0.001 ± 0.0	—	0.002 ± 0.0001	125
Cocoa butter	Ecuador	3	0.001 ± 0.0	—	0.003 ± 0.0001	125
Cocoa butter	Venezuela	3	0.001 ± 0.0	—	0.002 ± 0.0002	125
Molasses	Egypt		—	—	0.0003 ± 0.0001	38
Sugar cane juice	Egypt	3	—	—	0.004 ± 0.002	37
Sugar juice	Egypt		—	—	0.0001 ± 0.0	38

7.7.2 Barium

Concentration of barium (Ba) in plant foods is generally greater than that in animal products and ranged between 0.002 and 0.17 mg/100 g in vegetables, and from 0.001 to 0.08 mg/100 g in fruit juices (Table 7.22 and Table 7.25). Miscellaneous cereals contained from 0.03 to 0.08 mg Ba/100 g (Table 7.21). Higher levels of this element were observed in nuts containing up to 5.6 mg Ba/100 g (Table 7.24). It is interesting to note that the Brazil nut seems to be the richest source of barium because some kernels contained extremely high levels of this metal, i.e., up to 300 mg/100 g.[124] Tea leaves contained up to 3.88 mg Ba/100 g, whereas roasted coffee contained from 0.09 to 0.97 mg/100 g (Table 7.23). Low levels of barium are reported for fruit juices and alcoholic beverages (Table 7.25) such as beer (0.001–0.008 mg/100 g) and wine (0.003–0.01 mg/100 g).

7.7.3 Gallium

Concentration of gallium (Ga) in rice (0.0016 mg/100 g) is several orders of magnitude higher than that in vegetables (0.0000006–0.00017 mg/100 g). Beer and wine contain this element in the range of 0.000001 to 0.00004 mg/100 g (Table 7.25).

7.7.4 Germanium, Tantalum, and Caesium

Very little information is available on concentrations of germanium (Ge), tantalum (Ta), and caesium (Cs), elements sometimes called "electronic metals."[1] Different levels of these metals have been reported for vegetables — from 0.00005 to 0.28 mg/100 g for Ge, 0.000002 to 0.001 mg/100 g for Ta, and < 0.00002 to 0.0003 mg/100 g for Cs (Table 7.22). Alcoholic beverages such as wine and beer contain very low levels of caesium and germanium, from 0.000001 to 0.0001 mg/100 g

TABLE 7.19
Toxic Metals in Nuts, Seeds, and Oils (in mg/100 g w/w)

Name	Origin	n	Cd	Hg	Pb	Ref.
Nuts and Seeds						
Almond	Spain	6	—	—	0.03 0.03 – 0.04	78
Cashew	Spain	3	—	—	0.02 0.02 – 0.03	78
Chestnut	Spain	6	—	—	0.03 0.02 – 0.03	78
Hazelnut	Spain	5	0.001 0.0004 – 0.001	—	0.03 0.03 – 0.04	78
Mixed nuts (no peanuts)	U.S.		< 0.001	—	< 0.01	9
Nuts[b]	U.K.		0.005	0.0003	0.002	7
Nuts[b]	U.S.		0.005 ± 0.0004 < 0.0001 – 0.01	NA	< 0.005	9
Peanut	Spain	5	0.001 0.0003 – 0.001	—	0.03 0.03 – 0.03	78
Peanut	NE China	1	0.01	—	0.0004	74
Peanut	NE China	1	0.02	—	0.001	74
Peanut butter	U.S.		0.01	—	< 0.005	9
Pinenut kernel	Spain	7	—	—	0.03 0.03 – 0.04	78
Pistachio	Spain	8	—	—	0.02 0.02 – 0.03	78
Roasted salted corn	Spain	6	0.002 0.001 – 0.002	—	0.02 0.02 – 0.02	78
Sunflower seed	Spain	5	0.001 0.001 – 0.002	—	0.02 0.01 – 0.03	78
Walnuts	Spain	5	—	—	0.02 0.02 – 0.03	78
Oils						
Grape seeds oil		3	0.0004 ± 0.0001	—	0.001 ± 0.0004	91
Maize oil		5	0.001 ± 0.0002	—	0.001 ± 0.0002	91
Nuts oil		3	0.0002 ± 0.0001	—	0.005 ± 0.001	91
Oils and fats[b]	U.K.		0.002	0.0003	0.002	7
Oils and fats[b]	U.S.		< 0.002	NA	< 0.01	9
Peanut oil		6	0.0003 ± 0.0001	—	0.001 ± 0.0004	90
Rice oil		5	0.0001 ± 0.00004	—	0.001 ± 0.0002	90
Soy oil		5	0.0004 ± 0.0001		0.003 ± 0.001	90
Soy oil	Poland	1	—	< 0.0001	—	65
Sunflower oil		5	0.0002 ± 0.0002		0.002 ± 0.003	90

TABLE 7.20
Toxic Metals in Beverages (in mg/100 g w/w)

Name	Origin	n	Be	Cd	Hg	Pb	Ref.
Beverages[b]	U.K.		—	0.0001	0.00006	< 0.001	7
Beverages[b]	U.S.		—	< 0.0003	NA	< 0.002 – 0.002	9
Juices							
Tomato juice			—	0.002	—	< 0.002	9
Alcoholic Beverages							
Beer[d]	Poland	35	—	0.00001	0.00003	0.0003	91
				0.000002 – 0.00005	0.00001 – 0.0001	0.0001 – 0.001	
Martini	U.S.		—	—	—	< 0.002	9
Whiskey	U.S.		—	—	—	< 0.002	9
Whiskey[d]	Scotland	35	—	—	—	0.0003 ± 0.0004	50
						< 0.00001 – 0.003	
Dry table wine	U.S.		—	—	—	0.002	9
Wine[d]	Italy		0.000001 – 0.0001	0.000001 – 0.0001	—	0.001 – 0.04	53

TABLE 7.21
Other Metals in Miscellaneous Cereals (in mg/100 g w/w)

Name	Origin	n	Ag	Au	Ba	Bi	Cs	Ga	Ge	Ir	Li	Ref.
Bread												
Bread[b]	U.K.		—	0.0001	0.1	0.00003	—	—	0.0003	0.0001	0.001	7
Rice												
Rice	Different		—	—	0.03	—	0.001	0.002	—	—	—	13
Miscellaneous Cereals												
Miscellaneous cereals[b]	U.K.		—	0.0002	0.08	0.00003	—	—	0.0004	0.0002	0.002	7
Common buckwheat flour[c]	Slovenia		0.002 ± 0.0002	—	—	—	—	—	—	—	—	77
Tartary buckwheat fine flour[c]	Slovenia		0.0003 ± 0.0001	—	—	—	—	—	—	—	—	77
Tartary buckwheat flour[c]	Slovenia		0.001 ± 0.0003	—	—	—	—	—	—	—	—	77
Tartary buckwheat leaf flour[c]	Slovenia		0.003 ± 0.001	—	—	—	—	—	—	—	—	77
Soya flour			—	—	—	—	—	—	0.36	—	—	131
Common buckwheat bran[c]	Slovenia		0.003 ± 0.0004	—	—	—	—	—	—	—	—	77

Tartary buckwheat bran[c]	Slovenia		0.001 ± 0.0002	—	—	—	—	—	—	—	—	77	
Common buckwheat grain[c]	Slovenia		0.002 ± 0.0003	—	—	—	—	—	—	—	—	77	
Tartary buckwheat grain[c]	Slovenia		0.001 ± 0.0002	—	—	—	—	—	—	—	—	77	
Pearl barley			—	—	—	—	—	—	0.16	—	—	129	

Name	Origin	n	Pd	Pt	Rb	Rh	Ru	Sb	Sr	Tl	W	Y	Ref.
Bread													
Bread[b]	U.K.	—	0.0002	0.00001	—	0.00002	< 0.0002	0.0001	0.37	0.0001	—	—	7
Rice													
Rice	India, Pakistan	28	—		0.14 0.04–0.39				—		0.0001 0.00003 – 0.0004		12
Rice	Europe	25	—	—	0.14 0.02 – 0.88	—	—	—	—	—	0.00004 0.0 – 0.0002	—	12
Rice	Different		—	—	0.28	—	—	—	0.05	—	—	0.0005	13
Rice[c]	—	—	—	—	—	—	—	0.38 ± 0.08	—	—	—	—	14
Miscellaneous cereals[b]	U.K.	—	0.0001	0.00001	—	0.00001	< 0.0002	0.0004	0.13	0.0001	—	—	7
Common buckwheat flour[c]	Slovenia		—	—	0.82 ± 0.04	—	—	0.003 ± 0.0003	—	—	—	—	77
Tartary buckwheat fine flour[c]	Slovenia		—	—	0.15 ± 0.004	—	—	0.001 ± 0.0001	—	—	—	—	77

TABLE 7.21 (CONTINUED)
Other Metals in Miscellaneous Cereals (in mg/100 g w/w)

Name	Origin	n	Pd	Pt	Rb	Rh	Ru	Sb	Sr	Tl	W	Y	Ref.
Tartary buckwheat flour[c]	Slovenia		—	—	0.47 ± 0.005	—	—	0.001 ± 0.0001	—	—	—	—	77
Tartary buckwheat leaf flour[c]	Slovenia		—	—	1.36 ± 0.01	—	—	0.0005 ± 0.0001	—	—	—	—	77
Common buckwheat bran[c]	Slovenia		—	—	1.26 ± 0.01	—	—	—	—	—	—	—	77
Tartary buckwheat bran[c]	Slovenia		—	—	1.31 ± 0.01	—	—	0.0005 ± 0.0001	—	—	—	—	77
Quinoa[c]			—	—	—	—	—	0.001 ± 0.0001	—	—	—	—	77
Common buckwheat grain[c]	Slovenia		—	—	1.12 ± 0.011	—	—	—	—	—	—	—	77
Tartary buckwheat grain[c]	Slovenia		—	—	1.64 ± 0.001	—	—	—	—	—	—	—	77
Wheat grains[c]			—	—	—	—	—	< 0.0004	—	—	—	—	77

(Table 7.25). Among grain products, rice can contain up to 0.001 mg Cs/100 g, whereas concentrations of germanium in bread and miscellaneous cereals are often in the range of < 0.0003 to 0.0004 mg/100 g. Somewhat higher levels of germanium have been reported in soya flour (0.36 mg/100 g) and pearl barley (0.16 mg/100 g) (Table 7.21).

7.7.5 Gold

Gold (Au) is found in food crops in very low amounts, typically from 0.00001 to 0.001 mg/100 g in vegetables (Table 7.22), and 0.00004 to 0.001 mg/100 g in sugarcane plant and sugar (Table 7.23). Concentrations of gold in bread and miscellaneous cereals are in the range of 0.0001 to 0.0002 mg/100 g (Table 7.21).

7.7.6 Indium

Compilation of published data (Table 7.22 and Table 7.25) suggests that indium (In) contents in vegetables range from < 0.00001 to 0.0003 mg/100 g, whereas beer is characterized by extremely low levels of this element (< 0.000001 mg/100 g).

7.7.7 Lithium

Average concentrations of lithium (Li) in bread and miscellaneous cereals are estimated to be 0.001 and 0.002 mg/100 g, respectively, and < 0.0001 to 0.003 mg/100 g has been reported for vegetables (Table 7.21 and Table 7.22). Higher lithium levels are reported for Spanish potatoes, up to 0.30 mg/100 g. Fresh fruits and alcoholic beverages (wine) generally contain from < 0.0002 to 0.005 mg Li/100 g (Table 7.22 and Table 7.25).

7.7.8 Niobium

Concentrations of niobium (Nb) in vegetables are in the extremely low range of 0.000000004 to 0.0003 mg/100 g. Wines contain from 0.000001 to 0.00001 mg Nb/100 g (Table 7.22 and Table 7.25).

7.7.9 Platinum, Palladium, and Other Platinum Metals

Platinum (Pt) and palladium (Pd), as well as other platinum metals are present in food crops in very low amounts, ranging from 0.000001 to 0.001 mg/100 g. Concentrations of palladium in bread and miscellaneous cereals are an order of magnitude higher than those of platinum (Table 7.21). Concentrations. of platinum, palladium, iridium (Ir), rhodium (Rh), and ruthenium (Ru) in vegetables vary from 0.000001 to 0.001 mg/100 g, whereas wine contains palladium and rhodium in the range of 0.000001 to 0.0001 mg/100 g (Table 7.22 and Table 7.25). In Australia, elevated levels of platinum have been detected on roadside vegetation, suggesting the platinum released from catalytic converters in automobiles is deposited on food crops and is contaminating the human diet.[1]

TABLE 7.22
Other Metals in Vegetables, Mushrooms, and Fruits (in mg/100 g w/w)

Name	Origin	n	Ag	Au	Ba	Bi	Cs	Ga	Ref.
Vegetables									
Canned vegetables[b]	U.K.		—	0.00004	0.03	0.00005	—	—	7
Green vegetables[b]	U.K.		—	< 0.00004	0.04	0.0001	—	—	7
Onion conventional	Denmark		0.0001 ± 0.00003 0.00001 – 0.0002	0.0001 ± 0.0002 0.00002 – 0.001	0.01 ± 0.01 0.002 – 0.05	0.00001 ± 0.000005 0.000001 – 0.00003	0.00002 ± 0.00003 0.0 – 0.0001	0.00002 ± 0.00003 0.000001 – 0.0002	20
Onion organic	Denmark		0.0001 ± 0.00003 0.00002 – 0.0002	0.0001 ± 0.00004 0.000015 – 0.0002	0.01 ± 0.01 0.002 – 0.04	0.00002 ± 0.00001 0.000002 – 0.00004	0.00005 ± 0.0001 0.0 – 0.0003	0.00002 ± 0.00002 0.0 – 0.0001	20
Other vegetables[b]	U.K.		—	0.00005	0.05	0.00003	—	—	7
Peas conventional	Denmark		0.0001 ± 0.00005 0.000012 – 0.0003	0.0001 ± 0.00004 0.00001 – 0.0003	0.02 ± 0.01 0.003 – 0.07	0.00001 ± 0.00001 0.000001 – 0.00004	—	0.00004 ± 0.00002 0.00001 – 0.0001	20
Peas organic	Denmark		0.0001 ± 0.00004 0.00002 – 0.0003	0.0001 ± 0.0001 0.00003 – 0.0004	0.02 ± 0.02 0.005 – 0.07	0.00001 ± 0.00001 0.000001 – 0.0001	—	0.00003 ± 0.00002 0.00001 – 0.0001	20
Potatoes[c]	Idaho	342	—		0.03 ± 0.01 0 – 0.08	—	—	—	22
Potatoes[c]	Non-Idaho	266	—	—	0.04 ± 0.04 0 – 0.17	—	—	—	22
Potatoes[c]	U.K.		—	0.00004	0.02	0.00001	—	—	7
Mushrooms									
Bay bolete[c]	Poland	166	0.01 ± 0.004 0.0003 – 0.04	—	—	—	—	—	24
Fruits									
Fresh fruit[b]	U.K.		—	0.00004	0.04	0.00003	—	—	7
Fruit products[b]	U.K.		—	0.00004	0.03	0.00002	—	—	7

Name	Origin	n	Ge	Hf	In	Ir	Li	Nb	Ref.
Vegetables									
Carrot			0.06	—	—	—	—	—	129
Canned vegetables[b]	U.K.		< 0.0002	—	—	< 0.0001	0.001	—	7
Garlic			0.28	—	—	—	—	—	129
Green pepper			0.02	—	—	—	—	—	129
Green vegetables[b]	U.K.		< 0.0002	—	—	< 0.0001	0.001	—	7
Onion			0.03	—	—	—	—	—	129
Onion conventional	Denmark		0.001 ± 0.001 0.00005 – 0.003	0.0001 ± 0.0002 0.0 – 0.001	0.00004 ± 0.00003 0.0 – 0.0003	0.000003 ± 0.00001 0.0 – 0.00003	0.0001 ± 0.0002 0.0 – 0.0005	0.0001 ± 0.0001 0.0 – 0.001	20
Onion organic	Denmark		0.0001 ± 0.00004 0.0001 – 0.002	0.00004 ± 0.00005 0.0 – 0.0002	0.00003 ± 0.00005 0.0 – 0.0003	0.000001 ± 0.000003 0.0 – 0.00001	0.0 – 0.0002	0.0001 ± 0.0001 0.000000004 – 0.001	20
Other vegetables[b]	U.K.		0.0002	—	—	< 0.0001	0.003	—	7
Peas conventional	Denmark		0.01 ± 0.002 0.004 – 0.01	0.00003 ± 0.00004 0.0 – 0.0002	—	0.0 – 0.00001	—	0.0001 ± 0.0001 0.00002 – 0.0003	20
Peas organic	Denmark		0.01 ± 0.002 0.005 – 0.02	0.00004 ± 0.00005 0.0 – 0.0003	—	0.000001 ± 0.000005 0.0 – 0.00004	—	0.0002 ± 0.0001 0.000003 – 0.001	20
Potato			0.185	—	—	—	—	—	129
Potatoes	Spain		—	—	—	—	0.21 ± 0.13 0.11 – 0.30	—	21
Potatoes[b]	U.K.		< 0.0002	—	—	< 0.0001	0.001	—	7
Red pepper			0.05	—	—	—	—	—	129
Tomato			0.04	—	—	—	—	—	
Fruits									
Fresh fruit[b]	U.K.		< 0.0002	—	—	< 0.0001	0.0005	—	7
Fruit products[b]	U.K.		< 0.0002	—	—	< 0.0001	0.001	—	7

Name	Origin	n	Pd	Pt	Rb	Re	Rh	Ru	Ref.
Vegetables									
Canned vegetables[b]	U.K.		0.00004	< 0.00001	—	—	< 0.00001	< 0.0002	7
Green vegetables[b]	U.K.		0.0001	0.00001	—	—	< 0.00001	< 0.0002	7
Onion conventional	Denmark		—	0.00002 ± 0.00003 0.0 – 0.0002	0.04 ± 0.02 0.01 – 0.09	0.000001 ± 0.000003 0.0 – 0.00001	—	0.00002 ± 0.00002 0.0 – 0.0001	20
Onion organic	Denmark		—	0.00002 ± 0.00002 0.0 – 0.0001	0.07 ± 0.02 0.03 – 0.11	0.000001 ± 0.000002 0.0 – 0.00001	—	0.00001 ± 0.00002 0.0 – 0.00005	20
Other vegetables[b]	U.K.		0.00005	0.00001	—	—	< 0.00001	< 0.0002	7

TABLE 7.22 (CONTINUED)
Other Metals in Vegetables, Mushrooms, and Fruits (in mg/100 g w/w)

Name	Origin	n	Pd	Pt	Rb	Re	Rh	Ru	Ref.
Other vegetables[b]	U.K.		0.00005	0.00001	—	—	< 0.00001	< 0.0002	7
Peas conventional	Denmark		0.0005 ± 0.0004 0.0001 – 0.001	0.00001 ± 0.00002 0.0 – 0.0001	—	0.000001 ± 0.000004 0.0 – 0.00002	0.00004 ± 0.00003 0.00001 – 0.0002	—	20
Peas organic	Denmark		0.0004 ± 0.0002 0.0001 – 0.001	0.00001 ± 0.00002 0.0 – 0.0001	—	0.0000004 ± 0.000002 0.0 – 0.00001	0.00004 ± 0.00002 0.00001 – 0.0001	—	20
Potatoes	Spain		—	—	0.25 ± 0.16	—	—	—	21
Potatoes[b]	U.K.		0.00005	< 0.00001	—	—	< 0.00001	< 0.0002	7
Fruits									
Fresh fruit[b]	U.K.		0.00004	< 0.00001	—	—	< 0.00001	< 0.0002	7
Fruit products[b]	U.K.		0.00005	< 0.00001	—	—	0.00001	< 0.0002	7

Name	Origin	n	Sb	Sc	Sr	Ta	Te	Ref.
Vegetables								
Green vegetables[b]	U.K.		< 0.0001	—	0.16	—	—	7
Canned vegetables[b]	U.K.		0.0002	—	0.09	—	—	7
Onion conventional	Denmark		0.0002 ± 0.0001 0.00002 – 0.0005	0.0002 ± 0.0001 0.00002 – 0.0005	0.09 ± 0.05 0.02 – 0.31	—	0.0002 ± 0.0001 0.00005 – 0.0005	20
Onion organic	Denmark		0.0001 ± 0.0001 0.00003 – 0.0004	0.0001 ± 0.0001 0.00003 – 0.0004	0.06 ± 0.03 0.02 – 0.16	—	0.0002 ± 0.0001 0.00002 – 0.001	20
Other vegetables[b]	U.K.		0.0001	—	0.13	—	—	7
Peas conventional	Denmark		0.0001 ± 0.0001 0.00002 – 0.0003	0.0002 ± 0.0001 0.0001 – 0.001	0.05 ± 0.02 0.02 – 0.11	0.0001 ± 0.0001 0.000002 – 0.0004	0.0002 ± 0.0001 0.00001 – 0.0005	20
Peas organic	Denmark		0.0001 ± 0.0001 0.00002 – 0.0004	0.0002 ± 0.0001 0.0001 – 0.0005	0.05 ± 0.01 0.02 – 0.09	0.0001 ± 0.0001 0.00001 – 0.001	0.0002 ± 0.0002 0.00004 – 0.001	20
Potatoes[c]	Idaho	342	—	—	0.05 ± 0.03 0 – 0.14	—	—	22
Potatoes[c]	non-Idaho	266	—	—	0.03 ± 0.04 0 – 0.16	—	—	22
Potatoes[b]	U.K.		0.0001	—	0.08	—	—	7

Tomato fruits, rockwool high EC	Denmark		—	—	0.01	—	—	23
Tomato fruits, rockwool norm EC	Denmark		—	—	0.01	—	—	23
Tomato fruits, soil norm EC	Denmark		—	—	0.02	—	—	23
Fresh fruit[b]	U.K.		0.0001	—	0.08	—	—	7
Fruit products[b]	U.K.		0.0001	—	0.10	—	—	7

Name	Origin	n	Ti	Tl	W	Y	Zr	Ref.
Vegetables								
Green vegetables[b]	U.K.		—	0.0003	—	—	—	7
Canned vegetables[b]	U.K.		—	0.0001	—	—	—	7
Onion, conventional	Denmark		0.09 ± 0.06 0.01 – 0.25	0.0001 ± 0.0001 0.00001 – 0.0003	0.002 ± 0.001 0.001 – 0.004	0.00002 ± 0.00002 0.000003 – 0.0001	0.0003 ± 0.0003 0.00003 – 0.001	20
Onion, organic	Denmark		0.04 ± 0.03 0.01 – 0.14	0.0001 ± 0.00004 0.00002 – 0.0002	0.002 ± 0.001 0.0005 – 0.005	0.00001 ± 0.00001 0.000003 – 0.00003	0.0003 ± 0.0003 0.0001 – 0.002	20
Other vegetables[b]	U.K.		—	0.0001	—	—	—	7
Peas, conventional	Denmark		0.10 ± 0.04 0.04 – 0.19	0.00003 ± 0.00003 0.0 – 0.0002	—	0.00002 ± 0.00001 0.00001 – 0.0001	0.0002 ± 0.0001 0.00004 – 0.0005	20
Peas, organic	Denmark		0.06 ± 0.02 0.03 – 0.11	0.00004 ± 0.00004 0.00000001 – 0.0002	—	0.00002 ± 0.000005 0.00001 – 0.00004	0.0003 ± 0.0004 0.00005 – 0.003	20
Potatoes[b]	U.K.		—	0.0002	—	—	—	7
Fruits								
Fresh fruit[b]	U.K.		—	0.0001	—	—	—	7
Fruit products[b]	U.K.		—	0.0001	—	—	—	7

7.7.10 Rubidium

Ranges in the rubidium (Rb) content of grain products and vegetables typically are 0.02 to 1.36 mg/100 g and 0.01 to 0.25 mg/100 g, respectively (Table 7.21 and Table 7.22). Fruit juices contain from 0.01 to 1.1 mg Rb/100 g. Concentration of the element in wine (0.06–0.42 mg/100 g) is an order of magnitude higher than that in beer (0.01–0.04 mg/100 g) (Table 7.25). Tea leaves contain elevated levels of rubidium — from 2.31 to 13.8 mg/100 g (Table 7.23).

7.7.11 Silver

Levels of silver (Ag) in miscellaneous cereals and vegetables vary from 0.0003 to 0.003 mg/100 g and from 0.00001 to 0.0003 mg/100 g, respectively (Table 7.21 and Table 7.22). Mushrooms often show elevated levels of silver — up to 0.04 mg Ag/100 g. Extremely small amounts of this metal (0.000001–0.00002 mg/100 g) are present in beer (Table 7.25), whereas the richest dietary source are nuts with concentrations of 0.27 to 2.01 mg Ag/100 g (Table 7.24). Some vegetables, e.g., the *Brassica* species, have the ability to selectively accumulate silver to very high levels.[1]

7.7.12 Strontium

Concentrations of strontium (Sr) vary from 0.01 to 0.37 mg/100 g in vegetables and grain products, with the highest levels registered for rice (Table 7.21 and Table 7.22). Wines and fruit juices contain this element in the range of 0.004 to 0.26 mg/100 g, whereas the values for beers can vary from 0.01 to 0.07 mg/100 g (Table 7.25). Higher levels are reported for nuts (0.86 mg Sr/100 g) and especially tea leaves (2.22 mg Sr/100 g) (Table 7.23 and Table 7.24).

7.7.13 Tellurium

The few reported levels of tellurium (Te) in plant products are mostly below the detection limit of analytical measurements, i.e., < 0.0005 mg/100 g[1] (Table 7.22 and Table 7.25). Higher levels of up to 0.001 mg/100 g have been reported for onion. It should be noted that contamination of metal containers with tellurium may be its main source in foods.[1]

7.7.14 Thallium

Little is known about the levels of thallium (Tl) in food crops. Concentration of this element in vegetables is usually extremely small, ranging from < 0.000001 to 0.0002 mg/100 g (Table 7.22). Grain products, sugar, nuts, and oils show slightly higher levels, from 0.0001 to 0.0002 mg/100 g (Table 7.21, Table 7.23, and Table 7.24). Wine and beer have lower levels, in the range of 0.000001 to 0.00002 mg/100 g (Table 7.25). Thallium is readily accumulated from contaminated soil by various vegetables; concentrations of up to 4.2 mg/100 g have been reported in cabbage grown on heavily contaminated soil of Ghizhou Province in China.[126] Vegetables

TABLE 7.23
Other Metals in Tea, Coffee, and Confectionary Products (in mg/100 g w/w and in µg/100 g w/w[a])

Name	Origin	n	Ag	Au	Ba	Bi	Cs	Ge	Ir	Li	Pd	Ref.
Tea												
Tea	Africa	18	—	—	1.97 ± 1.07	—	0.05 ± 0.04	—	—	—	—	27
Tea	Asia	36	—	—	2.62 ± 0.89	—	0.03 ± 0.02	—	—	—	—	27
Tea	China	13	—	—	1.77 ± 1.1	—	0.05 ± 0.03	—	—	—	—	27
Tea	India, Sri Lanka	13	—	—	2.03 ± 0.81	—	0.02 ± 0.01	—	—	—	—	27
Tea[c]	China	39	—	—	1.82 ± 0.81 0.28 – 3.88	—	—	—	—	—	—	29
Beiqishen tea	China		—	—	2.84 ± 0.07	—	—	—	—	< 0.02	—	30
Instant tea[d] (<2% tea extract)		3	—	—	0.001 0.001 – 0.002	—	—	—	—	—	—	33
Instant tea[d] (100% extract)		2	—	—	0.001 0.001 – 0.002	—	—	—	—	—	—	33
Beiqishen tea extract[d]	China		—	—	0.01 ± 0.0004	—	—	—	—	< 0.00004	—	30
Black tea beverages[d]		23	—	—	0.003 ± 0.0 0.001 – 0.01	—	—	—	—	—	—	33
Green tea beverages[d]		20	—	—	0.002 ± 0.0004 0.001 – 0.01	—	—	—	—	—	—	33
Tea soft drinks[d]		8	—	—	0.002 0.001 – 0.003	—	—	—	—	—	—	33
Coffee												
Roasted coffee[c]	Mixture	18	—	—	0.45 ± 0.23 0.09 – 0.97	—	—	—	—	—	—	36
Green coffee[c]	Mixture	41	—	—	0.43 ± 0.15 0.17 – 0.76	—	—	—	—	—	—	84

TABLE 7.23 (CONTINUED)
Other Metals in Tea, Coffee, and Confectionary Products (in mg/100 g w/w and in μg/100 g w/w[a])

Name	Origin	n	Ag	Au	Ba	Bi	Cs	Ge	Ir	Li	Pd	Ref.
Confectionary Products												
Sugar	Egypt	2	0.0002 ± 0.0	0.5 ± 0.1[a]	—	—	—	—	—	—	—	37
Sugar	Egypt		—	—	—	—	—	—	—	0.0003 ± 0.001	—	38
Sugar cane plant	Egypt	3	0.4 ± 0.005[a]	0.001 ± 0.001	—	—	—	—	—	—	—	37
Molasses	Egypt		—	—	—	—	—	—	—	0.001 ± 0.001	—	38
Sugar cane juice	Egypt	3	0.0002 ± 0.0	0.001 ± 0.0002	—	—	—	—	—	—	—	37
Sugar juice	Egypt		—	—	—	—	—	—	—	0.001 ± 0.001	—	38
Sugar and preserves[b]	U.K.		—	0.00004	0.08	0.00003	—	0.0002	< 0.0001	0.001	0.00005	7

Name	Origin	n	Pt	Rb	Rh	Ru	Sb	Sc	Sr	Ti	Tl	Ref.
Tea												
Tea	Africa	18	—	6.15 ± 3.17	—	—	—	—	1.0 ± 0.58	2.58 ± 1.0	—	27
Tea	Asia	36	—	8.25 ± 3.29	—	—	—	—	2.09 ± 0.72	2.78 ± 0.7	—	27
Tea	China	13	—	8.17 ± 3.12	—	—	—	—	0.88 ± 0.4	2.63 ± 1.38	—	27
Tea	India, Sri Lanka	13	—	4.59 ± 1.74	—	—	—	—	1.1 ± 0.39	2.46 ± 0.53	—	27
Tea[c]	China	39	—	7.62 ± 2.92 2.31 – 13.8	—	—	—	—	1.13 ± 0.44 0.32 – 2.22	2.12 ± 1.2 0.37 – 6.46	—	29
Instant tea[d] (<2% tea extract)		3	—	—	—	—	—	—	0.001 0.001 – 0.001	—	—	33
Instant tea[d] (100% extract)		2	—	—	—	—	—	—	0.002 0.001 – 0.002	—	—	33
Black tea beverages[d]		23	—	—	—	—	—	—	0.004 – 0.02	—	—	33
Green tea beverages[d]		20	—	—	—	—	—	—	0.004 – 0.01	—	—	33
Tea soft drinks[d]		8	—	—	—	—	—	—	0.02 0 – 0.07	—	—	33

Coffee												
Roasted coffee[c]	Mixture	18	—	—	—	—	—	—	0.62 ± 0.18	—	—	36
									0.40 – 1.15			
Green coffee[c]	Mixture	41	—	—	—	—	—	—	0.55 ± 0.21	—	—	84
									0.12 – 1.12			
Confectionary Products												
Sugar	Egypt	2	—	—	—	—	—	—	0.003 ± 0.001	—	—	37
Sugar	Egypt	—	—	—	—	—	—	0.3 ± 0.2[a]	—	—	—	38
Sugarcane plant	Egypt	3	—	—	—	—	—	—	0.004 ± 0.002	—	—	37
Molasses	Egypt		—	—	—	—	—	0.002 ± 0.002	—	—	—	38
Sugarcane juice	Egypt	3	—	—	—	—	—	—	0.003 ± 0.0008	—	—	37
Sugar juice	Egypt		—	—	—	—	—	0.001 ± 0.001	—	—	—	38
Sugar and preserves[b]	U.K.		0.01[a]	—	< 0.01[a]	0.0002	0.0002	—	0.11	—	0.0001	7

TABLE 7.24
Other Metals in Nuts, Seeds, and Oils (in mg/100 g w/w)

Name	Origin	n	Ag	Au	Ba	Bi	Ge	Ir	Li	Pd	Pt	Rh	Ru	Sb	Sr	Tl	Ref.
Nuts and Seeds																	
Almond	Spain	6	0.83 0.44–2.01	—	—	—	—	—	—	—	—	—	—	—	—	—	78
Cashew	Spain	3	0.61 0.5–1.22	—	—	—	—	—	—	—	—	—	—	—	—	—	78
Chestnut	Spain	6	0.54 0.33–0.9	—	—	—	—	—	—	—	—	—	—	—	—	—	78
Hazelnut	Spain	5	0.94 0.5–1.22	—	—	—	—	—	—	—	—	—	—	—	—	—	78
Peanut	Spain	5	0.19 0.17–0.28	—	—	—	—	—	—	—	—	—	—	—	—	—	78
Pinenut kernel	Spain	7	0.51 0.4–0.62	—	—	—	—	—	—	—	—	—	—	—	—	—	78
Pistachio	Spain	8	0.45 0.3–0.52	—	—	—	—	—	—	—	—	—	—	—	—	—	78
Roasted salted corn	Spain	6	0.66 0.44–0.87	—	—	—	—	—	—	—	—	—	—	—	—	—	78
Sunflower seed	Spain	5	0.52 0.12–0.95	—	—	—	—	—	—	—	—	—	—	—	—	—	78
Walnuts	Spain	5	0.47 0.27–0.54	—	—	—	—	—	—	—	—	—	—	—	—	—	78
Hazelnut					1.66 ± 0.21 1.39–2.14												109
Nuts[b]	U.K.		—	0.00005	5.6	0.00004	0.0002	0.0001	0.001	0.0003	0.00001	0.0004	<0.0002	0.0001	0.86	0.0002	7
Oils																	
Oils and fats[b]	U.K.		—	0.0001	0.003	0.00002	0.0002	<0.0001	0.0003	0.00004	0.00002	<0.00001	0.0002	0.0002	0.01	0.0001	7

TABLE 7.25
Other Metals in Beverages (in mg/100 g w/w)

Name	Origin	n	Ag	Au	Ba	Bi	Cs	Ga	Ref.
Juices									
Grapefruit juice			—	—	0.001	—	—	—	45
Lemon juice			—	—	0.02	—	—	—	45
Lime juice			—	—	0.02	—	—	—	45
Mandarin juice			—	—	0.002	—	—	—	45
Orange juice	Australia	290	—	—	0.02 ± 0.001 0.001 – 0.04	—	—	—	45
Orange juice concentrate	Australia	83	—	—	0.01 ± 0.003 0.01 – 0.03	—	—	—	45
Orange juice concentrate	Brazil	42	—	—	0.03 ± 0.02 0.01 – 0.08	—	—	—	45
Pummelo juice			—	—	0.003	—	—	—	45
Seville orange juice			—	—	0.02	—	—	—	45
Tangello juice			—	—	0.03	—	—	—	45
Alcoholic Beverages									
Beer[d]		32	—	—	0.003 ± 0.001 0.001 – 0.01	—	—	—	49
Beer[d]	Poland	35	0.000004 0.000001 – 0.00002	—	—	0.000005 0.0000005 – 0.00005	0 0.00003 0.00001 – 0.0001	0.00001 0.000001 – 0.00002	91
Wine[d]	Italy		—	—	0.003 – 0.01	—	0.000001 – 0.0001	0.000004 – 0.00004	53
Beverages[d]	U.K.		—	0.00004	0.01	0.00001	—	—	7

Name	Origin	n	Ge	Hf	In	Ir	Li	Ref.
Juices								
Grapefruit juice			—	—	—	—	< 0.0002	45
Lemon juice			—	—	—	—	< 0.0002	45
Lime juice			—	—	—	—	< 0.0002	45
Mandarin juice			—	—	—	—	< 0.0002	45

TABLE 7.25 (CONTINUED)
Other Metals in Beverages (in mg/100 g w/w)

Name	Origin	n	Ge	Hf	In	Ir	Li	Ref.
Orange juice	Australia	290	—	—	—	—	0.0004 ± 0.0001 < 0.0002 – 0.004	45
Orange juice concentrate	Australia	83	—	—	—	—	0.0002 ± 0.0001 < 0.0002 – 0.001	45
Orange juice concentrate	Brazil	42	—	—	—	—	0.0001 ± < 0.0001 < 0.0002 – 0.0004	45
Pummelo juice			—	—	—	—	< 0.0002	45
Seville orange juice			—	—	—	—	< 0.0002	45
Tangello juice			—	—	—	—	< 0.0002	45
Alcoholic Beverages								
Beer[d]	Poland	35	—	—	0.000001 0.0000001 – 0.000002	—	—	91
Wine[d]	Spain	125	—	—	—	—	0.001 ± 0.0003 0.0002 – 0.002	52
Wine[d]	Italy		0.000001 – 0.0001	0.000001 – 0.00001	—	—	0.001 – 0.005	53
Beverages[d]	U.K.		0.0002	—	—	< 0.0001	0.0004	7

Name	Origin	n	Nb	Pd	Pt	Rb	Rh	Ru	Sb	Sr	Ref.
Juices											
Grapefruit juice			—	—	—	0.02	—	—	—	0.004	45
Lemon juice			—	—	—	0.07	—	—	—	0.04	45
Lime juice			—	—	—	0.17	—	—	—	0.02	45
Mandarin juice			—	—	—	0.07	—	—	—	0.01	45
Orange juice	Australia	290	—	—	—	0.09 ± 0.04 0.02 – 1.1	—	—	—	0.07 ± 0.01 0.01 – 0.26	45
Orange juice concentrate	Australia	83	—	—	—	0.06 ± 0.02 0.02 – 0.10	—	—	—	0.07 ± 0.02 0.04 – 0.15	45
Orange juice concentrate	Brazil	42	—	—	—	0.21 ± 0.05 0.09 – 0.31	—	—	—	0.04 ± 0.02 0.02 – 0.1	45

Name	Origin	n									Ref.
Pummelo juice			—	—	—	0.24	—	—	—	0.01	45
Seville orange juice			—	—	—	0.03	—	—	—	0.10	45
Tangello juice			—	—	—	0.04	—	—	—	0.16	45
Alcoholic Beverages											
Beer[d]		32	—	—	—	—	—	—	—	0.03 ± 0.02 0.01 – 0.07	49
Beer[d]	Poland	35	—	—	—	0.02 0.01 – 0.04	—	—	0.00004 0.00002 – 0.0001	—	91
Red wines[d]	Brazil	—	—	—	—	0.18–0.46	—	—	—	0.04–0.1	51
Red wines[d]	Portugal		—	—	—	0.08 – 0.4	—	—	—	0.07 – 0.15	51
Red wines[d]	Chile		—	—	—	0.25	—	—	—	0.08	51
White wines[d]	Brazil		—	—	—	0.13 – 0.3	—	—	—	0.03 – 0.03	51
White wines[d]	Chili		—	—	—	0.33	—	—	—	0.04	51
Wine[d]	Spain	125	—	—	—	0.19 ± 0.08 0.07 – 0.42	—	—	—	0.05 ± 0.02 0.02 – 0.13	52
Wine[d]	Italy		0.000001 – 0.00001	0.000001 – 0.0001	—	0.06 – 0.23	0.000001 – 0.00001	—	0.000001 – 0.0001	0.03 – 0.24	53
Beverages[d]	UK		—	0.00004	< 0.00001	—	0.00001	< 0.0002	< 0.0001	0.01	7

Name	Origin	n	Te	Ti	Tl	W	Zr	Ref.
Juices								
Grapefruit juice			—	0.0001	—	—	—	45
Lemon juice			—	0.0001	—	—	—	45
Lime juice			—	0.0001	—	—	—	45
Mandarin juice			—	0.0001	—	—	—	45
Orange juice	Australia	290	—	0.001 ± 0.0004 < 0.0001 – 0.002	—	—	—	45
Orange juice concentrate	Australia	83	—	0.0001 ± 0.0001 < 0.0001 – 0.001	—	—	—	45
Orange juice concentrate	Brazil	42	—	0.0003 ± 0.0001 < 0.0001 – 0.001	—	—	—	45
Pummelo juice			—	0.0001	—	—	—	45
Seville orange juice			—	0.0001	—	—	—	45
Tangello juice			—	0.0001	—	—	—	45

TABLE 7.25 (CONTINUED)
Other Metals in Beverages (in mg/100 g w/w)

Name	Origin	n	Te	Ti	Tl	W	Zr	Ref.
Alcoholic Beverages								
Beer[d]	Poland	35	—	—	0.000005 0.000002 – 0.00002	—	—	92
Red wines[d]	Brazil		—	0.02	—	—	—	51
Red wines[d]	Portugal		—	0.01 – 0.01	—	—	—	51
White wines[d]	Brazil		—	0.01	—	—	—	51
Wine[d]	Italy		0.000001 – 0.0001	0.00001 – 0.0001	0.000001 – 0.00001	0.000001 – 0.0001	0.000001 – 0.0001	53
Beverages[b]	UK		—	—	< 0.000	—	—	7

such as lettuce, kohlrabi, and broccoli cultivated in contaminated areas have been associated with levels of thallium that can exceed 0.01 mg/100 g.[1,127,128]

7.7.15 Titanium

Little information is available on levels of titanium (Ti) in food crops. Concentrations of ca. 0.2 mg Ti/100 g were detected in cereals and vegetables but these values may be questionable.[1] Recent studies[20] have reported more reliable data, with average concentrations in onions and peas given as 0.05 and 0.08 mg Ti/100 g, respectively (Table 7.22). The concentrations of titanium in fruit juices range from < 0.0001 to 0.002 mg/100 g, whereas the range for wines is 0.00001 to 0.02 mg/100 g (Table 7.25). According to Moreda-Piñeiro et al.,[27] tea leaves are characterized by high concentration of titanium, up to 2.63 ± 1.38 mg/100 g. Chinese tea may contain up to 6.46 mg Ti/100 g (Table 7.23).

7.7.16 Tungsten

There is little information on tungsten (W) in foods. Its concentration is low and range from 0.0005 to 0.005 mg/100 g in vegetables, and from < 0.00003 to 0.0004 mg/100 g in rice (Table 7.21 and Table 7.22). Wines contained from 0.000001 to 0.0001 mg W/100 g (Table 7.25).

7.7.17 Uranium and Thorium

Concentrations of uranium (U) and thorium (Th) in vegetables are very small and oscillated between 0.000000002 and 0.0001 mg U/100 g and between 0.000001 and 0.001 mg Th/100 g (Table 7.26). As is seen in Table 7.26 very low levels of these elements also occur in alcoholic beverages, i.e., wine (0.000001–0.00001 mg/100 g) and beer (0.000002–0.0001 mg/100 g). Rice contained higher levels of the elements, i.e., 0.003 mg/100 g (Table 7.26).

7.7.18 Zirconium, Scandium, Yttrium, and Hafnium

Very low levels of zirconium (Zr) were reported (Table 7.22 and Table 7.25) for vegetables (0.00003–0.003 mg/100 g) and wine (0.000001–0.0001 mg/100 g). Vegetables contained similarly small concentrations of scandium (Sc). Its content in confectionary products is within the range of 0.0003 to 0.02 mg/100 g (Table 7.23). Yttrium (Y) and hafnium (Hf) occur in plant food in low levels, i.e., from 0.000001 to 0.001 mg/100 g in vegetables and wines (Table 7.22 and Table 7.25).

7.7.19 Cerium and Other Rare Earth Elements

Concentration of cerium (Ce) in grain products represented by rice amounted to 0.002 mg/100 g, whereas vegetables contained from 0.0001 to 0.001 mg Ce/100 g (Table 7.26). Low levels of this element were also observed in wine, i.e., 0.000001 to 0.0001 mg/100 g (Table 7.26).

TABLE 7.26
Ce i REE in Food Crops and Beverages in μg/100 g w.w.

Name	Origin	n	Ce	Dy	Er	Eu	Gd	Ho	La	Lu	Ref.
Rice	India, Pakistan	28	—	—	—	—	0.02 0.0 – 0.1	0.0 0.0 – 0.01	—	—	12
Rice	Europe	25	—	—	—	—	0.01 0.0 – 0.4	0.0 0.0 – 0.01	—	—	12
Rice	Different		2	—	0.05	0.04	0.1	—	1	—	13
Onion conventional	Denmark		0.2 ± 0.2 0.0 – 1	0.002 ± 0.003 0.0 – 0.01	0.001 ± 0.002 0.0 – 0.05	0.03 ± 0.01 0.0 – 0.4	0.005 ± 0.005 0.0 – 0.03	0.0003 ± 0.0005 0.0 – 0.003	0.02 ± 0.02 0.003 – 0.1	0.0 – 0.003	20
Onion organic	Denmark		0.1 ± 0.2 0.0 – 1	0.005 ± 0.005 0.0 – 0.02	0.0 – 0.01	0.01 ± 0.05 0.0 – 0.4	0.01 ± 0.01 0.0 – 0.03	0.0002 ± 0.0005 0.0 – 0.002	0.01 ± 0.01 0.004 – 0.1	0.0002 ± 0.0001 0.0 – 0.003	20
Peas conventional	Denmark		0.0 – 0.2	0.001 ± 0.002 0.0 – 0.01	0.0005 ± 0.002 0.0 – 0.005	0.04 ± 0.04 0.0 – 0.2	0.01 ± 0.004 0.001 – 0.02	0.0001 ± 0.0004 0.0 – 0.002	0.02 ± 0.01 0.005 – 0.1	0.0 – 0.001	20
Peas organic	Denmark		0.1 ± 0.2 0.0 – 1	0.001 ± 0.001 0.0 – 0.004	0.002 ± 0.002 0.0 – 0.01	0.1 ± 0.1 0.01 – 0.4	0.01 ± 0.005 0.003 – 0.02	—	—	—	20
Beer[d]	Poland	35	—	—	—	—	—	—	—	—	91
Wine[d]	Italy		0.001 – 0.1	0.001 – 0.01	0.001 – 0.01	0.001 – 0.01	0.001 – 0.01	0.001 – 0.01	0.001 – 0.01	—	53

Name	Origin	n	Nd	Pr	Sm	Tb	Th	Tm	U	Yb	Ref.
Rice	India, Pakistan	28	—	—	—	—	—	—	—	—	12
Rice	Europe	25	—	—	—	—	—	—	—	—	12
Rice	Different		1	0.3	—	—	3	0.03	3	—	13
Onion conventional	Denmark		0.02 ± 0.02 0.0 – 0.1	0.003 ± 0.01 0.0 – 0.03	0.004 ± 0.004 0.0 – 0.02	0.001 ± 0.001 0.0 – 0.005	0.1 ± 0.1 0.03 – 0.5	0.0005 ± 0.0004 0.0 – 0.002	0.005 ± 0.005 0.000002 – 0.03	0.0005 ± 0.001 0.0 – 0.01	20
Onion organic	Denmark		0.01 ± 0.01 0.0 – 0.1	0.002 ± 0.004 0.0 – 0.01	0.004 ± 0.005 0.0 – 0.03	0.0001 ± 0.001 0.0 – 0.002	0.1 ± 0.2 0.03 – 1	0.0004 ± 0.001 0.0 – 0.003	0.002 ± 0.002 0.0 – 0.01	0.0002 ± 0.001 0.0 – 0.01	20
Peas conventional	Denmark		0.01 ± 0.01 0.002 – 0.05	0.003 ± 0.005 0.0 – 0.02	0.003 ± 0.003 0.0 – 0.01	0.0003 ± 0.001 0.0 – 0.002	0.03 ± 0.03 0.001 – 0.1	0.0003 ± 0.0002 0.0 – 0.001	0.005 ± 0.004 0.0 – 0.02	0.0001 ± 0.001 0.0 – 0.003	20
Peas organic	Denmark		—	—	0.002 ± 0.002 0.0 – 0.01	0.0001 ± 0.001 0.0 – 0.002	0.1 ± 0.1 0.003 – 0.5	0.0003 ± 0.0003 0.0 – 0.001	0.005 ± 0.01 0.000005 – 0.1	0.0001 ± 0.001 0.0 – 0.003	20

Beer[d]	Poland	35	—	—	—	—	0.01 0.002 – 0.03	—	0.03 0.004 – 0.1	—	91
Wine[d]	Italy		0.001 – 0.1	0.001 – 0.01	0.001 – 0.01	0.001 – 0.01	0.001 – 0.01	0.001 – 0.01	0.001 – 0.01	0.001–0.01	53

Concentrations of lanthanum (La), neodymium (Nd), praseodymium (Pr), holmium (Ho), thulium (Tm), erbium (Er), europium (Eu), and gadolinium (Gd) in grain products were very low, ranging from < 0.00001 to 0.001 mg/100 g. Concentration of these elements and those of dysprosium (Dy), lutetium (Lu), samarium (Sm), terbium (Tb), and ytterbium (Yb) in vegetables varied between values < 0.0000002 and 0.0004 mg/100 g, with the highest values for europium (Table 7.26).

REFERENCES

1. Reilly, C., *Metal Contamination of Food: Its Significance for Food Quality and Human Health*, Blackwell Science, Oxford, 2002.
2. Bennet-Chambers, M., Davies, P., and Knott, B., Cadmium in aquatic ecosystems in Western Australia; a legacy of nutrient-deficient soils, *J. Environ. Manage.*, 57, 283, 1999.
3. Smart, G.A. and Sherlock, J.C., Chromium in foods and diets, *Food Addit. Contam.*, 2, 139, 1985.
4. Borigiato, E.V.M. and Martinez, F.E., Iron nutritional status is improved in Brazilian preterm infants fed food cooked in iron pots, *J. Nutr.*, 128, 855, 1998.
5. Rojas, E. et al., Are metals dietary carcinogens?, *Mutation Res.*, 443, 157, 1999.
6. Coultate, T.P., *Food — The Chemistry of Its Components*, Royal Society of Chemistry, Cambridge, MA, 2002.
7. Ysart, G. et al., Dietary exposure estimates of 30 elements from the U.K. total diet study, *Food Addit. Contam.*, 16, 391, 1999.
8. Isserliyska, D., Karadjov, G., and Agelov, A., Mineral composition of Bulgarian wheat bread, *Eur. Food Res. Technol.*, 213, 244, 2001.
9. Capar, G.S. and Cunningham, W.C., Element and radionuclide concentrations in food: FDA total diet study 1991–1996, *J. AOAC Int.*, 83, 157, 2000.
10. Skibniewska, K.A. et al., Influence of starter culture and complex dough improver on in vitro digestibility of some minerals from bread, *Current Trends in Commodity Science, II*, Poznań 2002.
11. Ekholm, P. et al., Effects of natural chelating agents on the solubility of some physiologically important mineral elements in oat bran and oat flakes, *Cereal Chem.*, 77, 562, 2000.
12. Kelly, S. et al., The application of isotopic and elemental analysis to determine the geographical origin of premium long grain rice, *Eur. Food Res. Technol.*, 214, 72, 2002.
13. Al-Dayel Omar, A.F., Al-Kahtani Saad, A., and Hefne Jameel, A., Quantification of trace elements in rice by ICP-MS, *Asian J. Spectr.*, 6, 23, 2002.
14. Dębski, B. and Gralak, M.A., Komosa ryżowa-charakterystyka i wartość dietetyczna, *Żyw. Człow. Metab.*, XXVIII, 360, 2001.
15. Ranhotra, G.S. et al., Nutrient composition of spelt wheat, *J. Food Compos. Anal.*, 9, 81, 1996.
16. Kawashima, L.M. and Valente Soares, L.M. Mineral profile of raw and cooked leafy vegetables consumed in Southern Brazil, *J. Food Compos. Anal.*, 16, 605, 2003.
17. Mohamed, A.E., Rashed, M.N., and Mofty, A., Assessment of essential and toxic elements in some kinds of vegetables, *Ecotoxicol. Environ. Saf.*, 55, 251, 2003.
18. Rubio, C. et al., Mineral composition of the red and green pepper (*Capsicum annuum*) from Tenerife Island, *Eur. Food Res. Technol.*, 214, 501, 2002.

19. Martín-Belloso, O. and Llanos-Barriobero, E., Proximate composition, minerals and vitamins in selected canned vegetables, *Eur. Food Res. Technol.*, 212, 182, 2001.
20. Gundersen, V. et al., Comparative investigation of concentrations of major and trace elements in organic and conventional Danish agricultural crops. 1. Onions (*Allium cepa* Hysam) and Peas (*Pisum sativum* Ping Pong), *J. Agric. Food Chem.*, 48, 6094, 2000.
21. Padín, P.M. et al., Characterization of Galician (N.W. Spain) quality brand potatoes: a comparison study of several pattern recognition techniques, *Analyst*, 126, 97, 2001.
22. Anderson, K.A. et al., Determining the geographic origin of potatoes with trace metal analysis using statistical and neural network classifiers, *J. Agric. Food Chem.*, 47, 1568, 1999.
23. Gundersen, V., McCall, D., and Bechmann, I.E., Comparison of major and trace element concentrations in Danish greenhouse tomatoes (*Lycopersicon esculentum* Cv. Aromata F1) cultivated in different substrates, *J. Agric. Food Chem.*, 49, 3808, 2001.
24. Malinowska, E., Szefer, P., and Falandysz, J., Metals bioaccumulation by bay bolete, *Xerocomus badius*, from selected sites in Poland, *Food Chem.*, 84, 405, 2004.
25. Hardisson, A. et al., Mineral composition of the banana (*Musa acuminata*) from the island of Tenerife, *Food Chem.*, 73, 153, 2001.
26. Hakala, M. et al., Effects of varieties and cultivation conditions on the composition of strawberries, *J. Food Compos. Anal.*, 16, 67, 2003.
27. Moreda-Piñeiro, A., Fisher, A., and Hill, S. J., The classification of tea according to region of origin using pattern recognition techniques and trace metal data, *J. Food Compos. Anal.*, 16, 195, 2003.
28. Kumar, A., Availability of essential elements in Indian and U.S. tea brands, *Food Chem.*, 89, 441, 2005.
29. Xie, M. et al., Multielement analysis of Chinese tea (*Camellia sinensis*) by total-reflection X-ray fluorescence, *Z. Lebensm. Unters. Forsch. A*, 207, 31, 1998.
30. Blázovics, A. et al., In vitro analysis of the properties of Beiqishen tea, *Nutrition*, 19, 869, 2003.
31. Costa, L.M., Gouveia, S.T., and Nóbrega, J.A., Comparison of heating extraction procedures for Al, Ca, Mg, and Mn in tea samples, *Anal. Sci.*, 18, 313, 2002.
32. Malinowska, E. et al., Zawartość pierwiastków chemicznych w herbatach czerwonych oraz ocena procesu ługowania z liści do naparu, *Bromat. Chem. Toks.*, Supl., 395, 2003.
33. Fernández, P.L. et al., Multi-element analysis of tea beverages by inductively coupled plasma atomic emission spectrometry, *Food Chem.*, 76, 483, 2002.
34. dos Santos, E.J. and de Oliveira, E., Determination of mineral nutrients and toxic elements in Brazilian soluble coffee by ICP-AES, *J. Food Compos. Anal.*, 14, 523, 2001.
35. Anderson, K.A. and Smith, B.W., Chemical profiling to differentiate geographic growing origins of coffee, *J. Agric. Food Chem.*, 50, 2068, 2002.
36. Martín, M.J., Pablos, F., and González, A.G., Characterization of arabica and robusta roasted coffee varieties and mixture resolution according to their metal content, *Food Chem.*, 66, 365, 1999.
37. Mohamed, A.E., Environmental variations of trace element concentrations in Egyptian cane sugar and soil samples (Edfu factories), *Food Chem.* 65, 503, 1999.
38. Awadallah, R.M., Ismail, S.S., and Mohamed A.E., Application of multi-element clustering techniques of five Egyptian industrial sugar products, *J. Radioanal. Nucl. Chem.*, 196, 377, 1995.

39. Plessi, M. et al., Dietary fiber and some elements in nuts and wheat brans, *J. Food Compos. Anal.*, 12, 91, 1999.
40. Açkurt, F. et al., Effects of geographical origin and variety on vitamin and mineral composition of hazelnut (*Corylus avellana* L.) varieties cultivated in Turkey, *Food Chem.*, 65, 309, 1999.
41. Ozdemir, F. and Akinci, I., Physical and nutritional properties of four major commercial Turkish hazelnut varieties, *J. Food Eng.*, 63, 341, 2004.
42. Özdemir, M. et al., Evaluation of new Turkish hybrid hazelnut (*Corylus avellana* L.) varieties: fatty acid composition, α-tocopherol content, mineral composition and stability, *Food Chem.*, 73, 411, 2001.
43. Almazan, A.M. and Begum F., Nutrients and antinutrients in peanut greens, *J. Food Compos. Anal.*, 9, 375, 1996.
44. Stintzing, F.C., Schieber, A., and Carle, R., Phytochemical and nutritional significance of cactus pear, *Eur. Food Res. Technol.*, 212, 396, 2001.
45. Simpkins, W.A. et al., Trace elements in Australian orange juice and other products, *Food Chem.*, 71, 423, 2000.
46. Stuckel, J.G. and Low, N.H., The chemical composition of 80 pure maple syrup samples produced in North America, *Food Res. Int.*, 29, 373, 1996.
47. Fili, S.P., Oliveira, E., and Oliveira, P.V., On line digestion in a focused microwave-assisted oven for elements determination in orange juice by inductively coupled plasma optical emission spectrometry, *J. Braz. Chem. Soc.*, 14, 435, 2003.
48. Cámara, M., Díez, C., and Torija, E., Chemical characterization of pineapple juices and nectars: principal component analysis, *Food Chem.*, 54, 93, 1995.
49. Alcázar, A. et al., Multivariate characterization of beers according to their mineral content, *Talanta*, 57, 45, 2002.
50. Adam, T., Duthie, E., and Feldmann, J., Investigations into the use of copper and other metals as indicators for the authenticity of Scotch whiskies, *J. Inst. Brew.*, 108, 459, 2002.
51. Anjos, M.J. et al., Trace elements determination in red and white wines using total-reflection x-ray fluorescence, *Spectrochim. Acta Part B*, 58, 2227, 2003.
52. Conde, J.E. et al., Characterization of bottled wines from the Tenerife island (Spain) by their metal ion concentration, *Ital. J. Food Sci.*, 14, 375, 2002.
53. Marengo, E. and Aceto, M., Statistical investigation of the differences in the distribution of metals in Nebbiolo-based wines, *Food Chem.*, 81, 621, 2003.
54. Offenbachr, E.G. and Pi-Sunyer, F.X., Temperature and pH effects on the release of chromium from stainless steel into water and fruit juices, *J. Adv. Food Chem.*, 31, 89, 1983.
55. Saner, G., The metabolic significance of dietary chromium, *Nutr. Int.*, 2, 213, 1986.
56. Booth, C.K., Reilly, C., and Farmakalidis, E., Mineral composition of Australian ready-to-eat breakfast cereals, *J. Food Compos. Anal.*, 9, 135, 1996.
57. Crosby, N.T., Determination of heavy metals in food, *Proc. Institute of Food Science and Technology*, 10, 65, 1977.
58. Wenlock, R.W., Buss, D.H., and Dixon, E.J., Trace nutrients. 2. Manganese in British foods, *Br. J. Nutr.*, 41, 253, 1979.
59. Coughlan, M.P., The role of molybdenum in human biology, *J. Inherit. Metab. Dis.*, 6, S7, 1983.
60. Reilly, C., *The Nutritional Trace Metals*, Blackwell Publishing, Oxford, 2004.
61. Smart, G.A. and Sherlock, J. C., Nickel in foods and diets, *Food Addit. Contam.*, 4, 61, 1987.

62. Szefer, P., *Metals, Metalloids and Radionuclides in the Baltic Sea Ecosystem*, Elsevier Science B.V., Amsterdam, p. 764.
63. Capar, S.G. and Szefer, P., Determination and speciation of trace elements in foods, in *Methods of Analysis of Food Components and Additives*, Otles, S., Ed., CRC Press, Boca Raton, FL, 2005, chap. 6.
64. Kerdel-Vergas, F., The depilatory and cytotoxic action of 'Coco de Mono' (*Lecythis ollaria*) and its relation to chronic selenosis, *Econ. Bot.*, 20, 187, 1966.
65. Falandysz, J. and Kotecka W., Stężenia metali w wybranych produktach spożywczych Trójmiasta, *Bromat. Chem. Toks.*, 26, 143, 1993.
66. Díaz-Alarcón, J.P., Navarro-Alarcón, M., and López-García de la Serrana H., Determination of selenium in cereals, legumes and dry fruits from southeastern Spain for calculation of daily dietary intake, *Sci. Total Environ.*, 184, 183, 1996.
67. Mateos, C.J., Aguilar, M.V., and Martínez-Para, M.C., Determination of the chromium content in commercial breakfast cereals in Spain, *J. Agric. Food Chem.*, 51, 401, 2002.
68. Nriagu, J.O. and Lin, T.-S., Trace metals in wild rice sold in the United States, *Sci. Total Environ.*, 172, 223, 1995.
69. Kennedy, G. and Burlingame, B., Analysis of food composition data on rice from a plant genetic resources perspective, *Food Chem.*, 80, 589, 2003.
70. González, M., Gallego, M., and Valcárcel, M., Determination of nickel, chromium and cobalt in wheat flour using slurry sampling electrothermal atomic absorption spectrometry, *Talanta*, 48, 1051, 1999.
71. Bonafaccia, G. et al., Trace elements in flour and bran from common and tartary buckwheat, *Food Chem.*, 83, 1, 2003.
72. Jorhem, L., Sundström, B., and Engman, J., Cadmium and other metals in Swedish wheat and rye flours: longitudinal study, 1983–1997, *J. AOAC Int.*, 84, 1984, 2001.
73. Zieliński, H., Kozłowska, H., and Lewczuk, B., Bioactive compounds in the cereal grains before and after hydrothermal processing, *Innov. Food Sci. Emerg. Technol.*, 2, 159, 2001.
74. Zhang, Z.-W. et al., Lead and cadmium contents in cereals and pulses in north-eastern China, *Sci. Total Environ.*, 220, 137, 1998.
75. Malinowska, E. and Szefer, P., Distribution of some chemical elements in certain grains, seeds and cereal products, in *Proc. 2nd International IUPAC Symposium on Trace Elements in Food*, Kortsen, B., Bickel, M., and Grobecker, K.-H., Eds., European Commission, Directorate-General Joint Research Centre, Brussels, Belgium, 2004, p. 96.
76. Bellés, M. et al.., Reduction of lead concentrations in vegetables grown in Tarragona province, Spain, as a consequence of reduction of lead in gasoline, *Environ. Int.*, 21, 821, 1995.
77. Stalikas, C.D., Mantalovas, A.Ch., and Pilidis, G.A., Multielement concentrations in vegetable species grown in two typical agricultural areas of Greece, *Sci. Total Environ.*, 206, 17, 1997.
78. Cabrera, C. et al., Mineral content in legumes and nuts: contribution to the Spanish dietary intake, *Sci. Total Environ.*, 308, 1, 2003.
79. Anzano, J.M. et al., Zinc and manganese analysis in maize by microwave oven digestion and flame atomic absorption spectrometry, *J. Food Compos. Anal.*, 13, 837, 2000.
80. Tüzen, M., Determination of heavy metals in soil, mushroom and plant samples by atomic absorption spectrometry, *Microchem. J.*, 74, 289, 2003.

81. Michalak, A. and Buliński, R., Badania zawartości niektórych pierwiastków śladowych w produktach spożywczych krajowego pochodzenia. Cz. XVII. Zawartość wybranych pierwastków śladowych w niektórych mrożonkach owocowych, *Bromat. Chem. Toks.*, 28, 29, 1995.
82. Górecka, D. et al., Ocena jakości wybranych gatunków herbat różnego pochodzenia, *Bromat. Chem. Toks.*, XXXVII, 145, 2004.
83. Grembecka M. et al., Assessment of mineral composition of market coffee and its infusions, in *Proc. 2nd International IUPAC Symposium on Trace Elements in Food*, Kortsen, B., Bickel, M., and Grobecker, K.-H., Eds., European Commission, Directorate-General Joint Research Centre, Brussels, Belgium, 2004, p. 94.
84. Martín, M.J., Pablos, F., and González, A.G. Characterization of green coffee varieties according to their metal content, *Anal. Chim. Acta*, 358, 177, 1998.
85. Kannamkumarath, S.S. et al., HPLC–ICP–MS determination of selenium distribution and speciation in different types of nut, *Anal. Bioanal. Chem.*, 373, 454, 2002.
86. Grembecka, M., unpublished data, 2006.
87. Parcerisa, J. et al., Influence of variety and geographical origin on the lipid fraction of hazelnuts (*Corylus avellana* L.) from Spain: (III) Oil stability, tocopherol content and some mineral contents (Mn, Fe, Cu), *Food Chem.*, 53, 71, 1995.
88. Anzano, J.M. and Gónzalez, P., Determination of iron and copper in peanuts by flame atomic absorption spectrometry using acid digestion, *Microchem. J.*, 64, 141, 2000.
89. Kreft, I., Stibilj, V., and Trkov, Z., Iodine and selenium contents in pumpkin (*Cucubita pepo* L.) oil and oil-cake, *Eur. Food Res. Technol.*, 215, 279, 2002.
90. Dugo, G. et al., Determination of Cd(II), Cu(II), Pb(II) and Zn(II) content in commercial vegetable oils using derivative potentiometric stripping analysis, *Food Chem.*, 87, 639, 2004.
91. Wyrzykowska, B. et al., Application of ICP sector field MS and principal component analysis for studying interdependences among 23 trace elements in Polish beers, *J. Agric. Food Chem.*, 49, 3425, 2001.
92. Stijve, T. and Bourqui, B., Arsenic in edible mushrooms, *Deuts. Lebensm.-Runds.*, 87, 307, 1991.
93. Woolson, E.A., Arsenic in cotton seed byproducts, *J. Agric. Food Chem.*, 23, 677, 1975.
94. Rainey, C.J. et al., Daily boron intake from the American diet, *J. Am. Diet. Assoc.*, 99, 335, 1999.
95. Murray, F.J., A human health risk assessment of boron (boric acid and borax) in drinking water, *Regul. Toxicol. Pharmacol.*, 22, 221, 1995.
96. Schroeder, H.A. and Nason, A.P., Trace element analysis in clinical chemistry, *Clin. Chem.*, 17, 461, 1971.
97. Badmaev, V., Prakash, S., and Majeed, M., Vanadium: a review of its potential role in the fight against diabetes, *J. Altern. Compl. Med.*, 5, 273, 1999.
98. Hight, S.C. et al., Analysis of dietary supplements for nutritional, toxic and other elements, *J. Food Compos. Anal.*, 6, 121, 1993.
99. Anke, M. et al., Vanadium — an essential element for animals and humans?, in *Trace Elements in Man and Animals*, Roussel, A.M., Anderson, R.A., and Favier, A.E., Eds., Kluwer, New York, 2000, p. 221.
100. Myron, D.R., Givand, S.H., and Nielsen, F.H., Vanadium content of selected foods as determined by flameless atomic absorption spectroscopy, *J. Agric. Food Chem.*, 25, 297, 1977.
101. Koo, W.W.K., Kaplan, L.A., and Krug-Wispe, S.K., Aluminum contamination of infant formulas, *J. Parenter. Enteral. Nutr.*, 12, 170, 1988.

102. Bouglé, D. et al., Concentrations en aluminum des formules pour prématurés, *Arch. Pédiatr.*, 46, 768, 1989.
103. Baxter, M.J., Burrell, J.A., and Massey, R. C., The aluminum content of infant formula and tea, *Food Addit. Contam.*, 7, 101, 1990.
104. Biégo, G.H. et al., Determination of dietary tin intake in an adult French citizen, *Arch. Environ. Contam. Toxicol.*, 36, 227, 1999.
105. Board, P.W., The chemistry of nitrate-induced corrosion of tinplate, *Food Technol. Aust.*, 25, 16, 1973.
106. González, M.M., Gallego, M., and Valcárcel, M., Determination of arsenic in wheat flour by electrothermalatomic absorption spectrometry using a continuous precipitation-dissolution flow system, *Talanta*, 55, 135, 2001.
107. Alam, M.G.M., Snow, E.T., and Tanaka, A., Arsenic and heavy metal contamination of vegetables grown in Samta village, Bangladesh, *Sci. Total Environ.*, 308, 83, 2003.
108. Müller, M., Anke, M., and Illing-Günther, H., Availability of aluminium from tea and coffee, *Z. Lebensm. Unters. Forsch. A*, 205, 170, 1997.
109. Şimşek, A. et al., Determination of boron in hazelnut (*Corylus avellana* L.) varieties by inductively coupled plasma optical emission spectrometry and spectrophotometry, *Food Chem.*, 83, 293, 2003.
110. MacIntosh, D.L. et al., Dietary exposure to selected metals and pesticides, *Environ. Health Perspect.*, 104, 202, 1996.
111. Branca, P., Uptake of metal by canned food with length of storage, *Bull. Chim. Lab.*, 33, 495, 1982.
112. Shaper, A.G. et al., Effects of alcohol and smoking on blood lead in middle-age British men, *Br. Med. J.*, 284, 289, 1982.
113. Bourgoin, B.P. et al., Lead content in 70 brands of dietary supplements, *Am. J. Public Health*, 83, 1155, 1993.
114. Capon, C.J., Mercury and selenium content and chemical form in vegetable crops grown on sludge-amended soil, *Arch. Environ. Contam. Toxicol.*, 10, 673, 1981.
115. Bakir, F., Methyl mercury poisoning in Iraq, *Science*, 181, 230, 1973.
116. Förstner, U. and Wittmann, G.T.W., *Metal Pollution in the Aquatic Environment*, 2nd ed., Springer-Verlag, Berlin, 1983.
117. Phillips, D.J.H. and Rainbow, P.S., *Biomonitoring of Trace Aquatic Contaminants*, Elsevier Science, London, 1993.
118. Shimbo, S. et al., Cadmium and lead contents in rice and other cereal products in Japan in 1998–2000, *Sci. Total Environ.*, 281, 165, 2001.
119. Fang, G. et al., Spectrophotometric determination of lead in vegetables with dibromo-*p*-methyl-carboxysulfonazo, *Talanta*, 57, 1155, 2002.
120. Svoboda, L. et al., Leaching of cadmium, lead and mercury from fresh and differently preserved edible mushroom, *Xerocomus badius*, during soaking and boiling, *Food Chem.* 79, 41, 2002.
121. Svoboda, L., Zimmermannová, K., and Kalač, P., Concentrations of mercury, cadmium, lead and copper in fruiting bodies of edible mushrooms in an emission area of a copper smelter and a mercury smelter, *Sci. Total Environ.*, 246, 61, 2000.
122. Svoboda, L. and Kalač, P., Contamination of two edible *Agaricus* spp. mushrooms growing in a town with cadmium, lead, and mercury, *Bull. Environ. Contam. Toxicol.* 71, 123, 2003.
123. Segura-Muñoz, S.I. et al., Metal levels in sugar cane (*Saccharum* spp.) samples from an area under the influence of a municipal landfill and a medical waste treatment system in Brazil, *Environ. Int.*, 32, 52, 2006.

124. Lisk, D.J. et al., Absorption and excretion of selenium and barium from consumption of Brazil nuts, *Nutr. Rep. Int.*, 38, 183, 1988.
125. Mounicou, S. et al., Concentrations and bioavailability of cadmium and lead in cocoa powder and related products, *Food Addit. Contam.*, 20, 343, 2003.
126. Zhoud, X. and Liu, D.-N., Chronic thallium poisoning in a rural area of Ghizhou Province, China, *J. Environ. Health*, 48, 14, 1985.
127. Hislop, J.S. et al., An Assessment of Heavy Metal Pollution of Vegetables Grown near a Secondary Lead Smelter, Rep. No. AERE 2383, Atomic Energy Research Establishment, Harwell.
128. Richter, U., Thallium in food, *Ernahrungs-Umschau*, 46, 360, 1999.
129. McMahon, M., Regan, F., and Huges, H., The determination of total germanium in real food samples including Chinese herbal remedies using graphite furnace atomic absorption spectroscopy, *Food Chem.*, 97, 411, 2006.

8 Elemental Content of Wines

Smaragdi M. Galani-Nikolakaki and
Nikolaos G. Kallithrakas-Kontos

CONTENTS

8.1 INTRODUCTION

Wine is an ethanol-aqueous solution containing several organic and inorganic substances.[1] Although its elemental content can be a source of essential minerals and trace elements to human beings, there is also a universal concern for the heavy metals and other trace elements present in musts and wines. The elemental content of wines depends upon factors such as the type of ground and underground soil of the vineyard, the climatic conditions of the geographical region (temperature, sun exposure, proximity to sea, and amount of rainfall), the proximity of the vineyard to areas of high traffic and to areas overburdened with industrial activities, the agrochemical treatment of the vine plant, the vinification methods, the wine-processing equipment and, finally, the type of storage container, including the type of the cork used for bottling. Eschnauer and Neeb[2] have referred to about 50 inorganic species that have been identified in wine, including all nonmetals, the alkali metals, the alkaline earth metals, the metalloids boron, silicon and arsenic, many heavy metals, rare earths, and naturally-occurring radioactive elements. Trace and ultratrace elements are determined in wine in an effort to investigate the sources of contamination and possible methods of adulteration.

For decades, the adulteration of wines was easily accomplished due to its worldwide availability and chemical consistency (high alcohol content and low pH). In recent years, there has been an intense and systematic effort made to apply novel techniques for accurate determination of wine authenticity and detection of possible adulteration. For this purpose, certain elements in wine are being studied, together with other physicochemical parameters.[3–5] Several investigators have used specific elements such as Si, Mg, Ti, Mn, Mo, U, V, Al, Sb, Co, Zn, Sr, Rb, and the rare earth elements (REE) in order to discriminate between wine regions and wine types.[4,6–11] Although several countries have individually established maximum permissible levels for several elements, the International Office of Vine and Wine (OIV)[12] has established lower limits only for the following elements: As 0.2 mg/l, Cd 0.01 mg/l, Cu 1 mg/l, Pb 0.2 mg/l, and Zn 5 mg/l.

This chapter describes briefly the sources of the most important trace elements in wine. We also examine the role that each element plays in the wine-making and aging processes. Table 8.1 gives the concentration range in mg/l of 20 elements found in wine and lists those that are discussed in the text.

8.2 ALUMINUM

Aluminum toxicity and its effect on the stability of wines are two problems resulting from excessive aluminum. The contamination of wine with aluminum may result in spoilage due to haze formation and creation of an undesirable and unpleasant metallic taste.[13] The presence of aluminum can result in the formation of hydrogen sulfide by the reduction of sulfur dioxide and reduction of the color of white wines to an almost watery white condition.[13,14] Average concentrations usually are from 0.50 to 0.90 mg/l,[2] although in many countries the reported values have been higher. Stability problems have risen with concentrations greater than 10 mg/l, whereas the recommended upper limit is 3 mg/l.[15]

Grapes usually contain less than 1 mg/l,[13,16] and 90% of aluminum in grape juice is removed by yeast during fermentation.[16] Musts and wines are contaminated by contact with equipment made of aluminum or aluminum alloys and materials added to them.[2] The acidity of wine (low pH) results in the leaching of the inert aluminum oxide film that is formed on the outer side of all aluminum surfaces. This process is followed by the oxidation and corrosion of the aluminum, with the rate of aluminum dissolution in wine increasing, as the pH decreases.[15] Therefore, all aluminum surfaces that come in contact with wines must be treated with a permanent and durable inert coating.[13] McKinnon et al.[15] have reported that the addition of

TABLE 8.1
Range of Concentration in mg/l for the Analysis of 20 Elements in Wines

Element	Range	References	Element	Range	References
Al	0.2–9.5	8,9,11,67,68,70	Li	0.008–0.45	8,9,10,67,71,72,73,74,75
As	0–0.012	8,9,67,70,74,75	Mg	0.055–718	8,9,10,11,60,73,75
B	2–23.3	68,73,75	Mn	0–10	8,9,10,11,60,67,68,70,71,72,73,74
Ca	30–200	8,9,10,11,60,73	Na	5–213.8	9,10,11,60,71,72,73
Cd	0–0.066	8,9,67,69,70,75	Ni	0–0.21	8,9,67,69,71,72,74,75
Co	0.003–0.30	8,9,67,69,71,72,74,75	Pb	0.005–0.42	8,9,67,69,70
Cr	0–1.6	67,70,75	Rb	0.21–2.9	8,9,10,11,60,67,71,72,73,74,75
Cu	0.001–1.8	8,9,10,11,47,60,67,69,70,74,75	Sr	0–2.4	8,9,10,11,67,73,74,75
Fe	0.4–27.8	8,9,10,11,47,60,67,68,69,70,71,72,75	V	0–0.097	8,9,67,73,74,75
K	202–1860	10,11,60,71,72,73	Zn	0.10–31	8,9,10,11,47,67,68,69,70,71,72,73,74,75

Source: Data were obtained from recently published work.

bentonite in wines as a fining agent and the use of different filter aids can increase the aluminum content by at least twofold. The fining of white wines is mainly done with bentonite, so these wines may have higher aluminum values than red wines of the same region and producer. Interesse et al.[17] have also shown that an increase of aluminum concentration may be due to the leaching of the metal from glass containers after a period of 9 months. The replacement of lead caps on wine bottles with aluminum ones has resulted in the addition of aluminum in wine, due to the aluminum migration from caps to the wine.[18]

The major problem of the aluminum haze appearance, which usually ranges from faint opalescence to some type of precipitate, depends on the amount of the metal present in the particular wine. Maximum haziness occurs at pH 3.8.[13] Aluminum haze can be distinguished from iron and copper hazes because it dissolves on addition of hydrochloric acid but does not react with H_2O_2 or H_2S (ibid.).

8.3 ARSENIC

The use of arsenic compounds as pesticides and insecticides goes back to antiquity, and there are ancient references to arsenic compounds used for vineyard spraying.[19] Vines absorb arsenic from the soil in its most stable form of As(V) and about 60% of this amount is reduced during fermentation to the highly toxic trivalent state – As(III).[20] In wines, methylarsenic acid (MAA) and dimethylarsinic acid (DMAA) are usually in concentrations below detection limits of most analytical instruments. The reduction of As(V) is followed by a 10 to 30% precipitation of As(III) as As_2S_3, and the remaining amount stays dissolved in the wine.[21]

When vineyards are treated with arsenic pesticides, the corresponding wines appear to have relatively high arsenic concentrations (up to 1 mg/l).[20,22,23] Other factors that may influence the arsenic content of wines include soil composition and the possible corrosion of metallic caps.[21] The use of arsenic pesticides has been forbidden in all wine producing countries, but even after the prohibition of arsenic pesticides, the uptake of large amounts of arsenic in grapes from the residue in the soil remained a serious problem for a long time.

8.4 BORON

Boron can enter wines through two routes: (1) from the soil and (2) by application of boron compounds to vineyards, musts, and wines.[24] Boron fertilization delays grape maturity.[14] Although in some countries boron addition is illegal under wine legislation, boron compounds are added to wines in order to prevent tartrate precipitation or are used as antiseptics. The compounds used for treatment of wines as antiseptics include sodium borate, sodium borotartrate, or fluroborate.[25]

The OIV[12]-recommended upper level of boron is 80 mg/l. If the concentrations found (expressed as boric acid) exceed 100 mg/l, the wine samples must be checked further for possible illegal addition of boron or boron compounds.[25] The estimated intake of boron through drinks and food is around 40 mg/d, and for wine consumers, wine can be one of the main sources of intake.[25,26] It has been noted that the boron content increases from musts to fresh wine and then to aged wine. This tendency can be explained by the fact that high levels of boron can be found in grape skins, pips, and stalks and are extracted during the maceration period.[25]

8.5 CALCIUM

Calcium is a natural constituent of musts and wines, necessary for the normal course of alcoholic fermentation. The well-drained and therefore warm, calcaneous soils produce the best wines because warmth helps vine growth and grape ripening. Also, the grape vines cultivated in high calcium soils produce grapes with a higher potassium and magnesium concentration than those produced in neutral or slightly acidic soils.[14]

During fermentation, the Mg/Ca ratio increases as the calcium content decreases.[27] This decrease continues during the storage and the stabilization stages.[28] Elevated calcium levels influence the performance of wine yeasts, resulting in fermentation suppression. The problem can be solved by maintaining a high Mg/Ca ratio.[29]

Calcium sources for musts and wines include the soil, the treatment of the musts with calcium sulfate or calcium carbonate, filtering aids and pads, fining agents such as calcium bentonite, concrete tanks for storage, and ion exchange treatment.[30] When concrete tanks are used, it seems that the high acid wines extract more calcium than the low acid ones, especially when the tank is new.

When calcium reacts with the tartaric acid present in wine, calcium tartrate precipitates through a process that usually lasts during the first and the second winter of wine aging.[28] The precipitation reaction occurs best at pH 3.7. It is a slow process and occurs often after bottling.[27] The presence of calcium tartrate may not affect the real quality of wine, but its presence in bottled wines has a negative effect on prospective consumers. Although the addition of calcium carbonate reduces the excess acidity and stabilizes the wine, the calcium tartrate thus formed cannot be effectively removed by chilling as potassium tartrate can be.[14] McKinnon et al.[31] have shown that high concentrations of polyuronic acids present in wines can inhibit the calcium tartrate crystal formation.

8.6 CADMIUM

Although only traces of cadmium have been found in grapes, musts and wines can be contaminated by agrochemical products such as pesticides and fungicides from vineyards located near factory discharges, from atmospheric pollution, and from winery processing equipment, if it is made from alloys containing cadmium. Painted surfaces or metal joints of equipment are another source of contamination.[32] The natural cadmium content of musts is < 5 mg/l, and about two-thirds of this quantity precipitates out during fermentation.[33] During fermentation, extraction of cadmium from the must takes place and chemical factors such as alcohol content and acidity of the must help this type of extraction.[32]

8.7 CHROMIUM, NICKEL, COBALT, AND VANADIUM

The possibility of contamination of wine with toxic heavy metals has led to many attempts to improved wine-making technology. The increased use of stainless steel winery equipment and machinery has resulted in increased concentrations of metals such as Cr, Co, Ni, and V.[34] Chromium in wine is only present as the less toxic Cr(III).[35] If any Cr(VI) ions are introduced in wine, they would be immediately reduced to Cr(III).[2] The chromium content of wine also increases during the aging period because of the glass bottle leaching, eventually reaching mg/l levels.[17,35,36] Nickel and cobalt can also originate from the contact of wines with glass bottles if they are made from either nickel oxide or blue glass, which can contain up to 0.02–0.05 wt % cobalt.[21]

Wine is considered to be one of the possible routes that can contribute to the daily chromium intake.[1] Studies performed by Boulet and coworkers on French subjects have estimated a weekly intake of Cr through wine to be around 4 μg/d per person.[37]

8.8 COPPER

Trace amounts of copper in musts and wines play the role of catalyst for yeast metabolism. Copper originates partially from enzymes, especially oxidases, which are found in abundance on fresh grapes.[38] In musts and fresh wines, copper concentration is usually 0.1–0.3 mg/l.[30] For concentrations greater than 0.5 mg/l there is the possibility of haze formation.[39] For concentrations equal to or greater than 1 mg/l the organoleptical properties of wine change, and when copper exceeds 9 mg/l it becomes very toxic for the fermentation process and can either reduce its rate or stop it at all.[40]

The main sources of copper in wines include the winery equipment made of copper alloys,[21] fungicides such as copper sulfate, bordeaux mixture, and zinc thiocarbamate for spraying the vines,[9,41] and the addition of copper sulfate salts during the stage of vinification.[40] Elevated copper concentrations in wines are mainly due to the first or the second pathways.[40,42]

Although most dissolved copper can be converted to insoluble precipitates such as copper sulfide during fermentation, the residual dissolved copper plays an important role in the chemistry of wine. Copper complexes catalyze oxidation reactions more than iron complexes.[14] When the wine has suffered reduction there is mostly Cu^+ in wine leading to the appearance of copper turbidity or "browning."[39] Copper *casse* appears in bottled white wines, when they contain small amounts of dissolved copper together with sulfur dioxide, in the absence of oxygen and ferric ions.[14] This *casse* appears to have a bluish-white glistering hue and may add a bitter taste to the wine.[43] The exposure to light and the high temperatures accelerates the *casse* formation.[40] Copper can be removed from wines together with iron, but if a wine is poor in iron, there is never a complete removal of copper. In this case, the wine is treated with sodium sulfide, and copper sulfide is produced. The advantage of this method is that if there are traces of arsenic in the wine, they are removed as well.

The disadvantage is that it cannot be applied with wood barrels, where oxygen gradually converts cuprous atoms into Cu^{2+}.[39]

Copper *casse* can be prevented with levels of copper < 0.3 mg/l, with the addition of bentonite for protein removal from white wines and with limited SO_2 addition.[40] For many years, Cu (and sometimes Fe^{3+} and Mn^{3+}) reduction in wines was accomplished with the addition of potassium hexacyanoferrate (II). Recently, ion exchange techniques have been used for the removal of these metals from wines but, although the metal reduction is successful, the organoleptic properties may also be affected.[44]

8.9 IRON

The presence of iron is required, among other elements, for the initiation of wine fermentation.[14] Its evaluation in wines is of major importance because of the effect it may have on the stability of the wine, its oxidation, and its aging.[45] The iron content of wines is partly due to natural sources and partly to technological processes.[9] Natural sources include the soil, the percentage of iron in the dust covering the berries prior to vintage, and the atmospheric pollution during the same period. A natural source of iron in the berries is mitochondria, which are composed of iron–porphyrin–protein complexes, and they degrade during wine fermentation, releasing iron into the wine.[38]

A significant amount of the iron content of must is lost during fermentation because it is used by the yeast growth.[30] The percentage of iron lost through this route varies from 25 to 80%, depending on the amount of yeast present during fermentation, on the degree of aeration of the must, on the presence or absence of polyphenolic compounds, and the presence of coloring material. Fresh wines with no contact with equipment made from iron or iron alloys have iron content around 1 to 5 mg/l. For the cases where steel equipment has been used (pumps, tubes, or iron containers), normal iron concentrations vary from 10 to 30 mg/l.[30] An acceptable concentration of iron in wine is considered to be around 4 to 5 mg/l. Concentrations higher than 10 mg/l are more likely to cause cloudiness or color change.[14,39]

The yeast requires about 1 to 5 g/l of trace elements including iron. In such concentrations, iron plays a very important role for metabolism as enzyme initiator, stabilizer, and as necessary constituent of proteins. If the concentrations become a little higher (7 to 10 mg/l), the role of iron changes. It catalyzes oxidation of the wine, taking part in the creation of tannins and phosphoric salts that destroy certain characteristics of the wine such as clarity, flavor, and color.[40]

The oxidation state of iron in wine depends on the redox conditions of the wine. Under reducing conditions, as in the case of bottled wines, iron is mostly present as Fe(II) (80 to 95%), whereas Fe(III) is usually present as complex molecules with organic ligands and in much lower concentrations. On the other hand, the aerated wines exhibit higher concentrations of Fe(III).[39,45]

The technological sources that introduce iron in wines include the corrosion of vinification equipment and steel containers used for the transfer of berries or must.[38]

Excess iron in wine results in the formation of the white precipitate ($FePO_4.2H_2O$) with phosphate ions. Ferric phosphate is a colloidal substance that coagulates under the influence of calcium, potassium ions, and proteins and precipitates when its solubility product is exceeded. This precipitation is influenced by many different factors, besides the iron concentration such as the nature and concentration of the predominant acid, the pH and the oxidation-reduction potential of the system, the concentration of phosphates present, and the kind and amounts of tannins present.[14,39] This problem can be treated with the addition of citric acid that forms a citric complex of Fe(III).[30]

An additional second blue-green to blue-black cloudiness, called *browning*, often appears in red and white wines and is produced when wine comes in contact with oxygen in the air. Browning is due to the contact of the grape skins and must with ferrous iron, which is oxidized by air to the less soluble compounds of iron (III), followed by the reaction of tannic substances.[43] Browning is initiated by oxygen with polyphenols being the oxidizable matter, and the activators of the process are metals such as Fe, Cu, and Mn.[46,47] The method to solve the problem of browning involves the treatment of the wine with potassium hexacynoferrate (II) which eliminates part of the iron present, although there is the danger of production of the highly toxic cyanide, if there is unreacted excess of the reactant.[43,46] Recently, researchers have experimented with ion exchange techniques for the reduction of the excess metal content of wines, but it has been shown that, although the wines treated have demonstrated low susceptibility to browning, they have also undergone changes due to their organoleptic properties.[46]

8.10 LEAD

Wine is one of the main sources for the lead exposure of human beings. Lead poisoning from homemade wine has been reported even during the twenty-first century.[48] It was found that the absorbance of lead from food by adults is around 10% whereas the absorption rate from beer is about 20%. Assuming that the absorbance of lead from wine is the same as from beer, around 20%, we can estimate the amount of lead, that an adult absorbs weekly from consuming wine or beer. The possible increase of lead concentration in the wine or the increase of wine consumption becomes a significant source of lead intake.[49]

Lead should not be found in normal musts and wines, and its existence in either medium indicates one or more of three possible sources of contamination: (1) the soil, (2) the atmosphere, and (3) the technological sources due to the equipment used for the handling of these products during the harvesting, the pressing, the fermentation, and the bottling periods.[50] Such sources include soil and air polluted with lead from cars, lead sprays used in vineyards, unlined cement tanks, lead-based paints, lead-containing metals in several kinds of vinification equipment, lead in filter pads and in bentonite, several fining compounds, lead capsules in wine bottles, and lead-containing crystal decanters and glasses.[14,51]

Vines should not be cultivated in regions rich in lead minerals because it has been found that the corresponding wines contain lead in high concentrations.[51] The regions around industrial installations producing lead batteries or handling lead compounds and alloys should not be used for grapevines.[51] For many years, the use of certain agrochemicals such as insecticide lead arsenate, the fungicide copper sulphate (containing traces of lead), and the fertilizer superphosphate have been major sources of contamination for vines and grapes.[52] The following inferences have been made by Stockley et al:[52]

1. About 40% of the lead available for plant absorption in the top 30 cm of soil is accumulated in the 10 cm surface layer of the ground.
2. The concentration of lead in grape berries is about 400 times less than that of the corresponding soil.
3. The lead content of the grape juice is about 10 times less than that of the whole grape berries.
4. The concentration of lead in wine is approximately 20 times higher than that of the corresponding grape juice.[52]

In grape juice, lead may be distinguished as "external" (if it originates from atmospheric sources) and "internal" (if it comes from soil and groundwater). Grape seeds and the skin contain 65% of the total lead, even though their weight is only the 20% of the total weight. The weight of the pulp constitutes up to 80% of the total weight but it contains only 35% of the lead.[53] From the observations above, one can infer that: (1) the grapes should not be hard pressed because pressing helps the transformation of internal lead into the must, and (2) long maceration at high temperatures causes extraction of lead from skins and seeds and therefore should be avoided.[53] In general, in the same grapevine, red grapes contain more lead than white grapes.[51,54] The end result of the whole fermentation process is that most of the lead precipitates as insoluble PbS, when sulphur dioxide or lead tartrate are added.[54]

Most studies report total concentrations of lead in wine (mostly mineral). Teissedre et al.[55] studied the origin of the organolead compounds in wine and concluded that there was no methylation of mineral lead taking place during the grape fermentation, although there might be a slow degradation of the triethyl-lead present in the wine medium. This change occurs as the fermentation is progressing. No organolead compounds are found in wines coming from regions where lead has been banned from gasoline.[55] If a wine sample contains high levels of organolead compounds, a safe assumption is that the vineyard is located on a road with heavy traffic. On the other hand, the absence of organolead implies that either the wine originates from a country that has banned leaded gasoline or that it is a very recent European wine. "The relative proportion of trimethyl/triethyl lead indicates the year of production."[5] Also graphical representation of the ratios of the four lead isotopes can be used for wine distinction, whereas the isotopic determination of $^{206}Pb/^{204}Pb$ shows a variation depending on the year of production.[5]

Recent studies have also shown that lead is strongly complexed in wines, and therefore its bioavailability is expected to be low.[56] In white wines, residual proteins and procyanidins are the binding agents for lead, whereas for red wines the binding ligands are condensed tannins.[57]

For decades, the traditional sealing of wine bottles included a cork with a metal cap which was usually made out of lead.[58] It has been shown that the white powder found around the bottle mouth and the cork is due to the corrosion of the metallic cap and consists of lead carbonate hydroxide and its hydrated form. Both these compounds can be produced by prolonged contact of the cap with the wine, especially when the bottles are stored on their side. The pH of a wine, as well as the porosity of the cork, are two of the major parameters that influence the degree of the metal cap deterioration.[58] Such capsules were banned by the U.S. Food and Drug Administration (FDA) in 1996, but bottles sealed in this manner may still be around.

8.11 LITHIUM, RUBIDIUM

The lithium and rubidium concentrations in wines do not seem to be influenced by the production process. Their presence is due to the soil composition and the ability of the plant to absorb them. Lithium may be absorbed in wines from prolonged storage in glass bottles.[59] The importance of determining lithium and rubidium in wines has increased, because they can be employed for geographical classification of wines.[5]

8.12 MAGNESIUM

Magnesium is a natural constituent of musts. Magnesium plays a key role during several stages of the glycolytic cycle, together with K, Zn, Co, I, Fe, Ca, Cu, P, and S, over 20 enzymes, and 6 coenzymes. For growth alone, yeasts require Cu, Fe, Mg, K, P, and S, which they find in satisfactory amounts in musts.[14] The concentration of magnesium in wine depends on the grape variety,[60] the rate of pressing, the pH, the time and temperature at which the maceration process takes place, the addition of carbonates as disacidificants, the wine storage in concrete tanks, the relative concentration of alcohols and other constituents, and the use of ion-exchange resins.[14,27,42] The taste and the tartrate stability of wine are also affected by the amount of magnesium present.[27]

8.13 MANGANESE

Manganese is a natural wine constituent. The amount found in wines depends on the corresponding soil composition, the subsequent absorption by the vine, the atmospheric contamination, the use of herbicides, the process of wine making, and the fining agents added.[46] Manganese together with iron is responsible for the browning of white wines. When manganese concentrations are 0.8 mg/l, there is a strong possibility for browning formation to occur if there is also an increase in iron concentration.[46] Cacho et al.[61] have shown that manganese in wine helps the formation of acetaldehyde, whereas the iron present catalyzes the reaction of acetaldehyde with the polyphenolic compounds to form polymers that precipitate out. Benítez et al.[46] have found that in “fino” sherry wine, the two metals act synergistically. Iron acts as catalyst for the browning process in the presence of manganese. Elimination of browning can be achieved if there is a simultaneous elimination of the manganese and iron content.

Manganese is also one of the metals often used to differentiate wines according to their geographical origin.[5,8-10,62]

8.14 PLATINUM

The necessity for platinum analyses in wine became an issue with the widespread introduction of vehicles equipped with catalytic converters. Other anthropogenic pollution sources include fertilizers and therapeutics, active charcoal, and bentonite.[63] Concentrations of platinum found in wines vary from 0.0005 to 0.0024 μg/l.[63]

8.15 POTASSIUM

Potassium is required for yeast growth and fermentation and constitutes about 75% of the total cation content of wines.[14] Its concentration in musts and wines is influenced by soil, climate, grape variety, time of harvest, temperature of fermentation, storage conditions, percentage of alcohol, pH, ion-exchange resins, and fining agents.[30] Grapes grown on alkaline chalky soils contain more K and Mg than those from neutral or slightly acidic soils[14]

The potassium content of wines is increased by pressing, by addition of potassium caseinate used for wine fining,[14] and by addition of potassium metabisulfite or carbonate to the crushed grapes during vinification.[9] Red wines have a higher potassium content than white ones.[14]

Potassium reacts with tartaric acid to form potassium bitartrate, an unstable compound that precipitates under certain conditions during vinification and preservation.[30] Excess potassium ions can be removed almost completely by using ion exchange resins charged with sodium or hydrogen atoms.[14] The potassium content of wine can be used, together with other properties or characteristics, for the determination of the geographical origin and the variety of the wine.[5]

8.16 SODIUM

The interest of enologists in the sodium content of wines started when it was reported that people suffering from hypertension should be on low sodium diets. The sources of sodium in wine can be natural or industrial. Proximity to the sea constitutes a major natural source whereas artificial sources include addition of several sodium compounds in wine and use of ion exchangers.[30] Sodium bisulfite in dilute sulfuric acid is used for keeping the tanks or barrels in good condition, sodium caseinate as fining agent for clarification of wine, sodium carbonate for reducing the natural acidity of wines, and sodium metabisulfite for sterilizing and preserving wines, whereas the sodium salt of ascorbic acid is a sterilizing and preserving agent, which can also be used for inhibition of secondary fermentation and mold growth.[14]

Sodium ion-exchange resins are used for the removal of excess potassium, magnesium, calcium, and tartrate without changing the pH of the wine. When the wine is treated with the hydrogen form of the resin, sodium is also removed, and the wine becomes more acidic. Sometimes, a combination of the two types of resins are used to give the desired result.[14]

8.17 STRONTIUM

Strontium in wines is associated with the vineyard soil composition and used for assessing wine authenticity.[8,10,64] Almeida et al.[64] have shown that the release of strontium into the must from grape skins and seeds results in wines with higher strontium concentrations than the corresponding grape juice. They also observed that the $^{87}Sr/^{86}Sr$ isotope ratio was statistically identical in wines and soil and, hence, can be used as tracer of wine origin.[64]

8.18 TIN

Tin is seldom found in wines, and its presence, when detected, is due to contact of wine with tin utensils or tin-plated equipment or containers. The possibility of tin contamination should be avoided, because its presence at concentrations greater than 1 mg/l results in the formation of tin–albumin haziness. The protein precipitation appears also by heating the tin-containing wine. Tin can also react with sulfites present in wine, resulting in the formation of hydrogen sulfide and free sulfur.[14]

8.19 ZINC

In vines and musts, zinc originates from soil, from fungicides, and insecticides used, such as Bordeaux mixture and zinc thiocarbamates, and from vinification equipment.[9] Low zinc concentrations in wines play a vital role during fermentation, whereas high concentrations influence badly its organoleptic properties.[65] Zinc can create cloudiness if its concentration exceeds the recommended limit of 5 mg/l, established by OIV.[10] The haziness due to excess of zinc can be treated effectively with the careful addition of blue fining (potassium ferrocyanide).[66] Karadjova et al.[47] have found that less than 15% of the zinc present in wines is complexed with polyphenols, whereas more than 60% of the same metal is present as an active labile ion.

REFERENCES

1. Cabrera-Vique, C. et al., Determination and levels of chromium in French wine and grapes by graphite atomic absorption spectrometry, *J. Agric. Food Chem.*, 45 1808, 1997.
2. Eschnauer, H., and Neeb, R., Micro-element analysis in wine and grapes, in *Wine Analysis,* 6, Linskens, H.F. and Jackson, J.F., Eds., Springer-Verlag, Berlin, Heidelberg, New York, London, Paris, Tokyo, 1988, p. 67.
3. Moret, I., Scarpini, G., and Cescon P., Chemometric characterization and classification of five Venetian white wines, *J. Agric. Food Chem.*, 42, 1143, 1994.
4. Almeida, C.M.R. and Vasconcelos, M.T.S.D., Multielement composition of wines and their precursors including provenance soil and their potentialities as fingerprints of wine origin, *J. Agric. Food Chem.*, 51, 4788, 2003.

5. Arvanitoyannis, I.S. et al., Application of quality control methods for assessing wine authenticity: use of multivariate analysis (chemometrics), *Trends Food Sci. Technol.*, 10, 321, 1999.
6. Jakubowski, N., et. al., Analysis of wines by ICP-MS: is the pattern of the rare earth elements a reliable fingerprint for the provenance?, *Fresenius J. Anal. Chem.*, 364, 424, 1999.
7. Barbaste M. et al., Analysis and comparison of SIMCA models for denominations of origin of wines from de Canary Islands (Spain) builds by means of their trace and ultratrace metals content, *Anal. Chim. Acta*, 61, 472, 2002.
8. Taylor, V.F., Longerich, H.P., and Greenough, J.D., Multielement analysis of Canadian wines by inductively coupled mass spectrometry (ICP-MS) and multivariate statistics, *J. Agric. Food Chem.*, 51, 856, 2003.
9. Marengo, E. and Aceto, M., Statistical investigation of the differences in the distribution of metals in Nebbiolo-based wines, *Food Chem.*, 81, 621, 2003.
10. Díaz, C., Application of multivariate analysis and artificial neural networks for the differentiation of red wines from Canary Islands according to the island of origin, *J. Agric. Food Chem.*, 51, 4303, 2003.
11. Frías, S. et al., Classification of commercial wines from the Canary Islands (Spain) by chemometric techniques using metallic contents, *Talanta*, 59, 335, 2003.
12. O.I.V. Office International de la Vigne et du Vin Recueils des Methodes Interntionals d'Analyse des Vins et des Mouts, Paris, 1990.
13. Rankine, B., Aluminum haze in wine, *The Australian Grapegrower and Winemaker,* 18, 1983.
14. Amerine, M.A., Berg, H.W., and Cruess, W.V., *The Technology of Wine Making,* The AVI Publishing Co, INC, Westport, CT, 1972, chaps. 3, 5, 6, 15.
15. McKinnon, A.J., Cattrall, R.W., and Scollary, G.R., Aluminum haze in wine-its measurement and identification of major sources, *Am. J. Enol. Vitic.*, 43, 166, 1992.
16. Rankine, B., Aluminum haze in wine, *Aust. Wine, Brew. Spirit Rev.*, 15, 1962.
17. Interesse, F.S., Lamparelli F., and Allogio V., Mineral contents of some southern Italian wines, *Z. Lebensm. Unters. Forsch.*, 178, 272, 1994.
18. Larroque, M. and Cabanis, J.C., Determination of aluminum in wines by direct graphite furnace atomic absorption spectrometry, *J. AOAC Int.*, 77, 463, 1994.
19. Nriagu, J.O. and Azcue, J.M., Food contamination with arsenic in the environment, in *Food Contamination from Environmental Sources*, Nriagu, J.O. and Simmons, M.S., Eds., John Wiley & Sons, New York, 1990, chap. 5.
20. Crecelius, E.A., Arsenite and arsenate levels in wine, *Bull. Environ. Contam. Toxicol.*, 18, 227, 1977.
21. Eschnauer, H.R. and Stoeppler, M., Wine: an enological specimen bank, in *Hazardous Metals in the Environment,* Stoppler, M., Ed., Elsevier, Amsterdam, London, New York, Tokyo, 1992, chap. 4.
22. Noble, A.C. et al., Trace element analysis of wine by proton-induced x-Ray fluorescence spectrometry, *J. Agric. Food Chem.*, 24, 532, 1976.
23. Handson, D.P., Lead and arsenic levels in wines produced from vineyards where lead arsenate sprays are used for caterpillar control, *J. Sci. Food Agric.*, 35, 215, 1984.
24. Sanz, J. et al., Use of methyl borate generation-flame emission spectrometry combined technique for boron determination in wine, *Analusis*, 18, 279, 1990.
25. Hernández, G.G., de La Torre, and León, J.J.A., Boron sulphate, chloride and phosphate contents in musts and wine of the Tacoronte-Acentejo D.O.C. region (Canary Islands), *Food Chem.,* 60, 339, 1997.

26. Lutz, O., Ascertainment of boric acid esters in wine by 11B NMR, *Naturwissenschaften*, 78, 67, 1991.
27. Themelis, D.G. et al., Direct and selective flow injection method for the simultaneous spectrophotometric determination of calcium and magnesium in red and white wines using in line dissolution based on "zone sampling," *J .Agric. Food. Chem.*, 49, 5152, 2001.
28. Baluja-Santos, C., Gonzelez-Portal, A., and Bermejo-Martinez, F., Evolution of analytical methods for the determination of calcium and magnesium in wines, *Analyst*, 109, 797, 1984.
29. Birch, R.M., Ciani, M., and Walker, G.M., Magnesium, calcium and fermentative metabolism in wine yeasts, *J. Wine Res.*, 14, 3, 2003.
30. Amerine, M.A. and Ough, C.S., *Methods for Analysis of Musts and Wines,* John Wiley & Sons, New York, Chichester, Brisbane, Toronto, 1980, chap. 9.
31. McKinnon, A.J., Williams, P.J., and Scollary, G.R., Influence of uronic acids on the spontaneous precipitation of calcium L-(+)-tartrate in a model wine solution, *J. Agric. Food Chem.*, 44, 1382, 1996.
32. Mena, C. et al., Cadmium levels in wine, beer and other alcoholic beverages: possible sources of contamination, *Sci. Total Environ.*, 181, 201, 1996.
33. Danilatos, N. and Salaxa-Moutsopoulou, M., Amount of trace elements in Greek wines. I Lead and cadmium, *Hellenica Georgika Chronika*, 3, 87, 1983 (in Greek).
34. Eschnauer, H., Trace elements in must and wine primary and secondary contents, *Am. J. Enol. Vitic.*, 33, 226, 1982.
35. Carvalho, M.L. et al., Study of heavy metals in Madeira wine by total reflection x-ray fluorescence analysis, *X-Ray Spectrom.*, 25, 29, 1996.
36. Lazos, E.S. and Alexakis, A., Metal ion content of some Greek wines, *Int. J. Food Sci. Technol.,* 24, 39, 1989.
37. Boulet, D. et al., The development of behaviour of wine consumption in France, *Ovinis infos*, 26, 72, 1995.
38. Hsia, C.L., Plank R.W., and Nagel C.W., Influence of must processing on iron and copper contents of experimental wines, *Am. J. Enol. Vitic.*, 26, 57, 1975.
39. Tsakiris, A., *Oenology: From Grape to Wine,* Psihalou, Athens, 1994, p. 179 (in Greek).
40. Zoeklein, B.W. et al., *Production Wine Analysis,* Chapman and Hall, New York, London, 1990, chaps. 14–16.
41. Almeida, A. A., Cardoso, M. I., and Lima, J. L. F. C., Determination of copper in Port wine and Madeira wine by electrothermal atomization AAS, *At. Spectrosc.*, 73, 1994.
42. Pérez-Magariño, S., Ortega-Heras, M., and González-San José, M.L., Multivariate classification of rosé wines from different Spanish protected designations of origin, *Anal. Chim. Acta*, 458, 187, 2002.
43. Kontos, G., *Technology of Wine Making,* Lyhnos, Athens, 1980, p. 64 (in Greek).
44. Benítez, P. et al., Influence of metallic content of fino sherry wine on its susceptibility to browning, *Food Res. Int.*, 35, 785, 2002.
45. Costa, R.C.D.C. and Araújo, A.N., Determination of Fe(III) and total Fe in wines by sequential injection analysis and flame atomic absorption spectrometry, *Anal. Chim. Acta*, 438, 227, 2001.
46. Benítez, P., Castro, R., and Barroso, C.G., Removal of iron, copper and manganese from white wines through ion exchange techniques: effects on their organoleptic characteristics and susceptibility to browning, *Anal. Chim. Acta*, 458, 197, 2002.

47. Karadjova, I., Izgi, B., and Gucer, S., Fractionation and speciation of Cu, Zn, and Fe in wine samples by atomic absorption spectrometry, *Spectrochim. Acta Part B*, 57, 581, 2002.
48. Mangas, S., Visvanathan, R., and Alphen, M., Lead poisoning from homemade wine: a case study, *Environ. Health Perspect.*, 109, 433, 2001.
49. Smart, G.A., Pickford, C.J., and Sherlock, J.C., Lead in alcoholic beverages: a second survey, *Food Addit. Contam.*, 7, 93, 1990.
50. Edwards, M.A. and Amerine M.A., Lead content of wines determined by atomic absorption spectrophotometry using flameless atomization, *Am. J. Enol. Vitic.*, 28, 239, 1977.
51. Kourakou-Dragona, S., *Choices in Oenology,* 1st ed., Troxalia, Athens, 1997, 57 (in Greek).
52. Stockley, C.S. et al., The relationship between vineyard soil lead concentration of lead in grape berries, *Aust. J. Grape Wine Res.*, 3, 127, 1997.
53. Teissedre, P.L. et al., On the origin of organolead compounds in wine, *Sci. Total Environ.*, 153, 247, 1994.
54. Mena, C.M. et al., Determination of lead contamination in Spanish wines and other alcoholic beverages by flow injection atomic absorption, *J. Agric. Food Chem.*, 45, 1812, 1997.
55. Teissdre, P.L., Cabanis, M.T., and Cabanis J.C., Comparaison de deux méthodes de minéralisation en vue du dosage du plomb par spectrométrie d'absorption atomique électrothermique, Applicationà des échantillons de sols, feuilles de vignes, raisins, moûts, marcs et lies, *Analusis*, 21, 249, 1993.
56. Vasconcelos, M.T., Azenha, M., and De Freitas V., Electrochemical studies of complexation of Pb in red wines, *Analyst*, 125, 743, 2000.
57. McKinnon, A.J. and Scollary, G.R., Size fractionation of metals in wine using ultrafiltration, *Talanta*, 44, 1649, 1997.
58. Wai, C.M., Knowles, C.R., and Keely J.F., Lead caps on wine bottles and their potential problems, *Bull. Environ. Contam. Toxicol.*, 21, 2, 1979.
59. Zerbinati, O., Balduzzi, F., and Dell'Oro, V., Determination of lithium in wines by ion chromatography, *J. Chromatogr. A*, 881, 645, 2000.
60. Sauvage, L. et.al., Trace metal studies of selected white wines: an alternative approach, *Anal. Chim. Acta,* 458, 223, 2002.
61. Cacho J. et al., Iron, copper, and manganese influence on wine oxidation, *Am. J. Enol. Vitic.*, 46, 380, 1995.
62. Frías, S. et al., Metallic content of wines from the Canary Islands (Spain): application of artificial neural networks to the data analysis, *Nahrung/Food*, 46, 370, 2002.
63. Alt, F. et al., A contribution to the ecology and enology of platinum, *Fresenius J. Anal. Chem.*, 357, 1013, 1997.
64. Almeida, C.M.R. and Vasconcelos, M.T.S.D., Does the wine making process influence the wine $^{87}Sr/^{86}Sr$? A case study, *Food Chem.*, 85, 7, 2004.
65. Salvo, F. et al., Influence of different mineral and organic pesticide treatments on Cd(II), Cu(II), Pb(II) and Zn(II) contents determined by derivative potentiometric stripping analysis in Italian white and red wines, *J. Agric. Food Chem.*, 51, 1090, 2003.
66. Peynaud, E., *Knowing and Making Wine,* John Wiley & Sons, New York, Chichester, Brisbane, Toronto, Singapore, 1984, chaps. 26, 29.
67. Baxter, M.J. et. al., The determination of the authenticity of wine from its trace element composition, *Food Chem.*, 60, 443, 1997.

68. Brescia, M.A. et. al., Chemomertic classificationof Apulian and Slovenian wines using ^{1}H NMR and ICP-OES together with HPICE data, *J. Agric. Food Chem.*, 55, 21, 2003.
69. Buldini, P.L., Cavalli, S., and Sharma, J.L., Determination of transition metals in wine by IC, DPASV-DPCSV and ZGFAAS coupled with UV photolysis, *J. Agric. Food Chem.*, 47, 1993, 1999.
70. Galani-Nikolakaki, S., Kallithrakas-Kontos, N., and Katsanos, A.A., Trace element analysis of Cretan wines and wine products, *Sci. Total Environ.*, 285, 155, 2002.
71. Peña, R.M. et. al., Pattern recognition analysis applied to classification of Galician (NW Spain) wines with certified brand of origin Ribeira Sacra, *J. Sci. Food Agric.*, 79, 2052, 1999.
72. Rebolo, S. et al., Characterisation of Galician (NW Spain) Ribeira Sacra wines using pattern recognition analysis, *Anal. Chim. Acta*, 417, 211, 2000.
73. Thiel, G. and Danzer, K., Direct analysis of mineral components in wine by inductively coupled plasma optical emission spectrometry (ICP-OES), *Fresenius J. Anal. Chem.*, 357, 553, 1997.
74. Almeida, C.M.R. et. al., ICP-MS multi-element analysis of wine samples — a comparative study of the methodologies used in two laboratories, *Anal. Bioanal. Chem.*, 374, 314, 2002.
75. Almeida, C.M. and Vanconcelos, M.T.S.D., Advantages and limitations of the semiquantitative operation mode of an inductively coupled plasma-mass spectrometer for multi-element analysis of wines, *Anal. Chim. Acta*, 463, 65, 2002.

9 Heavy Metals in Food Packagings

Marcelo Enrique Conti

CONTENTS

9.1 INTRODUCTION AND REGULATION

The Codex Alimentarius Commission Procedure Manual (1997)[1] defines a contaminant as:

> ...any substance not intentionally added to food, which is present in such food as a result of the production (including operations carried out in crop husbandry, animal husbandry and veterinary medicine), manufacture, processing, preparation, treatment, packing, packaging, transport or holding of such food or as a result of environmental contamination. The term does not include insect fragments, rodent hairs and other extraneous matter.

Heavy metals, especially lead, cadmium, chromium, and mercury, specifically dealt with in this chapter, are environmental contaminants[2–4] found in most foods in the human food chain. If metal contaminants migrate into food, they can have negative effects on food safety and quality.

Apart from their chemical definition (which includes the stable metals or metalloids whose density is greater than 4.6 g/ml), heavy metals can also be classified on the basis of their toxicity in humans.[5] This classification identifies the following elements as exemplary heavy metals: As, Bi, Cd, Hg, In, Pb, Se, Sb, and Tl. A third, wider classification also considers their ecotoxicological properties, thereby adding Te and other rare elements to the list.

In general, any migration of chemical substances from packaging to food is the result of a series of diffusion processes subjected to both thermodynamic and kinetic controls.[6]

The regulations in the area of food contact materials and supporting scientific evidence are very complex and, at least in Europe, are constantly being revised. As always, a strong scientific basis is necessary to support any decision making processes.[7] The aim of EU directives is the harmonization of preexistent national rules so as to remove the legal barriers that may prevent a free circulation of foodstuffs and beverages.[8]

The EU Framework Directive 89/109/EEC[9] states that "food contact materials shall be safe and must not transfer constituents in quantities that could endanger human health or induce an unacceptable change in the foodstuffs composition." The directive establishes ten groups of materials and articles regulated as follows:

1. Plastics including varnishes and coatings
2. Regenerated cellulose
3. Elastomers and rubbers
4. Paper and board
5. Ceramic
6. Glass
7. Metals and alloys.
8. Wood, including cork
9. Textile products
10. Paraffin waxes and micro-crystalline waxes

The directive also establishes that all articles intended for food use shall be explicitly labeled "for food use" or bear the corresponding symbol (a glass and a fork) as stated by Directive 80/590/EEC.[10] There are specific directives for three groups of materials and articles: plastics, regenerated cellulose films, and ceramics. In addition, there are other prepromulgation rules issued by the Council of Europe (CoE) or by European Committee for Standardization (CEN). The CoE promulgates resolutions and guidelines that are not binding for member states but constitute a common legal reference. The CEN issued EU standards of reference to be included in the EU directives.

The Food and Drug Administration (FDA),[11] in Title 21, Parts 170–190, regulates substances intended to make contact with foodstuffs. They are classified as direct and indirect food additives. In the area of the indirect food additives (Parts 174–178) adhesives, coatings components, paper, paperboards and polymers are considered. The section highlights the chemical characteristics of substances and the authorized commodities with some use limitations (similar to EU positive lists) and the allowed analytical methods. Migration tests with liquid simulants, at different temperatures and time conditions, are expected for polymers and papers intended for food packaging applications.

9.2 HEAVY METALS IN PLASTIC PACKAGING MATERIALS

The recent Commission Directive 2002/72/EC[12] regulates plastics packaging. This directive consolidates the Directive 90/128/EEC[13] and its seven amendments. Generally, these amendments modify the list of authorized substances. The Directive 2002/72/EC establishes an overall migration limit (OML) of 60 mg (of substances) per kilogram of foodstuff or food simulants (or 10 mg/dm^2) for all substances migrating from a material into foodstuffs. The Directive presents also a positive list of authorized monomers and other starting substances with threshold limits on their use (and specific migration limits [SML] in food or food simulant) where applicable.

There are many analytical problems in the determination of migration levels of packaging contaminants.[14–16] Usually, migration tests in simulant liquids are used with the objective to represent foodstuff–packaging contact. A special Directive (82/711/EEC[17] and its amendments 93/8/EEC[18] and 97/48/EC[19]) set out procedures for migration tests, such as the conditions of contact. Generally, for plastic materials and articles intended for contact with all food types, the tests shall be carried out using the following food simulants under more severe test conditions (in terms of temperature and time): (1) 3% acetic acid (w/v) in aqueous solution, (2) 10% ethanol (v/v) in aqueous solution, and (3) rectified olive oil. There are many substances of toxicological interest, restricted for plastic materials and articles intended to come into contact with foodstuffs. These are restricted through the SML and QM (maximum permitted quantity of the residual substance in the material or article, expressed as mg/dm^2 of the surface in contact with foodstuffs). SML "is applied to individual authorized substances and is fixed on the basis of the toxicological evaluation of the substance." The SML is generally established according to the acceptable daily intake (ADI) or the tolerable daily intake (TDI) set by the Scientific Committee on Food (SCF). To set the limit, it is assumed that, every day throughout one's lifetime, a person of 60 kg eats 1 kg of food packed in plastics containing the relevant substance at the maximum permitted quantity.[8]

In the production of thermoplastic polymers, including polyethylene (PE), polypropylene (PP), polystyrene (PS), polyvinyl chloride (PVC), polyethylene terephtalate (PET), polycarbonate (PC), acrylonitrile-butadiene-styrene (ABS), and polyamide (PA), different catalysts may be used that can contain low levels of heavy metals. Generally, plastics used for food packaging contain low concentrations of metallic compounds. Lead and Cd are sometimes present in plastic packaging articles. The sources for Cd and Pb are impurities originating from inorganic pigments and stabilizers.[20]

A surveillance project carried out by the Ministry of Agriculture, Fisheries, and Food of the U.K. (MAFF)[21] has considered 39 samples of common "packaging grade" plastics. Samples were analyzed by inductively coupled plasma–mass spectrometry (ICP-MS) for a total of 72 elements. For this study, food–simulating liquids were selected for the standard EC tests for measuring migration. The analysis revealed that metals expected to be found as catalyst residues were present. Most elements and heavy metals of particular concern (i.e., Pb, Cd, and Hg) were not detected (below limits of detection [LOD]). These heavy metals were not expected to be present as catalyst residues, but it was important to demonstrate their absence as contaminants in plastics. Analogous studies were conducted by Fordham et al.[22] on digested food contact plastics, and migration experiments were carried out using three food simulants (olive oil, 3% acetic acid w/v, and 15% ethanol v/v). Results showed that residues of Al, Mg, Cr, and Zn were present at levels much lower than the OML value of 60 mg/kg established by Directive 2002/72/EC.[12] Generally, residues of elements can be attributed to the catalyst residues expected for each kind of polymer.

The additives in plastics represent less than 0.5% of the mass. The low concentrations make heavy metals quantification very difficult.[23] Antioxidants can contain Ni, whereas thermal stabilizers can contain Ni, Pb, and Sb. Metal deactivators used in food packaging materials prevent oxidative degradation caused by Cu or other metal catalysts. Metal deactivators form stable complexes with metals, in particular, with Cu ions. Thus, the stability of polymers in contact with Cu is largely improved by the use of metal deactivators.[24] Stabilization process of polymers is an important step because it increases the product shelf life. Some metallic compounds such as Sn and Cd in Ba/Cd carboxylates are used as stabilizers of PVC.[24] These stabilizers are being replaced because of serious toxicological concern about Cd. The Ba/Cd stabilizers typically contain from 1 to 15% Cd and usually constitute about from 0.5 to 2.5% of the final PVC mass. This PVC type is used, for instance, in water and drain pipes. Zinc in Ba/Zn stabilizer system is important because it is an alternative to cadmium–containing compounds in plasticized PVC. The Ca/Zn stearates and other carboxylates are widely used for stabilizing nontoxic PVC articles for manufacturing of food contact packaging materials. Also, lead compounds are often used as PVC stabilizers (especially the primary lead salts such as lead carboxylates). Most metal-free compounds used in the stabilization of PVC are costabilizers. A small fraction of lubricants used contains Pb.

Fordham et al.[25] studied trace element concentrations in a wide range of food contact polymers (low density polyethylene [LDPE], high density polyethylene [HDPE], linear low density polyethylene [LLDPE], PP, PS, PET, ABS, etc.) by ICP-MS (after digesting the samples by microwave digestion [MW]) and by direct neutron activation analysis (NAA). Results show that polymers contain low levels of trace elements; e.g., Pb was detected in the range of 0.24 to 4.1 mg/kg in 10 out of 33 polymers sampled, and Cr was not detectable in 31 samples out of 33. Limits of the detection (LOD) for metals in the polymers were roughly 1 mg/kg.

Perring et al.[26] have analyzed various food packaging materials (package films, HDPE, paperboard, paperboard and PE film, PE, PET, PP, and PS) using ICP-MS, inductively coupled plasma atomic emission spectrometry (ICP-AES), and cold vapor atomic absorption spectrometry (CV-AAS). Wet oxidation systems were used to mineralize samples, which then were analyzed for Cd, Cr, Hg, and Pb. All commercial packaging materials contained levels of Pb, Cd, and Cr below 1.5 mg/kg. Total Cr concentrations were in the range of 1.5 to 30 mg/kg. Authors set detection limits for ICP-MS 25- to 200-fold better than axial ICP-AES and CV-AAS and indicated that such low LODs are not important for heavy metals' determination in food packing materials. The EU Directive 94/62/EC[27] on packaging and packaging waste sets out a maximum limit of 100 mg/kg for all Pb, Cd, Cr(VI), Cd, and Hg in packaging materials. This limitation must be controlled in polymeric packaging materials and for calculating the possible heavy metal enrichment provoked by recycling processes of organic polymers.

Thus, reliable analytical routine methods have been proposed. These methods, as the above mentioned, are usually based on migration tests of elements from the packaging material into a contact solution and then on analysis of the extracted metals. On the other hand, it is also necessary to establish the metal concentrations in polymers in order to improve the quality of plastic articles.

A collaborative study was conducted with the aim to improve the analytical quality control of trace element determinations in polymer.[28,29] At present, there are available two certified reference materials (CRM 680 and 681). These CRMs are made from a polyolefinic base material that represents 48% of the total polymers production. Heavy metals such as Cd, Cr, Hg, Pb, and As (metalloid) were added to the base material to obtain the CRM. The preparation process of CRMs requires satisfactory accuracy; to ensure comparability of measurements over long time periods, CRMs need high standards of homogeneity and stability.[30–32]

A radiotracer method has been proposed for estimating the migration of trace elements from food contact packaging materials (PET and LDPE) into food simulants.[33] Also X-ray fluorescence spectrometry (XRF) can be used for the direct determination of trace elements in polymers.[25] This method does not require sample decomposition and needs to be calibrated using CRMs. Atomic absorption spectrometry (AAS) and inductively coupled plasma optical emission spectroscopy (ICP-OES) are the more common methods available among plastic manufacturers. AAS has the limitation that it is generally able to measure a single element at a time, and this method requires a difficult sample pretreatment process. ICP-MS became a very auspicious alternative in recent years because of its high sensitivity (at ppb levels) and the ability to obtain multielement data.[29,34] The wet oxidation by MW digestion techniques applied to polymers mineralization has led to a significant time reduction in the sample preparation.[25] Generally, modern digestion methods such as MW, the high pressure asher digestion method (HPA), and oxygen flask combustion procedures (OFC) are time saving, cost effective, and routinely used for metals analysis in polymers.[26,35–37] Heavy metals in polyolefins can also be analyzed by inductively coupled plasma isotope dilution mass spectrometry (ICP-IDMS) and thermal ionization isotope dilution mass spectrometry (TI-IDMS).[38]

Future trends in analytical spectroscopy of trace metals are toward direct introduction of solid samples into the atomization/ionization chamber and avoiding any sample treatment (e.g., new nebulizers design for ICP-OES and ICP-MS, and slurry techniques for graphite furnace atomic absorption spectrometry GFAAS).[34] NAA is a well-suited technique to measure trace metals impurities in polymer materials at trace and ultratrace levels. One of the critical problems is the high risk of contamination during sample preparation. Recently, Kil-Yong et al.[39] have conducted a study to improve the accuracy and sensitivity of NAA for analysis of metals in plastic materials. The authors proposed a technique of the sample crushing after treating with liquid N and reducing the capsule pressure by a pin hole. They also investigated irradiation and cooling optimal conditions of the method. NAA is a sensitive and nondestructive technique useful in obtaining quantitative multielement data.[25]

The level of migration of the metals into foodstuffs should be determined under conditions that represent the worst case scenario. It is recommended that the maximum potential release of metals into foodstuffs should be evaluated, according to protocols described in the Appendix of the Directive 97/48/EC[19] with the exception of the acidic foodstuffs. At present, there are many ongoing research programs for selecting the most appropriate test to replace the traditional (now outdated) acetic acid 3% test. Any procedure for the determination of heavy metals concentration in foodstuffs and food simulants has to be adequately validated to gain the status of "reference method."

9.3 HEAVY METALS IN PRINTING INKS FOR FOOD PACKAGING APPLICATIONS

Colorants are either pigments or dyes. Dyes are soluble at molecular level whereas pigments are insoluble in polymers. Pigments are particles whose dimension is within the 0.01 to 1 μm range size. It may generally be problematic to extract a list of colorants that represents the vast assortment of products that are used in food packaging applications. Nonetheless, pigments, dyes, and inks also contribute, although to a lesser extent, to the overall metal content of foods.

There is no EU directive that especially regulates colorants in food packaging. There are two CoE resolutions, one concerning the use of colorants in plastic materials[40] and the other concerning surface coatings that may contain colorants.[41] Also, there is a draft resolution on packaging inks[42] and a draft technical document for packaging ink raw materials.[43] Colorants that are integrated into the plastic materials are considered as plastic additives in the literature.[23] A typical formulation for printing inks consists of colorant (5 to 30%), binder (15 to 60%), solvent (20 to 70%), and additives (1 to 10%).[44] Several types of printing inks are used on cardboard for food packaging applications.[45]

Several types of inorganic pigments have been banned due to restrictions on the use of heavy metals like Cd and Pb.[46] The CoE resolution AP (89)[40] established the migration limits for Sb (< 0.05%), As (< 0.01%), Ba (< 0.01%), Cd (< 0.01%), Cr (< 0.1%), Pb (< 0.01%), Hg (< 0.005%), and Se (< 0.01%). Also, the employment of Cr(VI) pigments for food contact use is discouraged.

In the CoE draft resolution AP (2004),[42] packaging inks are defined as printing inks and varnishes intended to be printed on the nonfood contact surface of packaging and articles for foodstuffs. As described in the CoE draft Technical Document N.1,[43] the pigment colorants based on compounds of Sb, As, Cd, Cr(VI), Pb, Hg, and Se are excluded as raw materials for the manufacture of packaging inks. The exclusion list includes many other dangerous substances (i.e., carcinogenic, mutagenic, teratogenic, etc.). This document establishes a migration limit of 50 ppb for only toxic substances that are not listed; if the migration exceeds this limit, the substance should be rejected. Conversely, the FDA Code of Federal Regulations relating to indirect food additives is applied exclusively for packaging inks where there is direct food contact and does not cover packaging inks printed on the outer or intermediate layers.

In Europe, the EU Directive 91/338/EEC[47] has established a list of polymers in which Cd pigments cannot be used. In these older pigments, Cd contents ranged from 15 to 30% by weight for cadmium sulfide and cadmium sulfoselenide. At present, the main application of these compounds is in complex polymers that are processed at higher temperatures and require the unique durability and technical performance of a cadmium pigment. These pigments are usually incorporated in plastics (nonfood contact applications) at levels of 0.01 to 0.75% by weight. Generally, organic pigments are almost entirely metal free in their composition.[20] Many factors such as pH, heat levels, and extraction time, can affect Pb leaching from microwavable plastics with Pb-containing pigments. The correlations of these parameters with lead release were found to be strong, but the incidence of Pb leaching was marginal.[48]

The exclusion list of the European Council of Paint, Printing Ink, and Artists' Colours Industry (CEPE)[49] gives recommendations for substances that should be excluded from the manufacture of printing inks according to the criteria adopted by Dangerous Substance Directive 67/548,[50] which considers colorants based on Sb, As, Cd, Cr (VI), Pb, Hg, and Se as toxic pigments. Inorganic pigments free of heavy metals are available.[51] However, quality controls are strongly needed because some trace metal contamination can be introduced from raw materials, some additives, or equipments. Digestion procedures for inks are a critical step and an important error source in the analysis of inks for heavy metal.[52]

9.4 HEAVY METALS IN PAPER, PAPERBOARD AND RECYCLED PAPER FOR FOOD PACKAGING APPLICATIONS

There is no EU directive for paper and board packaging. Recently, QM restriction limits have been established for Cd, Pb, and Hg by CoE Resolution AP (2002)[53] on paper and board materials and articles intended to come into contact with foodstuffs. This document establishes that paper and paper board should be manufactured from cellulose-based natural fibers from bleached and unbleached fiber material. QM restriction quality limits (mg/dm^2 paper and board) are 0.002, 0.003, and 0.002 for Cd, Pb, and Hg, respectively. A restriction limit is also indicated for pentachlorophenol with a purity requirement of 0.15 mg/kg paper and board. The document establishes, also, the SML restrictions, to be reported in the "List of substances used in the manufacture of paper and board materials and articles intended to come into contact with foodstuffs" (document under preparation). At present, the SML restrictions can be tested (migration tests) using the conventional simulants reported in the plastic directives 82/711/EEC and its amendments.[17] The conventional ratio for food contact conditions is 1 kg of food to 6 dm^2 of paper.

Paper and board made from recycled fibers (fully or in part) should comply with the Resolution AP (2002) 1[53] restrictions and are subject to additional quality requirements to ensure their safety of use. These quality aspects are related to the presence in the feedstock of constituents of some substances, such as additives and residues containing printing inks, adhesives, or "stickies" that come from paper not used in food contact applications.

Guidelines on test conditions and methods of analysis for paper and board materials and articles intended to come into contact with foodstuffs, and paper and board materials and articles made from recycled fibers are given. The document also gives a guide for good manufacturing practice (GMP).

Paper and paper derivatives are used extensively in food packagings all over the world. Paper packagings are more inexpensive and have high quality standards of safety of use.[54] Like plastic materials, toxicological effects of heavy metals generally occur at higher concentrations than expected in most packaging materials (at μg/g levels). Analytical methods used for trace elements analysis of paper and paper board are similar to those of plastic materials (e.g., AAS and ICP) and have been the subject of many exhaustive reviews.[55–57] Several studies determine the total trace elements concentrations in food packaging paper. This method involves acid digestion to mineralize the matrix and then liberates the elements before their analysis.[55–58] Castle et al.[59] report a multielement screening by semiquantitative ICP-MS for 10 paper and board samples previously digested in a pressure bomb. Results show that As and Hg were not detected in all samples (LODs were 1.8 and 0.4 μg/g, respectively). Cadmium was detected in 2 samples only: teabag tissue and unbleached Kraft paper at 0.3 μg/g (LOD = 0.1 μg/g). There were 7 out of 10 samples in the range from 1.1 to 7.8 μg/g for total Cr (LOD = 1.0 μg/g). Lead was also detected in 9 out of 10 samples in the range of 0.3–6.6 μg/g (LOD = 0.1 μg/g). Also, Cd levels were too low in digested food packaging papers (candy wrappers, tea bags, flour sacks, kitchen towels, etc.), and it can be concluded that the health risk from exposure to this source of Cd is negligible.[60]

Other publications have described the heavy metals in paper and paperboard packaging after a migration test in a contact liquid (usually acetic acid 3%). Methods of preparation of paper and board samples were well described,[61–62] i.e., the content of heavy metals in food packaging paper boards should be determined only on the internal (back) side in direct contact with food.

Conti and Botrè[61] conducted heavy metals migration (3% acetic acid) tests in 7 samples of food packaging papers and reported the following results: n.d.–4.38; n.d.–0.11; n.d.–0.52, and 0.13–0.22 μg/g of Pb, Cd, total Cr, and Cr(VI), respectively. Conti[62] also reports data on migration (3% acetic acid) tests for 15 food packaging paper boards. The results were in the range: n.d.–2.37, n.d.–0.08, n.d.–0.51, and 0.024–0.033 μg/g of Pb, Cd, total Cr, and Cr(VI), respectively. Mercury was not detected for any samples of paper and paper boards. The LODs in Conti's[62] study were [Pb] < 4 μg/l, [Cd] < 0.1 μg/l, [Cr] < 4 μg/l, and [Hg] < 1 μg/l in the contact liquid. These studies show that substantial amounts of heavy metals are present in some of the assayed samples. Metals that have been detected in higher concentrations are primarily Pb and Cd whereas the levels of Cr are in most samples below the detection limits. Cr(VI) can be present at very low levels. It should be noted that where the product is made with 100% of virgin pulp, heavy metals are rarely detected (e.g., pizza samples).

At present, EU countries are promoting the recycling of cellulose fibers. According to European Declaration on Paper Recovery,[63] the amount of used paper that was recycled was 48.7% in 2000, and it is expected to reach 56% in 2005 in Europe. Paper pulp has a remarkable capability to absorb high quantities of heavy metals even during short time exposures. Recycled pulps usually contain higher heavy metals concentrations than primary pulps.[64] In a laboratory test,[65] the percentage of migrated cadmium into a virgin conifer's pulp was found to be between 30% and 90% of initial reference solutions. There is a need of analytical methods and kinetics of migration studies with the aim to identify and quantify residues of concern in paper and paperboard.[66,67] Odorous substances can be present in paperboard food packaging and can cause tainting of the packaged food products.[68] Also, microbiological quality should be a very important purity test of recycled fiber materials.[69] Sankey diagrams are useful tools to represent balances of heavy metal loads in paper making process that uses waste paper as raw material.[70]

Conti et al.[71] studied 14 samples of paper produced with "second use" fibers from different types of waste paper and migration tests were conducted (3% acetic acid). The Pb range was 0.8 to 37.2 μg/g in 7 samples out of 14. The Cd levels in 11 samples out of 14 were in the range of 0.01 to 0.37 μg/g. Total Cr concentration (3 samples out of 14) was in the range of 0.67 to 2.61 μg/g; these 3 samples showed remarkable Cr(VI) levels (0.25 to 0.34 μg/g). Mercury was not detected in any samples (LOD = 1 μg/l in the contact liquid). The authors concluded that samples did not come from good quality waste paper.

The de-inking process is often used in the recycling industry to enhance the quality of the recycled paper. The ink dispersion has a negative impact on the overall pulp brightness[72] and, as mentioned above, inks can contain certain levels of heavy metals. The increasing use of recycled fiber in a wide variety of paper applications requires that de-inked pulp must meet always higher quality standards.[73] However, little information is available on heavy metals in recycled fiber base materials and, to a greater extent, in food packaging papers. Several changes in EU laws and regulations in the past decade have probably influenced the very complex data interpretation obtained with many analytical methods. In view of this, it is also important to consider the difficulties of establishing limits of toxicological concern.

9.5 METALLIC AND OTHER FOOD CONTACT MATERIALS

9.5.1 Chromium and Other Metals

Chromium is an essential element for humans and is present in the diet mainly as Cr(III).[74–76] It has a role in carbohydrate, lipid, and protein metabolism related to improving the action of insulin. On the other hand, Cr is recognized as a food contaminant. Most trivalent Cr compounds are soluble in water at low pH values. Chromium (III) hydroxide precipitates generally at pH of 5 to 6.[77] Thus, Cr(III) cannot migrate at neutral pH in foodstuffs and, at the same time, its migration to foods with pH 5 or above is low.[78] Both solubility and oxidation state affect the potential for toxicity. Cr(III) is much less toxic than the hexavalent form; some Cr(VI) compounds may be carcinogenic.[79]

Some metals such as Cr, Fe, Cu, and Zn are used in alloy form. Stainless steels are iron-chromium alloys that contain high percentage of Cr (17 to 18 %) and are widely used in food contact materials.[78] Chromium has an important role in the formation and stabilization of the passive film which acts as a very thin (few Å thick) protective surface layer. This film confers the typical corrosion resistance of stainless steels. Stainless steels containing Cr are used in many food-related fields: for processing equipment (e.g., dairy, chocolate, and fruit industries), for containers (e.g., wine tanks, brew kettles, and beer kegs), processing of dry foods (e.g., cereals, flour, and sugar), utensils (e.g., blenders, and bread dough mixers), and kitchen appliances (e.g., knives, spoons, forks, electric kettles, cookware, etc.).[78]

Kuligowski and Halperin[80] have investigated the leaching of Cr(VI) of 6 saucepans that were more than 1 year old. Results show that 5% acetic acid did not cause Cr leaching from any of the saucepans. If the acid was boiled for 5 min in the saucepans, the concentration of Cr observed was not distinguishable from the analytical background (0.035 mg/l) in 3 cases, whereas it was 2-fold higher in 2 other cases and 8-fold higher (0.3 mg/l) in the 6th case. Offenbacher and Pi-Sunyer[81] have measured the release of Cr from stainless steel pots into acidified water. Cr-free water was acidified with HCl to pH 2.5 and 3.0. This acidity represents pH encountered in some foods. The Cr levels in leachate were in the range of 31 to 50 ng/g in the case of canned tomato juice, bottled pineapple juice, and lemon juice after 1-h boiling in a stainless steel pot; no Cr leaching was observed in unacidified water. It was concluded that Cr release from stainless steel was mainly dependent on pH of the food or drink.

Kumar et al.[82] have observed that there is no leaching of chromium into tea, coffee, milk, or fruit juice from either old or new stainless steel bowls and tumblers (with a Cr content 9.74 to 20.80%), whereas the leaching was 0.04 to 0.4 μg/g to curd or lemon pickle from new utensils and 0.03 to 0.3 μg/g from old utensils. Neither Cr nor Pb was released into water from twenty-six kettles sold on the Danish market in a study simulating regular household use. However, 10 out of 26 kettles released more than 50 μg/l of nickel into the water. Coffee machines similarly tested did not show significant liberation of Pb, Cr, Ni, and Al.[83]

The cooking process can induce migration of trace elements. The chromium content of crayfish hepatopancreas cooked in a stainless steel pan increased from approximately 0.05 to 0.15 μg/g fresh weight (f.w.). The cooking process for crayfish abdominal muscle did not reveal relevant changes in Cr concentrations compared with the raw fish.[84]

New saucepans on first use have the highest rates of Cr and Ni release.[85,86] After the first two cooking operations, the highest chromium release for apricots and rhubarb was 0.05 and 0.01 μg/g, respectively.[86] Generally, the contribution of Cr and Ni released from cooking utensils and glass pots to the average daily diet is negligible.[78,85–89] There does not appear to be significant differences in migration observed between ferritic and austenitic stainless steels and glass pots.[78,87]

Chromium and its compounds are also used in electroplating and in surface treatment of food cans. A "can" is defined as an hermetically sealed container in which food is subjected to a "canning" process, that is, heat treatment to increase shelf life.[90] Most cans are made from tinplate, that is, a composite packaging material consisting of a low carbon mild steel base. To prevent oxidation, the tinplate is "passivated" with a thin layer by the same process above described. Acidic foods can corrode the metal, thus, cans are frequently coated internally with a lacquer that is mostly a polymerized resin.[90] At present, in Europe, about 25,000 million food cans are produced every year; about 20% of these are unlacquered cans. The total worldwide production for food packaging is approximately 80,000 million cans.[91]

The performance of the lacquered food cans is greatly affected by the coating thickness. Low-aggressive foods such as apricots and beans require a thickness of 4 to 6 μm, whereas tomato concentrate needs a layer of 8 to 12 μm to prevent interaction between the can and its content.

9.5.2 Lead, Cadmium, Tin, Aluminium, and Other Metals

Lead and Cd are environmental pollutants,[5,92] and their presence in food contact materials is regulated by specific EU directives and CoE resolutions. The Joint FAO/WHO Expert Committee on Food Additives (JECFA) has established a Provisionally Tolerable Weekly Intake (PTWI) at 0.025 mg/kg body weight (bw) for Pb[93] and 0.007 mg/kg bw for Cd.[94]

A minor source of Pb in food cans comes as a small impurity in the tin and the coating. In the past, the lead solder on cans was an important source of Pb contamination of food. Lead exposure has declined over the years, having been reduced from an average of 0.2 μg/g in 1982 to 0.01 μg/g in 1990 in the U.K. Although lead can be found in tinplate (even in Grade A up to a maximum of 0.01%), the tin in cans dissolves preferentially due to the electrochemical properties of those 2 metals.[95] Foods in lacquered welded cans contain less Pb, Cr, and Sn than foods in unlacquered welded cans.[96] Lead leaching is prevented to a very high degree by lacquer coats.[97]

A survey was conducted on the Pb contents of 52 samples of grapefruit juice, 14 packaged in glass, 8 in waxed paperboard, and 30 in tin-coated carbon steel containers. Lead was detected only in juices packaged in metal cans with levels in the range of < 1-27 μg/l; the average was 7.7 μg/l.[98] Leaching studies, conducted on refrigerated juices for up to 30 d, showed that Pb content was highly increased in juices stored in tin-coated cans (from 3 to 90 μg/l after 6 d storage). Juices stored in glass, paperboard, or polyethylene containers did not show Pb increase. The authors concluded that the Pb in the cans could be attributed to Pb impurities in the tin coating.[98]

In the past, Pb and Cd pigments were used in ceramic glazes. The most important Pb pigment is white lead.[99,100] EU Directive 84/500/EEC[101] strictly regulates Pb and Cd leaching from ceramic materials. Colorants used in overglaze paints can contain pigments with Pb, Cd, Zn, and other heavy metals. The FDA has established leaching limits of 3 μg/ml for Pb and 0.5 μg/ml for Cd when dinnerware is filled with 4% acetic acid for 24 h at room temperature.[102]

Automatic dishwashings and scrubbings seem not have great influence on release of Pb from glazed ceramicware.[103] A study was conducted on ceramicware with initial Pb leach levels of less than 0.1 to 470 μg/ml that was cleaned in a dishwasher; no increase in lead release was noted. Another study reported a Pb release from ceramic pitchers (4% AcOH at 19°C for 24 h) of less than 0.01 μg/g.[104] Colored porcelain dishes can release Cd. Doemling[105] reported levels of more than 0.5 μg/g of Cd for 25% of the samples tested. At present, there are glazes that are free of Pb and Cd in their formulation.[106] An adherent glass coating for glass ceramic surfaces for cooking food free of Cd and Pb has recently been reported.[107]

The FDA has recently improved methods of determination of Pb and Cd extracted from ceramic foodware.[108,109] The flame atomic absorption spectrometry (FAAS) interlaboratory study reported excellent results in the quantification of the metals, and the data obtained among different laboratories (0.005 to 0.019 μg/ml for Pb and 0.0004 to 0.0019 μg/ml for Cd) were adopted as a First Action by AOAC International.[108] Since 1981, the International Organization for Standardization (ISO) issued international regulations on ceramic foodware in contact with food.[110] Instances of heavy metal poisoning from ceramic foodware surfaces remain rare, however.

At high exposure times (nearly 5 months), it was found that 2 μg/g of Pb was leached from a lead crystal wine glass (containing 24 wt.% lead oxide).[111] Guadagnino et al.[112] tested the Pb release from crystal glasses into some beverages (3 h of continuous contact) and found that the release increased in the order: cola drink > HOAc > whisky > white wine. If the number of contacts for both cola drink and wine was increased, the Pb levels showed a steep decrease. The study concluded that there were no significant health risks associated with ingestion of beverages in contact with crystalware. Lead migration is strongly correlated with the hydrolytic resistance of glass. There is a linear correlation of Pb migration after the first contact and the Na migration from the glass.[113] Many studies show that the Pb migration from beverages decreases significantly after the first contact, similar to the results of the 4% acetic acid tests.[114–116] Lead release in acetic acid (pH = 2.39) was higher than that for wine (pH= 3.14) because of higher pH and ethanol content of wine.[117] Coca Cola has an extraction ability similar to that of 4% HOAc.[118] Repeated-leaching experiments with Pb crystal wine glasses show that total Pb released (in micrograms) in 30 min decreased according to the following function:[117,119]

$$LR = a + b/L^2$$

where:

LR = lead concentration in the extractant
L = contact number
a = constant depending on glass composition
b = constant depending on the characteristics of surface layer

The EC directive 94/62 regulates tolerable concentration levels of Pb, Cd, Hg, and Cr(VI) in glass packaging.[27] Recent (December 2003) guidelines on Pb leaching from glass tableware into foodstuffs is derived from a draft resolution.[119] This document establishes fundamental principles for Pb leaching from glass hollowware and flatware and recommends washing the food equipment before first use because of the tendency for a high release of lead to occur at the very first contact. Lead release decreased by nearly 20% after washing of unused crystal wine goblets.[116] The weekly Pb intake attributable to lead extraction from tableware has been estimated as 35 μg/week, if one assumes a daily use of lead crystal stemware and a daily consumption of 5 dL beverage.[119] Glazes on the outside part of drinking glasses can release up to 300 mg Pb/dm^2 and up to 30 mg Cd/ dm^2 on 24-h extraction with 4% acetic acid.[120]

The data above, although still under evaluation, show that there is a need to improve the quality control of glass analysis in the food industry. At present, the CRM 664 is available at the Institute for Reference Materials and Measurements (IRMM). Elements certified include As, Ba, Cd, Co, Cl, Cr, Pb, Sb, and Se.[121,122] In the absence of a standard procedure, a method was proposed to evaluate the presence of hexavalent chromium in glasses. In this procedure, the glass sample is digested with H_2SO_4 and NH_4HF_2, then diphenylcarbazide is added, and a pink colored complex is obtained, which is measured with a spectrophotometer at 840 nm.[123,124] Mercury in glass packaging can be determined by a CV-AAS.[125]

The primary source of tin is associated with the can; [78,91,100,126] in the U.K., most dietary intake of Sn (94%) comes from canned fruit and vegetables.[127] Unprocessed foodstuffs contain generally less than 1 μg/g of Sn.[78,128] The PTWI for Sn is 14 mg/kg body weight,[129] and the recommended maximum permissible levels of Sn in food are typically 150 μg/g for canned beverages and 250 μg/g (200 μg/g for the U.K.[130]) in other canned foods. These levels of tin in foods can produce acute manifestations of gastric irritation in some individuals.[129] The typical organic acids found naturally in fruits and juices such as citric acid, malic acid, and their mixture with oxalic acid, can increase the rate of Sn liberation from tinplate cans.[131] Corrosion of tinplates is highly induced by the nitrate-containing acid foods.[132] Tin is used in alloys with Cu and Zn and to coat kitchen utensils.[78] Ceramic industry uses inorganic Sn compounds as pigments. The Cr-SnO_2 pigment has long been used in the ceramic industry.[133] Plastic materials containing chlorine are stabilized by organotin compounds.[100]

The MAAF[134] has conducted a survey on the levels of Sn in canned fruits and vegetables. All the 400 samples tested, with the exception of only 2 products, were found to contain levels of Sn below the legal limit of the U.K. (200 μg/g). The 234 food samples packaged in unlacquered cans showed an average Sn concentration of 59 μg/g and, overall, the Sn concentrations were lower than those found in earlier surveys.[134,135] Other studies reported that tin levels in foodstuffs stored in lacquered cans, in contrast to unlacquered cans (> 100 μg/g), were generally below 25 μg/g.[128] WHO[136] did not fix a threshold value in drinking water for inorganic Sn, possibly because of its low toxicity level.

Aluminum is a ubiquitous element naturally present in foods. It constitutes 8.13% of the earth's crust and appears in a large variety of minerals.[137] Foodstuffs and water are the main sources of Al intake in the diet. The JECFA has established a PTWI of 7 mg Al/kg of body weight.[138] The daily intake of Al from food and beverages in adults ranges between 2.5 and 13 mg. This is nearly 90 to 95 percent of total intake. Drinking-water may contribute roughly 0.4 mg/d according to present international guideline values, but it is more likely to be generally around 0.2 mg/d.[137]

Foodstuffs that contain Al concentrations greater than 1 μg/g of food are said to contain high values of Al;[139] alkaline and salty foodstuffs can further increase the Al intake. Usually, the levels of Al naturally present in food are very low,[140,141] with the exception of specific plants and crops like tea leaves that naturally bioconcentrate Al to levels of up to the 699 to 1943 μg/g found in dry leaves of some types of tea grown on acidic soils.[142–144] Mature tea leaves collected at Hong Kong accumulated very high concentrations of Al and F (15300 and 2070 μg/g, respectively).[143] A number of studies have demonstrated that Al leaching from food plants in the presence of fluoride is insignificant.[144,145] Ingestion of Al per cup through tea (100 ml capacity) can expose one to Al doses of between 1.1 and 1.4 mg.[144]

Aluminum is mainly used in packaging materials such as saucepans, pressure cookers, foils, wrappers, frozen dinner trays, cans, roasting pans, etc. Usually, aluminum packaging materials are covered with an intermediate resin-based coating. Thus, Al migration from coated food contact materials is negligible.[78] Generally, sauces and drinks packaged in Al coated cans at different storage times and temperature conditions do not leach out significant amounts of Al and thus do not represent a significant risk for human health.[146,147] Sauce samples stored at 50°C showed slightly higher Al contents than those stored at room temperature.[146] Acidic foodstuffs such as tomatoes, cabbage, soft fruits, etc., take up more Al from the containers than nonacidic foods. In particular, tomatoes and rhubarb in Al pans showed a significant increase in Al levels (0.5 μg/g wet weight of raw tomatoes to 3.3 μg/g wet weight of the cooked), whereas the accumulation in prepared rice or potatoes showed only a slight increase.[148] Aluminum cans can be corroded over time by canned beer. Beer has a pH of 4.15 and can extract Al despite the coating of the can.[149] Beer stored at room temperature (5 months at 23°C) showed Al levels that were significantly higher (546 and 414 μg/l in 2 brands) than those observed for refrigerated beers (50 and 117 μg/l for 5 months at 5°C). At the beginning of the experiment, Al levels ranged between 50 and 118 μg/l for the 2 sampled brands.[149] Cola and citrate–based drinks can also corrode Al coated cans. Ortophosphoric acid found in some soft drinks is more corrosive than citric acid.[150] Soft drinks from Al cans are a negligible source of dietary Al intake, however. The longer the storage period and higher the storage temperature, the more the leaching of Al from the packaging.[78,150,151]

Other contributors of Al to human diet are Al containing food additives, toothpaste, which can contain significant amounts of Al, above all, when packed in Al tubes, and some over-the-counter drugs such as antacids and buffered analgesics. In this context, Lione[152] has reported that the exposure to aluminum in antacids ranged from 840 to 5000 mg/d for antacids and 130 to 730 mg/d in buffered analgesics. This leads to an overall 2–3 fold increase with respect to the normal dietary intake.

Aluminum from cookware and uncoated articles in contact with foodstuff can migrate into foodstuffs; e.g., uncoated Al camping bottles can release up to 7 mg/l Al in tea acidified with lemon juice within 5 d storage.[153] The leaching of Al from aluminum cooking vessels in boiling test was 0.2 to 0.8 μg/g for boiling milk. Boiling of tap water in an aluminum pan can result in the migration of 0.54 to 4.3 mg/l Al.[154] Muller et al.[153] found levels of up to 2.6 mg/l after boiling tap water for 15 min in Al pans. In general, the rate of aluminum release depends on the water acidity and chemical compositions of the Al utensils.[155,156]

Some disorders (or their correlated events) have been associated with an excessive intake of Al in the human body such as dialysis, osteomalacia, fractures and high levels of bone Al, encephalomyelopathy, and Alzheimer's disease. There is some uncertainty on the definition of the exact cause of Alzheimer's disease. However, the IPCS[137] report has concluded that Al is not the cause of Alzheimer's disease. The storage of strongly acidic and salty liquid foodstuffs in uncoated aluminum utensils should be discouraged, and the manufacturers should be encouraged to label the Al uncoated utensils to prevent their contact with acidic or salty moist foods before or after cooking.[78]

Mercury is a highly toxic pollutant[157] and the main contributor in the diet is methyl mercury in fish where levels can reach 2 to 4 μg/g.[78] The PTWI for Hg is 0.005 μg/g bw.[158] Data on Hg migration for all types of packaging materials is very scarce; most reports show that the Hg levels in food packaging material are always below the LODs of the analytical techniques (see other sections of this chapter). Although Hg must be controlled in the food packaging processes, it should not be used in food contact materials, either.[78]

9.6 CONCLUSIONS

Food packaging materials have changed more in the past 15 years than in the previous 150 years, with a general, marked improvement of the quality, safety and nutritional content of packaged foods. The progress in understanding the toxicological relevance of the presence of heavy metals in food packaging is dependent primarily on the possibility of exactly determinating their content in selected matrices, and consequently on the further development of sensitive, accurate, fast, and precise analytical methods. The success of such analytical techniques is strictly correlated with both the simplification of sample treatment and the reduction of the corresponding costs.

In general, electrothermal atomization has been allowed to establish sufficient sensitivity limits for heavy metals determination in food simulants and food packaging articles. GFAAS and FAAS are the more widespread techniques used in analytical laboratories. More recently, advanced analyzers suitable for multielement analysis, such as ICP-MS, became available. There is also the need to have more suitable CRMs for the analytical quality control of packaging materials.

Further studies are also necessary to detail more precisely the toxicological relevance of heavy metals in food packaging, with the aim of refining the existing guidelines. In this view, it is mandatory to consider the rapidly growing area of metal speciation research. The major advances of this area are coming from coupling of the separation processes, carried out by universal techniques such as capillary electrophoresis or high-performance liquid chromotography (HPLC), with dedicated detectors for heavy metals analysis.

The improvement of analytical techniques should therefore correspond with a constant advance in the study of the role of heavy metals in food packagings, with special emphasis on health aspects. In this context, heavy metals limitations stated in EU and U.S. regulations can constitute a good starting point for similar guidelines in other countries, especially concerning those toxic heavy metals particularly relevant to the risk of contamination by food packagings.

REFERENCES

1. Codex Alimentarius Commission (1997), *Procedural Manual,* 10th ed., Joint FAO/WHO Food Standards Programme, Rome.
2. Nriagu, J.O. and Pacyna, J.M., Quantitative assessment of worldwide contamination of air, water and soils by trace metals, *Nature*, 333, 134, 1988.
3. Szefer, P., Metal pollutants and radionuclides in the Baltic Sea: an overview, *Oceanologia*, 44, 129, 2002.
4. Conti, M.E. and Cecchetti, G., A biomonitoring study: trace metals in algae and molluscs from Tyrrhenian coastal areas, *Environ. Res.*, 93, 99, 2003.
5. Fergusson, J.E., *The Heavy Elements: Chemistry, Environmental Impact and Health Effects*, Pergamon Press, Oxford, 1990.
6. Castle, L., Chemical migration from food packaging, in *Food Chemical Safety — Volume 1: Contaminants*, Watson, D.H., Ed., CRC Press, Woodhead Publishing Ltd., 2001, chap. 9.
7. Gilbert, J. and Rossi, L., European priorities for research to support legislation in the area of food contact materials and articles, *Food Addit. Contam.,* 17, 83, 2000.
8. EU Food Contact materials Resource Centre, http://cpf.jrc.it/webpack/, 2005.
9. Directive 89/109/EEC: European Community, Council Directive on the approximation of the laws of the Member States relating to materials and articles intended to come into contact with foodstuffs, December 21, 1988.
10. Directive 80/590/EEC: European Community, Council directive determining the symbol that may accompany materials and articles intended to come into contact with foodstuffs, June 9, 1980.
11. Code of Federal Regulations. Title 21 Parts 170–190, http://www.access.gpo.gov/cgi-bin/cfrassemble.cgi?title=200121.
12. Directive 2002/72/EC: European Community, Council directive relating to plastic materials and articles intended to come into contact with foodstuffs, August 6, 2002.
13. Directive 90/128/EEC: European Community, Council directive relating to plastic materials and articles intended to come into contact with foodstuffs, February 23, 1990.
14. O'Brien, A.P., Cooper, I., and Tice, P.A., Correlation of specific migration (Cf) of plastics additives with their initial concentration in the polymer (Cp), *Food Addit. Contam.*, 14, 705, 1997.
15. Tice, P.A. and Cooper, I., Migration tests for food packaging, *Food Sci. Rev.*, 2, 3, 1994.
16. Tice, P.A., Testing polymeric coatings on metal and paper substrates, *Food Addit. Contam.*, 11, 187, 1994.
17. Directive 82/711/EEC: European Community, Council directive laying down the basic rules necessary for testing migration of the constituents of plastic materials and articles intended to come into contact with foodstuffs, October 18, 1982.
18. Directive 93/8/EEC: European Community, Council directive amending Council Directive 82/711/EEC laying down the basic rules necessary for testing migration of constituents of plastic materials and articles intended to come into contact with foodstuffs, March 15, 1993.
19. Directive 97/48/EC: European Community, Council directive amending for the second time Council Directive 82/711/EEC laying down the basic rules necessary for testing migration of the constituents of plastic materials and articles intended to come into contact with foodstuffs, July 29, 1997.

20. Mark, F.E., Metals in source separated plastics packaging waste: a European overview — analytical methods and results, a report for the Association of Plastics Manufactures in Europe (APME), November 1996.
21. Ministry of Agriculture, Fisheries and Food of U.K. (MAAF), Joint Food Safety and Standards Group, Metallic Compounds in Plastics, Food Surveillance, Information Sheet, No 1, July 1993.
22. Fordham, P.J. et al., Element residues in food contact plastics and their migration into food simulants, measured by inductively-coupled plasma-mass spectrometry, *Food Addit. Contam.*, 12, 651,1995.
23. Zweifel, H., Ed., *Plastics Additives Handbook*, 5th ed., Hanser Gardner, Munich, Germany, 2001.
24. Zweifel, H., Stabilization of Polymeric Materials State-Of-The-Art, Scope and Limitations, www.sun.ac.za/unesco/PolymerED2000/Conf2000/ZweifelC(s).pdf.
25. Fordham, P.J. et al., Determination of trace elements in food contact polymers by semi-quantitative inductively coupled plasma mass spectrometry: performance evaluation using alternative multi-element techniques and in house polymer reference materials, *J. Anal. At. Spectrom.*, 10, 303, 1995.
26. Perring, L. et al., An evaluation of analytical techniques for determination of lead, cadmium, chromium, and mercury in food-packaging materials, *Fresenius J. Anal. Chem.*, 370, 76, 2001.
27. Directive 94/62/EC: European Community, Council directive on packaging and packaging waste, December 20, 1994.
28. Van Borm, W., Lamberty, A., and Quevauviller, P., Collaborative study to improve the quality control of trace element determinations in polymers, part 1: interlaboratory study, *Fresenius J. Anal. Chem.*, 365, 361, 1999.
29. Lamberty, A., Van Borm, W., and Quevauviller, P., Collaborative study to improve the quality control of trace element determinations in polymers, part 2: certification of polyethylene reference materials (CRMs 680 and 681) for As, Br, Cd, Cl, Cr, Hg, Pb, and S content, *Fresenius J. Anal. Chem.*, 370, 811, 2001.
30. Lamberty, A., Schimmel, H., and Pauwels, J., The study of the stability of reference materials by isochronous measurements, *Fresenius J. Anal. Chem.*, 360(3–4), 359, 1998.
31. Pauwels, J., Lamberty, A., and Schimmel, H., Homogeneity testing of reference materials, *Accred. Qual. Assur.*, 3, 51, 1998.
32. Linsinger, T.P.J. et al., Homogeneity and stability of reference materials, *Accred. Qual. Assur.*, 6(1), 20, 2001.
33. Thompson, D., Parry, S.J., and Benzing, R., The validation of a method for determining the migration of trace elements from food packaging materials into food, *J. Radioanal. Nucl. Chem.*, 217, 147, 1997.
34. Conti, M.E. et al., Biomonitoring of heavy metals and their species in the marine environment: the contribution of atomic absorption spectroscopy and inductively coupled plasma spectroscopy, *Trends Appl. Spectrosc.*, 4, 295, 2002.
35. Vollrath, A. et al., Comparison of dissolution procedures for the determination of cadmium and lead in plastics, *Fresenius J. Anal. Chem.*, 344, 269, 1992.
36. Besecker, K.D. et al., A simple closed-vessel nitric acid digestion method for a polyethylene/polypropylene polymer blend, *At. Spectrom.*, 19, 193, 1998.
37. Bonn, A. and Knezevic, G., A rapid method for analysis of metals in plastic, *Coating*, 30, 141, 1997.

38. Diemer, J. and Heumann, K.G., Development of an ICP-IDMS method for accurate routine analysis of toxic heavy metals in polyolefins and comparison with results by TI-IDMS, *Fresenius J. Anal. Chem*, 368, 103, 2000.
39. Kil-Yong, L. et al., An accurate and sensitive analysis of trace and ultratrace metallic impurities in plastics by NAA, *J. Radioanal. Nucl. Chem.*, 241(1), 129, 1999.
40. Resolution AP (89) 1: On the Use of Colorants in Plastic Materials Coming into Contact with Food, Council of Europe, September 13, 1989, www.coe.int/soc-sp.
41. Resolution AP (96) 5: On Surface Coatings Intended to Come into Contact with Food, Council of Europe, October 2, 1996. www.coe.int/soc-sp.
42. Partial Agreement in the Social and Public Health Field. Draft Resolution AP (2004). RD 8/1-42, On Packaging Inks Applied to the Non Food Contact Surface of Food Packaging Articles Intended to Come into Contact with Foodstuffs, 42nd session, Strasbourg, December 2–5, 2003, www.coe.int/soc-sp.
43. Partial Agreement in the Social and Public Health Field, Council of Europe Draft Technical Document No 1, RD 8/2-42, On requirements for the selection of packaging ink raw materials applied to the non food contact surface of food packaging articles intended to come into contact with foodstuffs, 42nd session, Strasbourg, December 2–5, 2003, www.coe.int/soc-sp.
44. Faigle, W., Lectures at the University of Printing and Media, Stuttgart/Germany, Dittmer, C. and Mayer D.T., Eds., 1997, www.hdm-stuttgart.de/projekte/printing-inks/inf_text.htm.
45. Davey, A.C., Printing inks for packaging applications, *Adv. Col. Sci. Technol.*, 4, 76, 2001.
46. Brede, C., Skjevrak I., and Fjeldal, P., Colors Substances in Food Packaging Material, SNT — Norwegian Food Control Authority, March 3, 2003, http://snt.mattilsynet.no/dokumentasjon/rapporter/2003/snt_arbeidsrapporter/200303.pdf.
47. Directive 91/338/EEC: European Community, Council Directive amending for the 10th time Directive 76/769/EEC on the approximation of the laws, regulations and administrative provisions of the Member States relating to restrictions on the marketing and use of certain dangerous substances and preparations, June 18, 1991.
48. Inthorn, D. et al., Factors affecting lead leaching from microwavable plastic ware made with lead-containing pigments, *J. Food Prot.*, 65, 1166, 2002.
49. European Council of Paint, Printing Ink and Artists' Colors Industry (CEPE), www.cepe.org/CEPE.htm, exclusion list for printing inks and related products, September 2001 (Brussels).
50. Directive 67/548/EEC: European Community, Council directive on the approximation of laws, regulations and administrative provisions relating to the classification, packaging and labeling of dangerous substances, June 27, 1967.
51. De Bruijn Boogaerd, F.H., Production of Inorganic Pigments Based on Non-Heavy Metals, Spanish Patent WO 9700918, *PCT Int. Appl.*, 1997.
52. DaRocha, M. et al., Analyzing for CONEG heavy metals in printing inks, *Am. Ink Maker (1923–2001)*, 73, 56, 58, 61–2, 64, 66, 68, 1995.
53. Resolution AP (2002) 1: On paper and board materials and articles intended to come into contact with foodstuffs, Council of Europe, September 18, 2002, www.coe.int/soc-sp.
54. Bureau, G. and Multon, J.L., *Food Packaging Technology*, Vol. 1–2, VCH, Weinheim, 1996.
55. Maeck, K., Determination of heavy metals in paper, *PTS-Manuskript*, (83), 7/1–7/17, 1999.

56. Toeppel, O., Griebenow, W., and Werthmann, B., Application of atomic absorption spectroscopy to the cellulose and paper fields. II. Developments and work towards the production and introduction of standardized AAS methods for the study and testing of cellulose and paper, *Papier* (Bingen, Germany), 31, 508, 1977.
57. Griebenow, W., Werthmann, B., and Toeppel, O., Application of atomic absorption spectroscopy to the cellulose and paper fields. I. Principles of atomic absorption spectroscopy and special application to the cellulose and paper fields, *Papier* (Bingen, Germany), 31, 503, 1977.
58. Knezevic, G. and Kurfuerst, U., Determination of heavy metals in papers — a comparison of methods (solid sampling and digestion analysis by graphite-tube AAS), *Fresenius Z. Anal. Chem.*, 322, 717, 1985.
59. Castle, L. et al., Migration studies from paper and board food packaging materials: 1. compositional analysis, *Food Addit. Contam.*, 14, 35, 1997.
60. Griebenow, W., Werthmann, B., and Schwarz, B., Cadmium contents of food packaging papers and paper in domestic use, *Papier* (Bingen, Germany), 39, 105, 1985.
61. Conti, M.E. and Botrè, F., The content of heavy metals in food packaging paper: an atomic absorption spectroscopy investigation, *Food Control*, 8, 131, 1997.
62. Conti, M.E., The content of heavy metals in food packaging paper boards: an atomic absorption spectroscopy investigation, *Food Res. Int.*, 30, 343, 1997.
63. European declaration on Paper Recovery, Confederation of European Paper Industries (CEPI), 2000, www.paperrecovery.org/files/English-164625A.pdf.
64. Jokinen, K. and Siren, K., Harmful residues in recycled fiber-metals and compounds, *Paperi ja Puu*, 77, 106, 109, 1995.
65. Garcia-Gomez, C., Carbonell, G., and Tarazona, J.V., Binding of cadmium on raw paper pulp: relationship between temperature and sorption kinetics, *Chemosphere*, 49, 533, 2002.
66. Triantafyllou, V.I., Akrida-Demertzi, K., and Demertzis, P.G., Migration studies from recycled paper packaging materials: development of an analytical method for rapid testing, *Anal. Chim. Acta*, 467, 253, 2002.
67. Song, Y.S., Park, H.J., and Komolprasert, V., Analytical procedure for quantifying five compounds suspected as possible contaminants in recycled paper/paperboard for food packaging, *J. Agric. Food Chem.*, 48, 5856, 2000.
68. Tice, P.A. and Offen, C.P., Odors and taints from paperboard food packaging, *Tappi J.*, 77, 149, 1994.
69. Sipilainen-Malm, T. et al, Purity of recycled fiber-based materials, *Food Addit. Contam.*, 14, 695, 1997.
70. Hamm, U. and Goettsching, L., Heavy metals in paper making process using waste paper as raw material: sources, depressions and valuation-criterions, *Papier* (Bingen, Germany), 43, V39, 1989.
71. Conti, M.E. et al., Heavy metals and optical whitenings as quality parameters of recycled paper for food packaging, *J. Food Process. Preservation*, 20, 1, 1996.
72. Seccombe, R., Brackenbury, K., and Vandenberg, D., Disperser bleaching with hydrogen peroxide — a tool for brightening recycled fibers, *Appita Annual Conference Proceedings*, 2002, p. 229, Rotorua (New Zealand).
73. Nada, A.M.A. and Hussein, A.Y., High efficiency recycling of newsprint. Cellulose and Paper Dep., National Research Centre, Cairo, Egypt., IPPTA, 13, 7, 2001.
74. Expert consultation WHO/FAO/IAEA: Trace Elements in Human Nutrition and Health, World Health Organization, Geneva, 1996.
75. Anderson, R.A., Chromium, glucose intolerance and diabetes, *J. Am. Coll. Nutr.*, 17, 548, 1998.

76. Krejpcio, Z., Essentiality of chromium for human nutrition and health, *Polish J. Environ. Stud.*, 10, 399, 2001.
77. Gauglhofer, J. and Bianchi, V., Chromium, in *Metals and Their Compounds in the Environment*, Merian, E., Ed., VCH, Weinheim, 1991, p. 853.
78. Partial Agreement in the Social and Public Health Field, Council of Europe, Policy statement concerning metals and alloys, Technical Document: Guidelines on metals and alloys used as food contact materials, February 13, 2002, www.coe.int/soc-sp.
79. Von Berg, R. and Liu, D., Chromium and hexavalent chromium, *J. Appl. Toxicol.*, 13, 225, 1993.
80. Kuligowski, J. and Halperin, K.M., Stainless steel cookware as a significant source of nickel, chromium, and iron, *Arch. Environ. Contam. Toxicol.*, 23, 211, 1992.
81. Offenbacher, E.G. and Pi-Sunyer, F.X., Temperature and pH effects on the release of chromium from stainless steel into water and fruit juices, *J. Agric. Food Chem.*, 31(1), 89, 1983.
82. Kumar, R., Srivastava, P.K., and Srivastava, S.P., Leaching of heavy metals (Cr, Fe, and Ni) from stainless steel utensils in food simulants and food materials, *Bull. Environ. Contam. Toxicol.*, 53, 259, 1994.
83. Berg, T. et al., The release of nickel and other trace elements from electric kettles and coffee machines, *Food Addit. Contam.*, 17, 189, 2000.
84. Jorhem, L. et al., Trace elements in crayfish: regional differences and changes induced by cooking, *Arch. Environ. Contam. Toxicol.*, 26, 137, 1994.
85. Flint, G.N. and Packirisamy, S., Systemic nickel: the contribution made by stainless-steel cooking utensils, *Contact Dermatitis*, 32, 218, 1995.
86. Flint, G.N. and Packirisamy, S., Purity of food cooked in stainless steel utensils, *Food Addit. Contam.*, 14, 115, 1997.
87. Haudrechy, P. et al., Innocuousness of stainless steels in contact with food or skin, in *Stainless Steels'96, Proceedings (2nd European Congress)*, Verein Deutscher Eisenhuettenleute, Düsseldorf, June 3–5, 1996, p. 228.
88. Accominotti, M. et al., Contribution to chromium and nickel enrichment during cooking of foods in stainless steel utensils, *Contact Dermatitis*, 38, 305, 1998.
89. Kawamura, Y. et al., Migration of metals from stainless steel kitchenware and tableware, *Shokuhin Eiseigaku Zasshi*, 38, 170, 1997.
90. Reilly, C., *Metal Contamination of Food*, 2nd ed., Elsevier Applied Science, Barking, England, 1991.
91. Blunden, S. and Wallace, T., Tin in canned food: a review and understanding of occurrence and effect, *Food Chem. Toxicol.*, 41, 1651, 2003.
92. Nriagu, J. O., A silent epidemic of environmental metal poisoning?, *Environ. Pollut.*, 50, 139, 1988.
93. JECFA, Evaluation of Certain Food Additives and Contaminants, Fifty-third report of the Joint FAO/WHO Expert Committee on Food Additives, World Health Organization, Technical Report Series 896, 53/81, 1999.
94. JECFA, Evaluation of Certain Food Additives and Contaminants. Fifty-fifth report of the Joint FAO/WHO Expert Committee on Food Additives, World Health Organization, Technical Report Series 901, 55/61, 2000.
95. Kent County Council, Trading Standards Report, September 2002, http://www.tradingstandards.gov.uk/kent/.
96. Jorhem, L. and Slorach, S., Lead, chromium, tin, iron and cadmium in foods in welded cans, *Food Addit. Contam.*, 4, 309, 1987.

97. Branca, P. and Pelizzone, G., Lead release from tinplated cans: theoretical and legal aspects and experimental results, *Bollettino dei Chimici dell'Unione Italiana dei Laboratori Provinciali*, 34, 325, 1983.
98. Stilwell, D.E. and Mustane, C.L., Lead content in grapefruit juice and its uptake upon storage in open containers, *J. Sci. Food Agric.*, 66, 405, 1994.
99. Herrmann, H.J., Ceramics, glass, enamel: release of harmful substances from everyday articles for foodstuffs, limit values, test, and analysis methods, *Keramische Z.*, 45, 267, 1993.
100. Beliles, R.P., The metals, in *Patty's Industrial Hygiene and Toxicology,* 4th ed., Clayton, G.D. and Clayton, F.E., John Wiley & Sons, New York, 1994, 2(Pt. C), 1879.
101. Directive 84/500/EEC: European Community, Council Directive on the approximation of the laws of the Member States relating to ceramic articles intended to come into contact with foodstuffs, October 15, 1984.
102. Sheets, R.W., Extraction of lead, cadmium and zinc from overglaze decorations on ceramic dinnerware by acidic and basic food substances, *Sci. Total Environ.*, 197, 167, 1997.
103. Gould, J. H. et al., Influence of automatic dishwashings and scrubbings on release of lead from glazed ceramicware, *J. Assoc. Off. Anal. Chem.*, 73, 401, 1990.
104. Ishiwata, H. et al., Determination of low levels of lead and cadmium released from ceramic ware into 4% acetic acid and grapefruit juice, *Shokuhin Eiseigaku Zasshi*, 32, 168, 1991.
105. Doemling, H. J., The extraction of cadmium from dishes, *Fresenius Z. Anal. Chem.*, 267, 118, 1973.
106. Clifford, J.F., Glaze Compositions. U.K. Patent, EP 509792, *Eur. Pat. Appl.*, 1992.
107. Mitra, Ina, S. et al., Abrasion-Resistant Adherent Glass Coating Compositions Free of Cadmium and Lead for Glass Ceramic Surfaces for Cooking Food, U.S. Patent 6525300, 2003.
108. Hight, S.C., Graphite furnace atomic absorption spectrometric determination of lead and cadmium extracted from ceramic foodware: collaborative study, *J. AOAC Int.*, 83, 1174, 2000.
109. Hight, S.C., Determination of lead and cadmium in ceramicware leach solutions by graphite furnace atomic absorption spectroscopy: method development and interlaboratory trial, *J. AOAC Int.*, 84, 861, 2001.
110. Lehman, R.L., International standards for lead and cadmium: release from ceramic foodware surfaces, *Ceramic Eng. Sci. Proc.*, 17, 129, 1996.
111. Seddon, A.B., and Whall, M.E., The extraction of lead from lead crystalware, *Glass Technol.*, 34, 71, 1993.
112. Guadagnino, E. et al., Estimation of lead intake from crystalware under conditions of consumer use, *Food Addit. Contam.*, 17, 205, 2000.
113. Guadagnino, E. et al., Parameters affecting lead migration from crystal glass in contact with 4% acetic acid, wine and brandy, *Rivista della Stazione Sperimentale del Vetro (Murano, Italy)*, 29(1), 5, 1999.
114. Barbee, S.J. and Constantine, L.A., Release of lead from crystal decanters under conditions of normal use, *Food Chem. Toxicol.*, 32, 285, 1994.
115. Frederes, K.P. and Varshneya, A.K., The leaching of lead and other metal ions from lead crystal glass, *Ceramic Trans. (Advances in Fusion and Processing of Glass)*, 29, 419, 1993.
116. Guadagnino, E. et al., Surface analysis of 24% lead crystal glass articles: correlation with lead release, *Glass Technol.*, 43, 63, 2002.

117. Hight, S.C., Lead migration from lead crystal wine glasses, *Food Addit. Contam.*, 13, 747, 1996.
118. Carelli, G. et al., Determination of lead in Coca Cola after leaching from crystal glasses by graphite furnace atomic absorption spectrometry (GFAAS), *Rivista della Stazione Sperimentale del Vetro (Murano, Italy)*, 29, 57, 1999.
119. 3Partial Agreement in the Social and Public Health Field. Council of Europe Draft Technical Document No 1. RD 5/1-42, Guidelines on Lead Leaching from Glass Tableware into Foodstuffs, 42nd session, Strasbourg, December 2–5, 2003, www.coe.int/soc-sp.
120. Doemling, H.J., Leaching of cadmium and lead from enamel and enamel colors: enamel utensils and glass drinking vessels, *Deutsche Lebensmittel-Rundschau*, 70, 439, 1974.
121. Guadagnino, E. and Quevauviller, Ph., Improvement of the quality control of glass analysis, part 1: interlaboratory studies, *Proc. Contr. Qual.*, 11, 147, 1998.
122. Quevauviller, Ph. and Guadagnino, E., Improvement of the quality control of glass analysis, part 2: certification of a glass reference material, *Proc. Contr. Qual.*, 11, 323, 1999.
123. Guadagnino, E. et al., Determination of hexavalent chromium as a requisite to comply with the packaging directive, *Rivista della Stazione Sperimentale del Vetro (Murano, Italy)*, 28, 63, 1998.
124. Guadagnino, E., Sundberg, P., and Corumluoglu, O., A collaborative study on the determination of hexavalent chromium in container glasses, *Glass Technol.*, 42, 148, 2001.
125. Guadagnino, E., Sundberg, P., and Heinrich, H.J., A collaborative study for the determination of mercury in glass packaging by cold vapor atomic absorption spectrometry, *Glass Technol.*, 42, 24, 2001.
126. Sherlock, J.C. and Smart, G.A., Tin in foods and the diet, *Food Addit. Contam.*, 1, 277, 1984.
127. Food Standards Agency (FSA), U.K., Tin in Canned Fruit and Vegetables, Report No. 29, 2002.
128. Codex, 1998, Position Paper on Tin Codex Committee on Food Additives and Contaminants, Thirtieth Session, The Hague Netherlands, March 1998. Joint FAO/WHO Food Standards Programme, CX/FAC/98/24.
129. JECFA, Evaluation of Certain Food Additives and Contaminants, Fifty-fifth Report of the Joint FAO/WHO Expert Committee on Food Additives, World Health Organization, Technical Report Series, 901, 55/69, 2000.
130. Ministry of Agriculture, Fisheries and Food of U.K. (MAAF), The Tin in Food Regulations, S.I. 496, HMSO, London, 1992.
131. Codex Alimentarius Commission, Doc. No. CX/FAC 96/17, Joint FAO/WHO Food Standards Programme, Codex General Standard for Contaminants and Toxins in Foods, 1995.
132. Albu-Yaron, A. and Semel, A., Nitrate-induced corrosion of tin plate as affected by organic acid food components, *J. Agric. Food Chem.*, 24, 344, 1976.
133. Juliàn, B. et al., A study of the method of synthesis and chromatic properties of the Cr-SnO2 pigment, *Eur. J. Inorg. Chem.*, 10, 2694, 2002.
134. Ministry of Agriculture, Fisheries and Food of U.K. (MAAF), Total Diet Study: Metals and Other Elements, Food Surveillance Information Sheet 131, HMSO, London, 1997.

135. Ministry of Agriculture, Fisheries and Food of U.K. (MAAF), Survey of Lead and Tin in Canned Fruits and Vegetables, Food Surveillance Information Sheet 122, HMSO, London, 1999.
136. World Health Organization, *Guidelines for Drinking-Water Quality: Recommendations*, Vol. 1, Geneva, 1993.
137. WHO, World Health Organization, International Programme on Chemical Safety (IPCS), Environmental Health Criteria, 194, Aluminum, 1997, www.inchem.org/documents/ehc/ehc/ehc194.htm.
138. JECFA, Evaluation of Certain Food Additives and Contaminants, Thirty-third Report of the Joint FAO/WHO Expert Committee on Food Additives, World Health Organization, Technical Report Series, 776, 33/26, 1988.
139. Birchall, J.D. and Chappell, J.S., Aluminum, chemical physiology, and Alzheimer's disease, *Lancet*, 2, 1008, 1988.
140. Pennington, J.A.T., Aluminum content of foods and diets, *Food Addit. Contam.*, 5, 161, 1988.
141. Pennington, J.A.T. and Jones, J.W., Dietary intake of aluminum, in *Aluminum and Health — A Critical Review*, Gitelman, H.J., Ed., Marcel Dekker, New York, 1989, p. 67.
142. Zhou, C.Y. et al., The behavior of leached aluminum in tea infusions, *Sci. Total Environ.*, 177, 9, 1996.
143. Fung, K.F. et al., Aluminum and fluoride concentrations of three tea varieties growing at Lantau Island, Hong Kong, *Environ. Geochem. Health*, 25, 219, 2003.
144. Rajwanshi, P. et al., Studies on aluminum leaching from cookware in tea and coffee and estimation of aluminum content in toothpaste, baking powder and paan masala, *Sci. Total Environ.*, 193, 243, 1997.
145. Savory, J., Nicholson, J.R., and Wills, M.R., Is aluminum leaching enhanced by fluoride?, *Nature*, 327, 107, 1987.
146. Joshi, S.P. et al., Detection of aluminum residue in sauces packaged in aluminum pouches, *Food Chem.*, 83, 383, 2003.
147. Sugden, J.K. and Sweet, N.C., A study of the leaching of aluminum ions from drink containers, *Pharm. Acta Helv.*, 64(5–6), 130, 1989.
148. Greger, J.L., Goetz, W., and Sullivan, D., Aluminum levels in foods cooked and stored in aluminum pans, trays and foil, *J. Food Prot.*, 48, 772, 1985.
149. Vela, M.M. et al., Detection of aluminum residue in fresh and stored canned beer, *Food Chem.*, 63, 235, 1998.
150. Seruga, M. and Hasenay, D., Corrosion of aluminum in soft drinks, *Z. Lebensm.-Unters. Forsch.*, 202, 308, 1996.
151. Seruga, M., Grgic, J., and Mandic, M., Aluminum content of soft drinks from aluminum cans, *Z. Lebensm.-Unters. Forsch.*, 198, 313, 1994.
152. Lione, A., The prophylactic reduction of aluminum intake, *Food Chem. Toxicol.*, 21, 103, 1983.
153. Muller, J.P., Steinegger, A., and Schlatter, C., Contribution of aluminum from packaging materials and cooking utensils to the daily aluminum intake, *Z. Lebensm.-Unters. Forsch.*, 197, 332, 1993.
154. Liukkonen-Lilja, H. and Piepponen, S., Leaching of aluminum from aluminum dishes and packages, *Food Addit. Contam.*, 9, 213, 1992.
155. Nagy, E. and Jobst, K., Aluminum dissolved from kitchen utensils, *Bull. Environ. Contam. Toxicol.*, 52, 396, 1994.
156. Mei, L. and Yao, T., Aluminum contamination of food from using aluminum ware, *Int. J. Environ. Anal. Chem.*, 50, 1, 1993.

157. Von Burg, R. and Greenwood, M.R., Mercury, in *Metals and Their Compounds in the Environment*, Merian, E., Ed., VCH, Weinheim, 1991, p. 1045.
158. JECFA, Evaluation of Certain Food Additives and Contaminants, Twenty-second Report of the Joint FAO/WHO Expert Committee on Food Additives, World Health Organization, Technical Report Series, 631, 22/26, 1978.

10 Pollutants in Food – Metals and Metalloids

Conor Reilly

CONTENTS

10.1 METALS AND METALLOIDS IN FOOD

Analysis of almost any plant sample will detect the presence of most of the metals and metalloids. Concentrations normally range from grams or milligrams down to almost undetectable trace amounts. It would be surprising if food did not contain a similar range because the tissues of plants can be expected to contain most of, if not all, the metals and metalloids that occur in soil.

Though food is likely to contain a wide range of metals and metalloids, many are at such low levels of concentration to hardly merit the designation of pollutant. However, there is a smaller group discussed in the following sections, which are known to be potentially serious contaminants and, as such, are included in the food standard regulations of many countries.

10.2 SOIL AS A SOURCE OF METALS AND METALLOIDS IN FOOD

Fertile soils contain, as well as the elements needed for plant growth, a wide range of other metals and metalloids in varying concentrations. These can range from as high as 700 to < 10,000 mg/kg of Al, 100 to 320,000 mg/kg of Ca, to lows of < 0.1 to 2.5 mg/kg of Ge and < 0.1 to 4.3 mg/kg of Se.[1] The amounts of the elements taken up by plants depend primarily on their concentrations in the soil. If any of those required for growth are in low concentrations, plants may suffer from deficiency. However, if concentrations are high, they may show signs of toxicity. Other elements also present may not harm the plants but could be a hazard to consumers.

10.2.1 Pollution of Food Plants Grown on Contaminated Soil

Amendment of agricultural soil can result in significant changes in its metal profile. This, in its turn, can result in changes in the pattern of uptake by crops. Fertilizers can sometimes contain high levels of toxic metals and cause contamination in crops.[2]

10.2.2 Metal Uptake by Foods from Agrochemicals

A number of chemicals used in agriculture can also cause problems. Copper fungicide used on grapes has resulted in contamination of wine. Organic mercurials, once widely used as antifungal seed dressings, have been responsible for Hg poisonings.[3] The use of Pb arsenate insecticides in orchards has been associated with high As levels in cider and perry.[4]

10.2.3 Metal Contamination of Food Caused by Sewage Sludge

The use of sewage sludge as a top dressing can result in contamination of crops. Sludge often contains a range of potentially toxic elements (PTEs), with, for example, up to 1,500 mg/kg Cd, 8,000 mg/kg Cu, 62,000 mg/kg Fe, and 49,000 mg/kg Zn.[5] Because of the dangers posed by crops grown on sludge-treated soil, restrictions have been placed in many countries on the practice.[6]

10.3 INDUSTRIAL CONTAMINATION AS A SOURCE OF METALS IN FOOD

Contamination of food by industrial activities is widespread. Wastewater was responsible for a buildup of Cd in rice in several mining regions in Japan in the 1960s.[7] High levels of Cd in vegetables grown in a U.K. village were traced to leakage of water from old mine workings.[8] Mercury contamination of domestic animals near gold mining operations has been reported in Brazil.[9]

Mercury used as a catalyst in a factory caused the notorious Minamata tragedy in Japan in the 1950s.[10] A U.K. study carried out found that one river in an industrialized area carried high levels of a wide range of potentially toxic metals, with resulting accumulation in fish.[11] A study of the effects of industrial emissions in the Ural region of Russia found significant contamination of crops, with levels of 3.6 μg/g As in 94% of potatoes and in 25% of other plant foods.[12]

10.4 CONTAMINATION OF FOOD FROM PROCESSING EQUIPMENT

Food processing equipment has long been recognized as a source of metal contamination, though in modern plants high-quality metals approved for use with foods normally prevent this. Some types of detergents used to clean equipment can result in leaching from stainless steel of As, Pb, Hg, and Cd.[13]

10.4.1 Metal Pick-Up during Canning

Ever since canning was introduced in the 19th Century, problems have arisen with regard to metal pickup by food, especially of Pb. So serious was this problem in the early days that there was a move in some countries to have the process banned.[14] Though the use of improved techniques has more or less solved the problem, contamination still occurs. Up to 10 mg/kg of Pb has been reported in canned fruit,[15] and 114 mg/kg in canned evaporated milk.[16] Uptake is greatest in acidic foods and is related to temperature and length of storage. Nitrate can also increase uptake.[17]

10.4.2 Contamination from Cooking Utensils

Tinned copper pans, formerly used in many kitchens, can still be a source of metal contamination where traditional catering methods are employed. Cabbage cooked in such utensils has been shown to increase its Pb content from 0.15 to 0.79 mg/kg, and Cu from 1.36 to 2.07 mg/kg.[18] Cast iron pots has been linked to incidence of iron overload in traditional communities in South Africa.[19] Indian Childhood Cirrhosis (ICC), in which excess Cu accumulates in the liver, may be related to the use of brass utensils to prepare infant food.

10.4.3 Ceramicware as a Source of Contamination

Poorly made ceramic utensils can leach a variety of potentially toxic elements (PTEs), unlike pottery made under controlled conditions in conformity with IOS (International Organisation for Standards) regulations. According to the FDA, lead-glazed ceramicware continues to be one of the major sources of dietary Pb in the US.[20]

10.4.4 Contamination from Plastic and Other Wrappings and Containers

Other domestic sources of contamination include decoration on food and beverage containers and wrappers. A Japanese study found high levels of Cd and Pb in the pigments used on drinking glasses.[21] High levels of Pb have been detected also on bread wrappers. Other metals, including Hg, Co, Cr, and Cd, are used to decorate confectionery wrappings.[22]

10.5 LEAD CONTAMINATION OF FOOD AND BEVERAGES

Lead has long been recognized as toxic. Writers in ancient Rome referred to a disease known as plumbism that afflicted shipbuilders who used white Pb in their work. A19th Century U.K. law prohibited pottery workers eating in areas where Pb glazes were used because of Pb poisoning. Today Pb in food is still a major concern.

10.5.1 Human Biology of Lead

Lead is found in every organ of the body, with a total of 100 to 400 mg in an adult male. Absorption from food in adults is about 10%, but in children it may be up to 50%, especially if the diet is low in calcium and protein and high in sugar.[23] Lead is transported around the body attached to blood cells. About 5% is placed in soft tissues where it forms an exchangeable compartment. The remainder is firmly sequestered in bone, though, under some conditions of stress, it can be released into the blood.[24] While unabsorbed Pb is excreted in the feces, about three quarters of the absorbed Pb will eventually leave the body in urine. In women some Pb appears in milk. Normally, the amount is low, between 2 and 5 μg/l, but under conditions of environmental contamination it may be much higher.[25]

10.5.1.1 Lead Toxicity

Though it has been claimed that Pb is essential for certain species of mammals,[26] there is little doubt that it has no nutritional benefits for humans and can cause both chronic and acute poisoning. It affects many different functions, with bone marrow, where hemoglobin synthesis is carried out, as its "critical organ" target.[27] Symptoms of acute inorganic Pb poisoning are relatively easily recognized. It usually manifests itself in gastrointestinal effects, accompanied by general weakness and malaise.[28] Encephalopathy is rare in adults but more common in children.

The effects of low-level chronic Pb poisoning are not so easily recognized. Symptoms attributed to it, such as tiredness and mild anaemia, can also be due to other causes. There is evidence that in children it interferes with neurophysiological development.[29]

The organic compounds of Pb are, in general, more toxic than inorganic compounds. They primarily affect the central nervous system. Tetraethyl Pb (TEL), for example, is rapidly distributed to various tissues, particularly brain, where it decomposes to triethyl Pb and inorganic Pb, resulting in toxic psychosis.

10.5.2 Lead in Food and Beverages

Lead is present in all foods and beverages primarily as a natural component but also as an accidental additive. Overall levels in foods throughout the world are surprisingly uniform, ranging from about 0.01 to 0.25 mg/kg, as seen in Table 10.1. There are, however, exceptions. In animals Pb may accumulate in the kidneys and other organs. Levels of 0.30 mg/kg have been detected in the kidneys in cows, and 0.52 mg/kg in pig, in Sweden.[30] In the U.K., levels in offal of 0.04 to 0.37 mg/kg have been recorded.[31]

10.5.2.1 Lead in Canned Foods

In recent years there has been a considerable reduction in levels in canned foods. However, they still are a major source of Pb in some diets.[32] The mean level in canned foods, determined in the U.S. Market Basket Survey of 1990–1991, was 0.01 mg/kg.[20] This is close to the 0.02 mg/kg found in the U.K. Total Diet Study.[33]

TABLE 10.1
Lead Content of Foods in Different Countries (mg/kg)

Food	Canada	Japan	U.K.
Cereals	0.012–0.078	0.092 (mean)	0.01–0.04
Meat/fish	0.011–0.121	0.186 (mean)	< 0.01–0.10
Dairy foods	0.001–0.082	0.032 (mean)	< 0.01–0.02
Vegetables	0.006–0.254	0.090–0.257	< 0.01–0.02

Source: Data adapted from Dabeka, R.W., McKenzie, A.D., and Lacroix, G.M.A., Dietary intake of lead, cadmium, arsenic and fluoride by Canadian adults: a 24 h duplicate diet study, *Food Addit. Contam.*, 4, 89–102, 1987; Ministry of Agriculture, Fisheries and Food, Lead, Arsenic and Other Metals in Food. Food Surveillance Paper No. 52, The Stationery Office, London, 1998; Teraoka, H., Morii, F. and Kobayashi, J., The concentration of 24 elements in foodstuffs and estimation of their daily intake, *Eiyo to Shokuryo*, 32, 221, 1981.

10.5.2.2 Lead in Water

Most natural waters contain about 5 μg Pb/l. Normally, municipal supplies contain well below this level, but there can be exceptions, especially in old urban areas where Pb plumbing is used. Surveys in the U.K. in the 1980s found that almost half the water samples in certain cities were above the WHO limit of 50 μg/l.[34]

Lead in domestic water can be a significant source of the metal in the diet. Infants in older houses in a U.K. city have been found to have an average intake of 3.4 mg/week, 0.4 mg higher than the WHO Provisional Tolerable Weekly Intake (PTWI) for adults. Intakes were even higher when food was prepared using water from the hot rather than the cold tap.[35]

10.5.2.3 Lead in Alcoholic Beverages

Lead levels of up to 100 μg/l have been detected in some wines. This is due, according to the U.S. Bureau of Alcohol, Tobacco, and Firearms (BATF) at least partly to tin-coated Pb foil caps on bottles.[36] In the European Community, Pb foil containing more than 1% Pb may not be used for this purpose.[37] Levels in alcoholic beverages can also be increased by storage in Pb crystal decanters[38] and in some glazed pottery vessels.[39]

10.5.2.4 Lead in Vegetables

Vegetables can be a source of high Pb intake when grown in urban areas. Soil in some urban areas in the U.K. contains as much as 1676 mg/kg (dry weight). Lead levels in vegetables grown on these soils have been found to exceed the statutory limit of 1 mg/kg, with, for example, 1.5 mg/kg in cabbage and 1.7 mg/kg in spinach.[31]

10.5.3 Dietary Intake of Lead

In recent decades a downward trend in dietary Pb has occurred in many countries. In the U.K., intakes between 1976 and 1994 decreased from 0.11 to 0.024 mg/day, pointing to the success of measures taken to reduce environmental Pb, such as replacement of soldered cans and phasing out of leaded petrol.[40] Not all countries have been so successful. In Japan, after an initial fall in the 1980s, dietary levels increased again, probably because of changes in eating habits such as increased wine consumption.[41] Significant decreases have occurred in the U.S. where intakes by children fell from 0.37 to 0.2 μg/kg b.w./d between 1986 and 1991.[43]

10.6 MERCURY CONTAMINATION OF FOOD AND BEVERAGES

To the ancient Greeks, Hg was known as *hydrargium* (liquid silver). It was valued for its medicinal properties and for other uses, such as silvering of mirrors. But its less welcome properties were also recognized. Those who worked with Hg could develop quicksilver disease,[43] a condition later made famous by the Mad Hatter of Lewis Carroll's *Alice in Wonderland*.[44] Mercury still causes problems, and WHO has raised concerns about its levels in foods and implications for health.[45]

One of the reasons for this concern is that up to 150,000 ton of Hg vapor are released annually into the atmosphere from the Earth's surface.[46] The vapor that returns to Earth in rainwater, and, especially, in lake and ocean sediments, is methylated by microorganisms. MethylHg enters the food chain via plankton feeders. At the end of the chain are large carnivorous fish such as marlin and shark that can accumulate considerable amounts of the metal and pose a risk to consumers.

10.6.1 Biological Effects of Mercury

Mercury is a cumulative poison. It is stored in the liver and certain other tissues. The amount taken up depends on the chemical form ingested. Elemental Hg and its inorganic compounds are poorly absorbed in contrast to the vapor that readily enters through the lungs.

Compounds of inorganic Hg, such as mercuric chloride (corrosive sublimate) in particular, are highly toxic, far more so than is elemental Hg itself. The greater part of Hg^{2+} is concentrated in the kidneys where it causes severe damage leading to kidney failure.

MethylHg has been listed as one of the six most dangerous environmental chemicals.[47] Up to 90% of most organic Hg compounds is absorbed from food, with the central nervous system as the central target. More than 95% of Hg in brain tissue is organic. MethylHg can cross the placenta and affect the fetal brain. Even after birth the infant can continue to be affected by the methylHg ingested by its mother, which appears in her milk.[48] In adults, clinical signs of poisoning include ataxia, tremor, slurred speech, blindness, loss of hearing, and death.[49]

TABLE 10.2
Mercury in Foods in the U.K.

Food Group	Mean Concentration (mg/kg Fresh Weight)
Bread, other cereal products	0.004
Meat, meat products	0.003
Offal	0.006
Fish	0.054
Vegetables (green)	0.002
Beverages	0.0006
Milk	0.0007
Nuts	0.003

Source: Data adapted from Ysart, G. et al., Dietary exposure estimates of 30 elements from the U.K. Total Diet Study, *Food Addit. Contam.*, 16, 391, 1999.

10.6.2 Mercury in Food and Diets

Mercury can be detected in most foods and beverages, at levels of < 1 to 50 μg/kg, as seen in Table 10.2.[41] Higher levels are often found in marine foods, as well as in offal.[50] Fish in polluted waters can contain as much as 1.61 mg/kg, and lettuce grown on sludge-amended soil 40 μg/kg.[51]

10.6.3 Dietary Intake of Mercury

Dietary intakes differ between countries, depending largely on the amount of fish consumed. In the U.K., the daily intake of fish by adults has been estimated to be on average 5 μg,[41] somewhat similar to the U.S. intake of 8 μg/d. In contrast, intakes of 78 μg have been recorded in Egypt.[52] Even though fish consumption is not high in the U.K., it still is a major contributor of Hg, with some 25% of total intake, three quarters of which is methylHg, coming from this source. Because of the high concentrations of Hg in deep-sea carnivorous fish, the U.K. Food Standards Authority has advised women who intend to become pregnant to avoid eating shark, marlin, and swordfish.[53] Similar advice has been given in the U.S., though the need to do so has been questioned by some experts.[54]

10.7 CADMIUM CONTAMINATION OF FOOD AND BEVERAGES

Cadmium has been described as "one of the most dangerous trace elements in the food and environment of man."[50] The danger to human health was brought to the world's attention by the itai-itai tragedy in post-WW II Japan, when rice fields were polluted with Cd contained in factory effluents.[55] The metal has again been attracting attention as evidence is growing that even accepted safe levels of intake may be too high.[56]

TABLE 10.3
Cadmium in Foods in the U.K.

Food Group	Mean Concentration (mg/kg Fresh Weight)
Bread	0.03
Meat	0.001
Offal	0.07
Fish	0.02
Vegetables (green)	0.006
Potatoes	0.03
Beverages	0.001
Milk	0.001
Nuts	0.05

Soucrce: Data adapted from Ysart, G. et al., Dietary exposure estimates of 30 elements from the U.K. Total Diet Study, *Food Addit. Contam.*, 16, 391, 1999.

10.7.1 Cadmium in Food and Beverages

Cadmium is found at low levels in most foods, as shown in Table 10.3. Similar levels have been reported in other countries.[57] Meat offal and seafoods are richer sources of Cd than other foods. A positive correlation has been found between fish consumption and blood Cd.[58]

The use of Cd-contaminated phosphate fertilizers has been responsible for high levels in potatoes. Cadmium released from industrial plants is believed to cause contamination of 9.5% of paddy fields in Japan.[59] Several areas in Eastern Europe have suffered from this problem.[60] Contamination of garden vegetables has occurred on reclaimed "brown field" sites in the U.K.[61]

10.7.1.1 Cadmium in Water and Other Beverages

Levels of Cd in domestic water are generally less than 1 μg/l. Contamination can occur when zinc-plated (galvanized) pipes and cisterns are used. Up to 21 μg/l in drinking water in a Scottish hospital were traced to this source. In Australia, where in rural areas rain is stored in galvanized tanks, levels of 3.6 μg/l have been detected in the water.[62] Soft drinks in vending machines with cadmium-plated parts have been found to contain up to 16 mg/l.[63]

10.7.2 Dietary Intake of Cadmium

Dietary exposure to cadmium in Europe ranges from 5 to 57 μg/d.[64] Intakes in the U.K., at 14 μg/d, and the U.S. at 18 μg/d, are in the lower quarter of this range.[41] Intakes by individuals may be much higher if they are smokers. Some cigarettes contain up to 2 μg Cd/g (d.w), which can be absorbed through the lungs. It is estimated that 20 cigarettes can contribute 2.0 μg of Cd to the body's burden.[65]

10.7.3 Metabolism and Health Effects of Cadmium

About 6% of Cd in food is absorbed from the gut. Phytate and some other food components can affect absorption. Cd is bound to albumin for transport in the blood. It is taken up by the liver where it induces synthesis of metallothionein (MT). The Cd–MT complex is released to the blood and travels to the kidneys where it is retained until the cells' synthetic capacity for MT is exceeded and the Cd is released. It can then cause irreversible renal damage.[66]

Ingestion of Cd initially causes nausea, vomiting, and abdominal cramp and, sometimes, shock. Long-term ingestion can cause kidney damage, anemia, and a severe loss of bone minerals resulting in extreme bone fragility. A diet low in protein and Ca may exacerbate these effects.[67]

There is some evidence that, at least in industrially exposed workers, a high Cd intake can increase the incidence of some cancers. There is no convincing evidence that Cd in food has the same effect. However, the finding of chromosome aberrations in victims of Cd poisoning suggests that Cd in food may have teratogenic effects.[68]

10.8 ARSENIC CONTAMINATION OF FOOD AND BEVERAGES

Arsenic is one of the most serious of the inorganic contaminants of the diet in some countries. According to the WHO, more than 20 countries have significant As contamination of groundwater, with millions of people at risk of poisoning.[69] Elsewhere, arsenic contamination of food and drink occurs only occasionally, though still sufficiently often to attract the vigilance of health authorities.[70]

10.8.1 Arsenic in Food and Beverages

Arsenic is found in most foods. Levels rarely exceed 1 mg/kg, except in seafoods, as shown in Table 10.4. Certain types of mushroom have been reported to contain high As levels. Some members of the widely consumed genus *Agaricus* contain more than 10 mg As/kg, d.w, and the less commonly used *Laccaria amethystina* up to 200 mg As/kg, d.w.[71]

In addition to free-swimming fish, other marine foods, such as shellfish and seaweed, can contribute significantly high levels of As to the diet. A U.K. study found 26 mg/kg in crab meat, and 170 mg/kg in prawns. There were lower levels in ocean haddock (1.0–6.0 mg/kg) and herrings (< 0.5–2.4 mg/kg), with higher levels in flounder (0.2–34 mg/kg) and sole (0.5–24 mg/kg) taken in a polluted estuary.[70]

Most of the As in marine products is in the organic form and is considerably less toxic than its inorganic compounds. In most countries, regulations that set the maximum permitted levels for As in foods refer only to the total amounts of the element. In the U.K. the maximum permitted level is 1 mg/kg in unspecified food, and between 0.1 and 10 mg/kg in certain specified foods. No distinction is drawn between organic and inorganic forms. A footnote to the regulations states that the limit does not apply to "fish, edible seaweed, or any product containing fish or edible

TABLE 10.4
Total Arsenic in Foods in the U.K.

Food Group	Mean Concentration (mg/kg Fresh Weight)
Bread	0.008
Meat	0.004
Offal	0.004
Fish	4.3
Vegetables (green)	0.003
Potatoes	0.005
Beverages	0.002
Milk	0.002
Nuts	0.009

Source: Data adapted from Ysart, G. et al., Dietary exposure estimates of 30 elements from the U.K. Total Diet Study, *Food Addit. Contam.*, 16, 391, 1999.

seaweed, where As is naturally present at levels above 1 ppm in that fish or seaweed."[72]

10.8.1.1 Arsenic in Water

Most potable waters contain some As, usually inorganic and usually in the lower reaches of a range of 0 to 200 μg/l.[73] Springs and spas can have higher levels than domestic water. This is recognized in the standards of some countries, such as the U.K., where higher levels are permitted in bottled mineral waters than in home supplies.[74]

Very high levels of As occur in groundwater in some countries. Up to 4 mg/l of inorganic As has been found in domestic water supplies in volcanic regions of South America.[75] In Taiwan the use of similarly contaminated water is believed to be responsible for "blackfoot disease."[76] Levels of up to 100 μg As/l, and in some cases even 1 mg/l, have been detected in water from deep tube wells in Bangladesh and neighboring parts of India. It is estimated that up to 40% of the wells in some regions of Bangladesh are contaminated.[77]

10.8.2 Other Sources of High Arsenic Intake

Industrial pollution is responsible for high levels of As in water and crops in some places. A coal-burning power station in the Czech Republic, which emitted more than 1 ton of As daily, seriously contaminated nearby surface waters and crops.[78] In the U.K., chronic As poisoning of farm animals has occurred in the vicinity of brick kilns and old mine workings.[79]

Accidental contamination of food with As has resulted in a number of serious poisonings. In the U.K., in the early 20th Century, As-contaminated beer poisoned more than 6000 drinkers.[70] The source of the contamination was As-containing sulphuric acid used to hydrolyze starch for fermentation. Contaminated sodium phosphate in infant formula poisoned 12,000 infants in Japan in the 1950s.[80]

10.8.3 Dietary Intake of Arsenic

The total U.K. population exposure to As in food, not including water, was estimated in 1997 at 63 μg/d, with a mean adult intake of 120 μg and an upper range of 420 μg.[41] The U.S. intakes were somewhat lower, with a mean intake of 38 μg/d in 1991.[81] Higher intakes have been reported in Spain (286 μg) and New Zealand (150 μg), apparently related to a higher intake of fish.[82]

10.8.4 Absorption and Metabolism of Arsenic

All As compounds are easily absorbed from food in the GI tract. It is then rapidly distributed to the different organs and subsequently redistributed to skin, nails, and hair, where it is tightly bound to keratin. After about 24 h the As begins to be released from the organs and excreted in urine, but is retained in the skin and hair.

The metabolism of inorganic As differs between animal species. In humans As is reduced and methylated in liver to generate less toxic organic compounds that are excreted in urine. Inorganic As species are large ligands, and their toxicity arises from their ability to attach themselves to sulphydryl groups of enzymes and block their action. Arsenic oxyanions, because their stereochemical structure resembles sulphate and phosphate, can enter mitochondria and compete with phosphate as a substrate to form high-energy compounds that are unstable and that uncouple oxidative phosphorylation. In contrast, organic As compounds are rapidily excreted unchanged, without any toxic effects.[83]

Symptoms of acute As poisoning include vomiting and hemorrhagic diarrhea, in some cases within minutes of ingestion. Hematuria and acute renal failure may follow. Chronic low-level As poisoning results in decreased appetite and weight loss, possibly with abdominal pain, facial oedema, respiratory difficulty, and obstructive jaundice.[84] Other symptoms include peripheral neuritis, skin problems with pigmentation and ulcers, and, possibly, higher-than-average levels of cancer.[85]

There is some evidence that As may play an essential role in mammals.[86] Deprivation of As in animals has been associated with depressed growth and abnormal reproduction. Arsenic has also been shown to activate some enzymes and may function as a substitute for phosphates in metabolism. However, whether these findings justify inclusion of As among essential elements is still uncertain.[87]

10.9 COPPER CONTAMINATION OF FOOD AND BEVERAGES

Copper was probably among the first of the metals to be used by humans because it is relatively easily extracted from ores and is sometimes found in its metallic state. For thousands of years it has been used for making vessels for cooking and storing food. This use was greatly extended when its ability to form a tough alloy with tin was discovered, and the Bronze Age began. In Asia, Cu also played a part in the preparation and storage of food and beverages, in the form of brass, its alloy with Zn. Today, though these uses have been greatly reduced, Cu and its alloys still play significant roles in food preparation and contamination.

TABLE 10.5
Copper in Foods in the U.K.

Food Group	Mean Concentration (mg/kg Fresh Weight)
Bread	1.6
Meat	1.4
Offal	40
Fish	1.1
Vegetables (green)	0.84
Potatoes	1.3
Beverages	0.10
Milk	0.05
Nuts	8.5

Source: Data adapted from Ysart, G. et al., Dietary exposure estimates of 30 elements from the U.K. Total Diet Study, *Food Addit. Contam.*, 16, 391, 1999.

10.9.1 Copper in Food and Beverages

Copper is present in all foods, with concentrations in most ranging from about 0.05 to 2.0 mg/kg. Levels are usually higher in foods such as meat offal and nuts, as shown in Table 10.5. The best food sources of Cu are, in addition to offal and nuts, cereals, especially wholegrain cereals. Natural grains can contain up to 4 mg/kg, whereas some ready-to-eat breakfast cereals have up to 10 mg/kg.[88]

10.9.1.1 Copper in Drinking Water

Levels in naturally occurring water supplies can vary widely. Public supplies should normally not exceed the WHO standard of 50 μg/l, but this can sometimes be exceeded in water that has passed through Cu pipes, especially if it is taken from the hot tap. A study in Australia found that the daily consumption of 1.5 l of cold tap water could give an intake of 80 μg of Cu, whereas from the hot tap the intake was 12.3 mg, equivalent to a concentration of 8.2 mg Cu/l.[89] Copper intoxication has been reported in U.S. children who habitually drank water that contained 1.6 mg Cu/l (higher than the U.S. upper limit of 1.0 mg/l).[90]

10.9.1.2 Adventitious Copper in Food

Food cooked in a Cu saucepan can contain twice as much Cu as the same food cooked in a stainless steel or Al utensil. The preparation of infant food in brass utensils has been associated with Indian childhood cirrhosis (ICC), in which excessive amounts of Cu accumulate in the liver.[91] A study in Austria found an apparent connection between the use of domestic water contaminated with Cu and outbreaks of Tyrolean childhood cirrhosis.[92]

10.9.2 Dietary Intake of Copper

Copper intakes reported in several countries range from about 1 to 3 mg/d. The U.K. total diet study (TDS) of 1994 found that adults consumed on average 1.4 mg/d, with an upper range of 3.0 mg.[42] In the U.S., intakes for adults varied little over a 9 year period from 1.21 mg/d in 1982 to 1.16 mg/d in 1991.[93]

Because Cu is an essential trace metal nutrient, it has been assigned dietary reference values in some countries. However, establishing a recommended intake has proved difficult, mainly because of lack of clarity about the metal's metabolic role. An ESADDI (estimated safe and adequate daily dietary intake) for adults of 1.5 to 3 mg Cu/d was proposed in the U.S. in 1989. In 2001 this was replaced by an RDA of 0.9 mg/d for adults, with a UL (tolerable upper intake level) of 10 mg/d.[94] The current RNI in the U.K. for adults is 1.2 mg/d.[87]

10.9.3 Absorption and Metabolism of Copper

The levels of Cu in the body are under homeostatic control over a wide range of intakes.[95] Between 30 and 60% of ingested Cu is absorbed, but the uptake can be affected by other components of the diet, such as phytate and other metals. Absorbed Cu is rapidly transferred in the blood to the liver, kidney, and brain. Much of the Cu circulating in blood is in the form of ceruloplasmin (Cp), a Cu protein synthesized in liver. Cp has important metabolic roles, such as regulation of transport of iron and other metals, and hemoglobin synthesis.

Most of the approximately 100 mg of Cu in an adult body is tightly bound to about 20 proteins, in addition to Cp. Several of these are metalloenzymes and include cytochrome *c* oxidase that is involved in ATP production, superoxide dismutase (SOD) that functions in oxygen metabolism, and tyrosinase that is necessary for the production of the pigment melanin. Copper deficiency can result in low levels of activity of these enzymes, and has been associated with anaemia, hypopigmentation, and abnormalities of glucose and cholesterol metabolism and heart function.[96] There are two hereditary diseases involving Cu: Menkes' syndrome, characterized by an inability to absorb Cu, and Wilson's disease, in which excessive amounts are accumulated.[97]

High intakes of Cu can be toxic, though because its effect is initially emetic, it is difficult to retain enough in the body to be fatal. Chronic intake can have long-term effects, including liver damage. However, doses of up to 0.5 mg/kg body weight/d are considered to be safe for adults.[98]

10.10 ZINC CONTAMINATION OF FOOD AND BEVERAGES

Zinc is another essential nutrient that can also be a food contaminant. Zinc deficiency in humans was first recognized in the 1960s when Zn-responsive dwarfish was detected in children in Egypt.[99] Zinc has now joined iodine and iron among trace elements, deficiencies of which are worldwide and urgently need to be addressed.[100]

TABLE 10.6
Zinc in Foods

Food Group	Mean Concentration (mg/kg Fresh Weight)
Bread	9.0
Meat	51
Offal	43
Fish	9.1
Vegetables (green)	3.4
Potatoes	4.5
Beverages	0.3
Milk	3.5
Nuts	31

Source: Data adapted from Ysart, G. et al., Dietary exposure estimates of 30 elements from the U.K. Total Diet Study, *Food Addit. Contam.*, 16, 391, 1999.

10.10.1 Zinc in Food and Beverages

In Western societies, upwards of 70% of Zn consumed comes from animal products.[101] It is found in most foods at levels normally of about 1 to 40 mg/kg. Higher levels are found in several foods, such as oysters, which can contain up to 1 g/kg. Other good sources are liver and other organ meats, as well as seeds, nuts, and wholegrain cereals. Fortified breakfast cereals may also contain high levels. Zinc levels in a variety of foods consumed in the U.K. are shown in Table 10.6.

10.10.2 Dietary Intake of Zinc

The average adult intake in the U.K. has been estimated to be about 9 to 12 mg/d, similar to those reported in the U.S. (10 to 15 mg/d).[102] In parts of the world, however, where meat does not form a major component of the diet, intakes can be considerably lower, in some cases to as little as 5 mg/d.[103]

10.10.3 Absorption and Metabolism of Zinc

Absorption of Zn from food ranges from about 10 to more than 90%, depending on the presence or absence of other components of the diet. Absorption can be restricted by phytate and fibre, as well as by competition for uptake by other elements, especially iron, Cu, and Cd. Calcium can also reduce uptake, but it is enhanced by animal protein. A diet rich in wholemeal bread, which contains several of these antagonists, has been shown to cause Zn deficiency.[104]

An adult body contains between 1.5 and 2.5 grams of Zn. Most of this is in the muscle, bone, liver, and other organs, with only about 2% in the plasma. High concentrations are found in the eye and, in men, in the prostate gland. In the blood, Zn is mainly loosely bound to metallothionein.[105] The body's Zn content is regulated homeostatically through control of absorption of exogenous Zn from the gut and by regulation of excretion of endogenous Zn in gastrointestinal secretions.[106] Plasma Zn levels can be affected by stress, physical exercise, and infection, and by other factors not necessarily diet-related.

Zinc plays many metabolic roles of considerable importance. It is a component of as many as 50 different enzymes in animals. In addition, it provides structural integrity in many proteins, for example in cell membranes. "Zinc finger protein" is involved in the processes of transcription factors that link with the double helix of the DNA to initiate gene expression.[107] Because of its multiple roles, an inadequate intake of the metal can result in a number of different clinical symptoms, not all of which are specific for Zn deficiency. These include growth retardation, loss of appetite, alopecia, skin lesions, reduced immunity, and slow wound healing. In parts of the Middle East during the 1960s, endemic deficiency brought about by consumption of a diet with inadequate available Zn caused poor growth, stunting, anaemia, hypogonadism, and delayed sexual maturity in young males.[108] In some poor countries such as Bangladesh, where Zn intake is often marginal, less dramatic but nevertheless serious symptoms of deficiency, such as depressed immunity and delayed growth,[109] are seen, especially in children.

10.10.3.1 Zinc Toxicity

Depending on the quantity consumed, Zn salts can cause irritation of the intestine, with nausea, vomiting, and abdominal pain. Consumption of about 1 g of a Zn salt is required to produce these effects. Ingestion of 75 to 300 mg/d, in the form of dietary supplements, has been shown to interfere with the absorption of other essential trace elements, especially Cu and iron. A fall in levels of the enzyme Cu-Zn superoxide dismutase has been observed after consumption of 50 mg of a Zn supplement daily for 6 weeks. Other effects of chronic high intake include microcytic anaemia, changes in immune function, and in lipoprotein metabolism.[110]

10.10.4 Zinc Requirements and Dietary Reference Values

As with Cu, problems have been experienced in trying to establish dietary requirements for Zn. These are largely due to the difficulty of assessing optimal Zn nutriture in humans because of the lack of reliable and specific biomarkers for Zn status. As a result there is some lack of uniformity between recommendations made in different countries. Thus, while the U.S. RDA for an adult male is 11 mg/d and 8 mg for a female, in the U.K. it is 5.3 and 4.0 mg/d. The U.S. upper intake level for all adults is 40 mg/d, and for infants 4 mg/d.

10.11 TIN CONTAMINATION OF FOOD AND BEVERAGES

Tin, and its alloys bronze and brass, have been used for making containers and cooking utensils for many thousands of years, a use that undoubtedly resulted in some contamination of food and beverages; yet there is no evidence that our predecessors suffered to any significant extent from tin poisoning. Today, in spite of a few expressions of concern, there is little evidence that tin is not a safe metal for use in connection with food and beverages.

10.11.1 Tin in Food and Beverages

Tin levels in most foods are generally low except where there has been contact with tinplate, as can be seen in Table 10.7. In France, levels in fresh foods of 0.03 ± 0.03 mg/kg, and 76.6 ± 36.5 mg/kg in canned foods have been reported.[111] The uptake of tin depends on the nature of the food as well as on whether cans are lacquered or not. Canned meat seldom contains more than trace amounts, whereas tomatoes may have up to 50 mg/kg. Colored fruits containing anthocyanins are highly aggressive towards tinplate, with 100 mg Sn/kg reported in blackcurrants.[112] The presence of nitrate can also increase uptake.[117] The age of the cans can be important. Tinned carrots that had been taken to the Arctic on the ill-starred Franklin expedition in 1837 and remained unopened for a century, contained 2.44 g Sn/kg.[113]

10.11.1.1 Organotin Compounds in Food

Organic compounds of Sn are found in some foods, especially fish. Their source appears to be industrial pollution rather than, as in the case of Hg, natural methylation of inorganic compounds. Dibutyltin salts are used as stabilizers in plastics and, until recently, tributyl Sn was used extensively as marine antifouling agents. High levels of organotin compounds have been reported in oysters in Australia,[114] salmon in the U.K.,[115] and several species of fish in the River Ganges in India.[116] Organotin has been detected in wastewater and sewage sludge in Switzerland.[117]

TABLE 10.7
Tin in Foods

Food Group	Mean Concentration (mg/kg Fresh Weight)
Bread	0.03
Meat	0.02
Offal	0.02
Fish	0.44
Vegetables (green)	0.02
Canned vegetables	44
Fresh fruit	0.03
Fruit products	17
Beverages	0.02
Milk	0.02
Nuts	0.03

Source: Data adapted from Ysart, G. et al., Dietary exposure estimates of 30 elements from the U.K. Total Diet Study, *Food Addit. Contam.*, 16, 391, 1999.

10.11.2 Dietary Intake of Tin

The total Sn intake in the U.K. in 1997 was 1.9 mg/d, with 94% of this contributed by canned foods.[118] US intakes were higher, at a mean of 2.7 mg/d.[119] A much lower intake of 0.644 mg/d has been reported in Japan, probably reflecting a lower use of canned foods.[120]

10.11.3 Absorption and Metabolism of Tin

Tin is poorly absorbed from food, with usually less than 1% crossing the gut wall. The rate of absorption depends on the chemical form, with Sn(II) compounds being taken up four times more readily than Sn(IV). Most of the absorbed Sn is rapidly excreted in urine, with only a small amount retained in bone and organs. There is little evidence that Sn plays any nutritional role in the body.

10.11.3.1 Toxicity of Tin

Inorganic Sn is generally considered to be nontoxic, though high intakes of more than 200 mg/kg of food can cause nausea and vomiting.[121] In rats, prolonged intakes of low levels have been associated with growth retardation and histological changes in the liver, and it may interfere with Fe and Zn metabolism.[122]

Organotin compounds are highly toxic. They can cause a variety of serious effects — from genotoxicity to mitochondrial damage and neurotoxicity. Other symptoms include hyperglycemia, changes in blood pressure, and immunotoxicity.[123]

10.12 ALUMINUM CONTAMINATION OF FOOD AND BEVERAGES

Aluminum is the most common metal in the Earth's crust. Though there is evidence that it was used in China as long ago as 300 A.D., it was not until the early 19th Century that it was discovered in the West. It took almost another 100 years before it began to be used commercially. Its use increased so dramatically that by the mid-20th Century Al had become the world's second most important metal.[124]

10.12.1 Aluminum in Food and Beverages

Aluminum is ubiquitous in the environment and is found in every food and beverage. Typical levels of concentration range from < 1.0 to about 5 mg/kg, as seen in Table 10.8. There are exceptions, and variations in levels of Al concentrations in foods can be considerable. A Swedish study found that levels in vegetables could vary by a factor of 10^{12}.[125] Levels of 0.1 to 53.0 mg/kg have been reported in meat offal in a number of countries. Certain plants, especially some herbs and spices, naturally contain relatively high levels of Al, with, for example, 56.50 mg/g (d.w.) in some Spanish herbs. The average levels in dried tea leaves has been reported as 50 to 1500 mg/kg (d.w.), with a highest level of 30,000 mg/kg.[126]

TABLE 10.8
Aluminum in Foods

Food Group	Mean Concentration (mg/kg Fresh Weight)
Bread	3.7
Miscellaneous cereals	78
Meat	0.49
Offal	0.35
Meat products	3.2
Fish	5.5
Vegetables (green)	1.8
Canned vegetables	1.1
Fresh fruit	0.57
Fruit products	1.0
Beverages	1.72
Milk	< 0.27
Dairy products	0.64
Nuts	11

Source: Data adapted from Ysart, G. et al., Dietary exposure estimates of 30 elements from the U.K. Total Diet Study, *Food Addit. Contam.*, 16, 391, 1999.

Processed foods are a major source, because Al compounds are used extensively in food processing. Permitted food additives and processing aids in many countries include sodium aluminosilicate and Al phosphate, Al stearate and sulphate, and many others. Food containing these additives include processed cheese, confectionery, dried milk, baked cereal goods, chewing gum, and extruded snack foods. Levels of Al detected in some products include 224.8 mg/kg in chocolate cake, 274 mg/kg in chewing gum,[127] and 78 mg/kg in bakery products.[41] Some beverages in Al cans have been found to be relatively rich in Al, with 3.24 mg/l in a cola-based drink,[128] and 10 mg/l in beer.[129]

10.12.1.1 Aluminum in Domestic Water

Natural waters are generally low in Al because many of its compounds are relatively insoluble. Concentrations in tap water are higher because it is normally treated to remove turbidity and improve clarity and color. One of the most commonly used coagulants is Al sulphate (alum). Reported levels of Al in treated water range from 10 to 2670 μg/l.[130] The WHO recommend a maximum of 0.2 mg/l in domestic water.[131] Accidental addition of coagulants can increase Al levels considerably, as occurred in the U.K., resulting in more than 100 mg/l.[132]

10.12.2 Dietary Intake of Aluminum

According to FAO/WHO, dietary Al intakes per day worldwide are 6 to 14 mg for adults and 2 to 6 mg for children.[133] Intakes in several countries have been falling in recent years. In the U.K., adult daily intakes fell from 11 mg in 1994 to 3.4 mg in 1997. The fall is believed to be due to a decrease in use of Al-containing additives, especially in cereal products.[134] A similar fall in the U.S. has been attributed especially to a reduction of levels in processed cheese.[130] An Australian study found that some people who are high consumers of tea and cakes can accumulate up to 4 mg/kg/b.w./d, which is equivalent to 200 mg/d for a 50-kg adult. There is some evidence of higher-than-average intakes by people who use Al cooking pots.[135]

10.12.3 Aluminum Absorption and Metabolism

Absorption of Al from food and drink is low, generally about 1%, though there can be variation between individuals.[136] The chemical form is an important factor in controlling absorption. Aluminum hydroxide and phosphate, the forms most often present in foods, are very insoluble, compared to the more easily absorbed citrate. In addition, both vitamin D and the parathyroid hormone appear to moderate absorption. Body stores of iron may also play a part, as well as health and age.[137]

The possible consequences of high dietary intakes of Al have been debated for many years, especially in relation to Alzheimer's disease. There is no doubt that Al is a neurotoxin. Renal failure has been caused by Al-contaminated water used in kidney dialysis machines.[138] High levels of Al have been reported in neurofibrillary tangles in the brains of Alzheimer victims.[139] High intakes have also been associated with osteomalacia and defective bone mineralization. However, whether such conditions can result from a normal dietary intake by people who do not suffer from chronic renal failure and are not on TPN or dialysis, and thus bypass the GI tract, the body's main barrier to Al absorption, is not clear.[140]

REFERENCES

1. Sparks, D.L., *Environmental Soil Chemistry*, Academic Press, New York, 1995, p. 24.
2. Williams, C.H. and David, D.J., Heavy metals in Australian soils, *Aust. J. Soil Res.*, 11, 43, 1973.
3. Bakir, F., Damluji, S.F., and Amin-Zaki, L., Methylmercury poisoning in Iraq, *Science*, 181, 230, 1973.
4. Ministry of Agriculture, Fisheries and Food, Survey of Arsenic in Food: the 8th Report of the Steering Group on Food Surveillance, the Working Party on the Monitoring of Foodstuffs for Heavy Metals, HMSO, London, 1982.
5. Pike, E.R., Graham, L.C., and Fogden, M.W., Metals in crops grown on sewage-enriched soil, *J. Am. Pediol. Assoc.*, 13, 19, 1975.
6. U.S. EPA, Clean Water Act, Section 503.58(2), U.S. Environmental Protection Agency, Washington, D.C., 1993.
7. Ashami, M.O., Pollution of soils by cadmium, in *Changing Metal Cycles and Human Health*, Nriagu, J.O. Ed., Springer-Verlag, Berlin, p. 9, 1984.

8. Morgan, H., *The Shipham Report: An Investigation into Cadmium Contamination and its Implications for Human Health*, Elsevier, London, 1988.
9. Palheta, D. and Taylor, A., Mercury in environmental and biological samples from a gold mining area in the Amazon region of Brazil, *Sci. Total Environ.*, 168, 63, 1995.
10. Futasuka, M., Kitano, T., and Shono, N., Health surveillance in the population living in a methyl mercury-polluted area over a long period, *Environ. Res.*, 83, 83, 2000.
11. Prater, B.E., Water pollution in the river Tees, *Water Pollut. Control*, 74, 63, 1975.
12. Skalney, A., Interelementary relationships and oncological morbidity in an As-polluted area, in *Trace Elements in Man and Animals –TEMA 8*, Anke, M., Meissner, D., and Mills, C.F., Verlag Media Touristik, Gersdorf, Germany, 1993, p. 794.
13. Witman, W.E., Interactions between structural materials in food plant, and foodstuffs and cleaning agents, *Food Progr.*, 2, 1, 1978.
14. Van Hamel, R., On tin in preserved articles of food, *Analyst*, 16, 195, 1891.
15. Page, G.G., Hughes, J.T., and Wilson, P.T., Lead contamination of food in lacquered and unlacquered cans, *Food Technol. N. Z.*, 9, 32, 1974.
16. Ramonaityte, D.T., Copper, zinc, tin and lead in canned evaporated milk, produced in Lithuania: the initial content and change in storage, *Food Addit. Contam.*, 18, 31, 2001.
17. Board, P.W., The chemistry of nitrate-induced corrosion of tinplate, *Food Technol. Aust*, 25, 16, 1973.
18. Reilly, C., Copper and lead uptake by food prepared in tinned-copper utensils, *J. Food Technol.*, 13, 71, 1978.
19. MacPhail, A.P. et al., Changing patterns of iron overload in black South Africans, *Am. J. Clin. Nutr.*, 32, 1272, 1979.
20. Bolger, P.M. et al., Identification and reduction of sources of dietary lead in the U.S., *Food Addit. Contam.*, 13, 53, 1996.
21. Watanabe, Y., Cadmium and lead on decorated drinking glasses, *Annu. Rep. Tokyo Metropolitan Lab. Public Health*, 25, 256, 1974.
22. Meranger, J.C., Cunningham, H.M., and Giroux, A., Plastics in contact with foods, *Can. J. Public Health*, 65, 292, 1974.
23. Bruening, K. et al., Dietary calcium intakes of urban children at risk of lead poisoning, *Environ. Health Perspect.*, 107, 431, 1999.
24. Baltrop, D. and Smith, A.M., Kinetics of lead interactions with human erythrocytes, *Postgrad. Med. J.*, 51, 770, 1985.
25. Snakin, V.V. and Prisyazhnaya, A.A., Lead contamination of the environment in Russia, *Sci. Total Environ.*, 256, 95, 2000.
26. Reichlmayer-Lais, A.M. and Kirchgessner, M., Depletionsstudien zur essentialität von blei an wachsenden ratten, *Arch. Tierennährg.*, 31, 731, 1981.
27. Rossi, E., Costin, K.A., and Garcia-Webb, P., Effects of occupational lead exposure lymphocyte enzymes involved in haem biosynthesis, *Clin. Chem.*, 36, 1980, 1990.
28. Pagliuca, A. et al., Lead poisoning: clinical, biochemical and haematological aspects of a recent outbreak, *J. Clin. Pathol.*, 43, 277, 1990.
29. Pocock, S.J., Smith, M., and Baghurst, P., Environmental lead and children's intelligence: a systematic review of epidemiological evidence, *Br. Med. J.*, 309, 1189, 1994.
30. Jorhem, L. and Sundström, B., Levels of lead, cadmium, zinc, copper, nickel, chromium, manganese, and cobalt in foods on the Swedish market, 1983–1990, *J. Food Contam. Anal.*, 6, 223, 1993.
31. Ministry of Agriculture, Fisheries and Food, *Lead, Arsenic and Other Metals in Food*. Food Surveillance Paper No. 52, The Stationery Office, London, 1998.
32. Food and Drug Administration, Lead-soldered food cans, *Fed. Reg.*, 58, 33860, 1993.

33. Ministry of Agriculture, Fisheries and Food, 1994 Total Diet Study: Metals and Other Elements, Food Surveillance Information Sheet No. 52, The Stationery Office, London, 1997.
34. Craun, G.F. and McCabe, L.J., Lead in domestic water supplies, *J. Am. Water Workers Assoc.*, 67, 593, 1970.
35. Sherlock, J.C. and Quinn, M.J., Relationship between blood lead concentrations and dietary lead intake by infants 0–1 year old, *Food Addit. Contam.*, 3, 167, 1986.
36. FDA, Tin coated lead foil capsules for wine bottles, *Fed. Reg.*, 58, 33860, Food and Drug Administration, 1992.
37. EC, Council Regulation EEC2356/91 amending Council Regulation EEC 2392/89, *Off. J. Eur. Commun.*, I.261, 1, 1991.
38. DeLacey, E.A., Lead-crystal decanters — a health risk? *Aust. Med. J.*, 147, 162, 1988.
39. Whitehead, T.P. and Prior, A.P., Lead poisoning from earthenware container, *Lancet*, I, 1343, 1960.
40. Ysart, G. et al., Dietary exposure estimates of 30 elements from the U.K. Total Diet Study, *Food Addit. Contam.*, 16, 391, 1999.
41. GEMS/Food, Summary of 1986–1988 Monitoring Data, World Health Organization, Geneva, 1991.
42. Gunderson, E.L., FDA Total Diet Study, July 1986–April 1991, dietary intakes of pesticides, selected elements, and other chemicals, *J. AOAC Int.*, 78, 1353, 1997.
43. Agricola, G., *De Re Metallica,* English translation by H.C. and L.H. Hoover, Dover, New York, 1950.
44. Waldron, H.A., Did the Mad Hatter have mercury poisoning?, *Br. Med. J.*, 287, 1961, 1983.
45. United Nations Environment Programme, International Labor Organization, World Health Organization, Mercury — environmental aspects, Environmental Health Criteria, 86, World Health Organization, Geneva, 1989.
46. World Health Organization, Environmental Health Criteria: Mercury, World Health Organization, Geneva, 1976.
47. Bennet, B.G., Six most dangerous chemicals named, Monitoring and Assessment Research Centre, London, on behalf of UNEP/PILO/WHO International Program on Safety, *Sentinel*, 1, 3, 1984.
48. Marsh, D.O. et al., Fetal methylmercury poisoning: relationship between concentration in single strands of maternal hair and child effects, *Arch. Neurol.*, 44, 1017, 1987.
49. Harada, M., Minamata disease: methyl mercury poisoning in Japan caused by environmental pollution, *Crit. Rev. Toxicol.*, 25, 1, 1995.
50. Vos, G., Hovens, J.P.C., and Delft, W.V., Arsenic, cadmium, lead and mercury in meat, livers and kidneys of cattle slaughtered in the Netherlands during 1980–1985, *Food Addit. Contam.*, 4, 73, 1987.
51. Capon, C.J., Mercury and selenium content and chemical form in vegetable crops grown on sludge-amended soil, *Arch. Environ. Contam. Toxicol.*, 10, 673, 1981.
52. Moharram, Y.G. et al., Mercury content of some marine fish from the Alexandria coast, *Nahrung*, 31, 899, 1987.
53. Food Standards Agency, Agency issues precautionary advice on eating shark, swordfish, and marlin, May 10, 2002, http://www.food.gov.uk/news/ressreleases/62503.
54. Clarkson, T.W. and Strain, J.J., Nutritional factors may modify the toxic action of methyl mercury in fish-eating populations, *J. Nutr.*, 133, 1S, 2003.
55. Asami, T., Cadmium pollution of soils and human health in Japan, in *Human and Animal Health in Relation to Circulation Processes of Selenium and Cadmium,* Lag, J., Norwegian Academy of Science and Letters, Oslo, 1991, p. 115.

56. Buchet, J.P., Lauwerys, R., and Roels, H., Renal effects of cadmium, *Lancet*, 336, 699, 1990.
57. Dabeka, R.W., McKenzie, A.D., and Lacroix, G.M.A., Dietary intake of lead, cadmium, arsenic and fluoride by Canadian adults: a 24 h duplicate diet study, *Food Addit. Contam.*, 4, 89–102, 1987.
58. Hovinga, M.E., Sowers, M., and Humphrey, H.E.B., Environmental exposure and lifestyle predictors of lead, cadmium, PCB, and DDT levels in great lakes fish eaters, *Arch. Environ. Health*, 48, 98, 1993.
59. Asami, M.O., Pollution of soil by cadmium, in *Changing Metal Cycles and Human Health*, Nriagu, J.O., Ed., Springer-Verlag, Berlin, 1984, p. 95.
60. Hoffmann, K. et al., The German environmental survey 1990/1992 (GerES II): cadmium in blood, urine and hair of adults and children, *J. Expo. Anal. Environ. Epidemiol.*, 10, 126, 2000.
61. MacIntosh, D.L. et al., Dietary exposure to selected metals and pesticides, *Environ. Health Perspect.*, 104, 202, 1996.
62. DeLaeter, J.R. et al., The cadmium content of rural tank water in Western Australia, *Search*, 8, 85, 1976.
63. Rosman, K.J.R., Hosie, D.J., and De Laeter, J.R., The cadmium content of drinking water in Western Australia, *Search*, 8, 85, 1977.
64. EC, Food Science and Techniques, report on Tasks for Scientific Cooperation, Dietary Exposure to Cadmium, EUR 17527, Office for Official Publications of the European Community, Luxumberg, 1997.
65. Ostergaard, K., Cadmium in cigarettes, *Acta Med. Scand.*, 202, 193, 1977.
66. Hammer, D.I., Finklea, J.F., and Creason, J.P., Cadmium exposure and human health effects, in *Trace Substances in Environmental Health*, Hempill, D.D., Ed., University of Missouri Press, Columbia, MO, 1971, p. 269.
67. Carruthers, M.M. and Smith, B., Evidence of cadmium toxicity in a population living in a zinc-mining area: pilot study of Shipham residents, *Lancet,* I, 663, 1979.
68. Piscator, M., Carcinogenicity of cadmium — review, presented at *Third Int. Conf. on Cadmium*, Miami, FL, February 3–5, 1981.
69. Biswas, B.K. et al., Detailed study report of Samata, one of the arsenic affected villages of Jessore district, Bangladesh, *Curr. Sci.*, 74, 134, 1998.
70. Ministry of Agriculture, Fisheries and Food, Survey of Arsenic in Food, HMSO, London.
71. Stijve, T. and Bourqui, B., Arsenic in edible mushrooms, *Deutsche Lebensmittel-Rundschau*, 87, 307, 1991.
72. Jukes, D.J., *Food Legislation in the U.K.: A Concise Guide,* Butterworth/Heinemann, Oxford, 1997, p. 137.
73. Bowen, H.J.M., *Trace Elements in Biochemistry*, Academic Press, London, 1966, p. 177.
74. Flowerdew, D.W., *A Guide to the Food Regulations in the U.K.,* British Food Manufacturing Industries Research Association, Leatherhead, Surrey, 1990, p. 156.
75. Queirolo, F. et al., Total arsenic, lead and cadmium levels in vegetables cultivated at the Andean village of Northern Chile, *Sci. Total Environ.*, 255, 75, 2000.
76. Shibata, A. et al., Mutational spectrum in the p53 gene in bladder tumors from the endemic area of blackfoot disease in Taiwan, *Carcinogenesis*, 15, 1085, 1994.
77. Dhar, R.K. et al., Groundwater arsenic calamity in Bangladesh, *Curr. Sci.,* 73, 48, 1997.
78. Wickstrom, G., Arsenic emission from the Novaky power station, *Work Environ. Health*, 9, 2, 1982.

79. Peach, D.F. and Lane, D.W., A preliminary study of geographic influence on arsenic concentrations in human hair, *Environ. Geochem. Health*, 20, 231, 1998.
80. Tsuchiya, K., Arsenic contamination of infant formula, *Environ. Health Perspect.*, 19, 35, 1977.
81. Gunderson, E.L., Total Diet Study, July 1986–April 1991: dietary intakes of pesticides, selected elements and other chemicals, *J. AOAC. Int.*, 78, 1353, 1995.
82. Vannort, R.W., Hannah, M.L., and Pickston, L., New Zealand Total Diet Study: Part 2. Contaminant elements, New Zealand Health, Wellington, New Zealand, 1995.
83. Clarkson, T.W., Molecular targets of metal toxicity, in *Chemical Toxicology and Clinical Chemistry of Metals*, Brown, S.S. and Savory, J., Eds., Academic Press, London, 1983, p. 211.
84. Lafontaine, A., Health effects of arsenic, in *CEC Trace Metals: Exposure and Health Effects,* Ferrante, E.D., Ed., Pergamon, Oxford, 1978, p. 107.
85. Chen, C.J., Kuo, T.L., and Wu, M.M., Arsenic and cancers, *Lancet,* I, 414, 1988.
86. Nielsen, F.H., Other trace elements, in *Present Knowledge of Nutrition,* Brown, M.I., Ed., International Life Sciences Institute, Washington, D.C., 1990, p. 294.
87. Department of Health, *Dietary Reference Values for Food Energy and Nutrients in the U.K.,* HMSO, London, 1991, p. 191.
88. Booth, C.K., Reilly, C., and Farmakalidis, E., Mineral composition of Australian ready-to-eat breakfast cereals, *J. Food Compos. Anal.*, 9, 135, 1996.
89. Reilly, C., The dietary significance of adventitious iron, zinc, copper and lead in domestically-prepared food, *Food Addit. Contam.*, 2, 209, 1985.
90. Spitalny, K.C. et al., Drinking-water-induced copper intoxication in a Vermont family, *Pediatrics*, 74, 1103, 1984.
91. Pandit, A. and Bhave, S.A., Present interpretation of the role of copper in the Indian childhood cirrhosis, *Am. J. Clin. Nutr.*, 63, 830S, 1996.
92. Müller, T. et al., Endemic Tyrolean infantile cirrhosis: an ecogenic disorder, *Lancet*, 347, 877, 1996.
93. Pennington, J.A.T., Intakes of minerals from diets and foods: is there a need for concern?, *J. Nutr.*, 126, 2304S, 1996.
94. Food and Nutrition Board National Academy of Sciences, Dietary Reference Intakes: Elements, 2003, http//www4.nationalacademies.org/10M/10Mhome.nst/pages/Food+and+Nutrition+Board.
95. Turnland, J.R. et al., Copper absorption and retention in young men at three levels of dietary copper by use of the stable isotope ^{65}Cu, *Am. J. Clin. Nutr.*, 49, 870, 1989.
96. Uauy, R., Olivares, M., and Gonzales, M., Essentiality of copper in humans, *Am. J. Clin. Nutr.*, 67, 952S, 1998.
97. Gitlin, J.D., The copper transporting ATPases in human disease, in *Trace Elements in Man and Animals 10*, Roussel, A.M., Anderson, R.A., and Favier, A.E., Eds., Kluwer/Plenum, New York, 2000, p. 9.
98. Food and Agricultural Organization/WHO, Evaluations of Food Additives, WHO Technical Reports Series No. 462, World Health Organization, Geneva, 1971.
99. Prasad, A.S., Discovery of human zinc deficiency and marginal deficiency of zinc, in *Trace Elements in Clinical Medicine*, Tomita, H., Ed., Springer-Verlag, Tokyo, 1990, p. 3.
100. Ranum, P., Zinc enrichment of cereal staples, *Food Nutr. Bull.*, 22, 169, 2001.
101. Welsh, S.O. and Marston, R.M., Zinc levels in the U.S. food supply: 1909–1980, *Food Technol.*, 36, 70, 1982.
102. Pennington, J.A.T. and Young, B., Iron, zinc, copper, manganese, selenium, and iodine in foods from the U.S. Total Diet Study, *J. Food Compos. Anal.*, 3, 166, 1990.

103. Osendarp, S.J.M. et al., Zinc supplementation during pregnancy and effects on growth and morbidity in low birthweight infants: a randomized placebo controlled trial, *Lancet*, 357, 1080, 2001.
104. Prasad, A.S., Human zinc deficiency, in *Biological Aspects of Metals and Metal-Related Diseases,* Sakar, A.B., Ed., Raven Press, New York, 1983, p. 107.
105. Jackson, M.J., Physiology of zinc: general aspects, in *Zinc in Human Biology*, Mills, C.F., Ed., Springer-Verlag, London, 1989, p. 1.
106. King, J.C., Shames, D.M., and Woodhouse, L.R., Zinc homeostasis in humans, *J. Nutr.*, 130, 1360S, 2000.
107. Berg, J.M. and Shi, Y., The galvanizing of biology: a growing appreciation of the role of zinc, *Science*, 271, 1081, 1996.
108. Prasad, A.S. et al., Zinc metabolism in patients with the syndrome of iron deficiency anemia, hepatosplenomegaly, dwarfism and hypogonadism, *J. Lab. Clin. Med.*, 61, 537, 1963.
109. Walsh, C.T. et al., Zinc health effects and research priorities for the 1990s, *Environ. Health Perspect.*, 102, 5, 1994.
110. Yadrick, M.K., Kenney, M.A., and Winterfeldt, E.A., Iron, copper and zinc status: responses to supplementation with zinc or zinc and iron in adult females, *Am. J. Clin. Nutr.*, 49, 145, 1989.
111. Biego, G.H. et al., Determination of dietary tin intake in an adult French citizen, *Arch. Environ. Contam. Toxicol.*, 36, 227, 1999.
112. Britton, S.C., *Tin versus Corrosion,* International Tin Research Institution Publication No.510, ITRI, Greenford, Middlesex, U.K., 1975.
113. Bartsiokas, A., The Franklin expedition and lead poisoning, *Eur. J. Oral. Sci.,* 108, 78, 2000.
114. Batley, G.E. et al., Accumulation of tributyltin in Sydney rock oysters, *Saccostrea commercialis, Aust. J. Mar. Freshwater Res.*, 40, 49, 1989.
115. Ministry of Agriculture, Fisheries and Food, Cadmium, Mercury and Other Metals in Food. Food Surveillance Paper No. 53, The Stationery Office, London, 1998.
116. Kannan, K., Senthilkumar, K., and Sinha, R.K., Sources and accumulation of butyltin compounds in Ganges river dolphin, *Platanista gangetica, Appl. Organometallic Chem.*, 11, 223, 1997.
117. Fent, K., Organotin compounds in municipal wastewater and sewage sludge: contamination, fate in treatment process and ecological consequences, *Sci. Total Environ.*, 185, 151, 1996.
118. Newman, P., Tin enters the free market era, *Metall.*, 50, 616, 1996.
119. Ministry of Agriculture, Fisheries and Food, MAFF U.K. 1997 — Total Diet Study, Food Surveillance Information Sheet No. 191, http://www.foodstandards.gov.uk.
120. Shimbo, S. et al., Use of food composition database to estimate daily dietary intake of nutrient or trace elements in Japan, with reference to limitation, *Food Addit. Contam.*, 13, 775, 1996.
121. DeGroot, A.P., Feron, V.J., and Til, H.P., Toxicology of tin compounds, *Food Cosmet. Toxicol.*, 11, 19, 1973.
122. Greger, J.L. and Johnson, M.A., Effect of dietary tin on zinc retention and excretion by rats, *Food Cosmet. Toxicol.*, 19, 163, 1981.
123. Baldwin, D.R. and Marshall, W.J., Heavy metal poisoning and its laboratory investigation, *Ann. Clin. Chem.*, 36, 267, 1999.
124. Tylecote, R.F., *A History of Metallurgy,* The Metal Society, London, 1976, p. 195.
125. Jorhem, L. and Haegglund, G., Aluminium in foodstuffs and diets in Sweden, *Z. Lebens. Unters. Forsch.*, 194, 38, 1992.

126. Ravichandran, R. and Parthiban, R., Aluminum content of South Indian teas and their bioavailability, *J. Food Sci. Technol.*, 35, 349, 1998.
127. Lione, A. and Smith, J.C., The mobilization of aluminum from three brands of chewing gum, *Food Cosmet. Toxicol.*, 22, 265, 1982.
128. Walton, J., Hams, G., and Wilcox, D., Bioavailability of Aluminum from Drinking Water: Co-exposure with Foods and Beverages, Research Report No. 83, Urban Water Research Association of Australia, Melbourne, Australia, 1994.
129. Williams, D., Aluminum in Beer, http://www.breworld.com/the-b/9603/br3.html.
130. Pennington, J.A.T., Aluminum content of foods and diets, *Food Addit. Contam.*, 5, 161, 1987.
131. World Health Organization, Evaluation of Certain Additives and Contaminants, 33rd Report of the Joint Expert Committee on Food Additives. WHO Technical Series 776. World Health Organization, Geneva, 1989.
132. Altmann, P. et al., Disturbance of cerebral function in people exposed to drinking water contaminated with aluminium sulphate: retrospective study of the Camelford water incident, *Br. Med. J.*, 319, 807, 1999.
133. Pennington, J.A.T. and Schoen, S.A., Estimates of dietary exposure to aluminium, *Food Addit. Contam.*, 12, 119, 1995.
134. Ministry of Agriculture, Fisheries and Food, Aluminum in Food. Food Surveillance Paper No. 39, HMSO, London, 1993.
135. Neelam, M., Bamji, S., and Kaladhar, M. Risk of increase aluminum burden in the Indian population: contribution from aluminum cookware, *Food Chem.*, 70, 57, 2000.
136. Alfrey, A.C., Physiology of aluminum in man, in *Aluminum and Health: A Critical Review*, Gitelman, H.J., Ed., Dekker, New York, 1989, p. 101.
137. Moore, P.B. et al., Absorption of aluminium-26 in Alzheimer's disease, measured using accelerator mass spectrometry, *Dementia Geriat. Cognit. Dis.*, 11, 66, 2000.
138. Platts, M.M., Goode, G.C., and Hislop, J.S., Composition of domestic water supply and the incidence of fractures and encephalopathy in patients on home dialysis, *Br. Med. J.*, 2, 657, 1977.
139. McLachlan, D.R., Aluminum and the risk of Alzheimer's disease, *Environmetrics*, 6, 233, 1995.
140. Doll, R., Review: Alzheimer's disease and environmental aluminum, *Age Ageing*, 22, 138, 1993.

11 Pollutants in Food – Radionuclides

Zitouni Ould-Dada

CONTENTS

11.1 INTRODUCTION

We have evolved in a naturally radioactive environment, and naturally radioactive elements have always been present in our environment. They are present in the air we breathe, the food we eat, the water we drink, and in all living organisms including man. Natural sources include cosmic rays, gamma rays from the earth, radon decay products in the air, and various radionuclides in food and drink. Artificial sources include fallout from the testing of nuclear weapons in the atmosphere, discharges of radioactive waste from the nuclear industry, and accidental releases.

TABLE 11.1
Symbols of Radionuclides Used in the Text

Element	Radionuclides
Caesium	^{134}Cs, ^{137}Cs
Strontium	^{89}Sr, ^{90}Sr
Tritium	^{3}H
Carbon	^{14}C
Iodine	^{129}I, ^{131}I
Technetium	^{99}Tc
Sulphur	^{35}S
Phosphorus	^{32}P
Uranium	^{233}U, ^{234}U, ^{235}U, ^{238}U
Plutonium	^{238}Pu, ^{239}Pu, ^{240}Pu, ^{241}Pu
Americium	^{241}Am

The most important radionuclides that contribute to human exposure via food are generally those that are mobile in the environment. Radionuclides with a potentially high environmental mobility are usually analogs of essential elements and include $^{134/137}Cs$, ^{90}Sr, ^{131}I ^{14}C, ^{3}H, ^{35}S, and ^{40}K. Those with low environmental mobility include radionuclides with large atomic weights such as P and Am.

Symbols of radionuclides used in the text are listed in Table. 11.1.

Food may become contaminated with radionuclides by a number of routes. These include uptake of radionuclides by plants from the soil or from deposition onto the upper parts of the plants. Radionuclides may be passed to humans directly from the consumption of these food crops or indirectly via foods derived from animals feeding on contaminated pastures or feedstuffs.

Natural and seminatural ecosystems (e.g., forests, uplands, heathlands, mountain pastures, marshlands) provide a variety of wild foodstuffs such as mushrooms, berries, honey, meat from game animals, and meat and milk from domestic ruminants (e.g., sheep, goats). These foods are known to concentrate radioactive elements, particularly ^{137}Cs. Radionuclides incorporated into forest food products such as mushrooms, berries, and grasses consumed by animals and humans can contribute significantly to the radiation dose to humans.

Because agricultural products constitute the basic diet of most populations, the fate and behavior of radionuclides in agricultural ecosystems are of primary importance when assessing radiation dose to humans from radioactivity released to the environment. It is essential to understand the factors affecting the behavior and transfer of radionuclides in the environment in order to accurately predict activity concentrations in plant- and animal-derived food products, interpret monitoring results, and develop appropriate and effective countermeasures.

11.1.1 Radiation Units

Becquerels (Bq) — Radioactivity is measured as the rate at which the atoms disintegrate. It is measured in becquerels, and one Bq is equal to one disintegration per second. When measuring radioactivity in the environment, it is usually measured relative to weight or volume, e.g., Bq/kg or Bq/l of substance.

Sievert (Sv) — A dose is a way of measuring how much radiation is delivered to a living body. The unit of radiation dose is the sievert. Because quite small amounts of radiation come from the environment, it is usual to measure it in either thousandths (milisievert — mSv) or millionths of a sievert (microsieverts — μSv).

The dose delivered by taking in radioactivity in food is worked out by measuring how much is in the food (Bq/kg), multiplying this by how much food is eaten over a period of time (e.g., kg/year), and then by a physical factor that tells us how much dose is caused by this number of Bq in the body.

11.2 SOURCES OF RADIOACTIVITY

11.2.1 Natural and Manmade

We have evolved in a naturally radioactive environment, subject to radiation from outer space, from the earth, from substances within the body, and even from burning wood and cultivating the ground. Radionuclides are found naturally in air, water, and land. Natural sources include cosmic rays, gamma rays from the earth, radon decay products in the air, and various radionuclides in food and drink. Artificial sources include fallout from the testing of nuclear weapons in the atmosphere, discharges of radioactive waste from the nuclear industry, and accidental releases.

Cosmic radiation comes from the sun, from our galaxy, and possible from even deeper in space. Radiation doses from cosmic rays increase with latitude and altitude. The average dose at ground level is about 0.26 mSv a year, with a range of 0.2 to 0.3 mSv. The Earth's crust is made up of materials that are naturally radioactive. Uranium, thorium, and ^{40}K, for instance, are dispersed throughout rocks and soil. Because building materials are extracted from the earth, they can be slightly radioactive, and people are irradiated indoors as well as outdoors. The radiation doses vary according to the rocks and soils of the area and the building materials in use.

Radon is a natural radioactive gas that comes from uranium, which is widespread in the Earth's crust. It is emitted from rocks or soil at the earth's surface and dispersed in the atmosphere. Radon decays to form other radioactive atoms which, when inhaled, can lodge in the lung and irradiate tissue. On average, the radiation dose from radon is far greater than the dose from any other source, natural or manmade (about 50% of the total dose). A recent paper by Thorne provides an excellent summary of background radiation, both natural and manmade.[1]

Since the beginning of the last century, humans have added to this natural background through civil and military uses of radiation. Under normal operation of nuclear power plants, the discharges of radioactive substances into the environment are controlled and have been continually reduced. The releases are kept low in order to protect humans and the environment, and the doses received from normal operation of nuclear plants are thus, in general, negligible.

In the U.K., for example, the annual effective dose, averaged over the whole population, is about 2.6 mSv in total (Table 11.2). Some 85% of this is from natural sources with over half from radon decay products in the home. Medical exposure

TABLE 11.2
Average Annual Doses to the U.K. Population from All Sources of Radiation

Source	Dose (mSv)
Natural	
Cosmic	0.26
Gamma rays	0.35
Internal	0.3
Radon	1.3
Artificial	
Medical	0.37
Occupational	0.007
Fallout	0.005
Products	0.0004
Discharges	0.0002
Total (rounded)	**2.6**

Source: National Radiological Protection Board (NRPB).

of patients accounts for 14% of the total, whereas all other artificial sources (e.g. fallout, discharges from nuclear industry) account for about 0.5% of the total value.

11.2.2 Planned and Accidental Discharges

Artificial radionuclides have been created by civil and military nuclear programs and are released into the environment as a result of nuclear weapons testing, releases from nuclear power stations and other sites (i.e., nonnuclear sites), and nuclear accidents. The discharges are made both in the atmosphere and in the aquatic environment. Discharges from nonnuclear sites include those made from premises such as hospitals, research establishments, and universities. Radionuclides detected in the effluents from hospitals and research establishments include ^{3}H, ^{14}C, ^{32}P, ^{35}S, and ^{131}I. The United Nations Scientific Committee on the Effects of Atomic Radiations (UNSCEAR) calculated the annual normalized production of radioisotopes used in medical, educational, and industrial applications in the developed countries and found the dominating contributors to the collective effective dose to be ^{14}C (90% of the dose), ^{131}I (9%) and ^{3}H (1.5%). Radionuclides discharged from nonnuclear sites generally have short radioactive half-lives compared with those discharged from nuclear sites. Also, the quantity of activity discharged from many nonnuclear sites is much less than that from the nuclear industry. Some nonnuclear sites, however, do discharge radionuclides with longer radioactive half-lives, such as tritium and carbon-14, and a few nonnuclear sites make discharges that exceed those from some of the smaller nuclear sites.

Radioactive wastes may be discharged to a freshwater, estuarine, or marine environment. In some countries such as the U.K., there are also discharges of radionuclides to the sewer system from hospitals, universities and research establishments, and a few nuclear sites.

A very small proportion of exposure of the population results from the discharge of airborne or liquid radioactive waste from operations such as power generation and fuel reprocessing. The average annual dose to the public is less than 0.001 mSv, although a small number of people receive doses of 0.15 to 0.2 mSv in a year. Radiation doses due to discharges from nuclear sites in the U.K. are reported annually for a range of exposure pathways in the Radioactivity in Food and the Environment (RIFE).[2] A study on collective doses to the U.K. population due to routine discharges from civil nuclear sites was recently carried out by The National Radiological Protection Board (NRPB).[3] This study has shown that the collective dose to the U.K. population due to routine discharges from U.K. civil nuclear activities peaked at around 130 man Sv in the late 1970s and fell to about 25 man Sv in 1985, falling further to less than 20 man Sv in the early 1990s.

Nuclear weapons tests injected a variety of radionuclides into the atmosphere from tritium to plutonium-241. Around 500 atmospheric explosions were carried out before the limited test ban treaty in 1963, with a few more until 1980. Currently, however, radionuclide concentrations in air, rain, and human diet are much lower than the peak values in the early 1960s. At present, the most important radionuclides in terms of human exposure are ^{14}C, ^{90}Sr, and ^{137}Cs. In the U.K., internal and external irradiation contribute nearly equally to the annual average effective dose of 0.006 mSv (peak of 0.125 mSv in 1963).

Radioactivity has also been widely dispersed as a result of nuclear accidents. The most significant accident was at the Chernobyl nuclear power plant in the Ukraine, causing the release of large amounts of radionuclides into the atmosphere and dispersal throughout Europe. In terms of doses to people, the most significant radionuclides were ^{131}I, ^{134}Cs, and ^{137}Cs. In the U.K., the average dose from Chernobyl contaminants was about 0.02 mSv in the first year after the accident. At present, it has declined to 0.001 mSv in a year. There are 386 farms under restriction in England, Scotland, and Wales, still affected by the Chernobyl accident.

Radionuclides may also be released to atmosphere or to the terrestrial environment from waste disposal sites. These include ^{14}C, ^{36}Cl, ^{99}Tc, ^{129}I, ^{137}Cs, ^{237}Np and daughters, and $^{239,240,242}Pu$ and daughters.[4] Because of its long half-life and mobility in soil, ^{36}Cl could be an important contributor to exposed-group doses in the far future.

Total effective doses from artificial radionuclides in the environment comprise about 77% from weapons test fallout, 20% from Chernobyl fallout, and 3% from discharges.

11.3 ENVIRONMENTAL TRANSFERS AND EXPOSURE PATHWAYS

11.3.1 Environmental Transfers

Radioactive material may be released into the environment as a result of normal operations, accidental releases, or following the disposal of solid radioactive waste. These releases can occur due to the use of radioactive material in nuclear power production, industrial operations, hospitals, and for research purposes. The most important radionuclides that contribute to human exposure via food are generally those which are mobile in the environment. Radionuclides with a potentially high environmental mobility are usually analogs of essential elements and include $^{134/137}Cs$, ^{90}Sr, ^{131}I ^{14}C, ^{3}H, ^{35}S, and ^{40}K. Those with low environmental mobility include radionuclides with large atomic weights such as P and Am.

In general, the processes involved in the terrestrial and aquatic transfers of radionuclides are well known, but they are subject to large variations. The mechanisms by which radioactive elements are incorporated into biological systems are basically the same as those by which plants and animals obtain their nutrients from the atmosphere, soil, water, and feed stuffs. Contamination of terrestrial foodstuffs occurs via a number of processes including:

- Foliar contamination by dry deposition and interception of contaminated rainfall
- Root uptake of radionuclides previously deposited onto the soil
- Contamination of plant surfaces by the dry deposition of resuspended soil that has been previously contaminated
- The splashing of contaminated surface soil onto plants by the action of rain

When radionuclides are depositing from the atmosphere, by dry or wet deposition processes, a fraction of the depositing material will be intercepted by vegetation, with the remainder reaching the ground. For the fraction intercepted by vegetation, environmental removal processes (e.g., wind, rain, snow, abrasion between leaves, leaf fall, shedding of cuticular wax) will combine with radioactive decay to reduce the quantity of this initial contamination. The growth of vegetation also causes a diluting effect on the retained contamination. Resuspension can occur due to the action of the wind or by mechanical disruption of a surface.[5,6] Resuspension can be quite high during agricultural activities (e.g., ploughing, harvesting). However, ploughing of fields reduces resuspension over a longer period as the surface layer is mixed with the lower uncontaminated soil.

Radionuclides deposited in the soil can be transferred to plants through the plant's root system or by direct contamination of the plant's surface with the soil (e.g., soil resuspension). Surface contamination of plants by soil is generally less important than root uptake because the material is less available for transfer within the plant and can be removed by the action of wind and rain. However, in the case of radionuclides with low soil-to-plant transfer factors (e.g., actinides), contamination by soil can have a significant effect on the activity in the plant.

Contamination of terrestrial ecosystems occurs by deposition (dry or wet) of radionuclides released to the atmosphere. Grazing of contaminated pasture by animals (and feeding animals contaminated hay and silage) will lead to the transfer of contamination to milk as well as animal tissues (e.g., meat, bone, liver). The extent to which animals eat contaminated food is one of the major factors affecting the contamination of animal products. Once ingested, some radionuclides are more absorbed in the body than others, e.g., I (100%), Cs (80%), Sr (20%), and plutonium (0.05%). Once absorbed, different radionuclides are accumulated in different animal tissues. For example, Cs accumulates in all soft tissues, radioiodine in the thyroid, and Sr in the bone. All these three radionuclides are also transferred to milk. Other radionuclides (e.g., Ru, Co) accumulate in offal (e.g., liver), but their rate of transfer is lower than that for Cs, I, and Sr due to their lower gut transfer rates and longer biological half-lives. The rate of uptake into animal tissues and its subsequent loss depends on the biological half-life. For example, mobile radionuclides such as Cs, Sr, and I have biological half-lives in milk of about 1 d.

The uptake of radionuclides from soil to plants via the root system can vary markedly depending on factors such as the type of radionuclide, crop type, soil characteristics, and time after the radionuclide enters the soil. With the exception of seminatural systems, the long-term migration of radionuclides in soil is of relatively low importance in radiological dose assessments for the transfer to terrestrial foods. The root uptake of radionuclides by vegetation depends on factors such as the physical and chemical characteristics of the radionuclides, the type and properties of the soil, the type and characteristics of vegetation, and the time after the radionuclide enters the soil. The most important factors controlling the extent to which radionuclides will migrate freely or be absorbed by plant roots are, in approximate order of importance: (1) type of radionuclide, (2) soil type and characteristics, and (3) agricultural practices, e.g., plowing and application of fertilizers.[7]

Environmental transfers for some radionuclides may continue despite substantial decreases in their discharges to the sea, e.g., ^{99}Tc and actinides. This is due to their remobilization from sediments. ^{99}Tc is released into the environment mainly as a result of nuclear fuel reprocessing. For example, in the U.K. such releases are made from the Sellafield plant into the Irish Sea. Soil-to-plant transfer factors for ^{99}Tc for some plants (e.g., lettuce, cabbage, spinach) can be an order of magnitude higher than that for ^{137}Cs.

11.3.2 The Importance of Speciation

The mobility, solubility, and fate of radionuclides in the environment are governed principally by their chemical forms (speciation). Little information is available in the literature on the chemical forms of radionuclides in routine emissions from nuclear facilities and on chemical speciation reactions affecting radionuclides in the atmosphere.[8] More information is available for routine emissions[9] than for accidental releases.[10] Volatile elements have the potential to be transported further and to undergo more and more complex reactions in the atmosphere than nonvolatile elements. Nonvolatiles, such as U and Pu, are likely to be deposited in insoluble forms such as oxides, whereas volatile elements, such as caesium and iodine, are more likely to be deposited in soluble forms. Radiocaesium is an alkali metal and is chemically analogous to other members of the same group such as K^+ and Na^+. It is relatively volatile and is likely to enter the atmosphere as an oxide. Following deposition, radiocaesium uptake by plants will be mostly by direct foliar absorption, with root uptake becoming increasingly important with time.[11] For example, the leaf-to-fruit translocation of soluble ^{134}Cs in grape vines was found to be the dominant process in the contamination of fruits compared to direct deposition.[12] Leaf-to-fruit translocation of radiocaesium varies with plant species, whereas root uptake is mostly affected by the type of soil. Experimental results on leaf-to-fruit translocation of ^{134}Cs[13] showed large differences in the amounts taken up by strawberries (22%), gooseberries (4%), and redcurrants (3.8%). Soil-to-plant transfer of ^{137}Cs was found to be the highest in blackcurrant, followed in decreasing order by strawberries, apple, gooseberries, rhubarb, and melon.[14] In the soil, Cs^+ interacts strongly with clay minerals and can become fixed indefinitely and essentially unavailable to biota.[15,16] Cs^+ uptake by plants, however, is enhanced in soils with a high organic matter content.[17]

Strontium is an element that belongs to the alkali earth group and exhibits many similar chemical properties to calcium. Strontium-89 behaves like ^{90}Sr in all respects except that, because of its short half-life, it is relatively more important as a foliar contaminant. Strontium tends to form soluble compounds and is therefore relatively mobile in most environmental systems. It is taken up by plants as a consequence of the plants' inability to discriminate between Sr^{2+} and Ca^{2+}. The extent of strontium uptake via roots is usually minimal when compared to foliar uptake immediately after contamination. Experimental results on the transfer of ^{134}Cs and ^{85}Sr to grape vines showed that Sr accumulates in leaves and growing shoots, whereas Cs accumulates in fruits.[18]

Emissions of ^{14}C from nuclear power production contribute a significant fraction of the radiation dose delivered to both local and global populations from this source.[2,19,20] High environmental mobility and long half-life (5730 years) are the main factors contributing to the radiological importance of ^{14}C. Once released into the atmosphere, $^{14}CO_2$ becomes incorporated into crops via photosynthesis, leading to enhanced levels of ^{14}C in crops.[21,22] Uptake may also occur via the root system as CO_2 diffuses into the soil.[23] ^{14}C (in the form of CH_4) can also diffuse into soils where it can be rapidly oxidized to CO by microorganisms.[23] CO_2 may be lost from plants due to respiration. In a review of the translocation and remobilization of ^{14}C in crops, it was reported that plant respiration within 24 h following exposure can account for up to 40% of assimilated ^{14}C with an average figure of around 25%. During subsequent growth (to maturity) the total loss may be as high as 60%.[21] Other studies, however, failed to detect any respiratory losses at all.[24–26]

The transfer processes of tritium between atmosphere, soil, and plant include dry and wet deposition, soil migration and diffusion, evaporation, uptake into plants, and evapotranspiration from plants. These processes are greatly influenced by factors such as the prevailing climatic conditions, the type and density of vegetation, and the soil type. Transfer of tritium into vegetation occurs through stomata on the leaf surface and uptake of soil water.[27,28] A large proportion of depositing tritium is rapidly returned to the atmosphere by evaporation from land surfaces and evapotranspiration processes from plants. Tritium may be incorporated in organic matter as an organically bound tritium (OBT) within crop tissues,[29] including fruit such as apples.[28]

Both elemental and organic iodine can enter plants by direct foliar absorption. In the case of ^{129}I, deposition and root uptake will depend on the forms in which iodine is present in the atmosphere and soil, and these are areas where further research is needed.[30] ^{131}I is radiologically significant only in the short term. In contrast, ^{129}I has a long half-life of 1.6×10^7 years and could present a long-term exposure risk.

Sulphur-35 may be emitted from CO_2-cooled reactors to atmosphere as carbonyl sulphide (COS) and hydrogen sulphide (H_2S). Atmospheric deposition is usually considered to be the primary uptake route by vegetation for ^{35}S, although there are few data on root uptake to support this assumption.[30] Information on speciation of ^{36}Cl might prove difficult to obtain because emissions of ^{36}Cl to the atmosphere are very difficult to measure. In postclosure assessment studies, the modeled transport of ^{36}Cl through the geosphere is assumed to be generally the same as that of ^{129}I.[31] Chlorine in soil is mainly present as the Cl ion,[2] and as a result of its high mobility in soils it is removed rapidly by leaching and by plants via root uptake.[32] Accumulation of Cl by plants from soil is virtually entirely through root uptake, and gaseous losses of Cl from soil are negligible.[33,34] Root-absorbed chlorine was found to be readily translocated to both stems and leaves of growing crops such as potatoes, peppermint, and tomatoes.[32] Because of its long half-life (3.01×10^5 years) and high mobility and bioavailability, ^{36}Cl warrants further research on its behavior on all components of both the terrestrial and aquatic ecosystems.

11.3.3 Exposure Pathways

Exposure pathways refer to the ways in which people can be exposed to radiation. Food may become contaminated with radionuclides by a number of routes (Figure 11.1). These include uptake of radionuclides by plants from the soil, or from deposition onto the above parts of the plants. Radionuclides may be passed to humans directly from the consumption of these food crops or indirectly via foods derived from animals feeding on contaminated pastures or feedstuffs. The relative importance of the various terrestrial foodstuffs pathways will vary depending on the amount and distribution of deposited activity at the location where the foodstuff is grown, the uptake and retention by plants and animals of the deposited radioactivity, and the individuals' consumption habits.

Deposition of radionuclides from the atmosphere to vegetation surfaces represent a direct pathway of plant contamination and is more important than root uptake in the short term following deposition. The uptake of radionuclides from soil is an indirect pathway of plant contamination following deposition to the atmosphere. It often results in a small fraction of the soil's radionuclide burden to plants. This is particularly the case for actinides (e.g., plutonium), where the contamination of external surfaces of plants is more important than root uptake. For animals, however, soil ingestion can provide a very important intake pathway for plutonium.[35,36] ^{99}Tc, on the other hand, is almost completely removed from soils by root uptake.

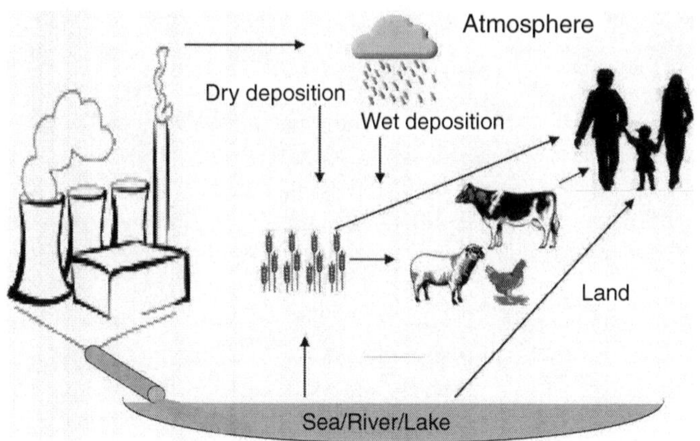

FIGURE 11.1 Exposure pathways.

Liquid radioactive effluents may be discharged into the marine environment, freshwater (lakes or rivers), or estuaries. Following the release to freshwater, the principal pathways leading to the irradiation of people are ingestion of foods from the river, drinking water from the river, drinking water used to irrigate crops and pasture and for animals' drinking, and external irradiation from sediments. In the case of releases into the marine environment, a number of pathways may lead to the irradiation of people, and these include ingestion of marine food, external irradiation from activity on beaches, and inhalation of sea spray. The ingestion of fish and shellfish constitutes an important pathway to humans for discharges of liquid effluent.

In some countries such as the U.K. there are also discharges of radionuclides to the sewer system from hospitals, universities and research establishments, and a few nuclear sites. This gives rise to possible routes of exposure to humans through the ingestion of crops and animal products.[37,38]

Radionuclides released into seawater are dispersed through advection and dispersion of the water masses and are also absorbed by sediments. In the case of discharges into the marine environment, the most exposed groups are likely to include those individuals who consume higher-than-average amounts of locally caught seafood, i.e., fish, crustaceans, and molluscs. Certain radionuclides discharged into sea become strongly associated with sediments (e.g., plutonium), which will be transported by tides and currents. Where pastures are inundated by seawater, some of these sediments can be deposited on the surface of the soil or vegetation. Animals grazing in such areas may therefore become contaminated with radionuclides as they inadvertently ingest sediment particles associated with vegetation. The intake of radionuclides by inhalation would generally be insignificant compared to the intake by ingestion, except possibly in winter when animals are given uncontaminated feed. Remobilization of radionuclides from sediments can be a source of contamination resulting from past authorized discharges and is likely to be increasingly important as authorized discharges decline.

Forests are an efficient reservoir for pollutants, and the long-term impact of contaminated forests can be important.[39,40] For example, radionuclides incorporated into forest food products such as mushrooms, berries, and grasses consumed by animals and humans can contribute significantly to the radiation dose to humans.

Interest in seaweed contamination has reemerged as a consequence of the release of ^{99}Tc from the Enhanced Actinide Removal Plant (EARP) plant at Sellafield. ^{99}Tc is highly concentrated by seaweeds such as *Fucus* sp. Seaweed is used as a fertilizer and soil conditioner in the Sellafield area, leading to increased concentrations of ^{99}Tc in soil and vegetable samples. In parts of northern Europe (e.g., northwest France) it is gathered commercially for use in the alginate industry and represents a potential pathway.

In addition to the internal irradiation from ingestion of contaminated food, radionuclides released into the atmosphere can give rise to internal irradiation to man following their inhalation, and external irradiation by photons and electrons emitted as a result of the radioactive decay process. Following their deposition onto the ground, radionuclides may lead to further irradiation of people by external irradiation, internal irradiation of resuspended activity, and inadvertent ingestion of soil.

11.4 RADIOACTIVITY IN TERRESTRIAL FOOD

11.4.1 Radioactivity in Crops

Following atmospheric fallout from nuclear weapons tests during the 1950s and 1960s, Cs and Sr were the main concern as contaminants of the food chain. Uptake of Cs and Sr by plants and animals represent an important pathway of radiation exposure to humans following the consumption of vegetables, fruits, and animal products. Plant surfaces can be an important source of contamination either directly on plants or indirectly through animals grazing the plants.

One of the major factors affecting the contamination of crops is the rate of transfer of radionuclides from soil to plants. For example, Cs remains mobile in organic soils but is strongly absorbed onto clay particles where it is gradually effectively immobilized. For ^{131}I, uptake by crops is not so important because it decays rapidly.

Some foodstuffs can accumulate high concentrations of natural radionuclides. For example, Brazil nuts can have high concentrations of ^{226}Ra. Plutonium is well known to be present at very low concentrations in plants and crops. For most people living in Europe, the majority of the dose currently received from ingestion of food is due to natural radionuclides, particularly ^{40}K in terrestrial food and ^{210}Po in fish and shellfish.

Annual doses from atmospheric discharges, indicative of those received by members of the critical groups residing near reprocessing plants, have been assessed.[41] The most important radionuclide (contributing 47% and 55% of the dose at 0.5 and 5 km, respectively, from the site) in 1987 at Sellafield is ^{129}I in milk, milk products and, to a lesser extent, in fruit. In 1996, ^{129}I was still important, contributing 49% and 66% of the dose at 0.5 and 5 km, respectively, from the site. For Cap de la Hague, ^{129}I was also an important radionuclide in 1996 (87% and 75% of the dose at 0.5 and 5 km, respectively, from the site). This is also true of ^{14}C, which contributed 8% and 19% of the dose at 0.5 and 5 km, respectively, from the site in 1996. At

Marcoule the annual dose received from discharges in 1996 was estimated at 100 μSv and 30 μSv at 0.5 and 5 km, respectively, from the site. This was dominated by discharges of ^{129}I.

Radiation dose received by humans due to the ingestion of radionuclides in food can be affected by a number of factors. For example, delays between production and consumption of food results in the radioactive decay of short-lived radionuclides (e.g., whether food is eaten fresh or frozen). Also, radionuclides may be removed from foods prior to consumption during food preparation and processing. Washing or peeling of vegetables grown in contaminated soils can significantly reduce the potential intake of radioactivity.

11.4.2 Radioactivity in Animal Products

A variety of animals provide foods for humans, such as milk, meat, and eggs. Cattle and sheep are considered particularly important because they are outdoors for much of the year and graze large areas of pasture. Other animals such as pigs and chickens are often housed indoors and fed on a variety of feeds, a significant proportion of which may not be locally produced. Some animals are important in some regions of Europe. For example, in Mediterranean countries the transfer of radionuclides to goat's milk has been found to be important. In sub-Arctic regions (e.g., Finland, Sweden, Norway) the transfer of radionuclides from lichen to caribou and reindeer has been found to be significant, but these animals are generally only consumed by a limited number of people. Other animals such as rabbits and horses are also consumed in parts of Europe. Animals can be contaminated by radioactivity via three routes: ingestion, inhalation, and through the skin. Ingestion of contaminated feed, soil, and water is the most important pathway to animals. Inhalation is potentially more important than skin absorption but is often not a major contamination pathway for agricultural animals.

Once absorbed in the body, different radionuclides are accumulated in different tissues. For iodine, the major storage organ in the body is the thyroid, and it is actively transferred into milk. For example, the development of thyroid cancers, particularly by young children, was the primary health effect arising from the Chernobyl accident. The major source of exposure to radioiodine was the ingestion of contaminated milk.[42,43]

Because ruminants are known to consume considerable quantities of soil as they graze, the inadvertent ingestion of soil can be a potentially important pathway by which radionuclides move through the food chain to humans.[35] Soil ingestion is particularly important for actinides and Cs than for other radionuclides that are readily taken up by plants (e.g., ^{90}Sr).

Following the Chernobyl accident, certain areas of the U.K. (Cumbria, north Wales, southwest Scotland, and Northern Ireland) received high levels of deposition of radionuclide (mainly ^{137}Cs) where the plume coincided with areas of high rainfall. As a result, restrictions were placed on the sale, movement, and slaughter of sheep grazing in the affected areas. To protect public health, any sheep that farmers wish to move off the restricted farms are live monitored, and if found to be over the action level of 1000 Bq/kg, are marked with indelible paint to block slaughter. Initially, restrictions covered 9036 farms with 4,278,000 sheep. Over the 18 years since the accident, the areas under restrictions have been reduced as contaminated levels have dropped. For example, in Northern Ireland restrictions were lifted in 2000.

It is essential to understand the factors affecting the behavior and transfer of radionuclides in animals in order to accurately predict activity concentrations in animal-derived food products, interpret monitoring results, and develop appropriate and effective countermeasures. A review and summary of the advances in animal radioecology since 1988 have been compiled by Howard and Beresford.[44]

11.4.3 Radioactivity in "Free" Food

The term *free food* used here refers to foodstuffs collected from the wild. Natural and seminatural ecosystems (e.g., forests, uplands, heathlands, mountain pastures, marshlands) provide a variety of wild foodstuffs such as mushrooms, berries, honey, meat from game animals, and meat and milk from domestic ruminants (e.g., sheep, goats). These foods are known to concentrate radioactive elements, particularly ^{137}Cs.[45] Mushrooms have the highest radiocesium activity concentrations among all the species in seminatural ecosystems. High radiocesium activities have been reported in a number of mushroom species both before and after the Chernobyl accident.[46,47] Forests, for example, are well known to be efficient collectors and reservoirs of pollutants, and the long-term impact of contaminated forests can be significant. For example, in contaminated forests, radioactivity can be transferred into food products such as mushrooms, berries, and plants. These food products can contribute significantly to the human radiation dose following consumption by humans and animals. Both experimental and modeling studies are difficult to conduct for forests because of the complexities of forest systems.

Contamination of game can also arise from "unusual" pathways. For example, in the U.K., pigeons visiting the Sellafield facility were found to be contaminated (largely by Cs) in excess of 100 kBq/kg.[48] These pigeons were habitually visiting part of the Sellafield facility and transferring significant amounts of radioactivity to the garden of a local resident who provided food for them. Wild boar is hunted in many central and eastern European countries. Data from Austria showed ^{134}Cs activity concentrations in wild boar reaching a maximum of 17.6 kBq in 1988.[49] Some high levels of radiocesium were reported in wildfowl during the first hunting season after the Chernobyl accident.[50]

Lichens form the major part of the diet of reindeer (> 75%), particularly in the winter period.[51] ^{137}Cs activity concentrations in reindeer during summer and early autumn tend to be less important and generally represent about 10 to 20% of those in winter. Reindeer can also have high concentrations of ^{210}Po.

Dairy goats graze in seminatural ecosystems in many parts of the world, and as for sheep, transfers of radiocesium to goat milk is higher than that to cow's milk.[52] In Europe, the average population intakes of goat milk are small as it is generally used for cheese production and by certain consumers such as those allergic to cow's milk. However, in some countries such as Norway, the more commonly consumed whey cheeses can have high levels of radiocesium as their preparation involves evaporation of the whey until dryness, which increases the radiocesium activity concentration by a factor of about 10.

The importance of the contribution of consumed natural food to the total radiation dose has been clearly demonstrated following the Chernobyl accident. For example, in Sweden the highest radiation doses were received by people living in high deposition areas who consume large amounts of lake fish, meat from game animals, and reindeer.[53] Although in most countries the contribution of natural food products to the total food consumption by humans is small, their contribution to the long-term ingestion dose can be significant. Also, natural foods can become very important under poor economic conditions. This was the case in Ukraine following the Chernobyl accident, where the economic difficulties resulted in the increased consumption of forest and natural foods (mushrooms, berries, meat of wild animals, and local fish) that were highly contaminated by ^{137}Cs.

11.5 RADIOACTIVITY IN AQUATIC FOOD

In addition to naturally occurring radionuclides in the marine environment, human activities such as coal mining, mining of phosphate rocks, and extraction of crude oil and natural gas all add to the natural radioactivity levels. Uranium, thorium, and potassium are the main elements contributing to natural radioactivity in the terrestrial and marine environments. From marine foodstuffs, by far the most important isotope is ^{210}Po.

High concentrations of tritium have been found in food and the environment near Cardiff where radiochemicals for research, medicine, and industry are produced. However, doses to high-rate seafood consumers were estimated to be relatively low at 0.031 mSv in 2002, similar to 0.036 mSv in 2001. Most of the dose was due to tritium and carbon-14 in fish from the Bristol Channel.[2] Technetium-99 was measured in one set of Scottish salmon and fish feed samples, and the results were 0.24 Bq/kg and 1.3 Bq/kg, respectively.

As a consequence of the low concentrations, estimated doses to high-rate consumers of farmed fish were also low, ranging from 0.017 mSv/year for adults to 0.038 mSv/year for 1-year-old infants. Most of these doses were due to naturally occurring lead-210 and polonium-210.[2]

The liquid effluents from fuel reprocessing plants are dominated by fission products such as ^{137}Cs, ^{106}Ru, ^{60}Co, and ^{90}Sr, and generally contain relatively significant quantities of actinides. The highest concentrations of anthropogenic radionuclides in seafood from the European seas are found in the Northeastern Irish Sea near Sellafield. Radionuclide concentration in fish and shellfish sampled in the vicinity of this area, however, peaked in the early to mid-1970s and declined thereafter as radioactive waste discharges continued to be reduced. In 2002 the highest concentrations of radioactivity in marine fish were found for ^{3}H at about 160 Bq/kg.[2] The concentrations of ^{3}H in local seawater was, however, less than 30 Bq/l, indicating that some bioaccumulation has taken place. It should be noted, however, that the radiotoxicity of ^{3}H is very low and the radiological importance of these concentrations is much less than that of other radionuclides.[2] Bioaccumulation of ^{3}H has also been observed in the Severn Estuary near Cardiff, U.K., where over 90% of the total

^{3}H in marine samples was found to be associated with organic matter. This form of ^{3}H is strongly bound to organic matter and sediment and has the potential to transfer through the marine foodchain from small organisms to accumulate in fish. Concentrations of ^{3}H near Cardiff recorded in 2002 were 12,000 Bq/kg (wet) in sole, 2000 Bq/kg (wet) in cod, and 14,000 Bq/kg (wet) in mussels.[2] The dose to the critical group of fish and shellfish consumers for organic H was about 0.031 mSv (including a contribution due to external radiation).

Technetium-99 has been the subject of considerable public concern in recent years. Shellfish, especially lobsters, are well known to concentrate ^{99}Tc.[54,55] Recent and current discharges of ^{99}Tc contributed around 15% of the dose to the Sellafield seafood consumers.[2] Transuranics are less mobile than radiocesium in seawater and have a high affinity for sediments. This was reflected in higher concentrations of transuranics in shellfish compared with fish, and a rapid reduction in their concentrations, particularly in shellfish, with distance from Sellafield. Mussels can accumulate high levels of radionuclides such as plutonium, americium, and polonium, which are highly particle-reactive. This can be attributed to the filter feeding nature of these organisms and the strong affinity of these radionuclides for fine particulate matter. In a 3-year study of radionuclides in winkles near Sellafield, retained sediment was found to carry major fractions of ^{137}Cs, $^{239+240}Pu$, and ^{241}Am.[56] The findings of this study suggested that removing the sediment prior to cooking would reduce the activity in these winkles by around 90%. The consumption of locally collected winkles has been a major contributor to the radiation dose received by a group of seafood consumers living near the Sellafield reprocessing plant in Cumbria, U.K. The magnitude of this dose may depend on the proportion of the activity associated with the sediments in the cooked winkles. Removing the sediments from these winkles prior to cooking (e.g., by soaking the live winkles prior to boiling) would substantially reduce the quantity of radionuclides consumed and hence the dose.

Atlantic salmon and rainbow trout are the main farmed fish species consumed in the U.K. A study on radioactivity in fishmeal fed to fish had suggested from a theoretical standpoint that farmed fish may contain enhanced levels of naturally occurring radionuclides as a result of the fishmeal they were fed.[57] Another study sought to test this thesis by monitoring fish directly in fish farms in England, Scotland, and Wales during 2001.[58] The results of this study showed low levels of radioactivity in farmed fish: ^{137}Cs (0.23 to 0.53 Bq/kg), ^{14}C (< 16 to 50 Bq/kg), ^{210}Pb (< 0.010 to 0.27 Bq/kg), and ^{210}Po (0.016 to 0.29 Bq/kg). ^{99}Tc was measured in one set of Scottish salmon and fish feed samples, and the results were 0.24 Bq/kg and 1.3 Bq/kg, respectively. As a consequence of these low concentrations, estimated doses to high-rate consumers of farmed fish were also low, ranging from 0.017 mSv/year for adults to 0.038 mSv/year for 1-year-old infants. Most of these doses were due to naturally occurring ^{210}Pb and ^{210}Po.

The radiological impact on the European Community (EC) populations from liquid discharges released into the Mediterranean Sea from European civil nuclear plants during the period 1980–1991 has been assessed under a project called MARINA-MED.[59] Results showed that ingestion of molluscs was the dominant pathway (60% of the collective dose) with ^{106}Ru (62%), ^{241}Am (22%), and ^{239}Pu (10%) being the main contributors to the dose. Ingestion of fish was also found to be a significant pathway (20% of the collective dose) with ^{90}Sr (51%) and ^{137}Cs (31%) being the most important contributors to the dose. Ingestion of crustaceans was found to be a negligible contributor to the collective dose (< 2%).

Discharges of radioactivity into the Mediterranean Sea have been more than two orders of magnitude lower than those made into the Northern European seas. Collective doses to the EC population due to liquid discharges into the Northern European sea waters were assessed in another EC project, called MARINA.[60] Results from this project showed that the total collective dose to the EC population was calculated to be more than nearly three orders of magnitude higher than that calculated for discharges into the Mediterranean.

Generally, doses to seafood consumers throughout Europe from naturally occurring radionuclides are approximately two orders of magnitude higher than those from anthropogenic radionuclides. For most people living in Europe, the majority of the dose currently received from ingestion of food is due to natural radionuclides, particularly ^{40}K in terrestrial food and ^{210}Po in fish and shellfish.

Collective and individual doses received as a result of routine radiological discharges from nuclear power plants and nuclear fuel reprocessing plants in the EC member states occurring between 1987 and 1996 has been assessed.[61] Doses to individuals residing near the Sellafield site have also decreased from 187 mSv in 1987 to 114 mSv in 1996. The contribution to the dose from various radionuclides and exposure pathways varied considerably over the study period. However, important radionuclides include ^{99}Tc in crustaceans, ^{14}C and ^{137}Cs in fish, and ^{106}Ru and ^{241}Pu in molluscs. For liquid discharges, however, the more significant exposures arise as a result of discharges from reprocessing plants.

11.6 MONITORING AND SURVEILLANCE

Member states of the European Union are subject to the terms of the Euratom Treaty. This makes provision for basic Community standards to protect the health of workers and the general public against the dangers arising from ionizing radiation. The Euratom Treaty also includes obligations for member states to monitor radioactivity in the environment (Article 35), to provide the European Commission with the results of such monitoring (Article 36), and to provide general data on any plan for the disposal of radioactive waste (Article 37).

Environmental monitoring programs in respect of normal disposal operations are carried out to satisfy one or more of the following objectives:

1. To comply with regulatory requirements
2. To ensure that regulatory requirements have been met
3. To check operators' results (for monitoring by regulators or local authorities)
4. To provide an independent means of surveillance for inadvertent or unrecorded discharges
5. To estimate public radiation exposures
6. To provide public reassurance
7. To detect any long-term trends

Operators of nuclear sites are required under their authorizations to carry out a regular program of environmental monitoring to ensure that no unacceptable levels of contamination appear in the local environment. Regulatory agencies also carry out their own programs of environmental monitoring as part of their enforcement responsibilities.

11.7 ASSESSMENT OF RADIATION DOSES TO CONSUMERS

11.7.1 Assessment Process

Radioactivity may be released into the environment as a result of normal operation due to the use of radioactive material in nuclear power production, hospitals, industrial operations, and for research purposes. An important part of the system of radiological protection is the assessment of the radiological impact of such releases on both individuals (individual dose) and population groups (collective dose). Individual exposure is used for comparison with the appropriate dose limits or constraints, whereas collective dose is used to estimate health detriment. Collective doses are also used to assess different process or discharge and disposal options (e.g., for the abatement of discharges).

Once released into the environment, radionuclides can be transferred through the soil and to plants and crops, animal products, and seafood. Mathematical models are used to describe the transfer of radionuclides in the environment through various media to humans. These include atmospheric dispersion and food chain models. In broad terms, the exposure pathways that have to be considered when evaluating radiation exposures of members of the public are ingestion, inhalation, and external exposure. The relative importance of these pathways depends on the radionuclide and the nature of the environmental media into which the deposition occurs.

It is essential to understand the factors affecting the behavior and transfer of radionuclides in the environment in order to accurately predict activity concentrations in crops and animal-derived food products, interpret monitoring results, and develop appropriate and effective countermeasures.

When radionuclides are released into the environment, they can lead to radiation doses to members of the public by different exposure pathways. When identifying exposure pathways to include in a dose assessment, it is important to take account of local conditions and people's habits. For example, in some areas the ingestion of radionuclides in wild food (e.g., mushrooms, berries, rabbits, reindeers) can be an important exposure pathway. Also, the ingestion of animal milk or meat from animals drinking river water could be a potential pathway to consider in a dose assessment. There are also some unusual exposure pathways that can lead to the transfer of radioactivity to humans via food. For example, crops farmed or animals grazed on land exposed to sea spray or reclaimed from the sea, and crops grown on soil conditioned with seaweed can be considered unusual, although their radiation doses are normally negligible.

Retrospective assessments are undertaken to estimate the actual dose received by consumers in specific circumstances. These are made based on measurements of radioactivity in food. Prospective assessments are carried out to measure the acceptability of planned radioactive discharges from nuclear sites and other users of radioactive material. This is based on computer modeling to predict levels of radioactivity in food. The "critical group" approach is used, where the dose to the group of consumers who would be most exposed is calculated for comparison with the dose criteria. Radiation dose assessments are also carried out for incidents involving unplanned releases of radioactivity into the environment and to respond to incidents where unexpected levels of radioactivity are found in foodstuffs.

Assessment of radiation doses should be as realistic as possible in order to avoid significant over- or underestimation and as close as possible to those that would actually be received by members of the public. In order to carry out a realistic dose assessment, it is important to have a good understanding of the local conditions around the site being assessed and to obtain as much site-specific information as possible. Such assessments will rely on the use of model parameter values and people's habits data to represent a realistic situation around the site being considered. For exposure pathways, the main focus will be on those pathways that contribute the highest doses to the reference group. Of course, the most realistic method for assessing the dose is to carry out an extensive monitoring of the main exposure pathways. This, however, will be costly and time consuming, in addition to the radionuclide concentration levels being below the detection limits. For these reasons, dose assessments involve a combination of measurement and modeled data. It is important that any models used should be robust, fit for the purpose, and have been validated against the measurement data.

The release of radioactive material into the environment can lead to the exposure of individuals by a variety of pathways. In an assessment of doses received by individuals and population groups, all of the important exposure pathways must be considered. However, in most situations involving the routine release of radionuclides, activity concentrations in environmental media are below detection limits and hence measurements cannot be used for calculating exposures. It is for this reason that mathematical models are often used to predict the transfer of radionuclides in the environment. A variety of models and data are required to predict the transfer of radionuclides through the environment and the resulting doses to people. For releases to atmosphere, dispersion models are used to estimate activity concentrations in air and the subsequent deposition of radionuclides to the ground. Other models are then used to predict the transfer of radionuclides through terrestrial foodchains, the behavior of radionuclides deposited on the ground and, where relevant, the resuspension of radionuclides from the ground back into the air.

In addition to computer models, a variety of data on people's habits are needed to assess radiation doses. These include intake rates of terrestrial and aquatic foods, water, and air together with occupancies of different environments, such as time spent indoors or near sediments. Generic or site-specific data are used, and both average and above-average habit data are used to assess doses. Dose coefficients are also needed in the dose assessment process. These relate the intake of activity (Bq) to the effective dose (Sv) and are available for a number of radionuclides and different age groups.

11.7.2 Uncertainties in Dose Assessment

Assessments of doses necessarily entail a series of assumptions about the identification and behavior of candidate critical groups and the transfer of radionuclides in the environment. It is widely recognized that environmental transfer models are imperfect representations of reality as they contain approximations and simplifying assumptions. They also contain a number of transfer processes that are not fully understood and parameter values, most of which are not accurately known.[62,63] Environmental transfer processes are also variable in both time and space. There is, therefore, an inevitable uncertainty associated with model predictions. There is also variability in habits and behavior, reflecting the genuine differences that occur between individuals within a group e.g., differences in food consumption rates or time spent indoors. Estimated doses will therefore vary according to the habits of the exposed population. Doses received by the critical group may cover a significant range, reflecting both the genuine variability of habits and the uncertainty due to the lack of information in the models used to assess the doses.[62–64]

11.8 CONCLUSION

Because radioactive materials occur everywhere in nature, it is inevitable that they get into food. Human activities have added to the natural background of radioactivity through civil and military uses of radiation. Under normal operation of nuclear power plants, discharges of radioactive substances into the environment are controlled and have been continually reduced. These releases are kept low in order to protect humans and the environment, and the doses received from the normal operation of nuclear plants are thus, in general, negligible.

For most people living in Europe, the majority of the dose currently received from ingestion of food is due to natural radionuclides, particularly ^{40}K in terrestrial food and ^{210}Po in fish and shellfish. In general, doses to consumers of seafood throughout Europe from naturally occurring radionuclides are approximately two orders of magnitude higher than those from anthropogenic radionuclides. Doses of radionuclides from natural origin in terrestrial and marine foodstuffs are currently much higher than those from artificial radionuclides. The highest concentrations of anthropogenic radionuclides in seafood from the European seas are found in the Northeastern Irish Sea near Sellafield.

It is essential to understand the factors affecting the behavior and transfer of radionuclides in the environment in order to accurately predict activity concentrations in crops and animal-derived food products, interpret monitoring results, and develop appropriate and effective countermeasures.

REFERENCES

1. Thorne, M.C., Background radiation: natural and man-made, *J. Radiol. Prot.*, 23, 29, 2003.
2. RIFE 4 to 9, Radioactivity in Food and the Environment (RIFE), Annual Reports produced by Food Standards Agency (previously MAFF), Environment Agency, Scottish Environment Protection Agency, and the Environment and Heritage Service of Northern Ireland, 1999–2003.
3. Bexton, A.P., Radiological impact of routine discharges from U.K. civil nuclear sites in the mid-1990s, National Radiological Protection Board, Chilton, NRPB-R312, 2000.
4. Naito M. and Smith G., Revised Sets of Example Assessment Contexts, BIOMASS Theme 1, Task group 3, Note 1, Version 1.0, 1998.
5. Nicholson, K.W., A review of particle resuspension. *Atmos. Environ.,* 22, 2639, 1988.
6. Ould-Dada, Z. and Baghini, N.M., Resuspension of small particles from tree surfaces, *Atmos. Environ.,* 35, 3799, 2001.
7. Shaw, G. and Bell, J.N.B, Transfer in agricultural and semi-natural environments, in *Radioecology: Radioactivity and Ecosystems*, Van der Stricht E. and Kirchmann R., Eds., International Union of Radioecology, Fortemps, Liege, Belgium, 2001.
8. Murdock, R.N., Chemical Speciation of Radionuclides in the Terrestrial Environment, Mouchel and Partners Ltd. LGM Report No. 48018.001-R1, 1996.
9. Bishop, G.P. and Cramp, T.J., The Effect of Chemical and Physical Form of Atmospheric Releases of Radionuclides on Environmental Behavior, Associated Nuclear Services Ltd. ANS Report No. 2132-1, 1988.
10. Warner, F. and Harrison, R.M., *Radioecology after Chernobyl: Biogeochemical Pathways of Artificial Radionuclides,* SCOPE 50, John Wiley & Sons, Chichester, 1993.
11. Sawidis T. et al., Caesium-137 accumulation in higher plants before and after Chernobyl, *Environ. Int.*, 16, 163, 1990.
12. Carini, F. et al., 134Cs foliar contamination of vine: translocation to grapes and transfer to wine, in *Proceedings of the International Symposium on Radioecology Ten Years Terrestrial Radioecological Research following the Chernobyl Accident*, Austrian Soil Science Society and Federal Environment Agency, Vienna, 163, 1996.
13. Kopp, P. et al., Foliar uptake of radionuclides and their distribution in the plant, in *Proceedings of the Environment Contamination following a Major Nuclear Accident*, IAEA, Vienna, 2, 37, 1990.
14. Green, N., Wilkins, B.T., and Hammond, D.J., Transfer of radionuclides to fruit, *J. Radioanal. Nucl. Chem.*, 226, 195, 1997.
15. Cornell, R.M., Adsoption of caesium on minerals: a review, *J. Radioanal. Nucl. Chem.*, 171, 483, 1993.
16. Dumat, C. et al., The effect of removal of soil organic matter and iron on the adsorption of radiocaesium, *Eur. J. Soil Sci.*, 48, 675, 1997.
17. Valcke, E. and Cremers, A., Sorption-desorption dynamics of radiocaesium in organic matter soils, *Sci. Total Environ.*, 157, 275, 1994.
18. Carini, F. and Lombi, E., Foliar and soil uptake of 134Cs and 85Sr by grape vines, *Sci. Total Environ.*, 207, 157, 1997.
19. McCartney, M., Baxter, M.S., and Scott, E.M., Carbon-14 discharges from the nuclear fuel cycle: 2. local effects, *J. Environ. Radioact.,* 8, 157, 1988.
20. UNSCEAR, United Nations Scientific Committee on the Effects of Atomic Radiations (UNSCEAR), Reports on Ionizing Radiation to General Assembly, New York, United Nations, 1982, 1988, 1993.

21. Collins, C.D., The Movement of C-14 and S-35 in Crops Following Deposition from the Atmosphere: A Review for the Ministry of Agriculture, Fisheries and Food, MAFF, London, 1991.
22. Collins, C. and Gravett, A., The deposition of 14C, 3H, and 35S to vegetation in the vicinities of a Magnox and an advanced gas cooled reactor, *Sci. Total Environ.*, 173–174, 399, 1995.
23. Morgan, J.E. and Beetham, C.J., Review of literature for radium, protactinium, tin and carbon, Nirex Safety Studies, BSS/R220, UK Nirex, Harwell, UK, 1990.
24. Moorby, J., The production, storage and translocation of carbohydrates in developing potato plants, *Ann. Bot.*, 34, 297, 1970.
25. Benjamin, L.R. and Wren, M.J., Root development and source sink relations in carrot, *Daucas carrota* L., *J. Exp. Bot.*, 29, 425, 1978.
26. Bell, C.J. and Incoll, L.D., The redistribution of assimilate in field grown winter wheat, *J. Exp. Bot.*, 41, 949, 1990.
27. Murphy, C.E., Tritium transport and cycling in the environment, *Health Phys.*, 65, 683, 1993.
28. Brudenell, A.J.P., Collins, C.D., and Shaw, G., Rain and the Effect of Washout of Tritiated Water (HTO) on the Uptake and Loss of Tritium by Crops and Soils, MAFF Report for Project RP 0175 produced by Imperial College, 1999.
29. Brudenell, A.J.P., Collins, C.D., and Shaw, G., Dynamic of tritiated water (HTO) uptake and loss by crops after short-term atmospheric release, *J. Environ. Radioact.*, 36, 197, 1997.
30. Nicholson, K.W. and Dore, C.J., Factors affecting the variability of radionuclides in terrestrial foodstuffs, MAFF Project Report produced by AEA Technology, Report No. AEAT-2935 Issue 1, 1998.
31. Johnson, L.H. et al., Radiological assessment of Cl in the disposal of used CANDU fuel, Atomic Energy of Canada Limited Report, AECL-11213, COG-94-527, 1995.
32. Coughtrey, P.J. and Thorne, M.C., *Radionuclide Distribution and Transport in Terrestrial and Aquatic Environments*, Vol. 1, A A Balkema, Rotterdam, 1983.
33. Sheppard, S.C. and Evenden, W.G., Response of some vegetable crop to soil applied halides, *Can. J. Soil Sci.*, 72, 555, 1992.
34. Sheppard, S.C., Eveden, W.G., and Amiro, B.D., Investigation of the soil-to-plant pathway for I, Br, Cl, and F, *J. Environ. Radioact.*, 21, 9, 1993.
35. Beresford, N.A. and Howard, B.J., The importance of soil adhered to vegetation as a source of radionuclides ingested by grazing animals, *Sci. Total Environ.*, 107, 237, 1991.
36. Hinton, T.G., Stoll, J.M., and Tobler, L., Soil contamination on plant surfaces from grazing and rainfall interactions, *J. Environ. Radioact.*, 29, 11, 1995.
37. Titley, J.G. et al., Investigation of the Sources and Fate of Radioactive Discharges to Public Sewers, Environment Agency R&D Technical Report P288, 2000.
38. Ham, G.J. et al., Partitioning of radionuclides with sewage sludge and transfer along terrestrial foodchain pathways from sludge-amended land — a review of data, National Radiological Protection Board, Report No. NRPB-W32, 2003.
39. ANPA, SEMINAT Long-Term Dynamics of Radionuclides in Semi-Natural Environments: Derivation of Parameters and Modelling. Agenzia Nazionale Per La Protezione Dell'Ambiente (ANPA), Final Report 1996–1999, Research contract No. FI4P-CT95-0022, European Commission Nuclear Fission Safety Programme, 2000.
40. Ould-Dada, Z., Dry deposition profile of small particles within a model spruce Canopy, *Sci. Total Environ.*, 286, 83, 2002.

41. EC, Radiation Protection 128: Assessment of the Radiological Impact on the Population of the European Union from European Union Nuclear Sites between 1987 and 1996. EC Luxembourg, 2002.
42. Kazakov, V.S., Demidchik, E.P., and Astakhova, L.N., Thyroid cancer after Chernobyl, *Nature*, 359, 21, 1992.
43. Likhtarev, I.A. et al., Thyroid dose assessment for the Chernigov region (Ukraine): estimation based on 131I thyroid measurements and extrapolation of the results to districts without monitoring, *Radiat. Environ. Biophys.*, 33, 149, 1994.
44. Howard, B. and Beresford, N.A., Advances in animal radioecology, in *Radioactive Pollutants Impact on the Environment*, Brechignac, F. and Howard, B. (Eds.), based on invited papers at the ECORAD 2001 International Conference, EDP Sciences 2001.
45. IAEA, Modeling of Radionuclide Interception and Loss Processes in Vegetation and of Transfer in Semi-natural Ecosystems, second report of the VAMP Terrestrial Working Group, Part of the IAEA/CEC Co-ordinated Research Programme on the Validation of Environmental Model Predictions (VAMP), International Atomic Energy Agency, Vienna, Report No. IAEA-TECDOC-857, 1996.
46. Gruter, H., Radioactive fission product 137Cs in mushrooms in West Germany during 1963–1970, *Health Phys.*, 20, 655, 1971.
47. Bakken, L.R. and Olsen, R., Accumulation of radiocaesium in fungi, *Can. J. Microbiol.*, 36, 704, 1990.
48. MAFF and SEPA, Radioactivity in Food and the Environment (RIFE) 1998, Annual report produced by the Ministry of Agriculture Fisheries and Food (MAFF) and the Scottish Environment Protection Agency (SEPA), 1999.
49. Tataruch, F., Schonhofer, F., and Klansek, E., Studies in levels of radioactivity in wildlife in Austria, in *Transfer of Radionuclides in Natural and Semi-natural Environments*, Desmet, G., Nassimbeni, P., and Belli, M., Eds., Elsevier Applied Science, London, 1990, p. 210.
50. Mascanzoni, D., Chernobyl's challenge to the environment: a report of Sweden, *Sci. Total Environ.,* 67, 133, 1987.
51. Rissanen, K. and Rhola, T., ^{137}Cs concentration in reindeer and its fodder plants, *Sci. Total Environ.,* 85, 199, 1989.
52. Hansen, H. S. and Hove, K., Radiocasium bioavailability: transfer of Chernobyl and tracer radiocaesium to goat milk, *Health Phys.*, 60, 665, 1991.
53. Mattsson, S. and Moberg, L., Fallout from Chernobyl and atmospheric nuclear weapons tests, Chernobyl in perspective, in *The Chernobyl fallout in Sweden*, Moberg, L., Ed., Swedish Radiation Protection Institute, Stockholm, 1991, p. 591.
54. Swift, D.J., The accumulation of ^{95m}Tc from sea water by juvenile lobsters (*Homarus gammarus* L.), *J. Environ. Radioact.*, 2, 229, 1985.
55. Knowles, J.F., Smith, D.L., and Winpenny, K., A comparative study of the uptake, clearance and metabolism of technetium in lobster (*Homarus gammarus*) and crab (*Cancer pagarus*), *Radiat. Prot. Dosimetry*, 75, 125, 1998.
56. McKay, W.A. and Fox, A.A., Particulate-associated nuclides in Cumbrian Winkles - implications for assessment of dose to man, *J. Environ. Radioact.,* 14, 1, 1991.
57. Smith, B.D. and Jeffs, T.M., Transfer of Radioactivity from Fishmeal in Animal Feeding Stuffs to Man Environmental Technical Note RL8/99. Centre for Environment, Fisheries, and Aquaculture Science, Lowestoft, UK (1999).
58. Smith, B.D., The Radiological Significance of Farmed Fish, Project R02015/C1289, RL 19/02, CEFAS, Lowestoft, U.K., 2002.

59. Chartier, M. and Despres, A., Radiological impact on EC members states of routine discharges into the Mediterranean sea, *Radiat. Prot. Dosimetry*, 75, 161, 1998.
60. CEC, The Radiological Exposure of the Population of the European Community from Radioactivity in North European Waters — Project MARINA, report by a Group of Experts Convened by the Commission of the European Communities, Report EUR 12483, Commission of the European Communities, Luxembourg, 1990.
61. EC, MARINA II project report Update of the MARINA project on the radiological exposure of the European Community from Radioactivity in North European Water, published by the European Commission at http://europa.eu.int/comm/environment/radprot, August 2002.
62. Smith, K.R. et al., Uncertainties in the Assessment of Terrestrial Food Chain Doses, NRPB-M922. NRPB, Chilton, Oxon, 1998.
63. Cabianca, T.R.A. et al., The Variability in Critical Group Doses from Routine Releases of Radionuclides to the Environment, Chilton, NRPB-M952, London, HMSO, 1998.
64. Ould-Dada, Z. et al., Assessment of prospective foodchain doses from radioactive discharges from BNFL Sellafield, *J. Environ. Radioact.*, 59, 273, 2002.

12 Assessment of Exposure to Chemical Pollutants in Food and Water

Peter J. Peterson

CONTENTS

12.1 INTRODUCTION

Food and water are the major exposure routes for most chemicals that adversely affect the health of nonoccupationally exposed people. Chemical contamination of food and hence the potential for exposure to a wide variety of hazardous substances can occur through the life cycle of foods, from cultivation through processing to consumption. Indeed, the need to address chemical contamination of food and water was recognized in Chapter Six of U.N. Agenda 21 as an important issue necessary for the protection and promotion of human health.[1] The World Summit on Sustainable Development, held 10 years after the Earth Summit, also called on governments, industry, and all stakeholders to use and produce chemicals in ways that do not lead to significant adverse effects on human health and the environment.[2] The principal aim of measuring concentrations of chemical contaminants in diets and dietary items — the determination of exposure — is to be able to reduce health risks and the social burden of food-borne diseases.

Substantial differences in dietary intake occur throughout the world. Consequently, exposure to chemical contaminants is diverse, based on the major ethnic differences, life style and behavioral actions, and cultural practices including the diet regimes of vegetarians, vegans, and groups who consume large amounts of raw foods including fish.[3]

Acute poisonings of humans in many developing countries have been reported from consumption of foods contaminated with a variety of pollutants that are hazardous and persistent chemical substances.[4] This situation has arisen in many countries because of inadequate chemicals management, lack of local awareness and education on the toxicity of chemicals, and general lack of food legislation and health care. Health impacts and social costs from exposure to contaminated foods can be serious as the bulk of the population in developing countries consists of malnourished poor people, often affected by chronic parasitic or infectious diseases and reduced immune systems, as well as being involved in heavy physical work and heat stress.[3]

In both developed and developing countries, chemical contaminants in foods can also involve long-term low-level exposure, giving rise to chronic effects that make it difficult to detect the health impact until long after the exposure has taken place.[5] The increasing information and growing consumer concerns about chemical residues and contaminants in foods[6] highlights the need for detailed chemical analyses, disclosure of results, and vigilance to ensure that dietary items are not a threat to human health. This chapter focuses on chronic long-term health risks from exposure to chemical contaminants.

Serious contamination of food that has occurred over the years has illustrated the negative effects of food contaminant problems on international trade.[6] For both economic and human health reasons, governments are responsible for ensuring that the specifications of food exports and imports comply with the Joint Food and Agriculture Organization of the United Nations and the World Health Organization (FAO/WHO) Codex Alimentarius Commission's Food Standards Programme,[7] including recommendations of the FAO/WHO Joint Expert Committee on Food Additives (JECFA) and the related FAO/WHO Joint Meeting on Pesticide Residues (JMPR). WHO's health assessments are undertaken by its International Programme on Chemical Safety (WHO/IPCS). Compliance with standards, guidelines, and recommendations of Codex means compliance with the international harmonization of food safety and control requirements of the World Trade Organization (WTO). Under WTO's Agreement on Sanitary and Phytosanitary Measures, all health and safety requirements for food must be based on sound scientific risk assessment. The JECFA and JMPR risk assessments performed for Codex comply with this agreement.

This Chapter focuses on dietary exposure to potentially toxic chemical substances in food and diets, such as lead, cadmium, mercury, arsenic, fluoride, and, to a lesser extent, boron, which arise via uptake processes from soil or water often contaminated by industrial processes.[5] Tens of thousands of national studies have been published over many decades dealing with these contaminants in foods and dietary intakes. In the last decade, emphasis has shifted to persistent toxic substances (PTS),[8] especially pesticides, many of which are addressed under the Stockholm Convention on Persistent Organic Pollutants (POPs).[9] Over 12,000 publications on

PTS in food and dietary intakes have been reported between 1990 and 2002, which illustrates how the current emphasis has shifted away from mineral components in foods. No international conventions have been adopted that address heavy metals, although major national actions have been implemented around the world addressing source reductions to reduce risk from such metals to human health.[10]

12.2 APPROACHES FOR ESTIMATING DIETARY EXPOSURE

Human exposure is defined as the contact over time and space between the chemical and the individual.[11] This definition of exposure is in agreement with the WHO concept of exposure, that is, the amount of a substance in contact with the outer boundary of the body.[5] The amount that passes the boundary by ingestion of food, the intake, is the potential dose, and the amount absorbed, the uptake, is the internal dose. The relationship between exposure, dose, and risk to human health is illustrated[12] in a simple manner in Figure 12.1. An expanded and more detailed version of the role of exposure in the environmental health paradigm has been published.[5]

The process of measuring and evaluating external exposure involves examining its magnitude, frequency, and duration.[5] Characterization of human exposure also involves differences with regard to age, gender, ethnicity, genetic characteristics, nutritional status, etc., of the individual or group. These elements of exposure are illustrated in Figure 12.2.

Dietary exposure, or dietary intake estimates that measure it, is also a complex issue involving measurements of the occurrence and concentration of the chemical in dietary items, changes in chemical concentration and speciation during food preparation, and examination of variations in dietary profiles across nations and regions.[13] The distinction of whether food was analyzed as prepared for consumption, such as having been washed, or was sampled as unprocessed food, can influence exposure measurements. Consequently a wide spectrum of exposure values have been measured and reported.[14]

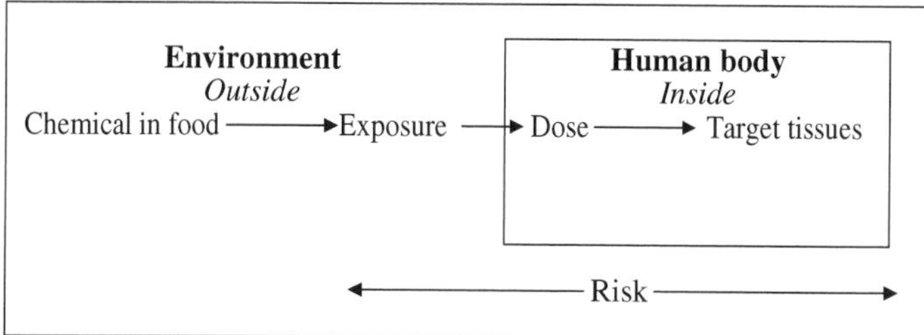

FIGURE 12.1 The relationship between chemicals in food, exposure, dose, and risk. (Modified from Huismans, J.W., Halpaap, A.A., and Peterson, P.J., *International Environmental Law: Hazardous Materials and Waste,* 2nd revised ed., United Nations Institute for Training and Research, Geneva, 2004.)

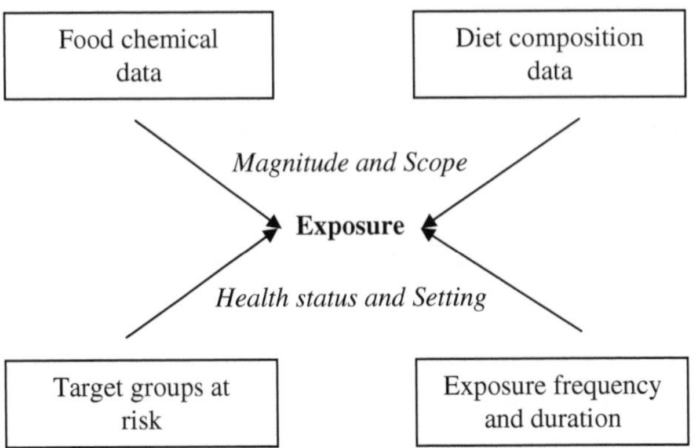

FIGURE 12.2 Important elements for determining exposure to contaminants in food.

The quantification of exposure to contaminant chemicals in food and water can be measured by three major approaches, each based on different data profiles, thus permitting the verification and validation of the information.[13] The approaches are:

- Chemical concentration scenario, i.e., food monitoring
- Point of contact method, i.e., duplicate diet studies
- Biological markers of exposure, i.e., analysis of biological samples.

In all three approaches the data used to determine exposure can be based either on single-point data, as an initial analysis, or involving probabilistic methods using data distributions for more refined evaluations.[15]

The first approach, the chemical concentration scenario,[13] involves measurements of the chemical in individual items including drinking water and the amounts consumed followed by estimates of the frequency and duration of exposure to such substances.

$$\text{Dietary exposure} = \text{Consumption} \times \text{Chemical concentration}$$

The method estimates exposure indirectly and considers standardized reference values for fluid intakes, activity patterns, and standard body weights of the groups of people concerned. Dietary intakes, i.e., consumptions, are calculated by summing the basic monitoring data of items in commonly consumed foods by using, for example, dietary recall questionnaires, food frequency questionnaires, food diaries, food habit questionnaires, or 24-h food recall, filled in by the participants of the study.[15] A second concept within the chemical concentration scenario involves the selection of a Total Diet Study (TDS) aimed at improving dietary exposure estimations. It relies on the direct analysis of a composite sample using a model diet or market basket representation of the average diet of the sample population. Total diets explicitly take into account kitchen preparations in order to assess the levels of contaminants in foods as consumed. The disadvantages of these two concepts are the assumptions made regarding which foods to analyze and calculations of their consumption rates, or, components to include within a total diet.[16] The approach's main strength is that concentrations of different chemicals can be multiplied by the amount of different food items consumed based on various exposure regimes, including those in high-risk groups. In the case of a total diet study, one major benefit is the small number of samples for analysis for the different populations studied. The approach, which is population-based, is discussed in detail in Subsection 12.4.1 and is the least expensive of the three.

The second approach, the point of contact method[13] in the case of food and drinking water, applies to duplicate meal studies in which a proportion of the customary meal is analyzed collectively whereas the remainder is eaten. Disadvantages of the method relate to the relatively short-term sampling period, differences in food items over the longer term, and the relationship between short-term exposure studies and long-term low-level exposures typical of chronic health effects. This approach also does not identify the contribution of particular contaminated dietary items to the total dietary exposure. Its main advantage is that it measures exposure directly over the period of time of the study. It is a household-based, or individual-based, approach. It is especially relevant for accurately evaluating exposure where major cultural differences in dietary patterns occur. The approach is expensive and is further discussed in Subsection 12.4.2.

The third approach involves estimations of exposure from the dose through the use of biological markers (biomarkers) of exposure.[13] It is a reconstructive method used to back-calculate dose and is individual-based. Biomarkers provide the link between external exposure and internal dosimetry and depend upon the relationship between the exposure and the health outcome.[11] Biomarkers can reflect chemical concentrations in tissues, organs, body fluids, excreta, etc., or what is bound to a specific target molecule.[5] They can represent past exposures (e.g., Pb in teeth), as well as current exposures (e.g., As in urine). Arsenic in urine reflects exposure over the last 1 to 3 d, and that in blood over a longer period, whereas contaminants in hair reflect exposure over 1 or 2 months.[5] Biomarkers of exposure can also help identify groups at risk where analytical values differ from background concentrations. The method is especially applicable for measuring the dietary intakes of contaminants by infants consuming breast milk.[17] A major disadvantage of breast milk is that it is an evaluation of all exposure routes for the mother, not just food and water. The method is expensive and is discussed further in Subsection 12.4.3.

Biomarkers are a specific group of indicators. In many cases, additional indicators have been developed that reflect either exposure to a wide range of chemicals or human health impacts (see Subsection 12.4.4).[18]

Each of the three scenarios outlined above, when compared with health outcomes, has provided useful data helping to set standards and guideline values designed to protect human health from chemical contaminants.[15] Exposure measurements are essential for the protection of populations and subgroups at high risk.[15]

12.3 DIETARY EXPOSURE AND RISK

A great many chemicals are circulating in the environment and occur in foods, so humans will continue to be exposed to varying degrees via their dietary intakes.

However risk, can only be estimated relative to measures of toxicity of specific chemical substances.[13] Simply stated:

$$\text{Risk} = \text{Hazard} \times \text{Exposure}$$

Toxicological methods of determining hazard are discussed elsewhere.[5,13]

Emphasis is placed in this chapter on determining exposure often involving three characteristic groups of people. These include the general population; important subgroups within the population including children,[19] pregnant women, and the elderly;[20] and the high end of the exposure distribution (90th percentile or higher). Within these subgroups, children are highly vulnerable to environmental toxicants because of their rapid growth and development associated with the high intake of food relative to body weight, their greater future years of life than most adults, and consequently their greater risk of chronic toxic effects than from exposure to chemicals occurring later in life.[21]

Possible risks to human health are the major consideration when evaluating exposure posed by chemical contaminants in foods and diets. Regulatory authorities are especially concerned with long-term intake and chronic public health effects, but they also have a responsibility to protect against short-term acute chemical impacts. Various approaches to measuring exposure are outlined in the following text, ranging from simple techniques to more accurate scientifically justifiable methods that can be used to evaluate risk to the general population.

12.3.1 Risk and Exposure via Food

Estimations or measurements of the actual dietary intake of chemical contaminants are essential for risk assessment. Protection of human health from such exposures has been through the establishment of exposure limits, standards, recommendations, guidance values, etc. Chronic dietary intakes are often evaluated in relation to the Acceptable Daily Intake (ADI), which is an estimate of the amount of a chemical that can be ingested daily over a lifetime without appreciable risk to health.[22] It is a health guideline based on the no-observed-adverse-effect-level (NOAEL) plus a safety factor usually of 100, and is made up of a factor of 10 for interspecific extrapolations and a further factor of 10 for individual variations in humans. It is assumed that the intake of chemical contaminants is "safe" if mean exposures are less than the ADI described from toxicological studies and expressed as mg/kg body weight/day. The Tolerable Daily Intake (TDI), Provisional Tolerable Weekly Intake (PTWI) and, in the case of pesticides, a health-based Maximum Residue Level (MRL) are used depending upon the toxic chemical in the food.

PTWIs have been established for metals e.g., Pb, Cd, Hg, etc., by the JECFA and are expressed on a body weight or per-person basis. For the protection of human health the amount of the contaminant should not exceed the PTWI or ADI over a prolonged period of time. Risk is therefore determined by comparing the actual intake as determined from the total diet study with the PTWI or ADI. As As, Cu, and Zn are not accumulated, a Provisional Tolerable Daily Intake (PTDI) is used instead of a PTWI. The word "provisional" is used because the upper limits for exposure can be reevaluated in the light of new knowledge. For chemicals that have been designated as carcinogens by the WHO's International Agency for Research on Cancer (IARC), no ADI is established.[5] The Acute Reference Dose (RfD) for humans is used in the U.S. and is comparable with the ADI, but it is weighted to large portions in dietary intake.[23]

The determination of human exposure (dietary intake) and its risks to human health, as mentioned above, all rely on accurate and reliable data quality on dietary intakes, identification of foods that contribute substantially to the intake, and the concentration of the chemical pollutants. Analytical Quality Assurance (AQA) and Analytical Quality Control (AQC) initiatives organized for many years under the GEMS/Food program are regularly published[24] and have contributed to improving national AQA capacity in many countries.[25] In addition to the need for adoption of standard AQA procedures, treatment of analytical data where the chemical is below the detection limit also can affect exposure calculations. Different approaches within total diet studies have been adopted, ranging from recording a zero value (often referred to as a "lower bound"), some percentage of the detection limit, or the actual detection limit ("upper bound"). Results may then be expressed as both the lower bound measures and upper bound measures. It may not be clear in some studies how samples that are below the detection limit were evaluated. These issues are addressed further in Chapter 3.

12.3.2 Risk and Exposure via Drinking Water

Contaminated drinking-water supplies contribute to the intake of chemical contaminants and may constitute the primary route of exposure, e.g., F, Pb, As, and B.[5] In developing countries, where most people have no access to treated and piped water, "raw" surface water, or well water, is the only source available. Ingestion of pollutants in drinking water primarily enters via two pathways, either direct ingestion from intrinsic water intake or secondarily through cooking in the contaminated water. Cultural differences in drinking-water intake have been reported.[26]

Health risks from drinking contaminated water have been evaluated on a substance-by-substance basis. Guideline values are commonly calculated as 10% of the ADI and based on a 60-kg person drinking 2 l water/d.[27] The guideline values indicate tolerable concentrations that should be as low as possible; they are not target values to aim at. Exposures of children and infants to contaminants in drinking water are adjusted to body weight and consumption. In the case of substances that are only slightly soluble in water, a dietary intake for adults from drinking water may be calculated as 1% of the ADI. As drinking water is not involved in international trade, there are no values comparable with the MRLs established by Codex for pesticide residues in food.

Some studies include drinking water in dietary exposure assessments, but they have not always been adopted.[15] The GEMS/Food dietary studies do recommend the inclusion of drinking water in their exposure estimates. As drinking water can be a major source of exposure to Pb, As, F, and B, whether or not it is directly included in the intake and whether or not domestic food is prepared/cooked in such water can influence dietary intakes. In one study in the U.K.,[28] drinking water was not included and food was prepared using distilled water, so the adult exposure to fluoride was 0.94 mg/d, which was compared with New Zealand and Canada total diet studies of 2.65 and 1.76 mg/d, respectively. Fluoride from tea contributed approximately 85% of the U.K. value. If drinking water was included in the U.K. diet at an average of 1 mg/l, and the consumer drank 2 l, then the actual exposure to F would be approximately 3 mg/d.

Boron can occur naturally in drinking water,[29] ranging from < 0.3 to >100 mg/l with average values of 0.5 mg/l. A daily drinking water intake of around 1 mg/d (drinking 2 l of water a day) can be compared with the B intake from food, also of approximately 1 mg/d, although specific country intakes can range to 10 mg/d, especially for food grown in areas of volcanic activity.[30]

Bottled water constitutes a special case. The WHO guidelines on drinking water apply to bottled drinking water, but bottled mineral waters are considered as food items and evaluated within Codex standards.[31]

12.4 CHEMICAL CONTAMINANTS AND EXPOSURE

Chemical contaminants in specific foods are discussed in detail, element by element, in Chapter 10. Speciation of chemical components in foodstuffs is discussed in Chapter 2. This section details and discusses the three approaches for evaluating exposure outlined in Section 12.2. Many exposure studies have been published that involve different dietary regimes and various groups of chemical pollutants.[5,13] Survey methods too may differ, including consumption periods from one day to one month, and are often based on different scientific and analytical criteria for estimating dietary exposures.

12.4.1 Chemical Concentration Scenario

Total diet studies and food monitoring programs provide the major approaches for exposure analysis. Results are presented as concentrations of individual pollutants in individual foods, or in total diets compared with the ADI or PTWI for those pollutants.[14,32] Comparisons between exposure studies in various countries may not be all that easy to evaluate, because analytical and sampling protocols may not have been standardized and quality controls are sometimes missing.[24] Drinking water may or may not have been included in the analysis of total dietary exposure.

Data on dietary exposure are too numerous to list (see Chapter 10), but much data for over 60 participating countries are included on the WHO Website, especially from the GEMS/Food programme:[33] (http://www.who.int/fsf). GEMS/Food has developed five regional diets (Middle Eastern, Far Eastern, African, Latin American, and European, the last including the U.S., Canada, and Australia) comprising 350 primary semiprocessed commodities as well as 13 consumption cluster diets based on the consumption of 36 basic foods and food groups.[34] In addition to the calculated median values, the 90th percentile has been calculated.

Total diet studies and estimated dietary intakes have been reported from many developed countries.[33] The data revealed differences in exposure between countries for a wide range of chemical contaminants and often mirror country-specific traditions and habits. Data are available on CD-ROM, e.g., from the U.S. Department of Agriculture.[35] In another example the 20th Australian total diet survey has been completed[36] and is available via http://www.anzfa.gov.au.

The Australian study[36] reported the range of mean estimated dietary exposure for metals, metalloids, several essential elements, and some pesticides, and compared them with the PTWI or ADI separately for women and men, for infants (9 months), children (2 years), teenagers (12 years), and adults (25–34 years). In all cases the dietary exposures to Pb, Cd, Hg, and As were below the PTWI/PTDI, although the highest mean exposure to Cd (13–69% PTWI), Pb (1–33%), Hg (1–35%), and As (12–48%) were recorded in infants mainly influenced by their high food consumption relative to body weight. Concern was expressed about the high potential Hg exposure for pregnant women consuming large amounts of fish, because of the sensitivity of the fetus to Hg.[37] An advisory of one half of the health standard for pregnant women compared with the general population was proposed in the Australian study.[36]

A comparable study — the 2000 U.K. Total Diet Study — has been reported[38] following analysis of 12 elements. It involved 119 categories of food collected from 24 towns and grouped into 20 similar foods for analysis, and the results were compared with the PTWI/PTDI. Diets were calculated to represent an elderly group of citizens (> 64 years), vegetarians, as well as toddlers (1.5–4.5 years), young people (4–18 years), and adults (16–64 years). The dietary exposure for toddlers for all elements analyzed were in general higher than those for other age groups as with the Australian study mentioned earlier, whereas the elderly had lower dietary exposures. Measuring dietary exposure of the elements compared with PTWI represent approximately 12% for Cd, 3% for Pb, and 12% for Hg, although high-level exposure groups were two or three times these percentages. A comparison with the annual U.K. population dietary exposures from 1976 to 2000 shows that concentrations of As, Cd, Hg, and Pb have consistently decreased over time.

The high concentration of Hg in fish, predominantly as methyl-Hg,[39,40] is the major contributor to dietary Hg exposure. Reported Hg exposure differences between countries are usually based on the amounts of fish consumed. Indigenous peoples living in some countries where gold-mining activities have discharged Hg into soil and rivers and who consume large amounts of fish,[41,42] have Hg intakes that exceed JECFA health guidelines.[37] This issue is discussed further in Subsection 12.4.3. Furthermore, in gold-mining areas, Hg in drinking water from boreholes and wells may exceed the WHO limit,[43] thus contributing to higher Hg exposures. Such results can be compared with Hg concentrations in drinking water from nonpolluted areas that are often below the limit of detection. Cd in drinking water is usually minimal, whereas Pb in water reticulated through Pb-pipes may contribute considerably to dietary exposure for many people including bottle-fed infants.[44,45] Exposures to As and F in drinking water are discussed in Subsection 12.4.4.

12.4.2 Point of Contact Method

Duplicate diets, in which a portion of the food prepared for consumption is analyzed for contaminants and the remainder is eaten provides a direct approach for evaluating human exposure to a variety of chemical substances.[14,32] In small-scale duplicate diet studies and analysis of fecal concentrations of Pb and Cd, food was shown to be the main exposure route, although in areas where concentrations of Pb in air was high, inhalation contributed significantly to exposure.[46] In another study,[47] large interindividual and day-to-day variations were shown with respect to exposure to Pb and Cd, which illustrates the usefulness of the duplicate diet approach. Concentrations of Cd in feces were again used to evaluate bioavailability of dietary Cd in women consuming a mixed diet low in shellfish, a diet higher in shellfish (shellfish accumulate Cd), or a vegetarian diet rich in fiber.[48] Despite differences in Cd intakes and a low absorption of dietary Cd, blood and urinary Cd values were similar.

A 7-d duplicate diet study of 12 elements and involving approximately 50 vegetarians in the U.K. has been reported[49] and compared with results from the 1997 general U.K. Total Diet Study.[50] As vegetarians do not eat fish, shellfish, or meat, Hg that is accumulated in these food items was not determined. Dietary exposure to As was lower in the vegetarian diet than in the general population (0.017 mg/d compared with 0.065 mg/d) as fish also contributes substantially to the As intake. With Cd, the intake was slightly higher in the vegetarian diet (0.015 mg/d compared with 0.012 mg/d), which may relate to the higher dietary intake of nuts, pulses, and vegetables (which contain high concentrations of Cd) thus offsetting Cd intakes by the general population from meat, offal, and shellfish that are also known to contain relatively high concentrations of Cd. With dietary exposure to Pb on the other hand, exposure of the vegetarians was lower than that of the general population (0.015 mg/d compared with 0.026 mg/d). Specifically, beverages (i.e., based on drinking water) that are major contributors to dietary exposures to Pb were not included in the diet of the vegetarians.

Vegetarians also have a higher intake of B than the general population via exposure from food because fruit, vegetables, pulses, legumes, and nuts have higher concentrations of B.[29] Average daily exposure of vegetarians to B, in EU countries for example, range from 2.4 to 7 mg B/d compared with an average diet of 1.6 to 4.5 mg B/d.[30]

The point-of-contact approach, however, has some limitations. The number of individuals participating in the studies is usually small, and errors with how the food is split between what is analyzed and what is consumed have been reported. These differences relate to behavioral differences and socioeconomic factors. In one study in the U.S. that lasted for 28 d, although only 7 d of food collections were consecutive, food consumed in social settings was not collected, and noncollection of meals and food increased after the third day of collection.[51] In the duplicate diet study of vegetarians mentioned earlier,[49] underreporting of foods consumed was also recorded based on diary records and mean daily energy intake calculations. As a result, more than a dozen dietary results were not used by the researchers to calculate dietary exposure.

Breast milk provides a rather different example of a duplicate diet approach. Breast milk, which forms the major dietary intake of infants, has also been chemically analyzed for a range of pollutants as mentioned in the following subsection.

12.4.3 Biological Markers of Exposure

Biomarkers can be of three types:[17,52]

- As indicators linking exposure of individuals to a chemical
- As indicators measuring biological response or effect
- As indicators measuring susceptibility to the chemical

Biomarkers have been used for various purposes,[53] including:

- Assessing exposure in terms of dose
- Setting standards relative to health effects
- Validating intake estimates of contaminants
- Calculating levels of risk
- Quantifying interindividual variability

Measured biomarker responses are typically noninvasive and can be examined in human fluids (commonly urine, feces, blood, and breast milk), and external tissues (commonly hair, nails, and shed teeth) as discussed earlier in Section 12.2.

Because some chemicals are excreted via the kidneys, urinary concentrations provide useful biomarkers[17,53] for F, As, and B. In the case of As, the species of chemical in the urine relates to the form of As in the dietary intake.[54] For example, although arsenate dominates in the urine of individuals exposed via drinking water, concentrations of the less toxic forms monomethylarsonic acid and dimethylarsinic acid also occur. With the intake of seafoods, As in urine can again be used to quantify As exposure, but the compound excreted is predominantly arsenobetaine, which is the principal As species in crustaceans.[55] Consequently, knowledge of the nature and concentration of As species in urine provides information on As exposure, the type of As compound ingested, and the methylating capacity of the individual.

Elevated concentrations of low-molecular-weight proteins in urine, such as β_2-microglobulin, α_1-microglobulin or retinal-binding protein, have been used as biomarkers of Cd exposure and indicate damage to the tubular protein absorption capability of the kidney.[56] Cd in feces is another useful biomarker as mentioned in the previous subsection, whereas Pb in blood is the principal method for estimating exposure to environmental Pb concentrations, although in this case it represents all routes of exposure, not just dietary intake.[14] Pb in whole blood reflects the absorbed dose, whereas Pb in blood plasma reflects the "active" fraction of Pb in blood and defines the relationship between blood-Pb and tissue- or organ-Pb accumulation and effects. Concentrations of Pb and Cd in diets over time have decreased in most countries, illustrating the successful implementation of regulatory controls to reduce exposure.[32]

Breast milk has been used extensively as a biomarker of exposure[17,53] for a great many substances in dietary intakes and can represent individual exposures or average values for a specific population group based on pooled samples. Data are expressed relative to a fat content of 3 or 3.5%. The concentrations of pollutant chemicals in milk and the levels of risk to infants are a function of various factors including not only the dietary intake but also nutritional factors.[57] Generally, chemicals enter breast milk by passive transfer from plasma, and their concentration in the milk is proportional to their solubility and lipophilicity.[58]

In the case of breast-milk-fed infants whose mothers consumed large amounts of fish, Hg concentrations were still below health guidelines.[59,60] Concentrations of Pb and Cd in breast milk were generally well below health guidelines.[32] Lead and Cd concentrations in breast milk are generally about 20% of the level in blood of the same person, which is attributed to their low lipid-solubility and high binding to erythrocytes.[61] In the case of F, exposure via breast milk was substantially lower than for formula-fed infants, which was largely determined by the concentration of F in the drinking water used to reconstitute the milk.[62]

Hair is another useful biomarker for certain pollutants, although external contamination can negate any dose–response relationship. Concentrations of As in bunched strands of hair have been shown to reflect retrospective exposure, especially to drinking water containing elevated concentrations of As, whereas patterns of variations of concentration along single strands relate to shorter-term exposure.[63]

Mercury in hair is a good indicator of Hg in dietary intakes and blood Hg concentrations.[37] Hg concentrations in hair have been reported to be higher in residents of areas[41] contaminated by Hg as well as in villagers living on the coast who eat fish compared with villagers living many kilometers inland.[64] Hence, measurements of methyl-Hg in hair reflect exposure to methyl-Hg in the diet being proportional to its simultaneous concentration in blood.[37] A blood–hair ratio for methyl-Hg has been calculated[37] as 1:250. Methyl-Hg exposure biomarkers as indicators of neurotoxicity in children have been discussed.[65] It has been considered by some authors[66] that although the JECFA PTWI for methyl-Hg (3.3 μg/kg body weight/week) is sufficiently protective for the general population, it may not be sufficiently protective for pregnant women due to the risk to the developing fetus. A lower value of 0.7 μg/kg body weight/week, which corresponds to the U.S. Environmental Protection Agency's (EPA) Reference Dose, is more appropriate. Again, as with As, concentrations of methyl-Hg have been shown to vary along single strands of hair, reflecting a longitudinal history of blood methyl-Hg.[67]

Lead concentrations in shed teeth of children have also been used as an indicator of exposure. Analyses of chemicals in fingernails and toenails have also been used to indicate exposure.[17]

12.4.4 Additional Indicators

In addition to the biomarkers mentioned above, various other groups of indicators have become widely used and play a significant role in trend analysis of exposures and chemical management response strategies.[18] Indicators summarize and condense large amounts of data into information more readily understandable than the original data. The significance of the indicator depends on the exposure concerned and the context of the issue or the problem being addressed. General exposure indicators can be used to relate to the proportion of a population exposed through dietary intakes to concentrations of methyl-Hg above background concentrations, or to the number of people exposed, that are above WHO guideline values.[37] Indicators can also be used to characterize exposure of specific at-risk groups, such as children or the poor, and management actions to ensure compliance with risk-reduction strategies.

Environmental health indicators too have been developed that relate exposure to health effects.[18] Various indicators have been reported that quantify exposure of people in China to high concentrations of As in drinking water.[68] Indicators that quantify dose–response relationships between exposure to As and skin lesions in farmers in Inner Mongolia have also been published.[69]

Because the predominant route of exposure to fluoride and development of dental and skeletal fluorosis diseases is via drinking water at locations on all continents,[70] indices (aggregate indicators) have been developed to measure past and present management actions to control exposure.[71] Different health outcomes for 14 provinces and autonomous regions in China were measured to evaluate health impacts and quantify exposure outcomes related to the local strategies needed to manage these diseases.

The use of indicators to evaluate human exposure and dietary intakes of chemical contaminants is not yet widespread unlike their extensive application for evaluating human exposure to urban air pollutants.[72]

12.5 DISCUSSION AND CONCLUSIONS

This chapter has summarized the methodology for evaluating exposure to chemical contaminants in foods and diets and has outlined selected results in terms of risk and risk management. But there is a recognition that more needs to be done on a range of issues.

One such issue is the changing population dietary patterns with time and hence the changing exposure scenarios in many countries. With increasing economic prosperity, especially in developed countries such as Japan, substantial changes in food composition have been recognized.[73] Dietary changes have also been reported in other countries associated with rice-based, wheat-based, and millet-based diets.[73] But in many countries the lack of base-line exposure data against which to measure trends is evident.

Risk from exposure to many individual chemical contaminants has been well documented for "general" populations in developed countries, using total diet studies and market-basket surveys. But how relevant are developed country average or median values for a general population exposed to a particular chemical when applied to developing country scenarios, even if using regional dietary data from GEMS/food? Much greater emphasis should be given to the high-risk exposure groups in developing countries, especially within the world's poorest countries. Conclusions from an international workshop stressed the need to conduct exposure assessments for different population subgroups.[74] Risks should also be evaluated in the light of poor nutrition and occurrence of infectious diseases, thus taking into account susceptible populations as mentioned in the Introduction Section 12.1. Indeed, one author has called for studies of effects of chemical contaminants on the poor.[3]

Combined low-level exposure to several pollutants, endocrine effects, and effects on the development of the fetal neural system, especially in high-risk groups, require further detailed study. Furthermore, how do the seasonal changes in availability and quality of locally grown food affect exposure of groups living only at the subsistence level? Biomarkers could be usefully applied to such groups to help identify and quantify the risks that would then be validated by the more usual dietary assessment.[52] Of course, solutions to such problems must be tailored to the dietary exposure and susceptibilities of the at-risk groups bearing in mind each country's unique circumstances.

Ethnic differences in dietary intake of sport-caught fish in the U.S. and risk from consuming methyl-Hg and other pollutants[75] provides a useful model for evaluating worst-case scenarios compared with the more "general" exposure.[37] Sport-fish consumption advisories have been issued in some circumstances in order to keep risks to an acceptable level. Yet, fish consumers may benefit from eating fish that contains omega-3 fatty acids.[76] Although sport-fish studies are highly specific, they illustrate the more general principle of the need to include risk–risk comparisons as well as risk–benefit comparisons.

Exposure to contaminants in traditional foods within specific ethnic groups provides a further example of worst-case scenarios, again involving methyl-Hg.[77] Inuit women eat large amounts of traditional foods, including, during pregnancy, fish, beluga, and seal meat, which contain high concentrations of methyl-Hg and other contaminants. Measurements of Hg in mothers' hair, plasma, breast milk, as well as children's cord blood showed levels consistent with cognitive deficits reported in other studies.

The addition of specific ethnic food items to a "general" diet in specific countries may markedly increase exposure of groups of people. Examples have been mentioned earlier of increased exposure to As, Hg, and Cd in seafoods, and specifically exposure of vegetarians. Yet there are other instances where exposure to a particular chemical such as F is less well known. For example, exposure to F is increased in endemic fluorosis areas in Kenya and Tanzania by the consumption of "Magadi" salt (a sodium bicarbonate encrustation on soda lakes), which is used in the cooking of vegetables.[70] Similarly, high exposure to F has also been reported from areas in China where consumption of Yushan salt (a high-F salt) and drinking of high-F "brick" tea is prevalent.[70] Consumption of high-F salt has also been reported from Thailand, Myanmar, and Vietnam, which has substantially contributed to increased F-exposure and associated fluorosis.

Populations in poor countries living on subsistence diets derived from one or two food staples, such as fava beans or cassava, can be further disadvantaged by the presence of potentially toxic substances. Cassava is the staple diet of around 400 million people worldwide but contains cyanogenic glycosides in the edible tubers and leaves.[78] Incompletely detoxified cassava products can readily lead to cyanide intoxication. Seeds and beans of various toxic plants are eaten by indigenous people throughout Africa, the Indian Subcontinent, China, etc., and can cause serious diseases such as lathyrism. Exposure to naturally occurring hazardous chemicals in foods adds to the chemical exposure of poor disadvantaged peoples.

How exposure and risk is to be communicated to decision-makers, let alone the public, and how dialogue can be fostered among all stakeholders, requires further attention.[79] A greater emphasis on assessing risks of vulnerable high-exposure groups by national scientists and managers is also clearly required. The overall aim is not just to evaluate the risk but to reduce chemical exposure and hence food-borne risks by taking appropriate management actions. One approach to highlight health issues requiring action was adopted by the WHO Regional Office for Europe, with their production of the book *Concern for Europe's Tomorrow*.[80] UNEP, too, has published regional studies on PTS as separate volumes to encourage more effective monitoring and generation of more representative dietary studies.[8] An extension of the GEMS/Food regional studies approach plus a series of regional publications would seem a useful way to publicize national data and highlight resultant national and regional needs.

As countries industrialize and their requirements for chemicals increase, they will need to shift chemical issues nearer the top of their priority concerns in order to reduce the social burden of food-borne diseases and to keep health risks low.[5] Food safety and compliance with Codex requirements are also of major economic importance for all nations involved in food trade. It can be concluded that studies of chemical exposures through food intakes and resultant risk evaluations are not a theoretical study; rather, they are an essential practical activity of local, national, and global importance.

REFERENCES

1. U.N., Agenda 21: The United Nations Programme of Action from Rio, New York, 1993.
2. WSSD, World Summit on Sustainable Development, Johannesburg, http://www.johannesburgsummit.org/html/documents/documents.html, 2002.
3. Goldman, L. and Tran, N., Toxics and Poverty: the Impact of Toxic Substances on the Poor in Developing Countries, World Bank, Washington D.C., 2002.
4. WHO, Major Poisoning Episodes from Environmental Chemicals, Report WHO/PEP/92, WHO, Geneva, 1992.
5. WHO/IPCS, Human Exposure Assessment, Environmental Health Criteria 214, WHO, Geneva, 2000.
6. WHO, WHO Global Strategy for Food Safety: Safer Food for Better Health, WHO, Geneva, 2002.
7. FAO/WHO, Codex Alimentarius Commission, Report of the Twenty-sixth Session, Rome, June 30–July 7, 2003.
8. UNEP, Regionally Based Assessment of Persistent Toxic Substance: Global Report 2003, UNEP, Geneva, 2003.
9. UNEP, Stockholm Convention on Persistent Organic Pollutants (POPs), UNEP, Geneva, 2001.
10. WHO, Health and Environment in Sustainable Development: Five Years after the Earth Summit, WHO, Geneva, 1997.
11. U.S. NRC, Frontiers for Assessing Human Exposures to Environmental Toxicants, U.S. National Research Council, National Academy Press, Washington D.C., 1991.

12. Huismans, J.W., Halpaap, A.A., and Peterson, P.J., *International Environmental Law: Hazardous Materials and Waste,* 2nd revised ed., United Nations Institute for Training and Research, Geneva, 2004.
13. WHO/IPCS, Principles for the Assessment of Risk to Human Health from Exposure to Chemicals, Environmental Health Criteria 210, WHO, Geneva, 1999.
14. Peterson, P.J., Assessment of exposure to chemical contaminants in water and food, *Sci. Total Environ.,* 168, 123, 1995.
15. WHO, Food Consumption and Exposure Assessment of Chemicals, Report of a FAO/WHO Consultation, Geneva February10–14, 1997.
16. EC, Report on Tasks for Scientific Co-operation: Improvement of Knowledge of Food Consumption with a View to Protection of Public Health by Means of Exchanges and Collaboration between Database Managers, Office for Official Publications of the European Commission, Luxembourg, 1997.
17. WHO/IPCS, Biomarkers and Risk Assessment: Concepts and Principles, Environmental Health Criteria 155, WHO, Geneva, 1993.
18. Von Shirnding, Y., Health in Sustainable Development Planning: the Role of Indicators, WHO, Geneva, 2002.
19. WHO/IPCS, Principles for Evaluating Health Risks from Chemicals during Infancy and Early Childhood, Environmental Health Criteria 59, WHO, Geneva, 1986.
20. WHO/IPCS, Principles for Evaluating the Effects of Chemicals on the Aged, Environmental Health Criteria 144, WHO, Geneva, 1992.
21. Suk, W. A., Beyond the Bangkok Statement: research needs to address environmental threats to children's health, *Environ. Health Perspect.,* 110, A284, 2002.
22. FAO/WHO, Summary of Evaluations by the Joint FAO/WHO Expert Committee on Food Additives (JECFA 1956–2001) First through Fifty-seventh Meeting, Internet Edition, FAO, Rome 2001.
23. U.S. EPA, Reference Dose (RfD): Description and Use in Health Risk Assessments, Background document 1A, U.S. EPA, Springfield, 1993.
24. WHO, GEMS/Food Analytical Quality Assessment Study 2000, report of the GEMS/Food Study of Food Analysis Performance of Heavy Metals, Report WHO/PHE/FOS/01.2, WHO, Geneva, 2001.
25. Weigert, P. and Müller, J., Analytical quality assurance in the German food contamination monitoring programme, *Fresenius Z. Anal. Chem.*, 736, 737, 1988.
26. Mushak, P. and Crocetti, A.F., Risk and revision in arsenic cancer risk assessment, *Environ. Health Perspect,.* 103, 684, 1995.
27. WHO, *Guidelines for Drinking Water Quality, Vol. 1: Recommendations*, WHO, Geneva, 1993.
28. FSA, 1997 Total Diet Study — Fluorine, Bromine, and Iodine, Food Survey Information Sheets 05/00, Food Standards Agency, London, 2000.
29. WHO/IPCS, Boron, Environmental Health Criteria 204, WHO, Geneva, 1998.
30. Mangas, S., Derivation of health investigation levels for boron and boron compounds, in *Health Risk Assessment of Contaminated Sites,* Langley, A. et al., Contaminated Sites Monograph Series 7, South Australian Health Commission, Adelaide, 1998, p. 229.
31. FAO, Codex Standards for National Mineral Water, Codex Alimentarius Commission XI, (III), FAO, Rome, 1994.
32. Jelinek, C.F., Assessment of Dietary Intake of Chemical Contaminants, WHO, Geneva, 1992.
33. WHO, GEMS/Food Total Diet Studies, Report of the Second International Workshop on Total Diet Studies, Brisbane, Australia, February 4–15, 2002.

34. WHO, GEMS/Food Regional Diets: Regional Per Capita Consumption of Raw and Semi-Processed Agricultural Commodities (revision September 2003), WHO, Geneva, 2003.
35. USDA, CSFII 1994–1996: 1998 Data Set, National Technical Information Service Accession No. PB 2000–500027, U.S. Department of Agriculture, Washington, D.C., 2000.
36. FSANZ, The 20th Australian Total Diet Survey, Food Standards Australia New Zealand, Canberra, 2003.
37. WHO/IPCS, Methyl mercury, Environmental Health Criteria 101, WHO, Geneva, 1990.
38. FSA, 2000 Total Diet Study of 12 Elements — Aluminum, Arsenic, Cadmium, Chromium, Copper, Lead, Manganese, Mercury, Nickel, Selenium, Tin and Zinc, Food Survey Information Sheets 48/04, Food Standards Agency, London 2004.
39. WHO/IPCS, Mercury: Environmental Aspects, Environmental Health Criteria 51, WHO, Geneva, 1989.
40. Boudou, A., and Ribeyre, F., Mercury in the food web: accumulation and transfer mechanisms, in *Mercury and Its Effects on Environment and Biology,* Sigel, A. and Sigel, H., Eds., Marcel Dekker, New York, 1997, p. 289.
41. Fréry, N. et al., Gold mining activities and mercury contamination of native Amerindian communities in French Guiana: key role of fish in dietary uptake, *Environ. Health Perspect.,* 109, 449, 2001.
42. Grandjean, P. et al., Methyl mercury neurotoxicity in Amazonian children downstream from gold mining, *Environ. Health Perspect.,* 107, 589, 1999.
43. Nyamah, D., Amonoo-Neizer, E.H., and Acheampong, K., Arsenic and mercury pollution at the mining environs of Obuasi, Ghana, *Ghana J. Chem.,* 1, 431, 2001.
44. FAO, Exposure of Infants and Children to Lead, FAO Food and Nutrition Paper 45, FAO Rome, 1989.
45. MAFF, Lead in Food: Progress Report, Food Surveillance Paper 27, Ministry of Agriculture, Fisheries and Food, Her Majesty's Stationary Office, London, 1989.
46. Vahter, M. and Slorach, S., GEMS Exposure Monitoring of Lead and Cadmium: An International Pilot Study within the WHO/UNEP Human Exposure Assessment Location (HEAL) Project, WHO Geneva and UNEP Nairobi, 1990.
47. Slorach, S., Jorhem, J., and Becker, W., Dietary exposure to lead and cadmium in Sweden, *Chem. Speciation Bioavail.,* 3, 13, 1991.
48. Vahter, M. et al., Bioavailability of cadmium from shellfish and mixed diets in women, *Toxicol. Appl. Pharmacol.,* 136, 332, 1966.
49. MAFF, Duplicate Diet Study of Vegetarians — Dietary Exposure to 12 Metals and Other Elements, MAFF Surveillance Information Sheet 193, Food Standards Agency, London, 2000.
50. MAFF, 1997 Total Diet Study — Aluminum, Arsenic, Cadmium, Chromium, Copper, Lead, Mercury, Nickel, Selenium, Tin and Zinc, Food Surveillance Information Sheet 191, The Stationary Office, London, 1999.
51. Thomas, K.W. et al., Testing duplicate diet sample collection methods for measuring personal dietary exposure to chemical contaminants, *J. Expo. Anal. Environ. Epidemiol.,* 7, 17, 1997.
52. Bingham, S.A., Biomarkers used to validate dietary assessments in human population studies, in *Biomarkers in Food Chemical Risk Assessment,* Crews, H.M. and Hanley, A.B. Eds., Royal Society of Chemistry, Cambridge, MA, 1995, p. 20,
53. WHO/IPCS, Biomarkers in Risk Assessment: Validity and Validation, WHO, Geneva, 2001.

54. Buchet, J.P. et al., Consistency of biomarkers to exposure to inorganic arsenic: review of recent data, in *Arsenic Exposure and Health Effects,* Chappell, W.R., Abernathy, C.O., and Calderon, R.L., Eds., Elsevier, New York, 1999, p. 31.
55. Donohue, J.M., and Abernathy, C.O., Exposure to inorganic arsenic from fish and shellfish, in *Arsenic Exposure and Health Effects,* Chappell, W.R., Abernathy, C.O., and Calderon, R.L., Eds., Elsevier, New York, 1999, p. 89.
56. Noonan, C.W. et al., Effects of exposure to low levels of environmental cadmium on renal biomarkers, *Environ. Health Perspect.,* 110, 151, 2002.
57. Sim, M.R. and McNeil, J.J., Monitoring chemical exposure using breast milk: a methodological review, *Am. J. Epidemiol.,* 136, 1, 1992.
58. Anderson, H.A. and Wolff, M.S., Environmental contaminants in human milk, *J. Expo. Anal. Environ. Epidemiol.*, 10, 755, 2000.
59. Skerfving, S., Mercury in women exposed to methyl mercury through fish consumption and their newborn babies and breast milk, *Bull. Environ. Contam. Toxicol.,* 41, 475, 1988.
60. UNEP, *Global Mercury Assessment,* UNEP, Geneva, 2002.
61. Needham, L.L. and Wang, R.Y., Analytic considerations for measuring environmental chemicals in breast milk, *Environ. Health Perspect.,* 110, A317, 2002.
62. Liteplo, R.G. et al., Inorganic fluoride: evaluation of risks to health from environmental exposure to fluoride in Canada, *Environ. Carcinog. Ecotoxicol. Rev.,* C12, 327, 1994.
63. Hindmarsh, J.T. et al., Hair arsenic as an index of toxicity, in *Arsenic Exposure and Health Effects,* Chappell, W.R., Abernathy, C.O., and Calderon, R.L., Eds., Elsevier, New York, 1999, p. 41.
64. Suzuki, T., *Advances in Mercury Toxicology,* Plenum Press, New York, 1991.
65. Grandjean, P. et al., Methylmercury exposure biomarkers as indicators of neurotoxicity in children aged seven years, *Am. J. Epidemiol.,* 150, 301, 1999.
66. COT, Statement on a Survey of Mercury in Fish and Shellfish, 2002/04, Committee on Toxicology of Chemicals in Food, Consumer Products and the Environment, Food Standards Agency, London, 2002.
67. Marsh, D.O. et al., Fetal MeHg poisoning: relationship between concentration in single strands of maternal hair and child effects, *Arch. Neurol.,* 44, 1017, 1997.
68. Peterson, P.J. et al., Development of indicators within different policy contexts for endemic arsenic impacts in the People's Republic of China, *Environ. Geochem. Health,* 23, 159, 2001.
69. Yang, L.-S. et al., The relationship between exposure to arsenic concentrations in drinking water and the development of skin lesions in farmers from Inner Mongolia, China, *Environ. Geochem. Health,* 24, 293, 2002.
70. Peterson, P.J. and Li, R.-B., Endemic Fluorosis: A Global Health Issue, Report UNEP/GEMS/92.H6, UNEP, Nairobi, 1992.
71. Yang, L.-S. et al., Developing environmental health indicators as policy tools for endemic fluorosis management in the People's Republic of China, *Environ. Geochem. Health,* 25, 281, 2003.
72. Kjellström, T. and Corvalán, C., Framework for the development of environmental health indicators, *World Health Stat. Q.,* 48, 144, 1955.
73. WHO, Diet, Nutrition, and the Prevention of Chronic Diseases, Technical Report 797, WHO, Geneva, 1990.

74. WHO, GEMS/Food Total Diet Studies, Document WHO/SDE/FOS/99.9, Report of Joint USFDA/WHO International Workshop on Total Diet Studies in Co-operation with the Pan American Health Organization, Kansas City, MO, July 26–August 6, 1999.
75. Burger, J., Gaines, K.F., and Gochfeld, M., Ethnic differences in risk from Hg among Savannah River fishermen, *Risk Anal.,* 21, 533, 2001.
76. Knuth, B. et al., Weighing health benefit and health risk information when consuming sport-fish, *Risk Anal.,* 23, 1185, 2003.
77. Muckle, G. et al., Determinants of polychlorinated biphenyls and methylmercury exposure in Inuit women of childbearing age, *Environ. Health Perspect.,* 109, 957, 2001.
78. Spencer, P.S. and Berman, F., Plant toxins and human health, in *Food Safety: Contaminants and Toxins,* D'Mello, J.P.F., Ed., Commonwealth Agricultural Bureaux International, Wallingford, 2003, p. 1.
79. FAO/WHO, The Application of Risk Communication of Food Standards and Safety Matters, Report of a Joint FAO/WHO Expert Consultation, FAO, Rome, February 2–6, 1998.
80. WHO, Concern for Europe's Tomorrow: Health and the Environment in the WHO European Region, Wiss. Verl.-Ges., Stuttgart, 1995.

13 Metal Contamination of Dietary Supplements

Melissa J. Slotnick and Jerome O. Nriagu

CONTENTS

13.1 INTRODUCTION

The safety of over-the-counter dietary supplements is an issue of growing concern given an increase in dietary supplement consumption, the vast range of products available to consumers of all ages, and the resulting regulatory challenges. The United States Food and Drug Administration (FDA) has estimated that over 29,000 different dietary supplements are available to consumers.[1] Dietary supplements are defined by the Dietary Supplement and Health Education Act of 1994 (DSHEA) as containing one or more of the following: vitamins, minerals, herbs or other botanicals, amino acids, dietary supplements meant to increase total dietary intake, or a combination of the above.[1,2]

The use of dietary supplements in the U.S. has been on the rise. Results from the 1999–2000 National Health and Nutrition Examination Survey (NHANES), conducted by the U.S. Centers for Disease Control and Prevention (CDC), indicate that 52% of the adult population reported using any type of dietary supplement, whereas 35% reported using a multivitamin or multimineral supplement.[3] This is over a 10% increase in dietary supplement use when compared with the previous NHANES survey, conducted between 1988 and 1994.[4] The increase in consumption corresponds to a rise in dietary supplement sales from $10 million in 1996 to over $17 million in 2000.[1] Furthermore, use of dietary supplements is not limited to adults; results from the 1988–1994 NHANES survey indicate that 51% of boys and 46% of girls aged 3 to 5 years use dietary supplements.[4]

Although vitamins or minerals appear to be the most commonly used dietary supplements, herbal supplement use is also prevalent in the U.S. In a 1999 survey of over 15,000 respondents enrolled in the Kaiser Permanents Medical Care Program of Northern California (KPMCP), 29.3% of the respondents indicated using herbal supplements in the past year.[5] When over 2500 adults in the 48 U.S. contiguous states and the District of Columbia were asked to report dietary supplement use during the 1-week period before the interview, the ten most commonly used herbal supplements were (in order from first to last): ginseng, *Gingko biloba* extract, *Allium sativum*, glucosamine, St. John's wort, *Echinacea augustifolia*, lechithin, chondroitin, creatine, and *Serenoa repens.*[6] The 10 most commonly used vitamins and minerals were: multivitamin, vitamin E, vitamin C, calcium, magnesium, zinc, folic acid, vitamin B_{12}, vitamin D, and vitamin A.[6] Dietary supplements in the U.S. are required to meet the standards put forth in the DSHEA but are not required to receive approval from the U.S. FDA. Under the DSHEA, (1) the manufacturer is responsible for ensuring labeling claims made are true, (2) the manufacturer must provide evidence that claims are truthful but is not required to submit evidence to the U.S. FDA, and (3) the manufacturer is responsible for ensuring quality and safety, but the U.S. FDA is responsible for proving it is not safe should a question of safety arise.[7] Therefore, much of the responsibility pertaining to product quality and safety lies on the shoulders of the manufacturer, with limited intervention by the federal government.

These issues have prompted awareness as to the safety of dietary supplements, and herbal supplements in particular.[7–9] Contamination of dietary supplements with toxic metals is a potential health hazard that has received increasing attention in recent years. Metal exposures from dietary supplements may cumulate with exposures from diet and other routes, increasing the potential of reaching elevated body burdens. Herein we review studies of trace metal content of mineral, vitamin, and herbal supplements, emphasizing future research needs.

13.2 METAL CONTENT OF DIETARY SUPPLEMENTS

13.2.1 Mineral Supplements

The majority of studies assessing the metals content of mineral supplements concern the lead content of calcium supplements (Table 13.1). In the early 1980s, the U.S. FDA cautioned consumers to restrict intake of calcium supplements, commonly consumed by pregnant women and children, due to elevated concentrations of lead.[11] At the same time, discussion of this potential contamination prompted researchers to investigate the scope of the problem. Early studies focused on analysis of lead and other heavy metals in bonemeal supplements, which are used primarily as calcium and phosphorus supplements.[19] Capar and Gould[19] analyzed Pb, Cd, F, Al, Cr, Cu, Fe, Mn, Mo, Ni, Ti, and Zn content of 20 brands of bonemeal supplements. From these analyses, lead concentrations were most concerning. The authors concluded that the supplement with the highest concentration of lead would have contributed approximately 22 μg of lead per day to overall lead intake.[19] Cadmium was also present in high levels in the bonemeal supplements, with one supplement

TABLE 13.1
Lead Concentrations in Calcium Supplements

Supplement Type	Lead Concentration (μg/g) Mean ± SD[a]	Range	Year	Authors	Study Location
Chelated	1.39 ± 0.06	n/a	1992	Bourgoin et al.[10]	Canada
	0.26 ± 0.36	0.03–1.21	1993	Bourgoin et al.[11]	Canada
	0.2 ± 0.0	n/a	1998	Amarasiriwardena et al.[12]	U.S.
	0.57 ± 0.54	0.04–2.8	2000	Scelfo and Flegal[13]	U.S.
	0.53 ± 0.02	ND[c]–1.81	2003	Kim et al.[14]	Korea
Refined[b] $CaCO_3$	0.34 ± 0.24[d]	0.04–0.92	1993	Bourgoin et al.[11]	Canada
	0.6 ± 0.28[d]	0.2–0.8	1998	Amarasiriwardena et al.[12]	U.S.
	0.73 ± 1.60	0–10.05	2000	Scelfo and Flegal[13]	U.S.
	0.26 ± 0.24[d]	ND–0.81	2000	Ross et al.[15]	U.S.
Dolomite	1.64 ± 0.94	0.5–2.75	1983	Roberts[16]	U.S.
	1.9 ± 0.5	n/a	1988	Boulos and von Smolinski[17]	U.S.
	1.11 ± 0.71	0.52–2.52	1993	Bourgoin et al.[11]	Canada
	0.94 ± 0.51	0.55–1.51	1994	Siitonen and Thompson[18]	U.S.
	3.37 ± 2.29[d]	1.9–6.0	1998	Amarasiriwardena et al.[12]	U.S.
	0.97 ± 0.49	0.39–1.56	2000	Scelfo and Flegal[13]	U.S.
Natural[e] $CaCO_3$	3.67 ± 0.23[d]	3.50–3.83	1992	Bourgoin et al.[10]	Canada
	2.11 ± 1.33	0.36–4.88	1993	Bourgoin et al.[11]	Canada
	0.67 ± 0.54[d]	0.17–1.26	1994	Siitonen and Thompson[18]	U.S.
	2.35 ± 1.34[d]	1.4–3.3	1998	Amarasiriwardena et al. [12]	U.S.
	0.88 ± 0.51	0.12–2.10	2000	Scelfo and Flegal[13]	U.S.
	0.28 ± 0.15[d]	ND–0.45	2000	Ross et al.[15]	U.S.
	0.39 ± 0.02	0.04–0.53	2003	Kim et al. [14]	Korea
	0.02 ± 0.001	ND–0.03	2003	Kim et al. [14]	Korea
Bonemeal	4.16 ± 2.05	1.5–8.7	1979	Capar and Gould[19]	U.S.
	8.86 ± 7.27	2–20	1983	Roberts[16]	U.S.
	2.67 ± 2.74	0.64–8.83	1993	Bourgoin et al.[11]	Canada
	4.27 ± 2.71[d]	1.21–6.39	1994	Siitonen and Thompson[18]	U.S.
	0.60 ± 0.39	0.21–1.38	2000	Scelfo and Flegal[13]	U.S.
	2.26 ± 0.16	0.05–6.72	2003	Kim et al.[14]	Korea

[a] Standard Deviation.
[b] Calcium carbonate produced in a laboratory.
[c] ND = Nondetectable.
[d] Mean and SD were derived from reported means. Levels below detection were assigned a value equal to half the detection limit.
[e] Calcium carbonate derived from limestone rock, oyster shells, or egg shells.

providing an additional 7.7 μg of cadmium to the daily intake; the other elements analyzed were not thought to be present at toxic levels.[19] Roberts[16] reported lead levels in 7 brands of bonemeal supplements as well as 6 brands of dolomite supplements. Results were averaged for comparison with other studies (Table 13.1). Boulos and von Smolinski[17] also analyzed one brand of dolomite tablets for lead, as well as Al, As, Cd, Cr, Cu, Mn, Se, and Zn. Lead concentrations were similar to previous studies (Table 13.1), but the low concentrations reported for other elements were well below the current daily intake guidelines set by various regulatory agencies (Table 13.2).[17]

Originally, lead levels were only thought to be elevated in dolomite and bonemeal-based supplements;[11] however, subsequent studies revealed lead contamination of other types of calcium supplements. Bourgoin et al.[10] measured lead and cadmium in three types of calcium supplements available in North America. The calcium supplements were either derived from calcium carbonate ($CaCO_3$) or were composed of chelate-bound calcium. The supplements were analyzed in four independent laboratories using four different types of instrumentation. Mean lead concentration ranged from 0.2 to 3.5 μg/g of supplement, and supplements derived from oyster shells ($CaCO_3$) contained the highest lead concentrations.[10]

In a later study, Bourgoin et al.[11] measured lead concentration in 70 different brands of supplements purchased in Canada and the U.S. The samples were categorized into five different types of calcium supplements: chelates, refined ($CaCO_3$ synthesized in laboratory), dolomite, natural source ($CaCO_3$ derived from limestone rock/oyster shells), and bonemeal. Dolomite supplements had significantly higher lead concentrations than chelates and refined supplements.[11] Supplements derived from bonemeal and natural $CaCO_3$ had the highest levels of lead, but concentrations were not significantly different from those of the dolomite supplements.[11] Of the 70 different brands of calcium analyzed, 25% of them exceeded the total tolerable daily intake for at risk populations of 6 μg lead/d.[11] Hight et al.[20] analyzed four types of calcium supplements derived from dolomite, egg shell, and oyster shell. In addition to lead, the supplements were analyzed for Al, Si, Sc, Tl, V, Fe, and Co.[20] Lead concentration ranged from 0.12 μg/g for the egg-shell based supplement to 1.9 μg/g for the oyster shell supplement.[20] These values are comparable to levels reported for other natural source $CaCO_3$ supplements analyzed around the same time period (Table 13.1).

In the following year, Siitonen and Thompson[18] published a report of lead content in calcium supplements classified as oyster shell $CaCO_3$, bone meal, dolomite, calcium lactate, and "other." Results from the first three types of supplements are reported in Table 13.1. Mean concentrations for the calcium lactate supplements ranged from nondetectable to 1.19 μg/g, whereas concentrations for the "other" types of supplements ranged from nondetectable to 0.82 μg/g.[18] Such continuation of emerging evidence began to raise public awareness regarding the lead content of calcium supplements.[21–24] Consequently, legislation was passed in the State of California that required health advisory warnings to be put on products with lead doses exceeding 1.5 μg/d.[13]

Given this growing concern over elevated lead concentrations, researchers have continued to monitor the lead content of calcium supplements. Amarasiriwardena et al.[12] reported lead concentration in 10 different brands of commercial calcium supplements classified as originating from dolomite, natural $CaCO_3$, synthetic processes, and chelated calcium. Means from three or four analyzed tablets were averaged, and the results are presented in Table 13.1. Only one brand of chelated calcium was analyzed; therefore, the results presented for chelated calcium are not averaged across brands.[12] Subsequently, Scelfo and Flegal[13] determined the lead concentration of 136 different brands of calcium supplements collected in 1996. The supplements included both natural sources of calcium (bonemeal, dolomite, or fossil oyster shell) and refined sources of calcium (calcium salts and chelate-bound calcium). Of the supplements purchased in 1996, two thirds exceeded the State of California guidelines for lead of 1.5 μg/d.[13] The refined calcium products had the lowest lead concentrations, although concentrations between supplement types were not significantly different.[13]

Ross et al.[15] analyzed the lead content of 21 brands of over-the-counter calcium supplements (7 natural-source $CaCO_3$, 14 refined $CaCO_3$) available in the U.S. Four of the seven natural-source $CaCO_3$ supplements had detectable levels of lead, with estimated daily doses ranging from 0.93 to 1.04 μg, given 800 mg of calcium supplementation per day.[15] Of the 14 refined $CaCO_3$ supplements, 4 had measurable lead concentrations, with estimated daily intakes of 0.96 to 1.83 μg.[15]

The lead content of calcium supplements available in North America appears to have declined over the past 20 years (Table 13.1). This decrease is apparent across all categories of supplements (chelated, refined $CaCO_3$, dolomite, natural $CaCO_3$, and bonemeal). In most studies analyzing multiple types of supplements, lead concentrations were highest in bonemeal supplements, followed by natural $CaCO_3$ and dolomite supplements. In recent years, however, natural $CaCO_3$ supplements had lead levels similar to refined or chelated calcium supplements (Table 13.1). Despite the decrease in the lead content of calcium supplements in North America, the potential for lead exposure from calcium supplements remains. This issue is of particular concern for sensitive populations (e.g., children, pregnant women) who also benefit from increased dietary calcium intake. Steps must be taken to ensure safety of calcium supplements for these populations and for sensitive populations around the world.

Metal contamination of dietary supplements is becoming a concern in other countries as use of dietary supplements increases.[14] Kim et al.[14] analyzed the lead content of 55 different brands of calcium dietary supplements available in Korea. Estimates of daily lead intake from calcium supplements ranged from 0.10 μg for calcium derived from eggshell, to 11.35 μg for calcium derived from bonemeal.[14] Lead levels in natural-source $CaCO_3$ supplements were comparable to levels currently observed in the U.S.; however, mean lead concentration in the bonemeal supplements was elevated (Table 13.1).

Mercury, cadmium, and arsenic concentrations of 55 Korean calcium supplements have also been reported.[25] Cadmium and mercury concentrations were low, contributing to less than 1% of the provisional tolerable daily intakes set by the Food and Agricultural Organization/World Health Organization (FAO/WHO).[25] Arsenic concentrations were also low, ranging from a mean total arsenic concentration of

TABLE 13.2
Elemental Intake from Commonly Consumed[a] Herbal Supplements

	Echinacea	Gingko biloba	Ginseng	St. John's Wort	Valerian	Intake Guidelines[b]
Al	n/a	n/a	101	n/a	9225	110,000[c]
			ND–203			
Sb	0.02	0.03	0.04	0.02	n/a	22[d]
	ND–0.03	ND–0.06	0.01–0.11	0.003–0.06		
As	0.28	0.62	0.30	0.20	n/a	16.5[d]
	0.03–0.91	0.03–3.08	0.06–0.70	0.02–0.83	0.02–0.01	
Cd	0.15	0.35	24.0	0.33	n/a	55[d]
	0.004–0.97	0.004–2.89	0.02–121	0.02–2.11	0.06–0.96	
Cr	4.83	4.21	2.66	2.43	n/a	165[d]
	0.13–9.37	0.05–12.88	0.73–5.64	0.001–6.05	0.001–0.006	
Co	n/a	n/a	0.37	n/a	3.38	550[c]
			0.17–0.56		1.13–2.93	
Cu	17.8	6.53	4.75	13.4	n/a	10,000[e]
	1.30–34.7	0.53–24.1	0.01–11.0	9.53–34.6		
Fe	n/a	n/a	324	n/a	5963	45,000[e]
			221–428		3375–6300	
Pb	0.56	2.37	5.70	0.60	n/a	75[f]
	0.02–2.90	0.02–12.5	ND–26.7	0.02–5.83	0.001–0.01	
Hg	0.05	0.03	0.12	0.08	n/a	5.5[d]
	ND–0.22	ND–0.13	ND–0.93	ND–0.25		
Mo	1.39	0.62	0.69	1.61	n/a	2,000[e]
	0.18–3.15	0.10–1.25	0.09–1.40	0.28–5.44		
Ni	n/a	n/a	2.37	n/a	n/a	1,000[e]
	0.003–0.008		ND–10.4	0.002–0.01	0.002–0.003	
Pd	0.52	0.42	0.44	0.24	n/a	n/a
	0.008–1.48	ND–1.26	0.06–1.15	0.06–0.59		

Sc	n/a	n/a	0.04	n/a	2.93	n/a
			ND–0.07		0.9– 2.48	
Tl	0.11	0.08	8.44	0.01	n/a	4.4[d]
	0.002–0.38	ND–0.32	ND–63.0	ND–0.03		
Sn	0.03	0.02	0.05	0.16	n/a	n/a
	0.002–0.09	0.003–0.05	0.01–0.11	0.01–0.64		
W	0.53	2.53	0.90	6.64	n/a	n/a
	0.03–1.72	0.04–8.80	0.26–1.8	0.09–36.0		
U	0.18	0.33	0.15	0.27	n/a	11[c]
	0.002–0.74	ND–1.46	0.01–0.65	0.01–1.35		
V	3.70	3.71	1.24	1.57	n/a	1,800[e]
	0.02–7.05	0.12–15.7	0.15–3.78	0.06–3.51	0.002–0.01	
Zn	29.8	46.0	14.0	35.1	n/a	40,000[e]
	3.20–79.7	3.46–141	0.05–27.7	16.8–75.8		

Note: Mean, range, and reference dose are reported (μg/day)

[a] From Kaufman, D.W. et al., Recent patterns of medication use in the ambulatory adult population of the U.S.: the Slone survey, *JAMA*, 287, 337, 2002.
[b] Intake Guidelines, special notes: Arsenic is specific to inorganic arsenic; mercury is specific to methyl mercury; chromium is specific to Cr(V); lead is for adults, and lower daily exposure levels are recommended for pregnant women (20 μg) and children under the age of six (6 μg).
[c] From ATSDR (Agency for Toxic Substances and Disease Registry), Minimum Risk Levels (MRLs) for Hazardous Substances, http://www.atsdr.cdc.gov/mrls.html, accessed December 3, 2005.
[d] From IRIS (Integrated Risk Information System), U.S. Environmental Protection Agency, http://www.epa.gov/iris/subst/0278.htm, accessed November 30, 2005.
[e] From Institute of Medicine, Dietary Reference Intakes for Vitamin A, Vitamin K, Arsenic, Boron, Chromium, Copper, Iodine, Iron, Manganese, Molybdenum, Nickel, Silicon, Vanadium, and Zinc. A report of the Panel on Micronutrients, Subcommittees on Upper Reference Levels of Nutrients and the Interpretation and Uses of Dietary Reference Intakes, and the Standing Committee on the Scientific Evaluation of Dietary Reference Intakes. Food and Nutrition Board, Institute of Medicine. National Academy Press, Washington, D.C., 2001.
[f] From Farley, D., Dangers of Lead Still Linger, U.S Food and Drug Administration, FDA Consumer, January–February 1998.

Source: From Raman, P., Patino, L.C., and Nair, M.G., Evaluation of metal and microbial contamination in botanical supplements, *J. Agric. Food Chem.*, 52, 7822, 2004; Dolan, S.P. et al., Analysis of dietary supplements for arsenic, cadmium, mercury, and lead using Inductively Coupled Plasma Mass Spectrometry, *J. Agric. Food Chem.*, 51, 1307, 2003; Huggett, D.B. et al., Organochlorine pesticides and metals in select botanical dietary supplements, *Bull. Environ. Contam. Toxicol.*, 66, 150, 2001; Hight, S.C. et al., Analysis of dietary supplements for nutritional, toxic, and other elements, *J. Food Compos. Anal.*, 6, 121, 1993; Khan, I.A., Determination of heavy metals and pesticides in ginseng products, *J. AOAC Int.*, 84, 936, 2001.

0.10 μg/g of supplement for chelated calcium supplements to 2.15 μg/g of supplement for calcium derived from algae.[25]

Cadmium has also been measured in calcium supplements from Canada.[10,11] Supplements derived from oyster shell calcium carbonate ($CaCO_3$) and supplements classified as organic chelate-bound calcium were analyzed for cadmium in four independent laboratories.[10] Mean cadmium concentrations ranged from nondetectable for the oyster shell $CaCO_3$ to approximately 3.5 μg/g for the chelate-bound calcium.[10]

Outside of calcium supplements, there are few data assessing the metal content of other mineral supplements. Krone et al.[26] analyzed the cadmium content of six different brands of zinc-containing mineral supplements obtained in the U.S. Cadmium levels in the supplements were estimated to contribute between 0.14 and 2.0 μg cadmium per day, a value that does not contribute significantly to the U.S. FDA's recommended maximum tolerable daily intake of 55 μg cadmium.[26] Hight et al.[20] also measured a variety of elements (Al, Si, Sc, Tl, V, Fe, Co, Pb) in four different vitamin and mineral products. The products included the following: a supplement with 7 minerals, 13 vitamins, and 50 other animal or botanical ingredients; a supplement with 9 minerals and fish liver oil; a product with 8 different trace elements; and zinc gluconate.[20] Concentrations of elements in these products were comparable to the herbal supplements analyzed by Hight et al.[20] and discussed in the following text, with the exception of thallium, which was reported to be 8400 μg/g in one supplement.

13.2.2 Vitamin Supplements

Studies analyzing the metal content of single vitamin supplements (e.g., vitamin C, vitamin E, vitamin D) are limited. Ponce de Leon et al.[27] measured the chromium (Cr), nickel (Ni), tin (Sn), and lead (Pb) content of two different types of synthetic vitamin E products. Mean concentrations (μg/g) for product 1 vs. product 2, respectively, were as follows: (Cr) 200, 162; (Ni) 60, nondetectable; (Sn) 9, 266; and (Pb) 45, nondetectable.[27] While this study demonstrates effectiveness of a new analytical method, further studies are necessary to better characterize concentrations of different metals in vitamin E supplements. Furthermore, additional data on metal contamination of other types and brands of single and multiple vitamin supplements is necessary to better understand risk assessment as it pertains to vitamin supplement intake. In particular, research efforts should focus on the most commonly used vitamins (multivitamins, vitamin E, vitamin C, vitamin B_{12}, vitamin D, and vitamin A).[6]

13.2.3 Herbal Supplements

Hight et al.[20] assessed elemental content (Al, As, B, Br, Ca, Cd, Cl, Co, Cr, Cs, Cu, F, K, Mg, Mn, Mo, Na, Ni, P, Pb, Rb, S, Sb, Sc, Se, Si, Sr, Tl, V, Zn) of 42 different types of dietary supplements purchased in the Washington D.C. area. The supplements included botanical products ranging from alfalfa to valerian, animal products such as bovine heart and brewer's yeast, vitamin and mineral products discussed in the previous subsection, and synthetic extracts and oils such as aloe vera and fish oil.[20] One to two whole tablets were analyzed, with the serving weight ranging from 2 to 2.5 g.[20] The authors estimated that, if the supplements were consumed as recommended on the product label, roughly half of the supplements analyzed had the potential to exceed the WHO 1989 PTTDI for lead.[20] One product in particular, bee propolis, was estimated to exceed the PTTDI for lead by a factor of ten.[20] Of the other elements analyzed, the supplements had the potential to exceed recommended daily intake of Zn, Mn, Mg, Fe, and Ca.[20]

While it is important to monitor the dietary supplement content for a wide spectrum of elements, as above, exposure to toxic metals is of particular concern. Huggett et al.[28] determined arsenic, cadmium, chromium, lead, and nickel concentrations in four different types of herbal dietary supplements (Valerian, St. John's wort, passion flower, and echinacea). Of these herbal supplements, three (Valerian, St. John's wort, and echinacea) are amongst the most commonly used herbal supplements.[6] A risk assessment considering the reported concentrations and calculated intake resulted in a low hazard index (< 0.2), suggesting that carcinogenic effects from metal intake from these particular supplements are unlikely.[28] It should be noted, however, that this risk assessment did not consider other exposure routes and is limited to the concentrations reported in this study. Further studies reporting concentrations in the context of risk assessment can help to better understand health risks associated with metal contamination of dietary supplements.

Khan et al.[29] reported chromium, mercury, arsenic, nickel, lead, and cadmium concentrations of 21 ginseng products purchased in the U.S., Europe, and Asia. Both solid (n = 13) and liquid (n = 8) supplements were analyzed. Chromium, mercury, and arsenic were not detected in any of the samples (detection limits were 0.1 mg/l, 50–100 ng/l, and 0.5 mg/l, respectively).[29] Nickel concentrations ranged from undetectable to 20 μg/l, lead concentrations ranged from undetectable to 62.2 μg/l, and cadmium concentrations ranged from 8.4 to 120.8 μg/l.[29] These data were used to calculate metal intake from the ginseng supplements, assuming the dose was 5 ml/d for tinctures, 120 ml/d for beverages, and 1000 mg/d for solid samples.[30] Average daily intake, carcinogenic risk, and a noncancer hazard index were calculated; from these calculations the authors determined that the metal concentrations from the analyzed ginseng supplements did not pose any significant risks.[30]

Given the vast number of herbal supplements available to consumers, it is important to continue monitoring a variety of different products. Dolan et al.[31] assessed arsenic, cadmium, mercury, and lead concentration of 95 different herbal supplements containing a variety of different extracts. Product components ranged from a single herbal extract (e.g., echinacea or St. John's wort) to combinations of multiple extracts (e.g., St. John's wort extract in combination with lecithin, kava kava rhizome and root extract, vitamins and minerals).[31] Concentrations reported for commonly consumed herbal supplements are included in Table 13.2; however, this table does not capture the complete range of concentrations found in all supplements analyzed. For example, intake of 11 of the products sampled would result in exposures that could exceed the tolerable intakes for sensitive populations, such as pregnant women and children.[31] Additionally, the highest arsenic concentration reported was 3.77 μg/g; this value was measured in a supplement containing red clover blossoms, echinacea, licorice root, buckhorn bark, burdock root, and other extracts, and therefore does not fall under the supplements classified as commonly consumed.[6] Given a recommended dose of 4.83 g/d,[31] this translates into 0.0182 mg of arsenic per day. This value slightly exceeds the chronic oral reference dose (RfD) of 0.0003 mg inorganic As/kg/d[34] for a 55-kg individual. This calculation assumes that the arsenic concentration measured is inorganic arsenic only, and therefore may be an overestimate of inorganic arsenic intake from this supplement; therefore, it highlights the importance of understanding both concentration and chemical form of metals in dietary supplements.

In addition to assessing the metal content of the commonly used[6] echinacea, *Gingko biloba*, ginseng, and St. John's wort, Raman et al.[32] analyzed the metal content for the following frequently used supplements: garlic, grape seed extract, kava kava, and saw palmetto. Different brands of the supplements were purchased from stores in Michigan, Illinois, and Indiana and analyzed for Pb, Cd, As, U, Cr, V, Cu, Zn, Mo, Pd, Sn, Sb, Tl, and W.[32] Concentrations of Cd, As, U, Cr, V, Cu, Zn, Mo, Sn, and Pb were less than the guidelines used (Minimum Risk Level, Recommended Dietary Allowance, No-Observed-Adverse-Effect Level); no reference values were available for Pd, Sb, Tl, and W.[32] It is important to remember, however, that additional exposure routes, such as diet or drinking water, may add to the total daily intake of the metals. Additionally, the authors reported a large degree of variability in lead concentrations for different batches of echinacea,[32] reiterating the importance of continued monitoring to ensure safety.

13.3 CONTRIBUTION TO DAILY INTAKE: COMMON HERBAL SUPPLEMENTS

Of the most commonly used herbal supplements,[6] echinacea, *Gingko biloba*, ginseng, St. John's wort, valerian, and glucosamine have been analyzed for metals and other elements.[20,28,29,31,32] For comparison, results reported in μg/g were converted to μg/day using either the dose[30,31] or midpoint capsule weight[20] reported by the authors. To estimate daily intake of metals and other elements from these supplements, reported values (means or individual product results) were averaged across studies (Table 13.2). Huggett et al.[28] only reported concentration ranges; while these values were not included in estimating an overall mean, the ranges were incorporated into Table 13.2. Instrument detection levels were not reported in all studies. For consistency, values below detection level (ND) were assigned a value of zero before means were averaged. Only one study reported mean intake levels (μg/day) for glucosamine (As = 0.04, Cd = 0.04, Pb = 0.15, Hg = 0.36).[31] Therefore, glucosamine results are not presented in Table 13.2.

Several guidelines proposed by various regulatory agencies and organizations in the U.S. were used for comparative purposes (Table 13.2). When available, the Upper Level Dietary Reference Intake (ULDRI) was used. The ULDRI is defined as "The highest level of daily nutrient intake that is likely to pose no risk of adverse health effects for almost all individuals in the general population."[33]

Oral chronic Reference Doses (RfDs) were reported for elements with no ULDRI. RfDs are established by the U.S. Environmental Protection Agency (U.S. EPA) and are defined as: "An estimate of a daily oral exposure for a given duration to the human population (including susceptible subgroups) that is likely to be without an appreciable risk of adverse health effects over a lifetime."[34] The RfDs are reported by the U.S. EPA in units of mg/kg/day, but, for ease of comparison in this review, were converted to µg/day based on a 55-kg individual.

For elements with no available ULDRI or RfD, the Minimum Risk Level (MRL) was used. The MRL is established by the U.S. Agency for Toxic Substances and Disease Registry (ATSDR), Centers for Disease Control, and is defined as the following: "... an estimate of the daily human exposure to a hazardous substance that is likely to be without appreciable risk of adverse noncancer health effects over a specified duration of exposure."[35] As with the RfDs, the MRLs were converted to µg/day, assuming a 55-kg individual. None of the above reference levels were available for lead. Instead, the tolerable daily diet intake recommended by the U.S. FDA[36] was used for comparison. It should be noted that the level listed in the table does not include sensitive populations; it is recommended that children under the age of 6 years and pregnant women limit exposures to 6 µg/d and 25 µg/d, respectively.[36] Intake guidelines were not available for Pd, Sc, Sn, and W.

Means and ranges are well below the intake guidelines for most elements (Table 13.2). Lead, cadmium, and thallium concentrations in some supplements, however, have the potential to exceed the daily reference dose for those metals. Specifically, the mean cadmium intake for ginseng, an average of 29 different products from three different studies, approximates half of the RfD set by the U.S. EPA (Table 13.2). Three products exceed the RfD, all of which were analyzed by Khan et al.[29] Although no values for lead exceed the reference value stated, several values come close to meeting or exceeding the tolerable daily diet intake for pregnant women (20 µg/d) and children (6 µg/d) (Table 13.2). Lastly, the mean thallium concentration (averaged from nine different ginseng products) is almost two times the RfD (Table 13.2). It should be noted that this mean appears to be driven by two extreme values (63 µg/d and 12.8 µg/d[20]) that far exceed the results for the other products. Therefore, additional sampling is necessary to determine the extent of thallium contamination in ginseng and other herbal supplements.

13.4 CONSIDERATIONS FOR RISK ASSESSMENT AND REGULATION

The results discussed herein raise several issues pertaining to quality and quantity of data that should be addressed in the context of risk assessment and regulation. First, the influence of batch variability on concentrations should be considered. Metal concentrations have been shown to vary by product batch. For example, Bourgoin et al.[11] found that lead levels differed significantly between two different lot samples of a bonemeal-based supplement. The variability might depend on production techniques or variability in concentrations in the biological material used from batch to batch. Multiple batches should be analyzed to gain a more accurate prediction of metal contamination. In addition to intrastudy variability, there is a high degree of variability in intake estimates between different studies for most elements (Table 13.2). Although mean concentrations do not exceed the intake guidelines, it is unknown how well the mean concentrations represent exposure for the general population. Improved monitoring is necessary to better characterize metal contamination of dietary supplements.

When estimating daily metal intake from dietary supplement consumption, it is also important to consider the chemical form of the metal of interest [e.g., inorganic vs. organic As, or Cr(III) vs. Cr(V)]. Therefore, future studies should focus on speciating the metals analyzed for comparison with reference doses. Although the average elemental intake discussed herein was compared with recommended daily doses of the elements analyzed, assessing risk in this manner does not consider other sources of metal intake, such as diet, drinking water, or occupational exposures. Additionally, we have assumed that individuals will consume the dose stated by the authors; in reality, many individuals may not take the recommended dose or may take several different supplements. According to the 1988–1994 NHANES study, 18.6% of the population (including all ages and sexes) reported consuming two or more types of supplements, and 14.4% reported consuming three or more dietary supplements.[4] Therefore, estimates of intake for risk assessment should be conservative to ensure inclusion and protection of the most at-risk populations.

The potential for chemical interaction between metal contaminants and the composition of the dietary supplement should also be considered. This issue is particularly relevant when assessing the risks associated with lead contamination of calcium supplements. Calcium and lead are known to interact metabolically; lead absorption decreases with increasing calcium concentrations.[37] Therefore, it is possible that the calcium supplement itself minimizes absorption of the lead contained in the supplement. These dynamics may depend on the form of the supplement as well as the metabolism of the individual. Gulson et al.[37] found that among adults aged 21 to 47 years, calcium carbonate supplements increased the $^{206}Pb/^{204}Pb$ ratio in blood samples, indicating that even though total blood lead concentration did not increase significantly, lead from the supplement contributed to blood lead. No change was observed for subjects taking a complex calcium supplement composed of carbonate, phosphate, and citrate, suggesting that supplement type may also affect lead absorption.[37] While these results are intriguing, the authors state the importance of future research focusing on sensitive subgroups, such as children. Furthermore, the benefits of calcium supplementation, such as the ability to increase bone mass and the potential to reduce preeclampsia in pregnancy,[37] should not be forgotten during risk assessment and public education efforts. This example highlights data complexities and uncertainties that further challenge the risk assessment process. Nonetheless, in the face of uncertainty, improved efforts should be made to minimize contamination of dietary supplements.

The above issues, in conjunction with the growing number of products available to consumers of all ages, pose many regulatory challenges. Continued testing and monitoring by manufacturers, governmental agencies, and independent researchers is necessary to ensure public safety. Specifically, a regulatory framework should be established to facilitate standardized monitoring procedures. Regardless of the future of dietary supplement regulation, increased public awareness as to the potential hazards of dietary supplements is critical, particularly for sensitive populations such as pregnant women and children. First and foremost, however, additional studies are necessary. Future investigations assessing the metal content of dietary supplements will help to better characterize the range of contamination and potential for adverse health impacts, information that is essential to ensure public health protection.

REFERENCES

1. Gibson, J.E. and Taylor, D.A., Can claims, misleading information, and manufacturing issues regarding dietary supplements be improved in the U.S.?, *J. Pharmacol. Exp. Ther.*, 314, 939, 2005.
2. Taylor, C.L., Regulatory frameworks for functional foods and dietary supplements, *Nutr. Rev.*, 62, 55, 2004.
3. Radimer, K. et al., Dietary supplement use by U.S. adults: data from the National Health and Nutrition Examination Survey, 1999–2000, *Am. J. Epidemiol.*, 160, 339, 2004.
4. Ervin, R.B., Wright, J.D., and Kennedy-Stephenson, J., Use of Dietary Supplements in the U.S., 1988–1994. U.S. Department of Health and Human Services, Centers for Disease Control and Prevention, National Center for Health Statistics. Hyattsville, MD, DHHS Publication No. (PHS), 1999, p. 99–1694.
5. Schaffer, D.M. et al., Nonvitamin, nonmineral supplement use over a 12-month period by adult members of a large health maintenance organization, *J. Am. Diet. Assoc.*, 103, 1500, 2003.
6. Kaufman, D.W. et al., Recent patterns of medication use in the ambulatory adult population of the U.S.: the Slone survey, *JAMA*, 287, 337, 2002.
7. DeSmet, P., Herbal remedies, *N. Engl. J. Med.*, 347, 2046, 2002.
8. Bent, S. and Ko, R., Commonly used herbal medicines in the U.S.: a review, *Am. J. Med.*, 116, 478, 2004.
9. Dorsch, K.D. and Bell, A., Dietary supplement use in adolescents, *Curr. Opin. Pediatr.*, 17, 653, 2005.
10. Bourgoin, B.P. et al., Instrumental comparison for the determination of cadmium and lead in calcium supplements and other calcium-rich matrices, *Analyst*, 117, 19, 1992.
11. Bourgoin, B.P. et al., Lead content in 70 brands of dietary calcium supplements, *Am. J. Public Health*, 83, 1155, 1993.
12. Amarasiriwardena, D., Sharma, K., and Barnes, R.M., Determination of lead concentration and lead isotope ratios in calcium supplements by inductively coupled plasma mass spectrometry after high pressure, high temperature digestion, *Fresenius J. Anal. Chem.*, 362, 493, 1998.
13. Scelfo, G.M. and Flegal, A.R., Lead in calcium supplements, *Environ. Health Perspect.*, 108, 309, 2000.
14. Kim, M., Kim, C., and Song, I., Analysis of lead in 55 brands of dietary calcium supplements by graphite furnace atomic absorption spectrometry after microwave digestion, *Food Addit. Contam.*, 20, 149, 2003.
15. Ross, E.A., Szabo, N.J., and Tebbett, I.R., Lead content of calcium supplements, *JAMA*, 284, 1425, 2000.
16. Roberts, J.H., Potential toxicity due to dolomite and bone meal, *South. Med. J.*, 76, 556, 1983.

17. Boulos, B.M. and von Smolinski, A., Alert to users of calcium supplements as antihypertensive agents due to trace metal contaminants, *Am. J. Hypertens.*, 1, 137S, 1988.
18. Siitonen, P.H. and Thompson, H.C., Analysis of calcium and lead in calcium supplements by inductively coupled plasma-atomic emission spectrometry and graphite furnace atomic absorption spectrophotometry, *J. AOAC Int.*, 77, 1299, 1994.
19. Capar, S.G. and Gould, J.H., Lead, fluoride, and other elements in bone meal supplements, *J. Assoc. Off. Anal. Chem.*, 62, 1054, 1979.
20. Hight, S.C. et al., Analysis of dietary supplements for nutritional, toxic, and other elements, *J. Food Compos. Anal.,* 6, 121, 1993.
21. Whiting, S.J., Safety of some calcium supplements questioned, Nutrition Reviews, 52, 95, 1994.
22. Consumer Reports, Calcium supplements: hold the lead, please, *Consumer Reports*, 64, 9, 1997.
23. Barnum, A., Environmental group sues over lead in calcium supplements, *The San Francisco Chronicle*, February 4, 1997.
24. Husted, A., Health watch: some calcium products may have excess lead, *The Atlanta Journal-Constitution*, January 29, 1997.
25. Kim, M., Mercury, cadmium and arsenic contents of calcium dietary supplements, *Food Addit. Contam.,* 21, 763, 2004.
26. Krone, C.A., Wyse, E.J., and Ely, J.T.A., Cadmium in zinc-containing mineral supplements, *Int. J. Food Sci. Nutr.,* 52, 379, 2001.
27. Ponce de Leon, C.A., Bayon, M.M., and Caruso, J.A., Trace element determination in vitamin E using ICP-MS, *Anal. Bioanal. Chem.*, 374, 230, 2002.
28. Huggett, D.B. et al., Organochlorine pesticides and metals in select botanical dietary supplements, *Bull. Environ. Contam. Toxicol.*, 66, 150, 2001.
29. Khan, I.A., Determination of heavy metals and pesticides in ginseng products, *J. AOAC Int.*, 84, 936, 2001.
30. Huggett, D.B. et al., Environmental contaminants in the botanical dietary supplement ginseng and potential human risk, *Hum. Ecol. Risk Assess.,* 6, 767, 2000.
31. Dolan, S.P. et al., Analysis of dietary supplements for arsenic, cadmium, mercury, and lead using inductively coupled plasma mass spectrometry, *J. Agric. Food Chem.*, 51, 1307, 2003.
32. Raman, P., Patino, L.C., and Nair, M.G., Evaluation of metal and microbial contamination in botanical supplements, *J. Agric. Food Chem.*, 52, 7822, 2004.
33. Institute of Medicine, Dietary Reference Intakes for Vitamin A, Vitamin K, Arsenic, Boron, Chromium, Copper, Iodine, iron, Manganese, Molybdenum, Nickel, Silicon, Vanadium, and Zinc, a report of the Panel on Micronutrients, Subcommittees on Upper Reference Levels of Nutrients and the Interpretation and Uses of Dietary Reference Intakes, and the Standing Committee on the Scientific Evaluation of Dietary Reference Intakes, Food and Nutrition Board, Institute of Medicine, National Academy Press, Washington, D.C., 2001.
34. IRIS (Integrated Risk Information System), U.S. Environmental Protection Agency, http://www.epa.gov/iris/subst/0278.htm, accessed November 30, 2005.
35. ATSDR (Agency for Toxic Substances and Disease Registry), Minimum Risk Levels (MRLs) for Hazardous Substances, http://www.atsdr.cdc.gov/mrls.html, accessed December 3, 2005.
36. Farley, D., Dangers of Lead Still Linger, U.S. Food and Drug Administration, FDA Consumer, January–February 1998.
37. Gulson, B.L. et al., Contribution of lead from calcium supplements to blood lead, *Environ. Health Perspect.,* 109, 283, 2001.

Index

A

B

C

Chewing gum 381

D

E

F

G

H

I

J

K

L

M

N

O

P

Q

R

S

T

U

V

W

X

Y

Z

Printed and bound by CPI Group (UK) Ltd, Croydon, CR0 4YY

29/10/2024

01780848-0004